Birkhäuser

Progress in Mathematics
Volume 297

Xianzhe Dai • Xiaochun Rong
Editors

Metric and Differential Geometry

The Jeff Cheeger Anniversary Volume

Birkhäuser

Editors
Xianzhe Dai
Department of Mathematics
University of California
Santa Barbara, New Jersey
USA

Xiaochun Rong
Department of Mathematics
Rutgers University
Piscataway, New Jersey
USA

ISBN 978-3-0348-0256-7 ISBN 978-3-0348-0257-4 (eBook)
DOI 10.1007/978-3-0348-0257-4
Springer Basel Heidelberg New York Dordrecht London

Library of Congress Control Number: 2012939848

Printed on acid-free paper

Springer Basel AG is part of Springer Science+Business Media (www.birkhauser-science.com)

Contents

Part III: Index Theory

Dedicated to

Jeff Cheeger

on his
65th birthday

Preface

The articles in this volume originates from lectures given at the international conference on *Metric and Differential Geometry*, May 11–15, 2009. Held at the Chern Institute of Mathematics, Tianjin and the Capital Normal University, Beijing, the main goal of the conference is to bring together leading experts to survey the recent advances in metric and differential geometry. Major topics covered in the conference include metric spaces, Einstein manifolds, Kähler geometry, geometric flows, index theory and hypoelliptic Laplacian, analytic torsions, and differential K-theory. This is reflected well in the collection in the volume. The conference was also used as an occasion to celebrate the 65th birthday of Jeff Cheeger, whose immense influence can be felt in many of these fields.

We acknowledge the financial support of Chern Institute of Mathematics and Capital Normal University for the realization of the conference and NSF for the preparation of these proceedings. We are very grateful for all the referees for their careful job in reviewing the contributions to this volume and for their thoughtful suggestions for improvements.

And we thank all the contributors to these proceedings. With submissions that they all could have sent for publication to excellent mathematical journals, they made possible the volume that we present here.

Xianzhe Dai and Xiaochun Rong
January 2012, Santa Barbara

Jeff and college friends, Pompei, Italy, summer after sophomore year in college (1962). From left: Steve, Jeff, Arnie, Jay. (credit: Jeff Cheeger)

On the beaches of Copacabana, Rio de Janeiro (1971), visit to IMPA. (credit: Jeff Cheeger)

Jeff, aged 2 years and
9 months; 1946, Brooklyn,
New York
(credit: Jeff Cheeger)

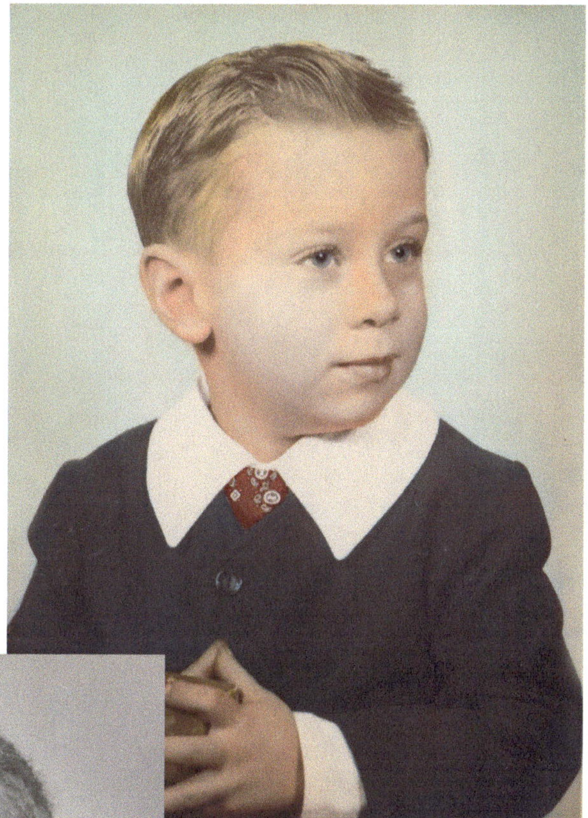

Jeff, high school graduation picture;
Brooklyn, New York, 1960.
(credit: Jeff Cheeger)

Jeff's parents, Thomas and Pauline Cheeger; Setauket, circa 1981. (credit: Jeff Cheeger)

Jeff and Emily; Setauket, 1981. (credit: Jeff Cheeger)

Calculations on spectral geometry of cones; Geneva 1976. (credit: Jeff Cheeger)

Deer Isle Maine, 1991. (credit: Jeff Cheeger)

Jeff and Detlef Gromoll, at Jim Simons house, Jim and Marilyn wedding reception, 1977 (credit: Jeff Cheeger)

Jeff, Werner Müller and Robert Schrader in East Berlin, circa 1982 (credit: Jeff Cheeger)

Jeff, Werner and Jean-Michel Bismut in Jerusalem, 2010 (credit: Jeff Cheeger)

Jeff and Misha Gromov, Conference "Differential Geometry, Mathematical Physics, Mathematics and Society" – 60th Birthday Conference Jean Pierre Bourguignon, IHES, 2007. (credit: Gert-Martin Greuel)

Jeff and Blaine Lawson, Conference "Differential Geometry, Mathematical Physics, Mathematics and Society" – 60th Birthday Conference Jean Pierre Bourguignon, IHES, 2007. (credit: Gert-Martin Greuel)

Jeff and students Dagang Yang, Xiaochun Rong and Xianzhe Dai at Chern Institute, 2009 (credit: Xiaochun Rong)

Curriculum Vitae of Jeff Cheeger

Jeff Cheeger, born December 1, 1943, Brooklyn, New York.

Education

- B.A., Harvard College, 1964
- M.S., Princeton University, 1966
- Ph.D., Princeton University, 1967

Career

- Princeton University, Teaching and Research Assistant, 1966–1967
- University of California, Berkeley, N.S.F. Postdoctoral Fellow and Instructor, 1967–1968
- University of Michigan, Assistant Professor, 1968–1969
- SUNY at Stony Brook, Associate Professor, 1969–1971
- SUNY at Stony Brook, Professor, 1971–1985
- SUNY at Stony Brook, Leading Professor, 1985–1992
- SUNY at Stony Brook, Distinguished Professor 1990–1992
- Courant Institute of Mathematical Sciences, NYU, Professor, 1989–
- Courant Institute of Mathematical Sciences, NYU, Silver Professor, 2003–

Honors and Prizes

- Sloan Fellow, 1971–1973
- Invited address, International Congress of Mathematicians, 1974, 1986
- Guggenheim Fellow, 1984–1985
- Max Planck Research Prize, Alexander von Humboldt Society, 1991
- National Academy of Sciences, elected 1997
- Finnish Academy of Science and Letters, foreign member, elected 1998
- Oswald Veblen Prize in Geometry, American Mathematical Society, 2001
- American Academy of Arts and Sciences, 2006

PhD Students

- 1976 Douglas Elerath, SUNY at Stony Brook,
  Dissertation: Nonegatively curved Manifolds Diffeomorphic to Eculidian Space
- 1978 Ping Charng Lue, SUNY at Stony Brook,
  Dissertation: Asymptotic Expansion of the Trace of the Heat Kernel on Generalized Surfaces of Revolution
- 1982 Arthur Chou, SUNY at Stony Brook,
  Dissertation: The Dirac Operator on Singular Spaces
- 1985 Aparna Dar, SUNY at Stony Brook,
  Dissertation: Intersection R-Torsion and Analytic Torsion for Pseudomanifolds
- 1985 Scott Hensley, SUNY at Stony Brook,
  Dissertation: Equivariant Reidemeister Torsion
- 1987 DaGang Yang, SUNY at Stony Brook,
  Dissertation: A Residue Theorem for Secondary Invariants of Collapsed Riemannian Manifolds
- 1989 Xianzhe Dai, SUNY at Stony Brook,
  Dissertation: Adiabatic Limit, Non-multiplicativity and Leray Spectral Sequence
- 1990 Xiaochun Rong, SUNY at Stony Brook,
  Dissertation: Collapsed 3-Manifolds and Rationality of Limiting n-invariants
- 1990 Shunhui Zhu, SUNY at Stony Brook,
  Dissertation: Bounding Topology by Ricci Curvature in Dimension Three
- 1991 Zhong-dong Liu, SUNY at Stony Brook,
  Dissertation: Nonnegative Ricci Curvature Near Infinity and Geometry of Ends
- 1996 Christina Sormani, Courant Institute, NYU,
  Dissertation: Noncompact Manifolds with Lower Ricci Curvature Bounds and Minimal Volume Growth
- 1998 Alireza Ranjbar-Motlagh, Courant Institute, NYU,
  Dissertation: Analysis on Metric-Measure Spaces
- 2001 Yu Ding, Courant Institute, NYU,
  Dissertation: Continuity of Some Analytic Objects under Measured Gromov-Hausdorff Convergence

The Mathematical Work of Jeff Cheeger, a Brief Summary

Xianzhe Dai and Xiaochun Rong

Jeff Cheeger's work, starting from his Ph.D. thesis, has been characterized by deep geometric insights, far-reaching consequences, and widespread influence. Some early work of Jeff Cheeger is exposed in Blaine Lawson's article in this volume, to which we refer for more discussions. Here we list some of the significant contributions of Jeff Cheeger to mathematics.

1. Cheeger-Gromov Compactness/Collapsing Theory. Cheeger's thesis [1] introduced finiteness theorems into Riemannian geometry. Subsequently, in a lecture at the Summer Institute on Global Analysis at Stanford (1973) Cheeger showed that the collection of such Riemannian manifolds is totally bounded in the Lipschitz topology. Hence, in the Lipschitz topology, one can actually take sublimits of sequences of such manifolds. The compactness theory was improved in important work of Gromov. It has had a very big impact on the subject.

Taking as point of departure Gromov's celebrated theorem on almost flat manifolds, Cheeger and Gromov developed their celebrated collapsing theory (F-structures) on the scale of the injectivity radius; [37, 46]). Later, with Fukaya, who had also done fundamental work on collapsing, they gave a theory (N-structures) which holds on a small but fixed scale; [53]. Collapsing theory has been one of the most important developments in Riemannian geometry in the past three decades.

2. Cheeger's Inequality for the Smallest Eigenvalue of the Laplacian. Cheeger proved a lower bound on the first nonzero eigenvalue of the Laplacian, which subsequently became known as "Cheeger's inequality" [6]. The lower bound is in terms of the "Cheeger constant", a certain isoperimetric constant. This had many applications in Riemannian geometry and geometric analysis. An analogous estimate for graphs (also known as Cheeger's inequality) has extraordinarily diverse applications in computer networks, computational complexity, computational geometry, random walks, the theory of error-correcting codes and so on.

3. Cheeger-Gromoll's Soul Theorem and Splitting Theorem. Cheeger's work with Detlef Gromoll [3, 13] on complete manifolds of nonnegative curvature followed soon after work of Gromoll-Meyer on complete noncompact manifolds of strictly positive curvature. The Soul Theorem states that such a manifold is diffeomorphic to the normal bundle of some close totally geodesic (in fact, totally convex) submanifold called a *soul*. In this terminology, Gromoll-Meyer had shown that if the curvature is positive, then the soul is a point. The Soul Theorem has been the subject of many research works; in particular, it is used in Perelman's resolution of the Poincaré conjecture.

By means of his triangle comparision theorem, V. Toponogov proved that if a complete manifold of nonnegative sectional curvature contains a *geodesic line*, then this line splits off as a isometric factor. In [13], by an argument based on Laplacian comparison for distance functions (roughly in the distribution sense) and the maximum principle, Cheeger and Gromoll extended the splitting theorem to the case complete manifolds of nonnegative Ricci curvature. It became a cornerstone of the subject. (As discussed below, much later, Cheeger and Colding extended the splitting theorem to the case of Gromov-Hausdorff limit spaces. This plays a basic role in their structure theory of Gromov-Hausdorff limits of manifolds with a uniform lower Ricci curvature bound.)

4. Cheeger-Simons Differential Characters. Shortly after the famous work of S.S. Chern and James Simons on transgression forms for principle bundles with connection (Chern-Simons invariants) in the early 1970s, Cheeger and Simons developed a corresponding theory of secondary invariants which live in the base space as elements of a new structure called *the ring of differential characters.* (This work was written up in notes distributed at the Stanford Summer Institute (1973) which were published only much later; see [34].) Differential characters are the first geometric model of what is now called differential cohomology, a refinement of ordinary integral cohomology. They have found interesting applications in theoretical physics.

5. The Ray-Singer Conjecture and the Cheeger-Müller Theorem. In [21], Cheeger proved the celebrated Ray-Singer conjecture which asserted that the analytic torsion is equal to the Reidemeister torsion. The proof involved considering two manifolds M_0, M_1 which differ by a surgery and interpolating between them a 1-parameter family M_t of manifolds with boundary which degenerates to M_0 at $t = 0$ and M_1 at $t = 1$. The proof is a true tour de force. It led in particular, to Cheeger's work on analysis on singular spaces. The Ray-Singer conjecture was proved independently and by a different method by Werner Müller. The Cheeger-Müller theorem has been used in Witten's study of two dimensional quantum gauge theories. More recently, it has been used to detect torsion cohomology classes of hyperbolic manifolds by Bergeron-Venkatesh and Müller.

6. Singular Spaces: Poincaré Duality, L_2-cohomology and Spectral Geometry. Cheeger realized that the technique developed in [21] for controlling heat kernels of degenerating families of manifolds with boundary, (he called it the "strong form of the method of separation of variables") gave rise to a functional calculus for the Laplacian on metric cones of arbitrary smooth cross-section. The key idea was to interpret the kernel $K_f(x_1, x_2, r_1, r_2)$ of a function $f(\Delta)$ of the Laplacian on the cone as a canonically associated family (parameterized by the radial variables r_1, r_2) of kernels of the Laplacian $\tilde{\Delta}$ of the cross-section. In this way, the functional calculus for $\tilde{\Delta}$ could be employed to gain control over the eigenfunction expansion of K_f (with respect to x_1, x_2) in regions where the expansion converged badly (or not at all). This led directly to his work on L_2-cohomology and Hodge theory on manifolds with (possibly iterated) conical singularities and to his study of the spectral geometry of piecewise flat pseudomanifolds; [20, 23, 28, 30, 31].

By this route, independently of Goresky and MacPherson, and in an entirely different context, Cheeger made the fundamental discovery that for pseudomanifolds (satisfying a certain condition on the middle dimensional cohomology of links) there exists a cohomology theory, L_2-cohomology, which satisfies Poincaré duality. He told this to Dennis Sullivan, who then informed him of the work of Mark Goresky and Bob MacPherson on intersection homology. Sullivan then conjectured that the two theories must be isomorphic and Cheeger proved that this was the case. Cheeger also introduced the notion of $*$-invariant ideal boundary conditions which can be used to restore Poincaré duality in certain cases in which the above mentioned condition on links fails to hold. On the topological side, this was done independently by Morgan (unpublished). Cheeger-Goresky-MacPherson made important conjectures (now established) asserting that the intersection homology of singular algebraic varieties satisfies properties analogous to those of the ordinary cohomology of smooth Kähler manifolds; [28].

Cheeger's theory of spectral geometry on piecewise flat spaces, which included a treatment of index theory via the heat equation method of McKean-Singer, had several remarkable applications. One was a purely combinatorial formula for the Euler number of a pseudomanifold in terms of certain angle defects which are defined in terms of products of dihedral angles. Another was an (analytically based) canonical local combinatorial formula for L-classes of PL-manifolds in terms of η-invariants of links. More generally, in this way, he defined L-classes (which are ordinary homology classes) for pseudomanifolds. Using intersection homology, a nonlocal topological definition in the spirit of Thom was given independently by Goresky-MacPherson.

Using the functional calculus for the Laplacian on cones, jointly with Michael Taylor, Cheeger gave a highly detailed treatment of diffraction of waves by cones of arbitrary cross section; [26, 27]. It was the first rigorous treatment of this classical problem and gave more information than earlier formal treatments. The first rigorous treatment of a special case of this diffraction problem dealt with diffraction by an edge (which can be viewed as diffraction by the cone on $[0, 2\pi]$ with Neumann boundary conditions). It was done by the physicist Arnold Sommerfeld in 1899 and subsquently rederived by a number of authors by means of different arguments, all involving nontrival summation formulas.

7. Regge Calculus and Lipschitz-Killing Curvatures for Piecewise Flat Spaces. The Lipschitz-Killing curvatures form a sequence of curvature measures (indexed by the degree of the polynomial in curvature). The sequence starts with scalar curvature times the volume measure and ends with the Chern-Gauss-Bonnet density. Cheeger, together with Werner Müller and Robert Schrader, studied the version of these curvatures for piecewise flat spaces which he had derived in his study of the index theory via the heat equation method; [25, 32, 39]. They showed that, for sequences of uniformly fat subdivisions of triangulations of smooth Riemannian manifolds, as the edge length goes to zero, the piecewise flat curvatures converge in the sense of measures to the smooth ones. For the case of the scalar curvature, this gives the first proof of the "classical limit theorem" for Regge calculus, which had long been widely assumed by physicists without rigorous justification.

8. η-Invariants, Families Index for Manifolds with Boundary and L_2-Index Theory. For a bundle with even dimensional fibre whose base space is a circle, Witten conjectured a formula relating the holonomy of an associated determinant line bundle to the adiabatic limit of the η-invariant of the total space. In [40, 41], Cheeger gave a proof of Witten's formula for signature operators. At about the same time, a different proof for general Dirac operators was given by Jean-Michel Bismut and Dan Freed. Subsequently, Bismut and Cheeger generalized the formula to base spaces and fibres of arbitrary dimension [43]. This involved their defining higher η-invariants (Bismut-Cheeger η-forms) which could also be viewed as Chern-Simons forms for certain bundles with infinite dimensional fibre. Cheeger had previously shown that the celebrated index formula of Atiyah-Patodi-Singer comes out naturally from the heat equation approach to the index formula in the conical setting. This is used in an essential way in the work of Bismut-Cheeger on the families index formula for manifolds with boundary [44, 45, 47].

L_2-index theory for infinite covering spaces of compact manifolds (based on the concept of Von Neumann dimension) was introduced by Atiyah and Singer. An extension to complete Riemannian manifolds with bounded curvature and finite volume was used in Cheeger and Gromov's study of integrals of characteristic forms over such manifolds; [35]. In this context, they also defined an L_2 version of η-invariants of compact manifolds and their associated ρ-invariants, which have had extensive applications in knot theory. The "good chopping" theorem, [48], whose generalization played a significant role in the work with Gang Tian on Einstein 4-manifolds ([74]), also arose in this context. (See also [36] for applications L_2-cohomology to group cohomology.)

9. Structure Theory for Limits of Manifolds with Lower Ricci Curvature Bounds and for Einstein Manifolds. In a series of trail blazing works [56, 60, 62, 64, 65], Cheeger and Tobias Colding gave a structure theory of Gromov-Hausdorff limit spaces of manifolds with lower Ricci curvature bounds. The fundamental tools are developed in [60], where some of the most important rigidity results such as the volume cone implies metric cone theorem, Cheng's maximal diameter theorem, and the Cheeger-Gromoll splitting theorem, are generalized quantitatively to almost rigidity theorems. Equivalently, the rigidity theorems are extended to the case of Gromov-Hausdorff limit spaces. An important tool in the discussion is their "segment inequality" which is a refinement of the Poincaré inequality.

For noncollapsed limit spaces Y^n, every tangent cone Y_y is shown to be a metric cone. A natural stratification $\mathcal{S}^0 \subset \mathcal{S}^1 \subset \cdots$ of singular set $\mathcal{S} \subset Y^n$ is defined in terms of the splitting properties of the tangent cones and the Hausdorff dimension bound dim $\mathcal{S}^k \leq k$ is proved. It is also shown that $\mathcal{S} = \mathcal{S}^{n-2}$ and that off a set of codimension 2, the limit space is bi-Hölder equivalent to a smooth connected Riemannian manifold. Moreover, the isometry group is a Lie group. (As had been conjectured by Cheeger-Colding and proved in spectacular recent work of Colding and Aaron Naber, this last statement holds in the collapsed case as well.) Even in the collapsed case, regular points have full measure with respect to any renormalized limit measure and (for such measures) up to a set of measure zero, the limit space is a countable union of measurable subsets, each of which is bi-Lipschitz equivalent to a subset of Euclidean space.

In work of (various subsets of) Cheeger, Colding and Tian, more refined regularity results were obtained in the noncollapsed case, when the members of the sequence are either Einstein, Kähler-Einstein [68] (possibly bounded L_p-norm of the full curvature tensor [69, 70]) or have special holonomy [73, 71].

In [74], the first results on Einstein manifolds whose hypotheses do not require a noncollapsing assumption are obtained. So far, they apply only in dimension 4.

9. Differentiation Theory for Metric Measure Spaces. Rademacher's theorem asserts the almost everywhere differentiability of Lipshitz functions $f : \mathbf{R^n} \to \mathbf{R^m}$. In a highly influential paper, [63], Cheeger formulated and proved a remarkable extension of Rademacher's theorem to spaces which might have fractional Hausdorff dimension and in particular, might not be Euclidean at the infinitesimal level. Let (X, d, μ) denote a doubling metric measure space that satisfies a p-Poincaré inequality for some $p \geq 1$. (In current terminology X is called a PI space.) Then (X, d, μ) has what is now known as a strong measurable differentiable structure (which, by definition, is unique). One way of expressing this is to say that X has a finite dimensional (measurable) cotangent bundle TX^* such that every Lipschitz function has a differential df which is a bounded measurable section of TX^*, such that the map $f \to df$ has the "usual" properties.

Cheeger found the following very generally applicable principle which underlies the whole discussion: If f (in this case a Lipschitz function) has finite energy (in this case, a Dirichlet energy) and the energy is lower semicontinuous then at the infinitesimal level almost everywhere, f will be a minimizer (i.e., harmonic) with constant energy density and hence linear in a generalized sense. Note that in $\mathbf{R^n}$, linear functions are precisely harmonic functions for which the norm of the gradient is a constant function.

One application of the theory is a general bi-Lipschitz nonembedding theorem. If a PI space X admits a bi-Lipschitz embedding into some finite-dimensional

Euclidean space, then at almost every point x, every tangent cone X_x is bi-Lipschitz equivalent to the fibre TX^*_x. (In general dim $X_x \geq TX^*$ and for many X, the inequality is strict.) Cheeger's theorem implies the known non-embedding results both for the Carnot-Carathéodory spaces and for Laakso spaces.

Cheeger and Bruce Kleiner gave an optimal extension of the differentiation theory to Banach space targets having the Radon-Nikodym Property; [79]. In [82, 83] they showed by a different method that the Heisenberg group $\mathbf{H}$ does not admit a bi-Lipschitz embedding into L_1. (The Banach space L_1 does not have the Radon-Nikodym Property; prior to [82, 83] there was no structure theory even for Lipschitz maps $\mathbf{R} \to L_1$.) Besides its significant intrinsic interest as a major advance at the intersection of metric embedding and differentiation theory, this result, when combined with work of Lee and Naor, gave a natural (qualitative) counterexample to the celebrated Goemans-Linial conjecture from theoretical computer science. By means of a highly nontrivial quantitative differentiation argument, this was made quantitative in [84], which provided an exponential improvement of the best previously known counterexample, due to Khot and Vishnoi (2005); for quantitative differentiation, see the discussion below.

11. Quantitative Differentiation. The simplest instance of "quantitative differentiation" arises in work of Peter Jones (1988) (who did not use this terminology). It makes the following assertion about functions $f : [0,1] \to \mathbf{R}$ with $|f'| \leq 1$. There is a natural measure on the collection of subintervals $J \subset [0,1]$ whose mass is infinite. However, for all $\epsilon > 0$, the measure of the collection of subintervals J such that $f \mid J$ fails to be ϵ-close to a linear function (in the appropriately scaled sense) is finite and has a bound in terms of ϵ (independent of the particular function f).

Through the work [63, 84] Cheeger discovered a surprisingly general, versatile, and powerful version of "quantitative differentiation"; see Section 14 of [84] and for a more precise formulation [88]. It can be viewed as the quantitative version of the above-mentioned general principle underlying [63]. When coupled with ϵ-regularity theorems, there are numerous applications to geometric analysis and partial differential equations. The technology for these applications (in which a key role is played by "cone structure") has been developed in (ongoing) work of Cheeger and Aaron Naber; see [86, 87]. A result from [86], which deals with non-collapsed Gromov-Hausdorff limits of Einstein manifolds, serves to illustrate the theme. Namely, the bound (from [68, 70]) on the $(n-4)$-dimensional Hausdorff measure of the singular set $\mathcal{S}$ is very significantly strengthened to the assertion that the volume of the set of points at which the C^2-harmonic radius is $\leq r$ is bounded by $c(n)r^4$.

12. Expository books and articles. Cheeger has written several influential expository works; see [17, 49, 67, 81, 71, 88]. In particular, [17], his book with David Ebin, "Comparison theorems in Riemannian Geometry" is a classic.

List of Publications of Jeff Cheeger

[1] Comparison and finiteness theorems for Riemannian manifolds, PhD Thesis, Princeton University, 1967.

[2] The relation between the diameter and the smallest eigenvalue of the Laplacian for manifolds of nonnegative curvature, Archiv. der Mathematik (1968) 558–560.

[3] The structure of complete manifolds of nonnegative curvature (with D. Gromoll) Bull. Amer. Math. Soc. (1968) 1147–1150.

[4] Pinching theorems for a certain class of Riemannian manifolds, Amer. J. Math., XCI, No. 3 (July 1969) 807–834.

[5] Infinitesimal isometries and Pontrjagin numbers (with P. Baum) Topology (1969) 173–193.

[6] A lower bound for the smallest eigenvalue of the Laplacian, Proc. of Princeton Conf. in Honor of Prof. S. Bochner (1969) 195–199.

[7] A combinatorial formula for Stiefel-Whitney classes, Proc. of Georgia Topology Conference (1969) 470–471.

[8] Counting topological manifolds, (with J. Kister) Proc. of Georgia Topology Conference (1969).

[9] Compact manifolds of nonnegative curvature, Proc. of Oberwolfach Conference on Differential Geometry (1969) 25–41.

[10] Finiteness theorems for Riemannian manifolds, Amer. J. Math. XCII, No. 1 (1970) 61–74.

[11] Homeomorphism types of topological manifolds (with J. Kister) Topology 9, (1970) 149–151.

[12] The splitting theorem for manifolds of nonnegative Ricci curvature (with D. Gromoll) J. Diff. Geom. 6, No. 1 (1971) 119–128.

[13] On the structure of complete manifolds of nonnegative curvature (with D. Gromoll) Ann. of Math. 96, No. 3. (1972) 413–443.

[14] Multiplication of differential characters, Proc. of Rome Conference on Geometry (1972) 441–445.

[15] Some examples of manifolds of nonnegative curvature, J. Diff. Geom. 8, No. 4 (1973) 623–628.

[16] Invariants of flat bundles, Proc. of International Congress of Mathematicians, Vancouver (1974) 3–6.

[17] Comparison theorems in Riemannian geometry, (book with D. Ebin) North-Holland (1975), reprinted Chelsea (2008).

[18] Analytic torsion and Reidemeister torsion, Proc. Nat. Acad. Sci. 74, No. 7 (1977) 2651–2654.

[19] Spectral geometry of spaces with cone-like singularities, (longer original preprint version of [20].)

[20] On the spectral geometry of spaces with cone-like singularities, Proc. Nat. Acad. Sci. 76 (1979) 2103–2106.

[21] Analytic torsion and the heat equation, Ann. of Math., 109, (1979) 259–322.

[22] On the lower bound for the injectivity radius of 1/4-pinched Riemannian manifolds (with D. Gromoll), (revised version of 1972 preprint) J. Diff. Geom. 15 (1980) 437–442.

[23] On the Hodge Theory of Riemannian pseudomanifolds, Amer. Math. Soc. Proc. Sym. Pur. Math. XXXVI (1980) 91–146.

[24] A lower bound for the heat kernel (with S.T. Yau) Com. Pur. Appl. Math. XXXIV (1981) 465–480.

[25] Lattice gravity of Riemannian geometry of piecewise linear spaces, (with W. Muller and R. Schrader) Proc. Heisenberg Symp., Munich (1981) 176–188.

[26] On the diffraction of waves by conical singularities I (with M. Taylor) Com. Pur. App. Math. XXV (1982) 275–331.

[27] On the diffraction of waves by canonical singularities II (with M. Taylor) Com. Pur. App. Math. XXXV (1982) 487–529.

[28] L^2-cohomology and intersection homology of algebraic varieties (with R. MacPherson and M. Goresky) Sem. Diff. Geom., S.T. Yau Ed. (Ann. Math. Study 102) Princeton Univ. Press (1982) 303–340.

[29] Finite propagation speed, kernel estimates for functions of the Laplace operator, and the asymptotic geometry of complete Riemannian manifolds (with M. Gromov and M. Taylor) J. Diff. Geom. 17 (1982) 15–53.

[30] Spectral geometry of singular Riemannian spaces, J. Diff. Geom. 18 (1983) 575–657.

[31] Hodge theory of complex cones, Analyse et topologie sur les espaces singuliers, (II-III) Asterisque 101-102 (1983) 118–133.

[32] On the curvature of piecewise at spaces (with W. Muller and R.Schrader) Commun. Math. Phys. 92 (1984) 405–445.

[33] Bounds on the Von Neumann dimension of L^2-cohomology and the Gauss Bonnet Theorem for open manifolds (with M. Gromov) J. Diff. Geom. 20 (1984) 1–34.

[34] Differential characters and geometric invariants (with J. Simons) mimeographed, Lecture Notes distributed at Amer. Math. Soc. Conference on Differential Geometry, Stanford (1973) and published in Geometry and Topology, Lecture Notes in Math. 1167, Springer-Verlag (1985) 50–80.

[35] Characteristic numbers of complete manifolds of bounded curvature and finite volume (with M. Gromov), (H.E. Rauch Memorial Volume I. Chavel and H. Farkas Eds.) Springer- Verlag (1985) 115–154.

[36] L_2-cohomology and group cohomology (with M. Gromov) Topology 24, No. 1.f (1985) 189–215.

[37] Collapsing Riemannian manifolds while keeping their curvature bounded (with M. Gromov) I, J. Diff. Geom. 23, No. 3 (1986) 309–346.

[38] A vanishing theorem for piecewise constant curvature spaces, Curvature and Topology of Riemannian Manifolds, K. Shiohama, T. Sakai and T. Sunada, eds. Lecture Notes in Math. 1201, Springer-Verlag (1986) 33–40.

[39] Kinematic and tube formulas for piecewise linear spaces (with W. Muller and R. Schrader) Indiana Math. J. 35, No. 4. (1986) 737–754.

[40] On the formulas of Atiyah-Patodi-Singer and Witten, Proc. International Congress of Mathematicians, Berkeley (1986) 515–521.

[41] Eta-invariants, the adiabatic approximation and conical singularities J. Diff. Geom. 26. (1987) 175–221.

[42] Invariants êta et indices des familles pour des variétés a bord (with J.-M. Bismut) C.R. Acad. Sci. Paris t. 305, Série 1. (1987) 127–130.

[43] Eta-invariants and their adiabatic limits (with J.-M. Bismut) J. Amer. Math. Soc. Vol. 2., No. 1.(1989) 33–70.

[44] The index theorem for families of Dirac operators on manifolds with boundary; superconnections and cones; I (with J.-M. Bismut) J. Funct. Anal. Vol. 89. No. 2 (1990) 313–363.

[45] The index theorem for families of Dirac operators on manifolds with boundary; superconnections and cones; II. (with J.-M. Bismut) J. Funct. Anal. Vol. 89 No. 3 (1990) 306–353.

[46] Collapsing Riemannian manifolds while keeping their curvature bounded, II. (with M. Gromov) J. Diff. Geom. Vol. 31, No. 4 (1990) 269–298.

[47] Remarks on the index theorem for families of Dirac operators on manifolds with boundary (with J.-M. Bismut) Pitman Press, (1990) 59–83.

[48] Chopping Riemannian manifolds (with M. Gromov) Differential Geoemtry, B. Lawson and K. Tenenblatt Eds., Pitman Press, (1990) 85–94.

[49] Critical points of distance functions and applications to geometry, Lecture Notes in Math, Vol. 1504, Springer Verlag, (1990) 1–38.

[50] Transgression de la classe d'Euler de $SL(2n;\mathbb{Z})$-fibrés vectoriels, limites adiabatiques d'invariants êta, et valeurs spéciales de fonctions L (with J.-M. Bismut), C.R. Acad. Sci. Paris. t. 312 Série 1 (1991) 399–404.

[51] Finiteness theorems for manifolds with Ricci curvature and $L^{n/2}$-norm of curvature bounded (with M. Anderson) GAFA, Geom. Funct. Anal.l. 1. No. 3 (1991) 231–252.

[52] C^α-compactness for manifolds with Ricci curvature and injectivity radius bounded below (with M. Anderson) J. Diff. Geom., Vol. 34, (1992) 265–281.

[53] Nilpotent structures and invariant metrics on collapsed manifolds (with K. Fukaya and M. Gromov), J. Amer. Math. Soc., Vol. 5, No. 2 (1992) 327–372.

[54] Transgressed Euler classes of $SL(2n;\mathbb{Z})$-vector bundles, adiabatic limits of eta invariants and special values of L-functions (with J.-M. Bismut), Ann. Scient Ec. Norm. Sup. 4e série. t.25 (1992) 335–391.

[55] On the cone structure at infinity of Ricci flat manifolds with Euclidean volume growth and quadratic curvature decay (with Gang Tian) Invent. Math. 118 (1994) 493–571.

[56] Almost rigidity of warped products and the structure of spaces with Ricci curvature bounded below (with T.H. Colding) C.R. Acad. Sci. Paris, t. 320, Série 1 (1995) 353–357.

[57] Collapsed riemannian manifolds with bounded diameter and bounded covering geometry (with X. Rong) GAFA, Vol. 5, N. 2 (1995) 141–160.

[58] Linear growth harmonic functions on complete manifolds with nonnegative Ricci curvature (with T. Colding and W. Minicozzi), GAFA, Geom. Funct. Anal., Vol. 5, No. 6, (1995) 948–954.

[59] Existence of polarized F-structures on collapsed manifolds with bounded diameter and curvature (with X. Rong) GAFA, Vol. 5 N. 3 (1996) 411–429.

[60] Lower bounds on Ricci curvature and the almost rigidity of warped products (with T.H. Colding) Annals of Math., 143 (1996) 189–237.

[61] Constraints on singularities under Ricci curvature bounds (with T. Colding and G. Tian) C.R. Acad. Sci. Paris, t. 324, Série I (1997) 645–649.

[62] On the structure of spaces with Ricci curvature bounded below; I (with T. Colding) J. Diff. Geom. 45 (1997) 406–480.

[63] Differentiability of Lipschitz functions on metric measure spaces, GAFA V. 9, N. 3 (1999) 428–517.

[64] On the structure of spaces with Ricci curvature bounded below; II (with T. Colding) J. Diff. Geom. 52 (1999) 13–35.

[65] On the structure of spaces with Ricci curvature bounded below; III (with T. Colding) J. Diff. Geom. 52 (1999) 37–74.

[66] Splittings and Cr-structures for manifolds with nonpositive sectional curvature (with J. Cao and X. Rong) Invent. Math. (2000).

[67] Degeneration of riemannian metrics under Ricci curvature bounds, Lezione Fermiane, Accademia Nazionale dei Lincei, Scuola Normale Superiore, Pisa (2001).

[68] On the singularities of spaces with bounded Ricci curvature (with T. Colding and G. Tian), GAFA Geom. Funct. Anal., V. 12 (2002) 873–914.

[69] L_p-curvature bounds, elliptic estimates and rectifiability of singular sets, C.R. Acad. Sci. Paris, t. 334, Série I (2002) 195–198.

[70] Integral bounds on curvature, elliptic estimates and rectifiablity of singular sets, GAFA Geom. Funct. Anal., V. 13 (2003) 20–72.

[71] Degeneration of Einstein metrics and metrics with special holonomy, Surveys in Differential Geometry, Volume VIII, S.-T. Yau ed., International Press (2004).

[72] Local splitting structures on nonpositively curved manifolds and semirigidity in dimension 3, (with J. Cao and X. Rong), Comm. Anal. and Geom. V. 12, N. 1, (2004) 389–415.

[73] Anti-self-duality of curvature and degeneration of metrics with special holonomy, (with G. Tian), Commun. Math. Phys. 255. (2005) 391–417.

[74] Curvature and injectivity radius estimates for Einstein 4-manifolds, (with G. Tian), J. Amer. Math. Soc., Vol. 19, N. 2. (2006) 487–525.

[75] Generalized differentiation and bi-Lipschitz nonembedding in L1, (with B. Kleiner), C.R. Acad. Sci. Paris 343 N. 5 (2006) 297–301.

[76] On the differentiability of Lipschitz maps from metric measure spaces to Banach spaces, (with B. Kleiner), Inspired by S. S. Chern – A Memorial Volume In Honor of A Great Mathematician, P. Griffths ed., World Scientific Press (2006) 129–152.

[77] L^2-cohomology of spaces with non-isolated conical singularities and nonmultiplicativity of the signature, (with X. Dai), Riemannian Topology and Geometric Structures on Manifolds, K. Galicki and S. Simanca, eds., Progress in Mathematics, Birkhäuser (2008) 1–24.

[78] Characterization of the Radon-Nikodym Property in terms of inverse limits (with B. Kleiner), Astérisque No. 321 (2008), 129–138.

[79] Differentiatiability of Lipschitz maps from metric measure spaces to Banach spaces with the Radon Nikodym Property, (with B. Kleiner), GAFA Geom. Funct. Anal. V. 19, N. 4 (2009) 1017–1028.

[80] A $(\log n)^{\Omega(1)}$ integrality gap for the sparsest Cut SDP, (with B. Kleiner and A. Naor), In Proceedings of 50th Annual IEEE Symposium on Foundations of Computer Science (FOCS 2009), 555–564, (2009).

[81] Structure theory and convergence in riemannian geometry, Milan Journal of Mathematics, V. 78, N. 1, (2010) 221–264.

[82] Differentiating maps into L^1 and the geometry of BV functions, (with B. Kleiner), Annals of Math. 171, (2010) 1347–1385.

[83] Metric differentiation, monotonicity and maps into L^1, (with B. Kleiner), Invent. Math. V. 182, N. 2, (2010) 355–370.

[84] Compression bounds for Lipschitz maps from the Heisenberg group to L^1 (with B. Kleiner and A. Naor), Acta Mathematica (to appear) arXiv:0910.2006.

[85] Realization of metric spaces as inverse limits and bilipschitz embedding in L^1 (with B. Kleiner), arXiv:1110.2406.

[86] Lower bounds on Ricci curvature and quantitative behavior of singular sets (with A. Naber), arXiv:1103.1819.

[87] Quantitative stratification and the regularity of harmonic maps and minimal currents (with A. Naber), arXiv:1107.3097.

[88] Quantitative differentiation, a general perspective, (to appear in) Com. Pur. App. Math.

The Early Mathematical Work of Jeff Cheeger

H. Blaine Lawson, Jr.

Introduction

This article is essentially a write-up of the talk I gave at the conference held in June of 2009 in Tianjing and Beijing in honor of Jeff Cheeger. The organizers had approached me about giving a lecture on Jeff's contributions to mathematics. I agreed in principle, but pointed out the impossibility to doing justice to his prodigious output in the space of an hour. So it was decided that I would restrict my remarks to his early work – "classical Cheeger", if you will. This actually made a nice topic to present because of the originality, sweep, and importance of his results in those years.

Like the lecture, these subsequent notes are informal and aimed at an audience of geometers. They are by no means historically complete, even in representing the early period. There are long and important collaborations (from the later period) which have been totally ignored. Notable among these are Jeff's collaborations with Jean-Michel Bismut, Toby Colding, Gang Tian, Xiaochun Rong, and most recently Bruce Kleiner. Had I begun to discuss any of these, I would not have known how to stop in reasonable time. I have also overlooked important, but less lengthy collaborations with: Shing-Tung Yau, Mike Anderson, Xianzhe Dai, Jianguo Cao, and most recently Assaf Naor. As you will see, there was still much left to say.

1. The thesis

Jeff wrote his thesis at Princeton under the direction of Salomon Bochner. In fact he was Bochner's last student. However, his more hands-on mentor was Jim Simons, who was at the Institute for Defense Analysis at that time. The work in Jeff's thesis was spectacular and gave promise of great things to come. One important result in [2] was his Finiteness Theorem, which he later refined, and which will be discussed below. A second basic result was the following.

Let X and $\bar{X}$ be compact simply-connected Riemannian n-manifolds with given points $x \in X$ and $\bar{x} \in \bar{X}$, and an isometry $I : T_x X \to T_{\bar{x}} \bar{X}$. To each broken geodesic $\gamma : [0, L] \to X$ there is a uniquely associated "congruent" broken geodesic $\bar{\gamma} : [0, L] \to X$ with $\bar{\gamma}'(0) = \gamma'(0)$, constructed as in the Cartan-Ambrose-Hicks Theorem (using parallel translations and to determine the new direction at each break in $\bar{\gamma}$).

Given $N > 0$, define

$$\rho_N(X, \bar{X}) \;=\; \sup_{\gamma} \left\{ \|R - I_\gamma^{-1}(\bar{R})\|, \|\nabla R - I_\gamma^{-1}(\nabla \bar{R})\| \right\}$$

where

$$I_\gamma \;=\; P_{\bar{\gamma}} \circ I \circ P_\gamma^{-1},$$

R denotes curvature, P_γ denotes parallel translation along γ, and the sup is taken over all broken geodesics with at most N breaks, each segment of which has length at most 1.

Theorem. *Fix X. Then there exists $N > 0$ such that for each $V > 0$ there is an $\epsilon > 0$ so that any $\bar{X}$ with*

$$\mathrm{vol}(\bar{X}) > V \quad \textit{and} \quad \rho_N(X, \bar{X}) < \epsilon$$

is diffeomorphic to X.

2. Pinching

For given Riemannian manifolds X and $\bar{X}$, Jeff introduced a notion of *relative pinching*. Fix $x \in X$, $\bar{x} \in \bar{X}$ and an isometry $I : T_x X \to T_{\bar{x}} \bar{X}$ as above. Let

$$r_d(X, \bar{X}) = \sup \|R - I_\gamma^{-1}(\bar{R})\|$$

where the sup is over broken geodesics γ with length $L(\gamma) \le d$. Take

$$r(X, \bar{X}) = r_{2d_c}(X, \bar{X})$$

where d_c = the maximum distance between conjugate points in X.

Theorem ([2], [3]). *Let X be a simply-connected symmetric space of rank 1. Then $\exists\, \delta > 0$ such that*

$$r(X, \bar{X}) < \delta \quad \Rightarrow \quad \bar{X} \textit{ is diffeomorphic to } X.$$

3. The soul theorem

About the time Jeff was writing his thesis, two young German mathematicians, Detlef Gromoll and Wolfgang Meyer, proved the following gorgeous result [39]. (In this section all manifolds are assumed connected.)

Theorem (Gromoll-Meyer). *Let X be a complete, non-compact Riemannian manifold of positive sectional curvature. Then X is diffeomorphic to Euclidean space.*

This leads to the obvious question: What happens if the curvature condition is relaxed to $K \geq 0$? The situation is certainly much more complicated as the next two examples show. Let G/H be a compact Riemannian homogeneous space (with the "Killing" metric) and consider the following Riemannian products.

Example 1. $X = (G/H) \times \mathbf{R}$.

Example 2. $X = (G/H)\times$ (a paraboloid of revolution).

Together Cheeger and Gromoll proved the following "Soul Theorem" [6].

Theorem (Cheeger-Gromoll). *Let X be a complete, non-compact Riemannian manifold of non-negative sectional curvature. Then there exists a compact totally convex (in particular totally geodesic) submanifold $S \subset X$ such that X is diffeomorphic to the normal bundle of S.*

Note. A subset $S \subset X$ is *totally convex* if every geodesic joining two points of S lies entirely in S. The subset S appearing in this theorem was called the *soul* of X.

A brief sketch of the proof

For each geodesic ray $c : [0, \infty) \to X$ (every segment minimizes length) they show that the *half-space*

$$H_c \equiv \bigcup_{t>0} B_t(\gamma(t))$$

is totally convex. The proof of this fact uses Toponogov's triangle comparison and is encapsuled in the diagram on page xxxiv of Wolfgang Meyer.

Intersection over such half-spaces is shown to give a compact totally convex set S.

Cheeger and Gromoll then proved a basic

Structure Theorem. *The set S is a smooth manifold with possibly non-smooth boundary.*

If the boundary of S is not empty, they contract down to a new totally convex set of smaller dimension. This new set consists of those points in S a maximal distance from the boundary.

The process ends with a soul.

Note. It is worthwhile to note that the Soul Theorem enters in at least two important ways in the recent proof of the Thurston Geometrization Conjecture by Grisha Perelman [49], [50], [51]. The first occurs in the "blow-up limit". This is when a bounded-time sequence $(M^3, g(t_k))$ with $t_k \nearrow t_0 < \infty$ under the Ricci flow, converges, after rescalings which tend to infinity, to a smooth limit (N, h). This limit has non-negative sectional curvature, and the Soul Theorem applies to severely restrict the possibilities. The second entry is in the latter part of the proof when one us studying the thick-thin decomposition and wants to prove that the thin part is a graph manifold (Collapsing Theory). Again limits under rescaling produce complete non-negatively curved 3-manifolds and the Soul Theorem applies in an important way.

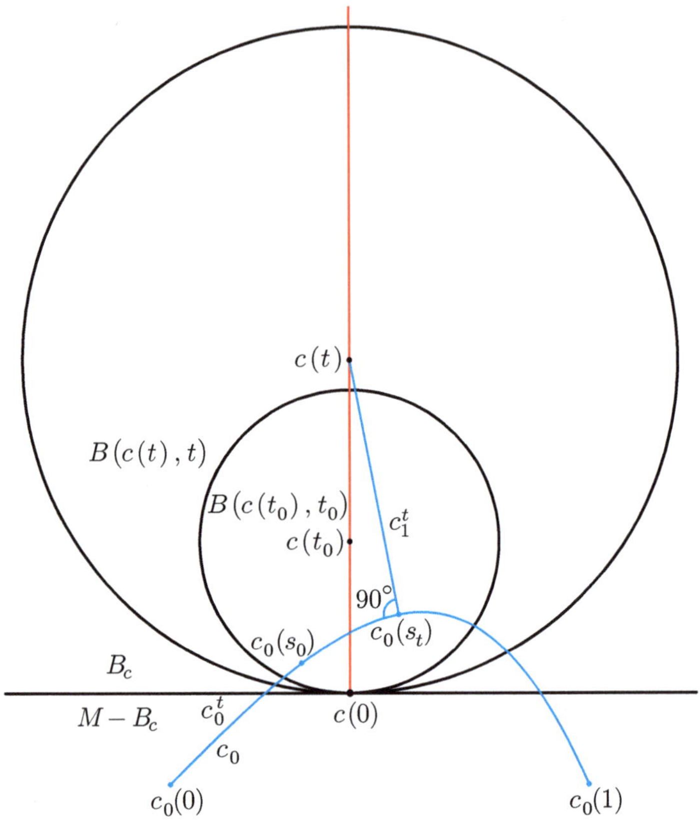

4. The splitting theorem

Jeff and Detlef then turned there attention from non-negative sectional curvature to the much wider and more interesting class of manifolds of non-negative Ricci curvature. They proved the following result [7] which has been fundamental in Riemannian geometry for the past 40 years.

Theorem (Cheeger-Gromoll). *Let X be a complete riemannian manifold of non-negative Ricci curvature. Then X is isometric to a riemannian product $\bar{X} \times \mathbf{R}^k$ where $\bar{X}$ contains no lines and $\mathbf{R}^k$ has the standard flat metric.*

Note. A *line* is an (arc-length) geodesic $\gamma : (-\infty, \infty) \to X$ such that every segment is length-minimizing, i.e.,

$$\mathrm{dist}(\gamma(a), \gamma(b)) \;=\; |b - a| \qquad \forall a, b \in \mathbf{R}.$$

Corollary 1. *Let X be compact with $\mathrm{Ric}_X \geq 0$. Then $\pi_1(X)$ contains a finite normal subgroup Γ such that $\pi_1(X)/\Gamma$ is a finite group extended by $\mathbf{Z}^k$, and the universal covering $\widetilde{X}$ splits isometrically as $\bar{X} \times \mathbf{R}^k$ where $\bar{X}$ is compact.*

Corollary 2. *Let X be complete with* $\mathrm{Ric}_X \geq 0$. *Then every finitely generated subgroup of* $\pi_1(X)$ *has polynomial growth of degree* $\leq \dim(X)$, *and there exists a subgroup for which equality holds iff X is compact and flat.*

The polynomial growth of π_1 with degree $\leq n$ was proved earlier by Milnor [46]. Corollary 2 establishes polynomial growth with degree $\leq n-1$ unless X is compact and flat.

Corollary 3. *Let X be complete and locally homogeneous with* $\mathrm{Ric}_X \geq 0$. *Then X is isometric to a flat vector bundle over a compact locally homogeneous space.*

A brief sketch of the proof

Consider a ray $c : [0, \infty) \to X$. For each $t \geq 0$ define

$$g_t(x) = \mathrm{dist}(x, c(t)) - t.$$

They show that as $t \to \infty$, $g_t \to g_c$ uniformly on compact subsets.

Basic Lemma. *The function g_c is superharmonic.*

Given a geodesic *line* $c : (-\infty, \infty) \to X$ there exist two such functions g_+ and g_-. They show that $g_+ = -g_-$ and therefore g_+ is harmonic. Bochner-type arguments then show that ∇g_+ is parallel.

The function g_+ then splits X isometrically as $\bar{X} \times \mathbf{R}$ using gradient lines and level surfaces.

Remark. In the Basic Lemma above, the main difficulty is dealing with the non-smooth points of the Busemann function g_c. The breakthrough was the realization that non-smoothness, if it occurred, actually helped the situation.

Interestingly, Jeff and Detlef were unaware at that time of the work of Calabi on the strong maximum principle [35], which dealt with a similar issue. In this paper Calabi introduced a notion of general "superhamonic" functions in terms of smooth comparison test functions. (This very much preceded modern viscosity theory, and Calabi is often given insufficient credit for his insight.) Later on, Eschenburg and Heintze realized the connection and gave a new more streamlined proof of the Splitting Theorem [37].

Remark. It should be noted that the splitting theorem for manifolds of non-negative sectional curvature was proved earlier by Toponogov [54], and reproved in [6].

Remark. It is interesting to note that twenty-five years later, in collaboration with Toby Colding, Jeff extended this Splitting Theorem to Gromov-Hausdorff limit spaces with Ricci curvature bounded below [33]. This has important applications to the infinitesimal structure of such limit spaces since it applies in particular to tangent cones.

5. Examples of manifolds with $K \geq 0$

In 1973 Jeff gave the first non-homogeneous examples of manifolds with non-negative sectional curvature and also metrics with non-negative Ricci curvature on certain exotic spheres [8]. Much work has subsequently been done in this area. One of the chief tools in the practitioner's kit is the technique of "Cheeger perturbations", introduced in [8].

The idea is the following. Let (M, g) have nonnegative curvature and suppose that the compact Lie group G acts isometrically on M^n. Equip G with a bi-invariant metric and $M \times G$ with the product metric. Then the diagonal action of G on $M \times G$ is isometric and induces by Riemannian submersion a new metric g_1 on M. The metric g_1 has nonnegative curvature, and if the plane section σ had positive curvature for g, then σ has positive curvature for g_1. Typically, some of the sections of zero curvature for g will also have positive curvature (although G will not act isometrically on (M, g_1)). These curvature assertions follow from O'Neill's fundamental equations for a Riemannian submersion[48], which imply that a Riemannian submersion is curvature non-decreasing on horizontal sections.

This technique is widely cited in the field.

6. The finiteness theorem

The following result was not just spectacular in its assertion, but the ideas and methods helped to dramatically change the course of Riemannian geometry.

Theorem (Cheeger [2], [5]). *Given $n, d, V, K > 0$, there exists at most finitely many diffeomorphism classes of compact n-manifolds X which admit a metric such that*

$$\operatorname{diam}(X) < d, \quad \operatorname{vol}(X) > V, \quad \text{and} \quad |K_X| < K.$$

Related to this was a finiteness result for δ-pinched manifolds.

Theorem (Cheeger) [2], [5]). *Given $n, \delta > 0$, there exists at most finitely many diffeomorphism classes of $2n$-manifolds X admitting a metric with $\delta \leq K_X \leq 1$.*

This was proved independently by Alan Weinstein [55] with the weaker conclusion of the finiteness of homotopy-types.

One of the key points in Jeff's proof of finiteness was the establishment of a lower bound on the injectivity radius, which comes down to finding a lower bound on the length of any smooth closed geodesic. The idea is that under the assumptions on curvature and diameter, a short closed geodesic forces the volume to be small.

Another important step was to show the existence of an atlas of charts of a definite size with definite bounds on the transition functions and the metric. The proof concludes with an Arzela-Ascoli argument.

7. The first eigenvalue theorem

The following is a result often quoted and used in many areas of mathematics. Suppose X is a compact Riemannian n-manifold, and $\lambda_1(X)$ denote the first non-zero eigenvalue of the Laplace-Beltrami operator Δ on functions. Let

$$\mathcal{C} \equiv \text{the set of compact hypersurfaces } S \subset X \text{ which divide } X \text{ into two non-empty submanifolds } X_1, X_2 \text{ with boundary } S.$$

and set

$$h(X) \equiv \inf_{\mathcal{C}} \left\{ \frac{A(S)}{\min\{\mathrm{Vol}(X_1), \mathrm{Vol}(X_2)\}} \right\}.$$

One has $h(X) > 0$ and the following result.

Theorem (Cheeger [4]).

$$\lambda_1(X) \geq \frac{1}{4} h^2(X).$$

This number $h(X)$ is known as the **Cheeger constant**, and it now appears in a serious way in an astonishing diversity of fields, including graph theory, dynamics, optimal networks, probability theory, computational vision, spectral theory, image segmentation, number theory and much more. Applying a Google search to this term produced 29,500 results, all referring, it appeared, to the number above.

8. The Ray-Singer conjecture

Let X be a compact Riemannian manifold with a flat Riemannian bundle

$$E \rightarrow X,$$

and consider the de Rham complex

$$\mathcal{E}^0(X,E) \xrightarrow{d} \mathcal{E}^1(X,E) \xrightarrow{d} \cdots \xrightarrow{d} \mathcal{E}^n(X,E).$$

Let Δ_k denote the Hodge Laplacian on $\mathcal{E}^k(X,E)$ with non-zero eigenvalues $\{\lambda_j\}_{j=1}^{\infty}$. The corresponding zeta function

$$\zeta_k(s) = \sum_{j=1}^{\infty} \lambda_j^{-s}$$

is well defined for $\mathrm{Re}(s) > \frac{n}{2}$, has a meromorphic continuation to the whole complex plane, and is regular at $s = 0$.

Definition. The *analytic torsion* is the quantity $T(X,E)$ where

$$\ln T(X,E) = \frac{1}{2} \sum_{k=1}^{n} (-1)^k k \cdot \zeta_k'(0).$$

The Ray-Singer Conjecture [52];

$$T(X,E) = \tau(X,E)$$

where $\tau(X,E)$ is a combinatorial invariant called the *Reidemeister Torsion*, defined below. This invariant depends on a choice of a volume forms on each $H^k(X,E)$. This is given by the induced metric on the harmonic forms.

Theorem (Cheeger [12], [13] and Müller [47]). *The Ray-Singer Conjecture is true.*

The proof, a *tour de force*, proceeds by showing the two quantities change in parallel under surgery.

Definition of Reidemeister torsion

Consider a complex of finite-dimensional real vector spaces

$$C^0 \xrightarrow{d_0} C^1 \xrightarrow{d_1} \cdots \xrightarrow{d_{n-1}} C^n \xrightarrow{d_n} 0$$

with $d_k \neq 0$ for $k \neq n$. We have short exact sequences

$$0 \longrightarrow B^k \longrightarrow Z^k \xrightarrow{\pi} H^k \longrightarrow 0.$$

Suppose we are given volume forms

$$\omega_k \text{ on } C^k \quad \text{and} \quad \mu_k \text{ on } H^k.$$

Let

$$t_k = \dim B^k \quad \text{and} \quad b_k = \dim H^k.$$

Choose

$$\rho_k \in \Lambda^{t_k}(C^k)^* \quad \text{with } \rho_k\big|_{B^k} \neq 0$$
$$\sigma_k \in \Lambda^{b_k}(C^k)^* \quad \text{with } \sigma_k\big|_{Z^k} = \pi^*(\mu_k).$$

Then

$$\rho_k \wedge d_k^*(\rho_{k+1}) \wedge \sigma_k \;=\; m_k\omega_k$$

for some scalar $m_k \neq 0$, and we set

$$\tau(C^*, \omega_*, \mu_*) \;\equiv\; \prod \frac{m_{2k}}{m_{2k+1}}.$$

Now let X be a compact Riemannian manifold and $E \to X$ a flat orthogonal bundle. Fix a smooth triangulation $\mathcal{T}$ of X, and note that each simplex σ in $\mathcal{T}$ determines a real vector space V_σ by taking the family of horizontal liftings of σ to E. These naturally form a chain complex $(C_*(X,E),\partial)$, since the boundary of a horizontal lift is a combination of horizontal lifts of the boundary faces. Note that $C_*(X,E)$ has a natural basis, which in turn gives a natural basis on the real cochain complex $C^*(X,E)$. Thereby, each space $C^k(X,E)$ carries a natural volume form ω_k. Now let μ_k be the metric on $H^k(X,E)$ coming from the L^2-metric on harmonic forms. With this information the torsion $\tau(C^*(X,E),\omega_*,\mu_*)$ is defined as above. The important fact is that this torsion remains unchanged when one passes to smooth subdivisions of the given triangulation of X. (See [Milnor] or [Cheeger].) Thus $\tau(X,E) = \tau(C^*(X,E),\omega_*,\mu_*)$ is a combinatorial invariant depending only on the smooth structure of X and volume forms on the $H^k(X,E)$ coming from harmonic theory.

9. Cheeger-Simons differential characters

In 1973 Cheeger and Simons introduced a theory of *differential characters* which greatly generalized the Chern-Simons invariants and had wide applications in geometry [10]. One may recall that the Chern-Simons forms were canonically defined on the total space of a smooth principal bundle with connection [36]. Differential characters arose as incarnations of these objects on the base. However, the reach of differential characters goes far beyond the study of principal bundles.

Fix a proper subgroup $\Lambda \subset \mathbf{R}$ (e.g., $\Lambda = \mathbf{Z}$ or $\mathbf{Q}$) and let $Z_k(X)$ denote the smooth singular k-cycles on X with $\mathbf{Z}$-coefficients.

Definition. A *differential character of degree* k mod Λ is a homomorphism

$$\rho : Z_k(X) \longrightarrow \mathbf{R}/\Lambda$$

whose coboundary is the mod Λ reduction of some smooth (necessarily closed) differential form φ. In other words for each smooth singular k-chain $c \in C_{k+1}(X)$

$$\rho(\partial c) \equiv \int_c \varphi \pmod{\Lambda}.$$

The set of such characters is denoted by $\widehat{H}^k(X)$.

Example 3 ($\Lambda = \mathbf{Z}$)**.** Let $L \to X$ be a $\mathbf{C}$ line bundle with unitary connection and let φ be its curvature 2-form. For each piecewise smooth closed curve γ in X define

$$\rho(\gamma) \in \mathbf{R}/\mathbf{Z} = S^1 \text{ to be the holonomy around } \gamma.$$

Suppose now that $\Sigma \subset X$ is a smooth surface with boundary (or more generally a smooth singular 2-chain). Then we have

$$\rho(\partial\Sigma) \equiv \int_\Sigma \frac{1}{2\pi}\varphi \pmod{\mathbf{Z}}.$$

Exact sequences

There are two fundamental exact sequences in the theory. The first is

$$0 \longrightarrow H^k(X;\mathbf{R}/\Lambda) \longrightarrow \widehat{H}^k(X) \xrightarrow{\delta_1} \mathcal{Z}_0^{k+1}(X) \longrightarrow 0$$

where $\delta_1(\rho) = \varphi$ and $\mathcal{Z}_0^{k+1}(Z)$ denotes the smooth $(k+1)$-forms with integral periods. The second is

$$0 \longrightarrow \widehat{H}^k(X)_\infty \longrightarrow \widehat{H}^k(X) \xrightarrow{\delta_2} H^{k+1}(X;\Lambda) \longrightarrow 0$$

where $\delta_2(\rho)$ is a class whose real reduction is the class of φ in $H^{k+1}(X;\mathbf{R})$. These two sequences can be assembled into a 2×2 grid. For $\Lambda = \mathbf{Z}$ it has the form:

$$\begin{array}{ccccccc}
 & 0 & & 0 & & 0 & \\
 & \downarrow & & \downarrow & & \downarrow & \\
0 \longrightarrow & \frac{H^k(X,\mathbf{R})}{H^k_{\text{free}}(X,\mathbf{Z})} & \longrightarrow & \widehat{H}^k_\infty(X) & \longrightarrow & d\mathcal{E}^k(X) & \longrightarrow 0 \\
 & \downarrow & & \downarrow & & \downarrow & \\
0 \longrightarrow & H^k(X,S^1) & \longrightarrow & \widehat{H}^k(X) & \xrightarrow{\delta_1} & \mathcal{Z}^{k+1}_0(X) & \longrightarrow 0 \\
 & \downarrow & & \delta_2 \downarrow & & \downarrow & \\
0 \longrightarrow & H^{k+1}_{\text{tor}}(X,\mathbf{Z}) & \longrightarrow & H^{k+1}(X,\mathbf{Z}) & \longrightarrow & H^{k+1}_{\text{free}}(X,\mathbf{Z}) & \longrightarrow 0 \\
 & \downarrow & & \downarrow & & \downarrow & \\
 & 0 & & 0 & & 0 &
\end{array}$$

Multiplicative structure and characteristic classes

Jeff defined a product $*$ on $\widehat{H}^*(X)$ making it a graded commutative ring for which the maps δ_1 and δ_2 are ring homomorphisms [11]. This was a substantial accomplishment.

With this structure Cheeger and Simons defined, for any complex vector bundle *with unitary connection*, refined Chern classes

$$\widehat{c}_k(E,\nabla) \in \widehat{H}^{k-1}(X)$$

such that

$$\delta_1(\widehat{c}_k(E,\nabla)) = \text{the Chern Weil form for } c_k(E,\nabla),$$
$$\delta_2(\widehat{c}_k(E,\nabla)) = \text{the integral class } c_k(E)$$

and the corresponding total Chern class satisfies

$$\widehat{c}_k(E\oplus F, \nabla^E \oplus \nabla^F) = \widehat{c}_k(E,\nabla^E) * \widehat{c}_k(F,\nabla^F).$$

9.1. Applications

There have been many important applications of the theory, including:

- The non-existence of conformal immersion of manifolds into Euclidean spaces.
- New invariants for flat bundles and foliations.
- A crucial role in modern forms of the geometric index theorem.
- New invariants for singularities of bundle maps.
- A vast generalization of the Chern-Simons invariants.

The theory of differential characters has been shown to have many quite different formulations, just as cohomology does (see [41], [42]). Recently the theory has axiomatized in the spirit of the Eilenberg-Steenrod Axioms for cohomology (see [53]). There have been wide-ranging generalizations of this theory (see for example [40] and [43]), with applications in physics (cf. [44]).

10. Hodge theory and spectral geometry on Riemannian pseudo-manifolds – intersection cohomology theory

In early 1976 Jeff began a deep study of Riemannian manifolds with iterated cone-like singularities [12], [13], [14], [15], [16], [17]. The basic idea was to study the L^2 theory of the Laplace operator on the complement of the singular locus. He used separation of variables on the deleted neighborhood of the vertex of a cone.

Among the many accomplishments in this body of work were the following. Jeff proved that:

(i) The harmonic forms represent the L^2-cohomology.

(ii) This cohomology satisfies Poincaré duality. In fact it was later shown that *this cohomology coincides with the intersection cohomology of Goresky-MacPherson* (in the middle perversity). Jeff is thus a co-founder of the theory of Poincaré duality for singular spaces.

(iii) The Laplace operator admits a spectral representation.

(iv) Analogues of Bochner's Vanishing Theorem hold for this cohomology.

(v) There is a good functional calculus for the Laplace operator. He then used it to construct a Greens operator and fundamental solutions of the heat and wave equations.

(vi) There exist generalizations of the Gauss-Bonnet Formula and the Hirzebruch Signature Theorem in this context.

(vii) There exists an asymptotic expansion of the trace of the heat kernel which involves a logarithmic term in t and spectral invariants of the base of the cone.

(viii) There is a combinatorial formula for the L-classes. This formula is in terms of the eta-invariants of the links in the iterated cone-structure, and it holds for the most general class of pseudo-manifolds. To date it is the only known combinatorial formula for the L-classes.

This foundational work led to a collaboration with Mike Taylor on the diffraction of waves by conical singularities, including in particular, cones of arbitrary cross-section in $\mathbf{R}^3$, a case of interest in applications. The main problem they address concerns singularities of the kernel of the operator $\sin\sqrt{t\Delta}/\sqrt{\Delta}$ on such manifolds.

The analysis is based on a functional calculus involving the Hankel transform, a technique developed and used in [13]. In the relevant case in [13] the cross-section was a round sphere. However, realizing the generality of his arguments led Jeff to investigate spaces with arbitrary (iterated) conical singularities. The rough idea is the following. Suppose that on a cone $C(X)$ with compact cross-section X, one wants to construct the kernel $K_f(r_1, x_1, r_2, x_2)$ of some function $f(\Delta)$ of the

Laplacian Δ on functions (for simplicity) on $C(X)$. Using separation of variables and the Hankel inversion formula, Jeff expressed this kernel as

$$K_f(r_1, x_1, r_2, x_2) = \sum_j h_f(r_1, r_2, \mu_j)\phi_i(x_1)\phi_j(x_2)$$

where the ϕ_j are eigenfunctions of the Laplacian $\widetilde{\Delta}$ with eigenvalue μ_j on the cross-section X, and the function h_f is determined by f. For a particular f, calculating h_f amounts to explicitly inverting a Hankel transform, and in all cases of interest, Jeff found by looking in Watson's classic book on Bessel functions, that the relevant integrals had already been calculated in the 19th century. Now the issue is the behavior of this kernel near the diagonal in $C(X) \times C(X)$ where the series may converge badly or not at all. Jeff's observation was that formally the right-hand side of this expression could be interpreted as a family (depending on r_1 and r_2) of kernels associated to the Laplacian $\widetilde{\Delta}$ on X. Synthesizing these kernels out of more standard kernels, Jeff was able to control the formal sum despite the lack of convergence. This method, which Jeff called "the strong form of the method of separation of variables", has been subsequently taken up by a number of geometric analysts.

This foundational work also led to important conjectures concerning the equivalence of L^2-cohomology and intersection cohomology on singular projective algebraic varieties (where generic singularities are not conical). These conjectures were developed and presented in [18]. They have been the focus of much study and many results.

11. The Cheeger-Gromov story

Back in approximately 1980 Jeff Cheeger and Misha Gromov began a long, fruitful and important collaboration. Their joint work has been transformational in geometry, and played a role in the recent resolution of the Thurston Conjecture.

Part one: Finite propagation speed and the geometry of complete Riemannian manifolds

Their first paper [21] was written in collaboration with Mike Taylor. In this work they gave deep estimates on the kernels of certain operators of the form $f(\Delta)$. Their idea was that the desired estimates for the wave kernel were a direct consequence of the finite propagation speed of that kernel. Building other kernels out of this one extended the results. The estimates were then used to establish a number of highly non-trivial results on the geometry of complete Riemannian manifolds.

The full power of one of the local geometric estimates in this paper, namely the local lower bound for the injectivity radius, was used in Perelman's proof of the Thurston Conjecture.

Jeff and Misha then went on to work on a number of other problems.

Part two: Characteristic numbers of complete manifolds of finite volume and bounded curvature

In [22] they studied the Gauss-Bonnet and Hirzebruch L-polynomial forms associated to such a metric. They showed that the integrals are well defined and give "geometric characteristic numbers". The question was:

When are these numbers homotopy invariants?

They assume that some neighborhood of infinity admits a covering which is either profinite or normal and has injectivity radius bounded below by a constant. With this assumption these two numbers are proper homotopy invariants. In fact the Gauss Bonnet number is shown to be a homotopy invariant, and when the manifold is of finite topological type, it is actually the Euler characteristic.

Part three: L^2-cohomology, von Neumann dimension and group cohomology

Another aspect of the work cited just above is the study of the L^2-Betti numbers of complete non-compact manifolds [22], [23]. One assumes $|K|$ and $\mathrm{vol}(M)$ bounded, and that on some normal covering $\tilde{M} \to M$ with covering group Γ, the injectivity radius is bounded below. For each degree k, M has a well-defined L^2-Betti number. It is defined as the trace of the kernel of the projection of L^2 k-forms onto the subspace $\mathcal{H}^k_{(2)}(M)$ of harmonic forms (defined to be those which are closed and co-closed). It can also be interpreted as the von Neumann dimension of the Γ-module $\overline{H}^k_{(2)}(\widetilde{M}) \equiv \ker d/\overline{\mathrm{Im} d}$. Their main result is that these L^2 Betti numbers are homotopy invariants.

Cheeger and Gromov then turn attention [24] to the more general case of a countable simplicial complex acted on simplicially by a countable group Γ. Taking L^2 cochains suitable limits leads to cohomology groups which are G-modules preserved by G-equivariant homotopies. The L^2 Betti numbers are defined to be the von Neumann Γ-dimensions of these modules.

Thus Cheeger and Gromov establish an invariant $b_k(\Gamma)$ as the kth L^2 Betti number of a contractible complex on which Γ acts freely. These invariants are shown to vanish in positive dimensions for amenable groups. A number of interesting facts are proved relating these invariants to geometry, topology and group theory.

One substantial achievement of [23] was the introduction of ρ-invariants for infinite coverings. Such invariants were first introduced by Atiyah-Patodi-Singer for finite coverings. In [23] they were generalized by using a von Neumann eta-invariant of the infinite covering. These objects, known as "Cheeger-Gromov invariants", have proved to have quite a few applications in knot theory.

Part four: Collapsing Riemannian manifolds while keeping their curvature bounded

They went on to write two important papers [25], [26] devoted to Riemannian manifolds with bounded curvature and small injectivity radius. Central idea is that of an **F-structure**.

Roughly an F-structure on X consists of a collection of tori (of varying dimensions) acting locally on finite covering spaces of subsets of X. These actions satisfy a compatibility condition which insures that X is partitioned into disjoint "orbits".

The F-structure has *positive rank* if the dimensions of all the orbits are positive.

Theorem (Cheeger-Gromov). *If X admits an F-structure, then X admits a family of Riemannian metrics g_δ with uniformly bounded curvature, such that as $\delta \to 0$, the injectivity radius i_p converges uniformly to zero at all points $p \in X$.*

Theorem (Cheeger-Gromov). *There exists a positive number $\epsilon(n)$, depending on dimension n, such that for a complete Riemannian n-manifold X with sectional curvature $|K_X| \leq 1$, there exists an open set $U \subset X$ so that*

1) *If $p \in X - U$, then $i_p > \epsilon(n)$, and*
2) *there exists an F-structure of positive rank on U.*

In particular, if $i_p < \epsilon(n)$ for all $p \in X$, then X has an F-structure of positive rank (and therefore collapses with bounded curvature).

Note. There is then later work with Fukaya. This paper [28] combines the approaches of Cheeger-Gromov and Fukaya, and in particular, gives a local picture of a manifold of bounded curvature on a fixed scale. They study the metric structure of the ϵ-collapsed part of a metric, and show there is a close approximation by a metric with *N-structure* – a sheaf of local nilpotent groups acting isometrically. It should be mentioned that Ghanaat, Min-Oo and Ruh [38] had independently found a local description of the geometry of a ball of a definite size in a manifold of bounded curvature, but they did not globalize it, (no N-structure) as was done in [28].

This paper [28] uses another work of Cheeger and Gromov [27] on "good choppings" (which refers to the possibility of surrounding a rough looking set with a manifold with boundary, for which the geometry of the boundary is suitably controlled.) Good choppings (and their compatibility with N-structures) played an an important role in Jeff's work with Gang Tian on Einstein 4-manifolds [34].

12. Work with Baum and with Müller and Schrader

There are still parts of Jeff's early work left to be discussed, but I will treat them only in broad terms. One contribution is his early paper with Paul Baum [1] which related characteristic numbers and Killing vector fields. Suppose X be a Riemannian manifold, V an infinitesimal isometry and Z is the set where $V = 0$. Baum and Cheeger established a formula which equates the integral of any characteristic form $p(\Omega)$ over X to an explicit residue integral over Z. This residue integral involves the universal polynomial p and the action of the Lie derivative L_V on the normal bundle of Z. This paper foreshadowed localization formulas in equivariant cohomology.

There is also a series of very nice papers written with Werner Müller and Robert Schrader [29], [30], [31]. These papers went deeply into the geometry of piecewise flat spaces. These papers studied the family of *Lipschitz-Killing curvatures* (which begins with scalar curvature and ends with the Gauss-Bonnet integrand). Among other things an important approximation theorem was proved and combinatorial formulas were derived. The formulas actually were derived in [12] where the contributions from singularities involved spectral invariants of the links. For these cases, the link invariants can actually be computed in terms of dihedral angles. (This is not true of the Cheeger formulas for the L-classes, in which the spectral invariants of the links are eta-invariants. These are global invariants for which there can be no canonical local formula, even in principle.)

Part of contribution from this collaboration was to give a rigorous foundation to the "Reggi calculus" often used in physics.

13. Some final remarks

As most readers will know, I have not began to touch on the profound contributions that Jeff has made in the last twenty some years. These include some long and very productive collaborations with Jean-Michel Bismut, Tobias Colding, Gang Tian, Xiaochun Rong, and most recently with Bruce Kleiner and with Kleiner and Assaf Naor.

References

[1] Paul Baum and Jeff Cheeger, *Infinitesimal isometries and Pontryagin numbers*, Topology **8** (1969), 173–193.

[2] Jeff Cheeger, "Comparison and Finiteness theorems for Riemannian Manifolds", Thesis (Ph.D.)-Princeton University, 1967.

[3] Jeff Cheeger, *Pinching theorems for a certain class of Riemannian manifolds*, Amer. J. Math. **91** (1969), 807–834.

[4] Jeff Cheeger, *A lower bound for the smallest eigenvalue of the Laplacian*, Problems in Analysis (Papers dedicated to Salamon Bochner, 1969), pp. 195–199, Princeton University Press, Princeton, NJ, 1970.

[5] Jeff Cheeger, *Finiteness theorems for Riemannian manifolds*, Amer. J. Math. **92** (1970), 61–74.

[6] Jeff Cheeger and Detlef Gromoll, *On the structure of complete manifolds of non-negative curvature*, Ann. of Math. **96** no. 2 (1972), 413–443.

[7] Jeff Cheeger and Detlef Gromoll, *The splitting theorem for manifolds of non-negative Ricci curvature*, J. Diff. Geom. **6** (1972), 119–128.

[8] Jeff Cheeger, *Some examples of manifolds of non-negative curvature*, J. Diff. Geom. **8** (1973), 623–628.

[9] Jeff Cheeger and Detlef Gromoll, *On the lower bound for the injectivity radius of* $\frac{1}{4}$*-pinched Riemannian manifolds*, J. Diff. Geom. **15** no. 3 (1980), 437–442.

[10] Jeff Cheeger and James Simons, *Differential characters and geometric invariants*, Geometry and Topology (College Park, Md. 1983–84), pp. 50–80, Lecture Notes in Math **1167**, Springer, Berlin, 1985.

[11] Jeff Cheeger, *Multiplication of differential characters*, Symposia Mathematica, Vol. XI (Convegno di Geometria, INDAM, Rome, 1972, pp. 441–445, Academic Press, London, 1973.

[12] Jeff Cheeger, *Analytic torsion and Reidemeister torsion*, Proc. Nat. Acad. Sci. U.S.A. **74** no. 7 (1977), 2651–2654.

[13] Jeff Cheeger, *Analytic torsion and the heat equation*, Ann. of Math. **109** no. 2 (1979), 259–322.

[14] Jeff Cheeger, *On the spectral geometry of spaces with cone-like singularities*, Proc. Nat. Acad. Sci. U.S.A. **76** no. 5 (1979), 2103–2106.

[15] Jeff Cheeger, *On the Hodge theory of Riemannian pseudomanifolds*, Geometry of the Laplace operator (Proc. Sympos. Pure Math., Univ. of Hawaii, Honolulu, Hawaii, 1979), pp. 91–146, Proc. of Sympos. Pure Math., XXXVI, Amer. Math. Soc., Providence, RI, 1980.

[16] Jeff Cheeger, *Spectral geometry of singular Riemannian spaces*, J. Diff. Geom. **18** no. 4 (1983), 575–657.

[17] Jeff Cheeger, *Hodge theory of complex cones*, Analysis and topology on singular spaces, II, III (Luminy, 1981), pp. 118–134, Astérisque, **101-102**, Soc. Math. France, Paris, 1983.

[18] Jeff Cheeger, Mark Goresky and Robert MacPherson, L^2*-cohomology and intersection homology of singular algebraic varieties*, Seminar on Differential Geometry, pp. 303–340, Ann. of Math. Studies **102** Princeton University Press, Princeton, NJ, 1982.

[19] Jeff Cheeger and Mike Taylor, *On the diffraction of waves by conical singularities, I*, Comm. Pure Appl. Math., **35** no. 3 (1982), 275–331.

[20] Jeff Cheeger and Mike Taylor, *On the diffraction of waves by conical singularities, II*, Comm. Pure Appl. Math., **35** no. 4 (1982), 487–529.

[21] Jeff Cheeger, Mikhael Gromov and Mike Taylor, *Finite propagation speed, kernel estimates for functions of the Laplace operator, and the geometry of complete Riemannian manifolds*, J. Diff. Geom. **17** no. 1 (1982), 15–53.

[22] Jeff Cheeger and Mikhael Gromov, *On the characteristic numbers of complete manifolds of bounded curvature and finite volume*, Differential Geometry and Complex Analysis, pp. 115–154, Springer, Berlin, 1985.

[23] Jeff Cheeger and Mikhael Gromov, *Bounds on the Von Neumann dimension of* L^2*-cohomology and the Gauss-Bonnet Theorem for open manifolds*, J. Diff. Geom. **21** no. 1 (1985), 1–34.

[24] Jeff Cheeger and Mikhael Gromov, *L^2-cohomology and group cohomology*, Topology **25** no. 2 (1986), 189–215.

[25] Jeff Cheeger and Mikhael Gromov, *Collapsing Riemannian manifolds while keeping their curvature bounded, I*, J. Diff. Geom. **23** no. 3 (1986), 309–346.

[26] Jeff Cheeger and Mikhael Gromov, *Collapsing Riemannian manifolds while keeping their curvature bounded, II*, J. Diff. Geom. **32** no. 1 (1990), 269–298.

[27] Jeff Cheeger and Mikhael Gromov, *Chopping Riemannian manifolds*, Differential Geometry, pp. 85–94, Pitman Monogr. Surveys Pure Appl. Math., Vol. 52, Longman Press, Harlow, 1991.

[28] Jeff Cheeger, Kinji Fukaya, and Mikhael Gromov, *Nilpotent structures and invariant metrics on collapsed manifolds*, J. Amer. Math. Soc. **5** (1992) no. 2, 327–372.

[29] Jeff Cheeger, Werner Müller and Robert Schrader, *Lattice gravity or Riemannian structure on piecewise linear spaces*, Unifies Theories of Elementary Particles (Munich, 1981), pp. 176–188, Lecture Notes in Physics, **160**, Springer, Berlin, 1982.

[30] Jeff Cheeger, Werner Müller and Robert Schrader, *On the curvature of piecewise flat spaces*, Comm. Math. Phys. **92** no. 3 (1984), 405–454.

[31] Jeff Cheeger, Werner Müller and Robert Schrader, *Kinematic and tube formulas for piecewise linear spaces*, Indiana Univ. Math.J. **35** no. 4 (1986), 737–754.

[32] Jeff Cheeger, *η-Invariants, the adiabatic approximation and conical singularities, I. The adiabatic approximation*, J. Diff. Geom. **26** no. 1 (1987), 175–221.

[33] Jeff Cheeger and Tobias Colding, *Lower bounds on Ricci curvature and the almost rigidity of warped products*, Ann. of Math. **144** (1996), no. 1, 189–237.

[34] Jeff Cheeger and Gang Tian, *Curvature and injectivity radius estimates for Einstein 4-manifolds*, J. Amer. Math. Soc. **19** no. 2 (2006), 487–525

[35] Eugenio Calabi, *An extension of E. Hopf's maximum principle with an application to Riemannian geometry*, Duke Math. J. **25** (1958), 45–56.

[36] Shiing-Shen Chern and James Simons, *Characteristic forms and geometric invariants*, Ann. of Math. **99** (1974), no. 2, 48–69.

[37] Jost Eschenburg and Ernst Heintze, *An elementary proof of the Cheeger-Gromoll splitting theorem*, Ann. Global Anal. Geom. **2** (1984) no. 2, 141–151.

[38] Patrick Ghanaat, Maung Min-Oo and Ernst Ruh, *Local structure of Riemannian manifolds*, Indiana Univ. Math. J. **39** (1990), 1305–1312.

[39] Detlef Gromoll and Wolfgang Meyer, *On complete open manifolds of positive curvature*, Ann. of Math. **90** (1969), 75–90.

[40] Mike Hopkins and Isador Singer, *Quadratic functions in geometry, topology and M-theory*, J. Diff. Geom. **70** (2005), no. 3, 329–452.

[41] Reese Harvey, Blaine Lawson and John Zweck, *The de Rham theory of differential characters and character duality*, Amer. J. Math. **125** (2003), no. 4, 791–847.

[42] Reese Harvey and Blaine Lawson, *From sparks to grundles – differential characters*, Comm. Anal. Geom. **14** (2004), no. 4, 673–695.

[43] Reese Harvey and Blaine Lawson, *D-Bar sparks*, Proc. London Math. Soc. **97** (2008), no. 1, 1–30.

[44] Alexander Kahle and Alessandro Valentino, *T-Duality and differential K-theory*, ArXiv:0912.2516.

[45] John Milnor, *Whitehead torsion*, Bull. A.M.S. **72** (1966), 348–426.

[46] John Milnor, *A note on curvature and the fundamental group*, J. Diff. Geom. **2** (1968), 1–7.

[47] Werner Müller, *Analytic torsion and R-torsion of Riemannian manifolds*, Advances in Math. **28** (1978), 233–305.

[48] Barrett O'Neill, *The fundamental equations of a submersion*, Michigan Math. J. **13** (1966), 459–469.

[49] Grisha Perelman, *The entropy formula for the Ricci flow and its geometric applications*, ArXiv:math/0211159.

[50] Grisha Perelman, *Ricci flow with surgery on three-manifolds*, ArXiv:math/0303109.

[51] Grisha Perelman, *Finite extinction time for the solutions to the Ricci flow on certain three-manifolds*, ArXiv:math/0307245.

[52] D.B. Ray and I.M. Singer, *R-torsion and the Laplacian on Riemannian manifolds*, Advances in Math. **7** (1971), 145–210.

[53] James Simons and Dennis Sullivan, *Axiomatic characterization of ordinary differential cohomology*, J. Topology **1** (2008), 45–56.

[54] V.A. Toponogov, *Riemannian spaces which contain straight lines*, Amer. Math. Soc. Transl. (2) **37** (1964), 287–290.

[55] Alan Weinstein, *On the homotopy-type of positively pinched manifolds*, Arch. Math. (Basel) **18** (1967), 523–524.

Part I

Differential Geometry

Boundary Value Problems for Metrics on 3-manifolds

Michael T. Anderson

Dedicated to Jeff Cheeger for his 65th birthday

Abstract. We discuss the problem of prescribing the mean curvature and conformal class as boundary data for Einstein metrics on 3-manifolds, in the context of natural elliptic boundary value problems for Riemannian metrics.

Mathematics Subject Classification (2000). Primary 53C42; Secondary 35J57.
Keywords. Conformal immersions, prescribed mean curvature.

1. Introduction

A question long of basic interest to geometers is the existence of complete Einstein metrics on manifolds. Any kind of theory for the existence or uniqueness of such metrics on compact manifolds is still far from sight. The only exception to this is the remarkable work of Perelman and Hamilton, which essentially gives a complete theory for closed 3-manifolds.

Instead of considering closed manifolds, it might be somewhat simpler to consider manifolds with boundary and look for a theory providing existence (and uniqueness) for geometrically natural boundary value problems. This has recently met with some success, in the context of complete conformally compact Einstein metrics, where one prescribes a conformal metric at conformal infinity [3], and in the context of a natural exterior boundary value problem for the static vacuum Einstein equations, [4].

In this note, we consider the simplest situation, namely boundary value problems for Einstein metrics in dimension 3, where the metrics are of constant curvature. Seemingly the simplest or most naive question one could ask in this context is the following:

Partially supported by NSF grant DMS 0905159.

QUESTION. Given a metric γ on a boundary surface $\partial M = S^2$ for instance, is there an Einstein metric (flat or constant curvature) on the 3-ball $M = B^3$ inducing γ on ∂M?

However, this is basically the isometric immersion problem for surfaces in $\mathbb{R}^3$, (or other space-forms), and is a notoriously difficult problem, also far from any current resolution. Note however that there are examples of smooth metrics on S^2 which do not isometrically immerse in $\mathbb{R}^3$, cf. [15], so the answer to the question is no in general.

The main difficulty here is that although the Einstein equations form an elliptic system of equations in a suitable gauge, Dirichlet boundary data for such a system never give rise to an elliptic boundary value problem. The Gauss constraint equation, (Gauss' Theorema Egregium), is an obstruction to such ellipticity. Thus, one should first consider what are the natural boundary value problems for the Einstein equations.

To describe this, let M be any 3-manifold with boundary ∂M which admits a metric of constant sectional curvature κ. We assume that

$$\pi_1(M, \partial M) = 0;$$

by elementary covering space arguments, this means that ∂M is connected and any loop in M is homotopic to a loop in ∂M, so that M is a three-dimensional handlebody.

Let $\mathcal{M}_\kappa = \mathcal{M}_\kappa^{m,\alpha}$ be the moduli space of metrics of constant curvature κ on M which are $C^{m,\alpha}$ up to ∂M, $m \geq 2$, $\alpha \in (0,1)$. This is the space of all such constant curvature metrics $\mathbb{M}_\kappa$ modulo the action of $\mathrm{Diff}_1^{m+1,\alpha}$ of diffeomorphisms of M equal to the identity on ∂M. In the case of $\kappa = 0$ for instance, the developing map gives an isometric immersion

$$D : (M, g) \to \mathbb{R}^3,$$

which induces an isometric Alexandrov immersion of $(\partial M, \gamma)$ into $\mathbb{R}^3$, where $\gamma = g_{\partial M}$. (An immersion of a surface in $\mathbb{R}^3$ is Alexandrov if it extends to an immersion of the bounding handlebody M). Similar remarks hold for all $\kappa \in \mathbb{R}$.

Let $\mathcal{C}^{m.\alpha}$ denote the space of conformal classes $[\gamma]$ of $C^{m,\alpha}$ metrics γ on ∂M, and let H denote the mean curvature of $\partial M \subset (M, g)$, with respect to the outward unit normal. It is proved in [5] that the moduli space $\mathcal{M}_\kappa = \mathcal{M}_\kappa^{m,\alpha}$ is a (C^∞) smooth Banach manifold. Moreover, setting as above $\gamma = g|_{T(\partial M)}$, the map

$$\begin{aligned} \Pi : \mathcal{M}_\kappa &\to \mathcal{C}^{m,\alpha}(\partial M) \times C^{m-1,\alpha}(\partial M), \\ \Pi([g]) &= ([\gamma],\ H), \end{aligned} \tag{1.1}$$

is a (C^∞) smooth Fredholm map, of Fredholm index 0. In fact the boundary data in (1.1) form elliptic boundary data for the Einstein equations. There are other elliptic boundary value problems for Einstein metrics, some of which are discussed in [5]. However, the data in (1.1) is geometrically the most natural so we restrict the discussion to this case.

It follows in particular that $Im\Pi$ is a variety of finite codimension in $\mathcal{C}^{m,\alpha}(\partial M) \times C^{m-1,\alpha}(\partial M)$. One would expect that generic metrics in $\mathcal{M}_\kappa$ are regular points for Π, in which case $Im\Pi$ would at least contain open domains in the target $\mathcal{C}^{m,\alpha}(\partial M) \times C^{m-1,\alpha}(\partial M)$; (a proof of this is still lacking however).

The result in (1.1) shows that one has a good local existence theory for this boundary value problem and it raises the global problem:

QUESTION. Given $([\gamma], H) \in \mathcal{C}^{m,\alpha}(\partial M) \times C^{m-1,\alpha}(\partial M)$, (possibly with some restrictions), does there exist a unique metric, (up to isometry), g on M such that

$$\Pi(g) = ([\gamma], H). \tag{1.2}$$

To the author's knowledge, it does not seem that this question, although clearly quite natural, has been studied previously. There are many previous works on the existence of surfaces of prescribed mean curvature in $\mathbb{R}^3$ for instance, cf. [18] for example, and [19] or [7] for further discussion and references. However, in these situations H is a given function on $\mathbb{R}^3$; moreover, there is no prescription of the conformal class. In the case of $\kappa = 0$ for example, the question can be rephrased as the question of the existence and uniqueness of an Alexandrov immersion of a surface $F : \Sigma = \partial M \to \mathbb{R}^3$ with prescribed conformal class $[\gamma]$ and prescribed mean curvature H, i.e.,

$$[F^*(g_{\mathrm{Eucl}})] = [\gamma], \quad H(F(x)) = H(x). \tag{1.3}$$

Note that the diffeomorphism group of ∂M acts non-trivially on both parts $[\gamma]$ and H of the boundary data: if $\varphi \in \mathrm{Diff}(\partial M)$, then the immersion $F \circ \varphi$ has the same image as F, but is a reparametrization of F. For different but related studies on surfaces of prescribed mean curvature, see for example [9], [11] and [12].

To address a global question as above, the basic issue is whether the boundary map Π is proper. In analogy to the simpler method of continuity commonly used in PDE, this is the closedness issue; one requires a priori estimates or compactness properties for spaces of solutions. For very simple reasons, the map Π is not proper in general, and one first needs to sharpen the problem to account for this. Thus, for example smooth bounded domains in $\mathbb{R}^3$ may degenerate from the "inside", in that the injectivity radius within M may go to 0 near ∂M, causing the boundary to develop self-intersections and the domain M is no longer a manifold. This behavior can be ruled out via the maximum principle, (see also Lemma 2.4 below), under the assumption that $H > 0$. Thus, let

$$\mathcal{M}_\kappa^+ = \Pi^{-1}(\mathcal{C}^{m,\alpha} \times C_+^{m-1,\alpha}),$$

be the space of constant curvature metrics with $H > 0$ at ∂M. Clearly, $\mathcal{M}_\kappa^+$ is an open submanifold of $\mathcal{M}_\kappa$ and one may consider the associated (restricted) boundary map

$$\Pi_+ : \mathcal{M}_\kappa^+ \to \mathcal{C}^{m,\alpha} \times C_+^{m-1,\alpha}. \tag{1.4}$$

Next, recall by the uniformization theorem that the space of metrics $\mathrm{Met}(S^2)$ on S^2 equals $\mathrm{Diff}(S^2) \times C_+(S^2)$; any metric γ is of the form $\gamma = \varphi^*(\lambda^2\gamma_{+1})$, where

γ_{+1} is the round metric of radius 1. (Similarly for surfaces Σ of higher genus, $\mathrm{Met}(\Sigma)$ is a bundle over the Riemann moduli space with fiber $\mathrm{Diff}(\Sigma) \times C_+(\Sigma)$.) The group $\mathrm{Diff}(S^2) \times C_+(S^2)$ thus acts transitively on $\mathrm{Met}(S^2)$ and has stabilizer at γ_{+1} equal to the group of essential conformal transformations $\mathrm{Conf}(S^2)$ of $S^2(1)$. It follows that one has a natural identification

$$\mathcal{C}^{m,\alpha}(S^2) \simeq \mathrm{Diff}^{m+1,\alpha}(S^2)/\,\mathrm{Conf}(S^2). \tag{1.5}$$

The conformal group also acts on the space $C_+^{m-1,\alpha}$ of mean curvature functions: $H \to H \circ \varphi$, for $\varphi \in \mathrm{Conf}(S^2)$. It is easy to verify that this action is free and proper, except on the functions $H = \mathrm{const}$, which are the fixed points of the action; this is because the flow of the conformal vector fields contracts or expands all of $S^2 \setminus \{pt\}$ to a point.

It follows then that at the special values $([\gamma], c)$ where $H = c$, the map Π_+ in (1.4) is not proper. The non-compact conformal group $\mathrm{Conf}(S^2)$ fixes this data, but acts nontrivially (and faithfully) on $\mathcal{M}_\kappa^+$; if $\Pi_+(g) = ([\gamma], c)$, then also $\Pi_+(\varphi^*(g)) = ([\gamma], c)$, for any $\varphi \in \mathrm{Conf}(S^2)$ extended to a diffeomorphism of M. On the other hand, this is the only value where $\mathrm{Conf}(S^2)$ acts non-properly. (Note this issue arises only for S^2, not for boundaries of higher genus.)

There are two ways to deal with this issue. First, one may just study the behavior of Π_+ away from the "round" metrics $H = c$, i.e., consider the global behavior of the map

$$\Pi' : \mathcal{M}'_\kappa \to \mathcal{C}^{m,\alpha} \times (C_+^{m-1,\alpha})', \tag{1.6}$$

where $(C_+^{m-1,\alpha})' = C_+^{m-1,\alpha} \setminus \{\text{constants}\}$ and $\mathcal{M}'_\kappa = \Pi_+^{-1}((C_+^{m-1,\alpha})')$. This map is again smooth and Fredholm, of index 0.

Alternately, one may include the round metrics, but divide out by the action of $\mathrm{Conf}(S^2)$. Briefly, as is standard, choose a fixed marking to freeze the action of the conformal group on S^2. Thus, fix three points p_i, $i = 1, 2, 3$ on $S^2(1)$ with

$$\mathrm{dist}_{\gamma_{+1}}(p_i, p_j) = \pi/2. \tag{1.7}$$

Let $\mathcal{N}_\kappa^+$ be the marked moduli space of constant curvature metrics on $M = B^3$ consisting of metrics g such that (1.7) holds on $\partial M = S^2$ with $\gamma = g|_{\partial M}$ in place of γ_{+1}. The condition (1.7) can always be realized by changing g by a conformal diffeomorphism, so that

$$\mathcal{N}_\kappa^+ \simeq \mathcal{M}_\kappa^+/\,\mathrm{Conf}(S^2),$$

the condition (1.7) giving a slice for the action of $\mathrm{Conf}(S^2)$ on $\mathcal{M}_\kappa$. (Of course this marking is not necessary in case $\chi(\partial M) \leq 0$ so that $\mathcal{N}_\kappa^+ \simeq \mathcal{M}_\kappa^+$ in such cases). The map Π_+ in (1.4) clearly restricts under the slicing (1.7) to a smooth map

$$\Pi_+ : \mathcal{N}_\kappa^+ \to \mathcal{C}^{m,\alpha}(\partial M) \times C^{m-1,\alpha}(\partial M). \tag{1.8}$$

However, the index of this map is now -3, since $\mathrm{Conf}(S^2)$ is three-dimensional and so one must also divide the target space by the remaining action of $\mathrm{Conf}(S^2)$

on the space $C_+^{m-1,\alpha}$ of mean curvature functions. Let then $\mathcal{B}[\mathcal{C}^{m,\alpha} \times C_+^{m-1,\alpha}]$ be the quotient space

$$\mathcal{B}[\mathcal{C}^{m,\alpha} \times C_+^{m-1,\alpha}] = [\mathcal{C}^{m,\alpha} \times C_+^{m-1,\alpha}]/\operatorname{Conf}(S^2) = \mathcal{C}^{m,\alpha} \times (C_+^{m-1,\alpha}/\operatorname{Conf}(S^2)).$$

The map Π_+ in (1.8) then descends to a smooth Fredholm map

$$\Pi_+ : \mathcal{N}_\kappa^+ \to \mathcal{B}[\mathcal{C}^{m,\alpha} \times C_+^{m-1,\alpha}], \tag{1.9}$$

of Fredholm index 0. The two formulations (1.6) and (1.9) are basically equivalent.

Now if Π_+ in (1.9), (or Π' in (1.6)), is proper, then by work of Smale [16], it has a well-defined degree $\deg \Pi' \in \mathbb{Z}_2$, (and most likely a $\mathbb{Z}$-valued degree if the spaces can be given an orientation). Elementary degree theory implies that if

$$\deg \Pi_+ \neq 0,$$

then Π_+ in (1.6) is surjective, answering at least the existence part of the question above.

In fact, if Π_+ is proper, then one has

$$\deg \Pi_+ \neq 0 \text{ for } \partial M = S^2, \text{ but } \deg \Pi_+ = 0 \text{ for } \partial M \neq S^2. \tag{1.10}$$

This follows from the Alexandrov-Hopf rigidity theorems, [1], [8]. Namely, any metric on a surface $\Sigma = \Sigma_g$ of genus g which is Alexandrov immersed in a space-form with $H = \text{const}$ is necessarily a round sphere. This uniqueness also holds infinitesimally, showing that the "round" conformal class $([\gamma_{+1}], H = c)$ is a regular value of Π. Since the Hopf theorem implies that the inverse image of this round regular value is unique, it follows that $\deg \Pi_+ \neq 0$ for $\Sigma = S^2$. For Σ_g with $g \neq 0$, the same argument shows that Π_+ in (1.9) is not surjective, which implies $\deg \Pi_+ = 0$.

Thus, at least in the case of S^2 the existence question above has been reduced to the properness of Π_+. This issue will be discussed in detail in the next section; we will show however that Π_+ or Π' is in fact not proper, so that further modifications are necessary to understand the global behavior of these boundary maps.

2. Analysis of the Boundary map Π

We begin by filling in some details from the discussion in §1. Since the full curvature is determined by the Ricci curvature in 3-dimensions, any metric $g \in \mathbb{M}_\kappa$ satisfies the Einstein equation

$$\operatorname{Ric}_g - 2\kappa \cdot g = 0. \tag{2.1}$$

We wish to view (2.1) as an elliptic equation for g. This is not possible due to the diffeomorphism invariance of (2.1), and so one needs to choose a gauge to break this invariance. Let $\widetilde{g} \in \mathbb{M}_\kappa$ be a fixed but arbitrary (constant curvature) background metric. The simplest choice of gauge is the Bianchi-gauge, with the associated Bianchi-gauged Einstein operator, given by

$$\begin{aligned} \Phi_{\widetilde{g}} &: \operatorname{Met}(M) \to S_2(M), \\ \Phi_{\widetilde{g}}(g) &= \operatorname{Ric}_g - 2\kappa g + \delta_g^* \beta_{\widetilde{g}}(g), \end{aligned} \tag{2.2}$$

where $(\delta^* X)(A,B) = \frac{1}{2}(\langle \nabla_A X, B\rangle + \langle \nabla_B X, A\rangle)$ and $\delta X = -\operatorname{tr}\delta^* X$ is the divergence and $\beta_{\widetilde{g}}(g) = \delta_{\widetilde{g}} g + \frac{1}{2} d \operatorname{tr}_{\widetilde{g}} g$ is the Bianchi operator with respect to $\widetilde{g}$.

Clearly g is Einstein if $\Phi_{\widetilde{g}}(g) = 0$ and $\beta_{\widetilde{g}}(g) = 0$, so that g is in the Bianchi-free gauge with respect to $\widetilde{g}$. Using standard formulas for the linearization of the Ricci and scalar curvatures, cf. [6] for instance, one finds that the linearization of Φ at $\widetilde{g} = g$ is given by

$$L(h) = 2(D\Phi_{\widetilde{g}})_g(h) = D^* D h - 2R(h). \tag{2.3}$$

The zero-set of $\Phi_{\widetilde{g}}$ near $\widetilde{g}$,

$$Z = \{g : \Phi_{\widetilde{g}} = 0\}, \tag{2.4}$$

consists of metrics $g \in \operatorname{Met}(M)$ satisfying the equation $Ric_g - 2\kappa g + \delta_g^* \beta_{\widetilde{g}}(g) = 0$.

Given $\widetilde{g}$, consider the Banach space

$$\operatorname{Met}_C(M) = \operatorname{Met}_C^{m,\alpha}(M) = \{g \in \operatorname{Met}^{m,\alpha}(M) : \beta_{\widetilde{g}}(g) = 0 \text{ on } \partial M\}. \tag{2.5}$$

Clearly the map

$$\Phi : \operatorname{Met}_C(M) \to S^2(M),$$

is C^∞ smooth. Let Z_C be the space of metrics $g \in \operatorname{Met}_C(M)$ satisfying $\Phi_{\widetilde{g}}(g) = 0$, and let $\mathbb{M}_C = \mathbb{M}_\kappa \cap Z_C$ be the subset of constant curvature metrics g, $\operatorname{Ric}_g = 2\kappa g$ in Z_C. It is proved in [5] that Z_C is a smooth Banach manifold and

$$\mathbb{M}_C = Z_C,$$

so that any metric $g \in Z_C$ near $\widetilde{g}$ is necessarily constant curvature, with $Ric_g = 2\kappa g$, and in Bianchi gauge with respect to $\widetilde{g}$. This result also holds at the linearized level. The spaces Z_C are smooth slices for the action of the diffeomorphism group $\operatorname{Diff}_1^{m+1,\alpha}$ on $\mathbb{M}_\kappa$ and it follows that the quotient $\mathcal{M}_\kappa$ is a smooth Banach manifold.

Next consider elliptic boundary data for the operator Φ in (2.2). Dirichlet or Neumann boundary data are not elliptic; this follows by inspection from the Gauss constraint equation (2.12) below, (or from the proof below). The following result is proved in [5]; we give the main details of the proof, since it is useful to compare this with the discussion in §3.

Proposition 2.1. *The Bianchi-gauged Einstein operator Φ with boundary conditions*

$$\beta_{\widetilde{g}}(g) = 0, \quad [g^T] = [\gamma], \quad H_g = h \ \text{ at } \ \partial M, \tag{2.6}$$

is an elliptic boundary value problem of Fredholm index 0.

Proof. It suffices to show that the leading order part of the linearized operators at the Euclidean metric forms an elliptic system. The leading order symbol of $L = D\Phi$ is given by

$$\sigma(L) = -|\xi|^2 I, \tag{2.7}$$

where I is the 3×3 identity matrix. In the following, the subscript 0 represents the direction normal to ∂M in M, and Latin indices run from 1 to 2. The positive roots of (2.7) are $i|\xi|$, with multiplicity 3. Writing $\xi = (z, \xi_i)$, the symbols of the leading order terms in the boundary operators are:

$$-2izh_{0k} - 2i\sum \xi_j h_{jk} + i\xi_k \operatorname{tr} h = 0,$$
$$-2izh_{00} - 2i\sum \xi_k h_{0k} + iz \operatorname{tr} h = 0,$$
$$h^T = (\gamma')^T \mod \gamma, \quad H'_h = \omega,$$

where h is a 3×3 matrix. Ellipticity requires that the operator defined by the boundary symbols above has trivial kernel when z is set to the root $i|\xi|$. Carrying this out then gives the system

$$2|\xi|h_{0k} - 2i\sum \xi_j h_{jk} + i\xi_k \operatorname{tr} h = 0, \tag{2.8}$$
$$2|\xi|h_{00} - 2i\sum \xi_k h_{0k} - |\xi| \operatorname{tr} h = 0, \tag{2.9}$$
$$h_{kl} = \varphi\delta_{kl}, \quad H'_h = 0. \tag{2.10}$$

where φ is an undetermined function.

Multiplying (2.8) by $i\xi_k$ and summing gives

$$2|\xi|i\sum \xi_k h_{0k} = 2i^2\xi_k^2 h_{kk} - i^2\xi_k^2 \operatorname{tr} h.$$

Substituting (2.9) on the term on the left above then gives

$$2|\xi|^2 h_{00} - |\xi|^2 \operatorname{tr} h = -2\sum \xi_k^2 h_{kk} + |\xi|^2 \operatorname{tr} h,$$

so that

$$|\xi|^2 h_{00} - |\xi|^2 \operatorname{tr} h = -\sum \xi_k^2 h_{kk} = -\varphi|\xi|^2.$$

Using the fact that $\operatorname{tr} h - h_{00} = \sum h_{kk} = n\varphi$, it follows that $\varphi = 0$ and hence $h^T = 0$.

A simple computation shows that to leading order, $H'_h = \operatorname{tr}^T(\nabla_N h - 2\delta^*(h(N)^T))$, which has symbol $iz\sum h_{kk} - 2i\xi_k h_{0k}$. Setting this to 0 at the root $z = i|\xi|$ gives

$$\sum(|\xi|h_{kk} + 2i\xi_k h_{0k}) = 0.$$

Since $h^T = 0$, this gives $\sum \xi_k h_{0k} = 0$, which, via (2.9) gives $h_{00} = 0$ and hence via (2.8), $h = 0$.

This proves that the boundary data (2.6) are elliptic for Φ. The proof that the Fredholm index is 0 is given in [5]. □

We now turn to the main issue, the properness of the map Π_+ in (1.9) or Π' in (1.6). This amounts to proving (a priori) estimates for metrics $g \in \mathcal{M}_\kappa$ in terms of the boundary data $([\gamma], H)$. The main result in this direction is the following:

Proposition 2.2. *Let $\mathcal{K}$ be a compact set in the space of boundary data $\mathcal{C}^{m,\alpha} \times C_+^{m-1,\alpha}$. Then for any $K < \infty$, the space of metrics $g \in \mathcal{N}_\kappa^+$ such that*

$$\Pi_+(g) \in \mathcal{K} \quad \text{and} \quad a = \operatorname{area}(\partial M) \leq K, \tag{2.11}$$

is compact.

This result shows that Π_+ is proper, under the assumption of an upper bound on a. The proof of this result follows below, organized into several lemmas.

To begin, we recall the constraint equations at ∂M, i.e., the Gauss and Gauss-Codazzi equations:

$$|A|^2 - H^2 + 2K_\gamma = s_g - 2\,\mathrm{Ric}_g(N,N) = 2\kappa, \tag{2.12}$$

$$\delta(A - H\gamma) = -\,\mathrm{Ric}(N,\cdot) = 0, \tag{2.13}$$

where A is the second fundamental form and N is the outward unit normal.

One of the most important issues is to obtain a bound on $|A|$.

Lemma 2.3. *There is a constant $C_0 < \infty$, depending only on $\mathcal{K}$ and K in Proposition 2.2, such that*

$$|A| \leq C_0. \tag{2.14}$$

Proof. The proof is by contradiction, by means of a blow-up argument. To begin, integrating the Gauss constraint (2.12) and using the Gauss-Bonnet theorem gives

$$\int_{\partial M} |A|^2 = \int_{\partial M} H^2 - 4\pi\chi(\partial M) + 2\kappa \cdot \mathrm{area}(\partial M). \tag{2.15}$$

By assumption, $a = \mathrm{area}(\partial M)$ is uniformly bounded,

$$\mathrm{area}(\partial M) \leq K < \infty.$$

This and (2.15) give an a priori bound on the scale-invariant quantity $\int |A|^2$:

$$\int_{\partial M} |A|^2 \leq C. \tag{2.16}$$

Now choose a point x on ∂M where $|A|$ is maximal, and rescale the metric so that $|A|(x) = 1$, with $|A|(y) \leq 1$ everywhere, so $\bar{g} = \lambda^2 g$ where $\lambda = |A|(x)$. It follows directly from the constraint equation (2.12) that the intrinsic curvature $K_{\bar{\gamma}}$ of $\bar{\gamma}$ is also uniformly bounded. The family of such metrics is compact in the pointed $C^{1,\alpha}$ topology, by the Cheeger-Gromov compactness theorem for instance; this means that modulo diffeomorphisms of ∂M, the metric γ itself is uniformly controlled in $C^{1,\alpha}$, (in suitable local coordinates and within bounded distance to x).

Now by assumption, the conformal class $[\gamma]$ of γ is uniformly controlled. It follows that the diffeomorphisms above, (in which $\bar{\gamma}$ is uniformly controlled), are themselves controlled modulo the group of conformal diffeomorphisms, cf. (1.5). Thus, passing if necessary from $\bar{\gamma}$ to $\gamma' = \varphi^*(\bar{\gamma})$, where φ is a conformal diffeomorphism, it follows that the metric γ' is uniformly controlled in $C^{1,\alpha}$, (locally, within bounded distance to x). Together with the uniform bound on $|A|$ above, it follows from Proposition 2.1 and elliptic regularity that the metric g' is controlled in the stronger $C^{m,\alpha}$ norm, up to its boundary.

Suppose then g_i is a sequence where $\max |A| \to \infty$. By rescaling as above one may pass to a smoothly convergent subsequence of $\{g_i'\}$ to obtain a smooth limit g'. The smooth ($C^{m,\alpha}$) convergence implies on the one hand that the limit is not flat, since the condition $|A|(x) = 1$ passes continuously to the limit. The estimate

(2.16) also holds on the limit. Since $H \to 0$ in the rescalings, it follows that the limit is a complete immersed minimal surface in $\mathbb{R}^3$ with finite total curvature

$$\int_\Sigma |A|^2 < \infty.$$

Moreover, since the conformal classes $[\gamma_i]$ of g_i on ∂M are uniformly controlled, the sequence γ_i' has a uniformly controlled (large scale) atlas of conformal coordinates. Hence the limit is conformally isometric to $\mathbb{R}^2$, i.e., the limit minimal surface is pointwise conformal to $\mathbb{R}^2$. Finally, these minimal surfaces are in fact embedded; this follows from Lemma 2.4 below. However, it is well known that the only such surfaces are flat planes, (cf. [14] for instance), and hence $A = 0$ in the limit. This contradiction establishes the bound (2.14). □

Next we show that the normal exponential map has injectivity radius bounded below

$$\mathrm{inj}_N \geq i_0$$

where i_0 depends only on an upper bound for $|A|$. This follows from the following Lemma.

Let here N be the inward unit normal to ∂M in M and consider the associated normal exponential map to ∂M, $tN \to \exp_p(tN)$, giving the geodesic normal to ∂M at p. This is defined for t small, and let $D(p)$ be the maximal time interval on which $\exp_p(tN) \in M$, (so that the geodesic does not hit ∂M again before time $D(p)$). Thus, $D : \partial M \to \mathbb{R}^+$.

Lemma 2.4. *Given $H > 0$, suppose $|A| \leq C_0$. Then there is a constant t_0, depending only on C_0 and κ, (and the lower bound on H when $\kappa < 0$), such that*

$$D(p) \geq t_0. \tag{2.17}$$

Proof. This is a well-known result in Riemannian geometry, essentially due to Frankel, and follows from the 2nd variational formula for geodesics. First, given bounds on $|A|$ and κ, by standard comparison geometry one has a lower bound on the distance to the focal locus of the normal exponential map $\exp(tN)$, i.e., a lower bound d_0 on the distance to focal points. Suppose then

$$\min D < d_0.$$

If the minimum is achieved at p, then the normal geodesic to ∂M at p intersects ∂M again at a point p', and the intersection is orthogonal to ∂M at p'. Denoting this geodesic by σ, and letting $\ell = D(p)$ be the length of σ, the 2nd variational formula of energy gives

$$E''(V,V) = \int_0^\ell (|\nabla_T V|^2 - \langle R(T,V)V,T\rangle)dt - \langle \nabla_V T, V\rangle|_0^\ell, \tag{2.18}$$

where $T = \dot{\sigma}$ and V is any variation vector field along σ orthogonal to σ. By the minimizing property, one has $E''(V,V) \geq 0$, for all V. Choose then $V = V_i$ to be parallel vector fields e_i, running over an orthonormal basis at $T_p(\partial M)$. The first term in (2.18) then vanishes, while the second sums to $-\mathrm{Ric}(T,T) = -2\kappa$. The boundary terms sum to $\pm H$, at p and p'. Taking into account that T points into M at p while it points out of M at p', this gives

$$0 \leq -2\kappa\ell - (H(p) + H(p')).$$

Since $H > 0$, this gives immediately a contradiction if $\kappa \geq 0$, and also gives a contradiction if $\kappa < 0$, if H is bounded below, depending only on the size of κ (if ℓ is sufficiently small). This proves the estimate (2.17). □

Finally we show that Lemma 2.3 implies that the intrinsic geometry of $(\partial M, \gamma)$ is controlled.

Lemma 2.5. *There is a constant C_1, depending only on C_0 in* (2.14) *and $\mathcal{K}$, such that*

$$|K_\gamma| \leq C_1, \tag{2.19}$$

Moreover, the metric γ is uniformly controlled, modulo conformal diffeomorphisms, by C_1, (and a).

Proof. As in the proof of Lemma 2.3, via the Gauss constraint equation (2.12), a bound on $|A|$ implies a bound on K_γ, giving (2.19). A standard simple analytic argument then gives control on the metric γ itself when $\chi(\partial M) \leq 0$. Namely, write $\gamma = \lambda^{-2}\gamma_0$, where γ_0 is the conformal metric with constant curvature σ and σ is chosen so that areaγ = areaγ_0. The formula for the behavior of Gauss curvature under conformal changes then gives

$$\lambda^2 \Delta_{\gamma_0}(\log \lambda) = -\lambda^2\sigma - K_\gamma.$$

The maximum principle implies an upper bound on λ and hence, by elliptic regularity, one has uniform $C^{1,\alpha}$ control on λ and so, (via standard bootstrap arguments), λ is controlled in $C^{m,\alpha}$.

This argument does not work when $\chi(\partial M) > 0$, (since the minimum or maximum principle does not hold). In this case, one can use the same argument as that given in the proof of Proposition 2.2. Thus, the bound $|K_\gamma|$ implies that the metric is controlled modulo diffeomorphisms, by the Cheeger-Gromov compactness theorem. Here we use the fact that the length of the shortest closed geodesic, and hence the injectivity radius of γ, is bounded below, since $|A|$ is bounded above. Since the conformal class $[\gamma]$ is assumed to be controlled, the diffeomorphisms are themselves controlled, modulo the group of conformal diffeomorphisms.

Note that (2.15) shows that $a = \text{area}(\partial M)$ is bounded below, and hence the diameter of $(\partial M, \gamma)$ is also bounded above and below. This proves the result. Note also that since $\text{diam}(M, g) \leq \text{diam}(\partial M, \gamma)$, this also gives a uniform upper bound on the diameter of (M, g). □

The results above prove Proposition 2.2. This result implies that the "enhanced" boundary map

$$g \to ([\gamma], H, a) \tag{2.20}$$

is proper. While this map is Fredholm, it is Fredholm of index -1 and so does not have a well-defined degree; for this, one needs the Fredholm index to be non-negative.

There are two ways in which one may try to proceed at this point.

(I). One may try to prove that $a = \text{area}(\partial M)$ is controlled by the boundary data $([\gamma], H)$, which would then prove that Π itself, (i.e., Π_+ or Π'), is proper.

However, this is false. It follows from the proof of Proposition 2.2 that counterexamples must closely resemble the helicoid (in a suitable scale), since the helicoid is the unique complete embedded minimal surface in $\mathbb{R}^3$ conformally equivalent to $\mathbb{R}^2$, (besides the plane), cf. [13]. In fact conversely, one may use the helicoid to construct examples of metrics g_i where $([\gamma_i], H_i)$ are uniformly bounded but

$$a_i \to \infty. \tag{2.21}$$

To see this, consider the helicoid $\mathcal{H} = \mathcal{H}_L$,

$$x = \rho\cos\theta, \quad y = \rho\sin\theta, \quad z = L^{-1}\theta,$$

$L = L(\varepsilon) >> 1$, wrapping around z-axis arbitrarily many times in the interval $z \in [-\varepsilon, \varepsilon]$; assume here $\rho \in [-1, 1]$. Consider an almost horizontal $S^1 \subset \mathcal{H}$ formed by connecting two line segments parallel to the x-axis in $\mathcal{H}$ at height $\pm\varepsilon$ by a circular arc joining their endpoints along a helix in $\mathcal{H}$. This S^1 bounds a disc $D^2 \subset \mathcal{H}$. Now form the vertical z-cylinder over this boundary S^1, and take it to a fixed height, say $z = \frac{1}{2}$ and then cap off the circular boundary at $z = \frac{1}{2}$ by a horizontal disc. This gives first an immersed S^2, which is also Alexandrov immersed, since it may be perturbed to an embedding. This S^2 may also be perturbed so that $H > 0$ everywhere. Namely, one may first deform the helicoid very slightly to a surface with $H > 0$, and in fact with H uniformly bounded away from 0 and ∞, cf. [17] for instance. The vertical cylinder has $H > 0$ and one can bend the top flat disc to $H > 0$. Finally, the corners of S^2 may also be smoothed to $H > 0$ everywhere.

The conformal class of the helicoid is fixed under arbitrary rescalings, (i.e., variations of L), and the gluing process above is also uniformly controlled; hence the conformal class of the collection of surfaces above is uniformly controlled. It is clear from the construction that (2.21) holds as the number of wrappings of the helicoid is taken to infinity.

Recall the Hopf uniqueness theorem: if Σ is a sphere immersed in a space-form of constant curvature with $H = \text{const}$, then Σ is umbilic and so locally isometric to a round sphere. The examples above seem to indicate or suggest that the rigidity associated to the Hopf theorem cannot be weakened to an "almost rigidity" theorem; thus we expect given any $\varepsilon > 0$, there exist surfaces $\Sigma_\varepsilon \subset \mathbb{R}^3$ diffeomorphic to S^2 such that

$$2 - \varepsilon \leq H_{\Sigma_\varepsilon} \leq 2 + \varepsilon,$$

which are not close to a round sphere $S^2(1) \subset \mathbb{R}^3$. Of course, one must have $\text{area}(\Sigma_\varepsilon) \to \infty$ as $\varepsilon \to 0$. It would be interesting to know the answer to this question.

(II). Instead, referring to the context of (2.20), one may add an extra scalar variable λ to the domain to obtain a map of Fredholm index 0. [Alternately, one may restrict the domain $\mathcal{M}'_\kappa$ in (1.6), or $\mathcal{N}^+_\kappa$ in (1.9) to surfaces where $a = 1$; correspondingly, one must then divide the target space by an $\mathbb{R}^+$ action. There is no essential difference between these so we discuss only the former].

Thus, extend for example the domain $\mathcal{M}'_\kappa$ to $\mathcal{M}'_\kappa \times \mathbb{R}^+$, and consider the following typical examples:

$$(g, \lambda) \to ([\gamma], H, a + \lambda). \tag{2.22}$$

$$(g, \lambda) \to ([\gamma], (H/H_{\min}) - 1 + \lambda, a), \tag{2.23}$$

where $H_{\min}$ is the minimum value of H.

These maps are Fredholm, of Fredholm index 0. However, neither map is proper; in (2.22), one may have $\lambda \to 0$ while in (2.23) one may have $H_{\max} \to \infty$, both within compact sets of boundary data. Consider next shifting the λ-variable also to the space $\mathcal{C}$. For example, let ψ_λ be a curve of diffeomorphisms of ∂M with $\psi_\lambda \to \infty$ as $\lambda \to 0$ and consider

$$(g, \lambda) \to ([\psi^*_\lambda(\gamma)], H, a + \lambda). \tag{2.24}$$

As before, this behaves well in the second two factors, but it is not clear if there exists a curve ψ_λ for which (2.24) is proper; in any case we have not been able to find a construction to make this map proper.

Next consider enlarging the domain by adding a scale factor, when $\kappa \neq 0$. Thus, let $\mathcal{M}' = \cup_{\kappa<0}\mathcal{M}'_\kappa$, and define

$$\begin{aligned} \mathcal{M}' &\to \mathcal{C} \times C_+, \\ g &\to ([\gamma], H, a). \end{aligned} \tag{2.25}$$

This gives of course control of both H and a, and by (2.15), one deduces uniform control on $|\kappa|$, when $\kappa < 0$. However, one cannot prevent the possibility that $\kappa \to 0$, so that again its not clear if this map can be made proper. (Including the spaces $\mathcal{M}'_\kappa$ with $\kappa > 0$ also does not seem to help.)

Consider finally the following modification of (2.23):

$$\widetilde{\Pi} : (g, \lambda) \to ([\gamma], (H/H_{\min}) - 1 + \lambda, a + H_{\max}). \tag{2.26}$$

The map $\widetilde{\Pi}$ is Fredholm, of index 0, and is now proper by Proposition 2.2, since control of the data in the target space gives control on H, a and the conformal class $[\gamma]$. It thus has a well-defined degree.

However, the Hopf rigidity theorem now shows that

$$\deg \widetilde{\Pi} = 0. \tag{2.27}$$

Namely, consider the case $\partial M = S^2$ and $\kappa = 0$. When restricted to the "round" metrics where $H = \text{const}$, by the constraint equation (2.12) one has $H^2 a = 16\pi$, so that $H = H_{\max} = \sqrt{16\pi/a}$. This gives

$$\beta(a) \equiv a + H_{\max} = a + \sqrt{16\pi/a},$$

which, for a given value of $\beta(a) = c$ has two positive real solutions $a > 0$. The function β is a simple fold map $\mathbb{R}^+ \to [(4\pi)^{1/3}, \infty)$. This implies (2.27). (Although the round metric is not in $\mathcal{M}'_\kappa$, the discussion above remains valid for data near the round metric.)

There is another, quite different argument showing that $\widetilde{\Pi}$ is not onto, and hence has degree 0. Namely on any $g \in \mathcal{M}_\kappa$ with boundary data (γ, H), one has

$$\int_{S^2} X(H)dV_\gamma = 0, \tag{2.28}$$

where X is any conformal Killing field on S^2. This follows from the constraint equation (2.13). Namely, pairing (2.13) with a vector field X and integrating over $(\partial M, \gamma)$ gives

$$\int_{\partial M} \langle \delta A, X \rangle = -\int_{\partial M} \langle dH, X \rangle.$$

The left side equals $\int_{\partial M} \langle A, \delta^* X \rangle$, and for X conformal Killing, $\int_{\partial M} \langle A, \delta^* X \rangle = \frac{1}{n} \int_{\partial M} H \operatorname{div} X$. On the other hand, $-\int_{\partial M} \langle dH, X \rangle = \int_{\partial M} H \operatorname{div} X$, which gives (2.28), since $\dim \partial M = n \neq 1$. The result (2.28) is essentially due to [2], although the proof given here is much simpler. The condition (2.28) is of course reminiscent of the Kazdan-Warner type obstruction [10] for the prescribed Gauss curvature problem. As in the Gauss curvature problem, note that the condition (2.28) is not conformally invariant, i.e., it is not a well-defined condition on the target space $\mathcal{C}^{m,\alpha} \times C_+^{m-1,\alpha}$.

The "balancing condition" (2.28) implies for instance that any H which is a monotone function of a height function on $S^2(1)$ is not the mean curvature of a conformal immersion in a space-form. On the other hand, although (2.28) formally represents 3 independent conditions on H, it does not imply that $\operatorname{Im} \Pi$ has codimension 3, (or any other codimension), in $\mathcal{C}^{m,\alpha} \times C_+^{m-1,\alpha}$, again since it is not defined on this target space.

Although we have succeeded in constructing a proper map $\widetilde{\Pi}$, it is not at all clear what $\operatorname{Im} \widetilde{\Pi}$ is, or what the images of the closely related maps Π_+ and Π' in (1.9), (1.6) are. For instance, can $\operatorname{Im} \Pi_+$ be described as the locus where a finite number of real-valued functions on the target are positive? Can one explicitly identify such functions characterising the boundary values of metrics in $\mathcal{M}_\kappa$?

Finally, regardless of the surjectivity issue, the discussion in (I) above suggests that Π_+ is infinite-to-one, so highly non-unique. In sum, it would be interesting to understand these issues better, which seem on the whole much easier than the existence and uniqueness question for the isometric immersion problem discussed in §1.

3. Generalization

Let (M_κ, g_κ) be any complete Riemannian 3-manifold of constant curvature κ; thus, up to scaling, M_κ is one of $\mathbb{R}^3$, $\mathbb{H}^3$ or $\mathbb{S}^3$ or a quotient of one of these spaces. Let $f : \partial M \to M_\kappa$ be an Alexandrov immersion, and let F denote an extension of f to M. Then since $\pi_1(M, \partial M) = 0$, the metric $F^*(g_\kappa)$ is uniquely determined by the immersion f on ∂M, modulo $\mathrm{Diff}_1(M)$. Thus, the map Π_+ in (1.4) is equivalent to a map

$$\Pi_+ : \mathrm{Imm}_A(\partial M) \to \mathcal{C}^{m,\alpha}(\partial M) \times C_+^{m-1,\alpha}(\partial M).$$

This suggests that one could replace the space of metrics $\mathcal{M}_\kappa$ by the space of immersions of $\Sigma = \partial M$ into a space-form M_κ or more generally into an arbitrary complete Riemannian manifold (N, g_N). This is in fact the case:

Proposition 3.1. *Let $\Sigma = \Sigma_g$ be a compact surface of genus g and let (N, g_N) be any complete Riemannian 3-manifold. Let $\mathrm{Imm}^{m+1,\alpha}(\Sigma, N)$ be the space of $C^{m+1,\alpha}$ immersions of $\Sigma \to N$. Then the map*

$$\begin{aligned} \Pi : \mathrm{Imm}^{m+1,\alpha}(\Sigma, N) &\to \mathcal{C}^{m,\alpha}(\Sigma) \times C^{m-1,\alpha}(\Sigma), \\ \Pi(f) &= ([f^*(g_N)|_{\partial M}], H(f(x))), \end{aligned} \tag{3.1}$$

is a smooth Fredholm map of Fredholm index 0.

Proof. The space $\mathrm{Imm}(\Sigma, N)$, is a smooth Banach manifold; the tangent space is given by the space of vector fields v along a given immersion $f : \Sigma \to N$. The differential $D\Pi$ of Π in (3.1) is given by

$$([(\delta^* v)^T]_0, H'_{\delta^* v}), \tag{3.2}$$

where $(\delta^* v)^T$ is the restriction of $\delta^* v$ to $T(\partial M)$. The Fredholm property then follows by showing that the data (3.2) form an elliptic system of equations for v. We do this following the proof of Proposition 2.2.

Thus, write $v = v^T + fN$, where v^T is tangent and N is normal to $T(\partial M)$. Then

$$\delta^* v = \delta^* v^T + fA + df \cdot N, \tag{3.3}$$

so that $(\delta^* v)^T = \delta^* v^T + fA$. The second term is lower order in v and so does not contribute to principal symbol. The principal symbol σ of the first term is thus

$$\sigma([(\delta^* v)^T]_0) = \xi_i v_j - (\xi_i v_i / 2)\delta_{ij}, \tag{3.4}$$

where i, j are indices for ∂M. For the mean curvature, one has $H'_{\delta^* v} = -\Delta f + v(H)$, so that the leading order term is just $-\Delta f$ with symbol $|\xi|^2 f$. Hence one has elliptic data for the normal component f of v.

For the tangential part of v, (3.4) gives

$$\xi_1 v_2 = \xi_2 v_1 = 0 \quad \text{and} \quad \xi_1 v_1 = \xi_2 v_2.$$

Since $(\xi_1, \xi_2) \neq (0, 0)$, it is elementary to see that the only solution of these equations is $v_1 = v_2 = 0$, which proves ellipticity. It is straightforward to verify further that Π has Fredholm index 0. □

References

[1] A.D. Alexandrov, A characteristic property of spheres, Ann. Mat. Pura Appl., **58**, (1962), 303–315.

[2] B. Ammann, E. Humbert and M.O. Ahmedou, An obstruction for the mean curvature of a conformal immersion $S^n \to \mathbb{R}^{n+1}$, Proc. Amer. Math. Soc., **135**, (2007), 489–493.

[3] M. Anderson, Einstein metrics with prescribed conformal infinity on 4-manifolds, Geom. & Funct. Analysis, **18**, (2008), 305–366.

[4] M. Anderson and M. Khuri, The static vacuum extension problem in General Relativity, (preprint, 2009), arXiv: 0909.4550 (math.DG).

[5] M. Anderson, On boundary value problems for Einstein metrics, Geom. & Topology, **12**, (2008), 2009–2045.

[6] A. Besse, Einstein Manifolds, Springer Verlag, NY, (1987).

[7] D. Gilbarg and N. Trudinger, Elliptic Partial Differential Equations of Second Order, Second Edition, Springer Verlag, NY, (1983).

[8] H. Hopf, Differential Geometry in the Large, Lect. Notes in Math., **1000**, Springer Verlag, Berlin (1983).

[9] G. Kamberov, Prescribing mean curvature: existence and uniqueness problems, Elec. Res. Ann. Amer. Math. Soc, **4**, (1998), 4–11.

[10] J. Kazdan and F. Warner, Curvature functions for compact 2-manifolds, Annals of Math., **99**, (1974), 14–47.

[11] K. Kenmotsu, The Weierstrass formula for surfaces of prescribed mean curvature, Math. Annalen, **245**, (1979), 89–99.

[12] H.B. Lawson and R. Tribuzy, On the mean curvature function for compact surfaces, Jour. Diff. Geom., **16**, (1981), 179–183.

[13] W. Meeks III and H. Rosenberg, Uniqueness of the helicoid, Annals of Math, **161**, (2005), 727–758.

[14] W. Meeks III, A. Ros and H. Rosenberg, The Global Theory of Minimal Surfaces in Flat Spaces, Lect. Notes in Math., **1775**, Springer Verlag, Berlin, (2002).

[15] E. Poznyak, Examples of regular metrics on the sphere and circle which are not realized on twice continuously differentiable surfaces, Vestnik Moskov. Univ. Ser. I, Mat. Meh., **1960**, (1960), 3–5.

[16] S. Smale, An infinite dimensional version of Sard's theorem, Amer. Jour. Math., **87**, (1965), 861–866.

[17] G. Tinaglia, Multi-valued graphs in embedded constant mean curvature discs, Trans. Amer. Math. Soc., **359**, (2007), 143–164.

[18] A. Treibergs and W. Wei, Embedded hyperspheres with prescribed mean curvature, Jour. Diff. Geom., **18**, (1983), 513–521.

[19] S.T. Yau, Problem Section, Seminar on Differential Geometry, Annals of Math. Studies, **102**, Ed. S.T. Yau, Princeton Univ. Press, Princeton, NJ, (1982).

Michael T. Anderson
Dept. of Mathematics
Stony Brook University
Stony Brook, NY 11794-3651, USA
e-mail: `anderson@math.sunysb.edu`

Space of Kähler Metrics (V) – Kähler Quantization

Xiuxiong Chen and Song Sun

Dedicated to Jeff Cheeger for his 65th birthday

Abstract. Given a polarized Kähler manifold (X, L). The space $\mathcal{H}$ of Kähler metrics in $2\pi c_1(L)$ is an infinite-dimensional Riemannian symmetric space. As a metric space, it has non-positive curvature. There is associated to $\mathcal{H}$ a sequence of finite-dimensional symmetric spaces $\mathcal{B}_k (k \in \mathbb{N})$ of non-compact type. We prove that $\mathcal{H}$ is the limit of $\mathcal{B}_k$ as metric spaces in certain sense. As applications, this provides more geometric proofs of certain known geometric properties of the space $\mathcal{H}$.

Mathematics Subject Classification (2000). 53C55.

Keywords. Quantization, Kähler metrics, geodesic distance.

1. Introduction

Let (X, ω, J) be an n-dimensional compact Kähler manifold. By [9] this gives rise to two infinite-dimensional symmetric spaces: the Hamiltonian diffeomorphism group $\mathrm{Ham}(X, \omega)$ and the space $\mathcal{H}$ of smooth Kähler potentials under the natural (Weil-Petersson type) L^2 metric. The relation between these two spaces is analogous to that between a finite-dimensional compact group G and its noncompact dual $G^{\mathbb{C}}/G$, where $G^{\mathbb{C}}$ is a complexification of G. As in the finite-dimensional case, at least formally, $\mathrm{Ham}(X, \omega)$ has non-negative sectional curvature, while $\mathcal{H}$ has non-positive sectional curvature. It is proved by E. Calabi and the first author in [2] that $\mathcal{H}$ is non-positively curved in the sense of Alexandrov. On the other hand, it is well known in the literature of geometric quantization that $\mathcal{H}$ is the limit of a sequence of finite-dimensional symmetric spaces $\mathcal{B}_k (k \in \mathbb{N})$ when X is (Kähler) polarized. Indeed, the polarization is given by an ample line bundle L over X, and we consider the space $H^0(X, L^k)$ of holomorphic sections of L^k for large enough k.

The first author was partially supported by NSF grant.

Here k^{-1} plays the role of Planck constant $\hbar$, while $k \to \infty$ should correspond to the process of taking the classical limit. Denote by N_k the dimension of $H^0(X, L^k)$. By the Riemann-Roch theorem this is a polynomial in k of degree n for sufficiently large k. Then $\mathcal{B}_k$ is by definition the space of positive definite Hermitian forms on $H^0(X, L^k)$. It could also be viewed as the symmetric space $GL(N_k; \mathbb{C})/U(N_k)$. The spaces $\mathcal{B}_k$ are related to $\mathcal{H}$ through two naturally defined maps: $\mathrm{Hilb}_k : \mathcal{H} \to \mathcal{B}_k$ and $FS_k : \mathcal{B}_k \to \mathcal{H}$. (The precise definition of these maps will be given in the next section.) In [24], G. Tian proved that given any $\phi \in \mathcal{H}$, $FS_k \circ \mathrm{Hilb}_k(\phi) \to \phi$ in C^4 topology. Using Tian's peak section method and canonical coordinates, W.-D. Ruan proved the C^∞ convergence in [20]. S. Zelditch ([26]) beautifully generalized Tian's theorem and derived C^∞ convergence from the asymptotic expansion of the Bergman kernel. If we denote by $\mathcal{H}_k$ the image of $FS_k \circ \mathrm{Hilb}_k$, then these results tell us that

$$\mathcal{H} = \overline{\bigcup_{k \in \mathbb{N}} \mathcal{H}_k},$$

where the closure is taken in C^∞ topology.

In [10], S.K. Donaldson suggests that the geometry of $\mathcal{B}_k$ should also converge to that of $\mathcal{H}$. In particular, the geodesics in $\mathcal{H}$ should be approximated by geodesics in $\mathcal{H}_k$. In [3], the first author proved the existence of $C^{1,1}$ geodesics and it subsequently lead to many interesting applications in Kähler geometry (see [7], [4] for further references). The limitation of $C^{1,1}$ regularities is purely technical. In a very interesting paper([18]), Phong-Sturm proved that the $C^{1,1}$ geodesics in $\mathcal{H}$ are the weak C^0 limits of Bergman geodesics, assuming the existence of $C^{1,1}$ geodesics. It would provide a canonical smooth approximation of $C^{1,1}$ geodesics if one could show the convergence is in $C^{1,1}$ topology. More evidence comes from a series of beautiful works by S. Zeltdich and his collaborators, see Song-Zelditch ([23]), Rubinstein-Zelditch ([22]). They proved that on toric varieties both geodesics and harmonic maps in $\mathcal{H}$ (automatically smooth) are C^2 limits of corresponding objects in $\mathcal{H}_k$. Recently, J. Fine ([14]) proved a remarkable result that the Calabi flow in $\mathcal{H}$ could be approximated by balancing flows in $\mathcal{B}_k$. In this paper, we shall prove the following convergence of geodesic distance:

Theorem 1.1. *Given any $\phi_0, \phi_1 \in H$, we have*

$$\lim_{k \to \infty} k^{-\frac{n+2}{2}} d_{\mathcal{B}_k}(\mathrm{Hilb}_k(\phi_0), \mathrm{Hilb}_k(\phi_1)) = d_{\mathcal{H}}(\phi_0, \phi_1).$$

Remark 1.2. It is easy to identify the scaling factor by simply taking $\phi_1 = \phi_0 + c$, then $d(\phi_0, \phi_1) = c$, while $d(\mathrm{Hilb}_k(\phi_0), \mathrm{Hilb}_k(\phi_1)) = c \cdot k\sqrt{N_k}$. This scaling also indicates that the limit can not have bounded curvature, which can also be speculated from the expression of the infinitesimal curvature of $\mathcal{H}$:

$$R_\phi(\phi_1, \phi_2) = -\frac{1}{4} \frac{\|\{\phi_1, \phi_2\}_\phi\|_{L^2}^2}{\|\phi_1\|_{L^2}^2 \|\phi_2\|_{L^2}^2}.$$

Remark 1.3. This theorem indicates that the L^2 metric defined in [9] is in a sense canonical. We notice that there are also natural affine structures on both $\mathcal{H}$ and $\mathcal{B}_k$, given by embedding them as convex subsets of the affine spaces $C^\infty(X)$ and the space of all Hermitian forms on $H^0(X, L^k)$ respectively. It would be interesting to see whether the affine structures on $\mathcal{H}$ and $\mathcal{B}_k$ have nice relations with each other as the above metric geometric structures do.

From the proof it follows that the convergence is uniform if both potentials vary in a C^l compact neighborhood for large l. So an easy corollary is the convergence of angles:

Corollary 1.4. *Given three points ϕ_i $(i = 1, 2, 3)$ in $\mathcal{H}$, let $H_{k,i} = \mathrm{Hilb}_k(\phi_1)$. Then*

$$\lim_{k\to\infty} \angle H_{k,1}H_{k,2}H_{k,3} = \angle \phi_1\phi_2\phi_3.$$

This corollary leads to that $\mathcal{H}$ is non-positively curved in the sense of Alexandrov, as was originally proved in [2].

Theorem 1.1 together with Theorem 5 in [14] imply the following corollary, because the downward gradient flow of a geodesically convex function on a finite-dimensional manifold is distance decreasing:

Corollary 1.5 (Calabi-Chen [2], [4]). *Calabi flow in $\mathcal{H}$ decreases geodesic distance.*

The "quantization" approach, namely using the approximation by $\mathcal{B}_k$ to handle problems about the Kähler geometry of (X, ω, J), turns out to be quite powerful and intriguing. It was shown in [9] that the existence and uniqueness problems in Kähler geometry are related to the geometry of $\mathcal{H}$. More precisely, the uniqueness of extremal metrics is implied by the existence of smooth geodesics connecting any two points in $\mathcal{H}$, and the non-existence is conjectured to be equivalent to the existence of a geodesic ray in $\mathcal{H}$ where the K-energy is strictly decreasing near infinity. The technical problem comes from the lack of regularity of geodesics in $\mathcal{H}$. There are two ways at present to circumvent this problem. The first way is to study the geodesic equation directly. One can use the continuity method, as in [3]. In [3], the first author constructed a continuous family of ϵ-approximate geodesics converging in weak $C^{1,1}$ topology to a $C^{1,1}$ geodesic. This gives the proof of the uniqueness of extremal metrics when $c_1(X) \leq 0$ and the other conjecture of Donaldson that $\mathcal{H}$ is a metric space. In [7] a new partial regularity was derived from studying the complex Monge-Ampère equation associated to the geodesic equation. Solutions with such regularity already have fruitful applications. The second way is to exploit the approximation of $\mathcal{H}$ by $\mathcal{B}_k$. This requires a polarization. The essential thing is to quantize the whole problem. Many problems have been settled using either of the two approaches: the uniqueness of cscK metrics ([7], [9]), the lower bound of Mabuchi energy assuming the existence of cscK metrics([7], [12]), the lower bound of Calabi energy ([5], [13]), etc.

Due to [12], the Mabuchi functional (K-energy) E has a nice quantization defined on $\mathcal{B}_k$, which we denote by Z_k. More precisely, up to a constant, given any $\phi \in \mathcal{H}$, we have $Z_k(\mathrm{Hilb}_k(\phi)) \to E(\phi)$. It is well known that E is convex along smooth geodesics in $\mathcal{H}$, but it is not known that E is convex along $C^{1,1}$ geodesics, although it is well defined (cf. [6]). One good thing is that Z_k is convex on a finite-dimensional space $\mathcal{B}_k$, which is geodesically complete. This makes it possible to derive some weak convexity of E, such as an alternative proof of two inequalities of the first author. Namely

Corollary 1.6 ([4]). *For any $\phi_0, \phi_1 \in \mathcal{H}$, let $\phi(t)(t \in [0,1])$ be the $C^{1,1}$ geodesic connecting them. Then*

$$dE_{\phi_0}(\dot{\phi}(0)) \leq dE_{\phi_1}(\dot{\phi}(1)).$$

Corollary 1.7 ([4]). *For any $\phi_0, \phi_1 \in \mathcal{H}$, we have*

$$E(\phi_1) - E(\phi_0) \leq d(\phi_0, \phi_1) \cdot \sqrt{Ca(\phi_1)}.$$

Both corollaries are useful in the study of Kähler geometry by the space of Kähler metrics. By Theorem 3.1 in [2], Corollary 1.6 immediately gives Corollary 1.5. Using these corollaries, the first author proved the sharp lower bound of the Calabi energy in any given Kähler class. The original proof of these two corollaries depends heavily on the delicate regularity results of Chen-Tian [7]. Our proof in the algebraic case is more geometric and substantially simpler.

Organization

In Section 2 we briefly recall the geometry of the space of Kähler metrics and the associated finite-dimensional spaces $\mathcal{B}_k$. In Section 3 we prove the convergence of geodesic distance, using the existence of $C^{1,1}$ geodesics in $\mathcal{H}$, the non-positivity of curvature of $\mathcal{B}_k$, and the convergence of infinitesimal geometry. In Section 4 the weak convexity of K-energy is discussed and we prove Corollary 1.6 and Corollary 1.7. The idea is that the K-energy could be approximated by convex functions Z_k on $\mathcal{B}_k$, while the norm of the gradient of Z_k approximates the norm of gradient of K-energy, which is the Calabi functional. So Corollary 1.7 follows. To prove Corollary 1.6, we need to estimate the difference between the initial direction of an "almost" geodesic γ (i.e., $|\ddot{\gamma}|$ is small) in $\mathcal{B}_k$ and that of the genuine geodesic connecting the two end points $\gamma(0)$ and $\gamma(1)$. This is done in Lemma 4.5. In the last section, we discuss some further problems.

2. Preliminaries

Let (X, ω, J) be a Kähler manifold. Here we always assume it is polarized, i.e., there is a holomorphic line bundle L whose first Chern class is $[\omega]/2\pi$. Fix the base metric ω and define the space of Kähler potentials as

$$\mathcal{H} = \{\phi \in C^\infty(M) | \omega + \sqrt{-1}\partial\bar{\partial}\phi > 0\}.$$

$\mathcal{H}$ can also be identified with the space of Hermitian metrics on L with positive curvature form. In the following we will not distinguish between these two meanings. We write ω_h to denote $\sqrt{-1}$ times the curvature of h for a Hermitian metric h, and $\omega = \omega_{h_0}$. Thus,

$$\omega_\phi := \omega_{e^{-\phi}h_0} = \omega + \sqrt{-1}\partial\bar{\partial}\phi.$$

Denote

$$d\mu_h = d\mu_\phi = \frac{\omega_h^n}{(2\pi)^n n!}.$$

There is a Weil-Petersson type metric defined on $\mathcal{H}$ by

$$(\delta_1\phi, \delta_2\phi)_\phi = \int_X \delta_1\phi\delta_2\phi d\mu_\phi,$$

for any $\delta_1\phi, \delta_2\phi \in T_\phi\mathcal{H} = C^\infty(X)$. Due to [9], $\mathcal{H}$ is formally an infinite-dimensional symmetric space, and the sectional curvature of the Levi-Cività connection is given by:

$$R_\phi(\delta_1\phi, \delta_2\phi) = -\frac{1}{4}\|\{\delta_1\phi, \delta_2\phi\}_\phi\|^2.$$

It was confirmed in [2] that $\mathcal{H}$ satisfies the triangle comparison theorem of a non-positive Alexandrov space, by using so-called ϵ-approximate geodesics. The geodesic equation in $\mathcal{H}$ is

$$\ddot{\phi} - |\nabla_\phi\dot{\phi}|_\phi^2 = 0, \tag{1}$$

where we have used the notation of complex gradient. The following theorem is proved in [3] by the first author:

Lemma 2.1 (Geodesic Approximation Lemma). *Given any $\phi_0, \phi_1 \in \mathcal{H}$, there is a positive number ϵ_0, and a one-parameter smooth family of smooth curves $\phi_\epsilon(\cdot) : [0,1] \to \mathcal{H}$ ($\epsilon \in (0, \epsilon_0]$), such that the following holds:*

(1) *For any $\epsilon \in (0, \epsilon_0]$, ϕ_ϵ is an ϵ-approximate geodesic, i.e., it solves the following equation:*

$$(\ddot{\phi}_\epsilon - |\nabla\dot{\phi}_\epsilon|^2)\omega_{\phi_\epsilon}^n = \epsilon \cdot \omega^n,$$

and $\phi_\epsilon(0) = \phi_0$, $\phi_\epsilon(1) = \phi_1$.

(2) *There exists a $C > 0$, such that for all $\epsilon \in (0,1]$ and $t \in [0,1]$, we have*

$$|\phi_\epsilon(t)| + |\dot{\phi}_\epsilon(t)| \le C,$$

and

$$|\ddot{\phi}_\epsilon(t)| \le C.$$

(3) *$\phi_\epsilon(\cdot)$ converges to the unique $C^{1,1}$ geodesic connecting ϕ_0 and ϕ_1 in the weak $C^{1,1}$ topology when $\epsilon \to 0$.*

(4) *We have uniform estimates when ϕ_0 and ϕ_1 varies in a C^k compact set for some large k.*

The space $\mathcal{H}$ has a global flat direction given by addition of a constant. The de Rham decomposition theorem turns out to be true in this case. There is a functional I giving the isometric decomposition

$$\mathcal{H} = \mathcal{H}_0 \oplus \mathbb{R}.$$

I is only defined up to an addition of a constant, and its derivative is given by:

$$\delta I = \int_X \delta\phi d\mu_\phi.$$

So I is linear along $C^{1,1}$ geodesics in $\mathcal{H}$. $\mathcal{H}_0$ is then an arbitrary level set of I, and it can also be regarded as the space of Kähler metrics cohomologous to ω. An interesting thing to notice is that $\mathcal{H}$ is embedded as a convex subset into an affine linear space $C^\infty(M)$. It is easy to show that I is indeed convex along any linear path in $\mathcal{H}$.

There is a well-known K-energy functional E on $\mathcal{H}$, defined up to an additive constant, with its variation given by:

$$\delta E = -\int_X (S - \underline{S})\delta\phi d\mu_\phi. \tag{2}$$

Here $\delta\phi$ is the infinitesimal variation of ϕ, and S is the scalar curvature of ω_ϕ. From the point of view of [8], this is a natural convex functional associated to a moment map of a Hamiltonian action (for infinite-dimensional manifold). Indeed, by a straightforward calculation, if $\phi(t)$ is a smooth geodesic in $\mathcal{H}$, then

$$\frac{d^2}{dt^2}E(\phi(t)) = \int_X |\mathcal{D}\dot{\phi}|^2 d\mu_\phi \geq 0,$$

where $\mathcal{D}$ is the Lichnerowicz-Laplacian operator. By studying the explicit expression of E, it is shown in [6] that E is well defined for $C^{1,1}$ Kähler potentials, but it is not obvious from the definition that E is convex along $C^{1,1}$ geodesics.

There is a sequence of finite-dimensional symmetric spaces $\mathcal{B}_k$ consisting of all positive definite Hermitian forms on $H^0(X, L^k)$. This can be identified with $GL(N_k; \mathbb{C})/U(N_k)$ by choosing a base point in $\mathcal{B}_k$, where $N_k = \dim H^0(X, L^k)$. We shall review some basic facts about these symmetric spaces. $U(N)$ is a compact Lie group and admits a natural bi-invariant Riemannian metric given by

$$(A_1, A_2)_A = -Tr(A_1A^{-1} \cdot A_2A^{-1}),$$

for any $A_1, A_2 \in T_AU(N)$. The sectional curvature is given by:

$$R(A_1, A_2) = \frac{1}{4}\|[A_1, A_2]_A\|^2.$$

The geodesic equation is

$$\ddot{A} = \dot{A}A^{-1}\dot{A}.$$

It is then clear that the geodesics all come from one parameter subgroups given by the usual exponential map. Now the non-compact dual of $U(N)$ is $GL(N; \mathbb{C})/U(N)$ – the space of positive definite $N \times N$ Hermitian matrices. We can explicitly write down the metric on it. For any $H_1, H_2 \in T_H(GL(N; \mathbb{C})/U(N))$, define

$$(H_1, H_2)_H = Tr(H_1H^{-1} \cdot H_2H^{-1}).$$

Then the sectional curvature becomes non-positive:

$$R(H_1, H_2) = -\frac{1}{4}\|[H_1, H_2]_H\|^2.$$

A path $H(t)$ is a geodesic if it satisfies the following equation:

$$\ddot{H} = \dot{H}H^{-1}\dot{H}. \tag{3}$$

This looks very similar to equation (1), except that it is a nonlinear ODE instead of a nonlinear PDE. In our setting, we get a sequence of equations (3) depending on k. For this reason, we may say that this sequence of equations "quantizing" equation (1). It is easy to see that all the geodesics in this space are of the form

$$H(t) = H(0)^{\frac{1}{2}} \exp(tA) H(0)^{\frac{1}{2}},$$

for some initial point $H(0)$ and initial tangent vector A.

Similar to the infinite-dimensional case, $\mathcal{B}_k$ also admits an isometric splitting given by the function $I_k = \log\det$, defined up to an additive constant. It is easy to see that I_k is linear along geodesics in $\mathcal{B}_k$ and convex along a linear path when we regard $\mathcal{B}_k$ as a convex subset of the affine linear space of all $N_k \times N_k$ complex matrices. The splitting corresponds to the following isometry:

$$GL(N;\mathbb{C})/U(N) \simeq SL(N;\mathbb{C})/SU(N) \times \mathbb{R}.$$

We will see in Section 4 that I_k indeed "quantizes" the I functional.

To "quantize" functionals on $\mathcal{H}$, there are two natural maps relating $\mathcal{H}$ and $\mathcal{B}_k$, where we have adopted the notation of [12]:

$$\mathrm{Hilb}_k : \mathcal{H} \to \mathcal{B}_k;$$
$$FS_k : \mathcal{B}_k \to \mathcal{H}.$$

Explicitly, given $h \in \mathcal{H}$, and $s \in H^0(X, L^k)$, we define

$$\|s\|^2_{\mathrm{Hilb}_k(h)} = \int_X |s|_h^2 d\mu_h.$$

For $H \in \mathcal{B}_k$, pick an orthonormal basis $\{s_\alpha\}$ of $H^0(X, L^k)$ with respect to H. Define

$$FS_k(H) = \frac{1}{k} \log \sum_\alpha |s_\alpha|^2_{h_0^k},$$

where h_0 is the base metric in $\mathcal{H}$. In particular, we have

$$\sum_\alpha |s_\alpha|^2_{e^{-k\cdot FS_k(H)} h_0} \equiv 1.$$

Notice that $\omega_{FS_k(H)}$ coincides with the pull-back of the Fubini-Study metric on $\mathbb{CP}^{N_k-1}$ under the projective embedding induced by $\{s_\alpha\}$, while $FS_k(H)$ coincides with the induced metric on the pull-back of $\mathcal{O}(1)$. Let $\mathcal{H}_k$ be the image of FS_k, then the well-known Tian-Yau-Zelditch theorem says that

$$\overline{\bigcup_k \mathcal{H}_k} = \mathcal{H}.$$

To be more precise, for any $\phi \in \mathcal{H}$, and $k > 0$, we choose an orthonormal basis $\{s_\alpha\}$ of $H^0(X, L^k)$ with respect to $\mathrm{Hilb}_k(\phi)$, and define the density of state function:

$$\rho_k(\phi) = \sum_\alpha |s_\alpha|^2_{h^k}.$$

Then we have the following C^∞ expansion:

Lemma 2.2 ([26], [15]).

$$\rho_k(\phi) = k^n + A_1(\phi)k^{n-1} + A_2(\phi)k^{n-2} + \cdots, \tag{4}$$

with $A_1(\phi) = \frac{1}{2}S(\phi)$, where S denotes the scalar curvature. Moreover, the expansion is uniform in that for any l and $R \in \mathbb{N}$,

$$\left\| \rho_k(\phi) - \sum_{j \le R} A_j k^{n-j} \right\|_{C^l} \le C_{R,l} k^{n-R},$$

where $C_{R,l}$ only depends on R and l.

Remark 2.3. Later we will need a generalization of the previous expansion theorem. We want to differentiate the expansion of the density of state function several times along a smooth path $\phi(t)$ in $\mathcal{H}$, and still to get a uniform expansion. This could be done following the arguments of Z. Lu and G. Tian, see [16].

Now if we let $\phi_k = FS_k \circ \mathrm{Hilb}_k(\phi)$, then

$$\phi_k - \phi = \frac{1}{k} \log \rho_k(\phi) \to 0,$$

as $k \to \infty$.

To quantize the K-energy, following [12], we define

$$\mathcal{L}_k = I_k \circ \mathrm{Hilb}_k + \frac{kd_k}{V} I,$$

and

$$Z_k = \frac{kd_k}{V} I \circ FS_k + I_k + d_k(\log V - \log d_k),$$

where $d_k = \dim H^0(X, L^k)$, and $V = \mathrm{Vol}_{[\omega]}(X)$. Notice that the definition of both functionals requires choices of base points in both $\mathcal{H}$ and $\mathcal{B}_k$. From now on, we always fix the same base points for defining these two functionals. By a straightforward calculation,

$$\delta \mathcal{L}_k = \int [\Delta \rho_k - k\rho_k]_\phi \delta\phi d\mu_\phi,$$

where we define for any function f, $[f]_\phi := f - \frac{1}{V}\int f d\mu_\phi = f - \underline{f}$. From (4), we see that

$$[\Delta \rho_k - k\rho_k]_\phi \to -\frac{1}{2} k^n (S - \underline{S}).$$

Thus, there are constants c_k, such that

$$\frac{2}{k^n} \mathcal{L}_k + c_k \to E,$$

where the convergence is uniform on $C^l (l \gg 1)$ bounded subsets in $\mathcal{H}$. From now on, we will denote the left-hand side by $\mathcal{L}_k$, and $Z_k + c_k$ by Z_k.

In [12] the following relation between $\mathcal{L}_k$ and Z_k was shown:

$$\mathcal{L}_k(FS_k(H)) - Z_k(H) = \log\det(\mathrm{Hilb}_k \circ FS_k(H)) - \log\det H - d_k \log \frac{V}{d_k} \le 0;$$

and

$$Z_k(\mathrm{Hilb}_k(\phi)) - \mathcal{L}_k(\phi) = \frac{kd_k}{V}(I(FS_k \circ \mathrm{Hilb}_k(\phi)) - I(\phi)) + d_k \log \frac{V}{d_k} \leq 0.$$

Thus, given any $\phi \in \mathcal{H}$, we have

$$\mathcal{L}_k(\phi_k) = \mathcal{L}_k(FS_k \circ \mathrm{Hilb}_k(\phi)) \leq Z_k(\mathrm{Hilb}_k(\phi)) \leq \mathcal{L}_k(\phi).$$

Since ϕ_k converges to ϕ smoothly, and $\mathcal{L}_k$ converges to E uniformly on $C^l(l \gg 1)$ bounded subsets in $\mathcal{H}$, we immediately obtain:

Lemma 2.4. *Z_k quantizes E in the sense that given any $\phi \in \mathcal{H}$, we have as $k \to \infty$,*

$$Z_k(\mathrm{Hilb}_k(\phi)) \to E,$$

and the convergence is uniform in $C^l(l \gg 1)$ bounded subsets of $\mathcal{H}$.

In Section 3 we will investigate more about the relation between Z_k and E.

The following is proved essentially in [12]:

Lemma 2.5 ([12]). *$\mathcal{L}_k$ is convex on $\mathcal{H}$, and Z_k is convex on $\mathcal{B}_k$.*

3. Convergence of geodesic distance

In this section we shall prove Theorem 1.1. The following lemma about the derivative of Hilb_k and FS_k follows from a simple calculation:

Lemma 3.1. *For $\phi \in \mathcal{H}$, and $\delta\phi \in T_\phi\mathcal{H}$, we have for any $s_1, s_2 \in H^0(X, L^k)$, that*

$$d_\phi \ \mathrm{Hilb}_k(\delta\phi)(s_1, s_2) = \int (s_1, s_2)_\phi(-k\delta\phi + \Delta\delta\phi)d\mu_\phi. \tag{5}$$

For $H \in \mathcal{B}_k$, and $\delta H \in T_H\mathcal{B}_k$, we have

$$d_H \ FS_k(\delta H) = -\frac{1}{k}\sum_{i,j} \delta H(s_i, s_j) \cdot (s_j, s_i)_{FS_k(H)}, \tag{6}$$

where $\{s_i\}$ is an orthonormal basis of H.

The following theorem proves the convergence of infinitesimal geometry.

Theorem 3.2. *Given a smooth path $\phi(t) \in \mathcal{H}(t \in [0,1])$, after normalization, the length of the induced path $H_k(t) = \mathrm{Hilb}_k(\phi(t))$ converges to that of $\phi(t)$ as $k \to \infty$, where the corresponding metrics on $\mathcal{H}$ and $\mathcal{B}_k$ are used. More precisely,*

$$\lim_{k\to\infty} k^{-n-2}\|\dot{H}_k\|^2 = \|\dot{\phi}\|^2 = \int_X \dot{\phi}^2 d\mu_\phi,$$

for any $t \in [0,1]$.

Proof. W.L.O.G, assume $t = 0$. Let $\{s_i(t)\}$ be an orthonormal basis of $H_k(t) = \mathrm{Hilb}_k(\phi(t))$. Then by Lemma 3.1,

$$\dot{H}_{ij} = \dot{H}(s_i(t), s_j(t)) = \int_X (s_i, s_j)(-k\dot{\phi} + \Delta\dot{\phi})d\mu_\phi,$$

where for convenience we drop the subscript k of H_k. Diagonalize it at $t = 0$, we get

$$\dot{H}_{ij} = \delta_{ij} \cdot \int_X |s_i|_0^2(-k\dot{\phi} + \Delta\dot{\phi})d\mu_\phi.$$

Now let $\phi_k(t) = FS_k(H_k(t))$, then

$$\dot{\phi_k} = -\frac{1}{k}\sum_i \dot{H}_{ii}|s_i|_{\phi_k}^2.$$

Therefore,

$$\begin{aligned}
\|\dot{H}\|^2 &= \sum_i |\dot{H}_{ii}H_{ii}^{-1}|^2 \\
&= \int_X \sum_i \dot{H}_{ii}|s_i|^2(-k\dot{\phi} + \Delta\dot{\phi})d\mu_\phi \\
&= -k\int_X \dot{\phi_k}(-k\dot{\phi} + \Delta\dot{\phi})\rho_k(\phi)d\mu_\phi
\end{aligned}$$

By Lemma 2.2, we know

$$\rho_k(\phi) = k^n + \frac{1}{2}S \cdot k^{n-1} + O(k^{n-2}),$$

and

$$\dot{\phi_k} - \dot{\phi} = \frac{1}{k} \cdot \frac{\dot{\rho}_k(\phi)}{\rho_k(\phi)} = \frac{\dot{S}}{2k^2} + O\left(\frac{1}{k^3}\right).$$

Hence,

$$\|\dot{H}\|^2 = k^{n+2}\left(\int_X \dot{\phi}^2 d\mu_\phi + O\left(\frac{1}{k}\right)\right).$$

□

Remark 3.3. Actually we have proved that for any smooth ψ, and $\{s_i\}$ an orthonormal basis of H_k,

$$\lim_{k\to\infty} k^{-n-2}\sum_{i,j}\left|\int_X (s_i, s_j)_\phi(-k\psi + \Delta\psi)d\mu_\phi\right|^2 = \int_X \psi^2 d\mu_\phi.$$

Indeed, the convergence is uniform for ψ varying in a C^l compact set. So the following holds:

$$\lim_{k\to\infty} k^{-n}\sum_{i,j}\left|\int_X (s_i, s_j)_\phi \psi d\mu_\phi\right|^2 = \int_X \psi^2 d\mu_\phi. \tag{7}$$

Now we can prove one side inequality of Theorem 1.1.

Corollary 3.4. *Given two metrics $\phi_1, \phi_2 \in \mathcal{H}$, and denote*

$$H_{k,i} = \mathrm{Hilb}_k(\phi_i), \quad i = 1, 2,$$

then we have

$$\limsup_{k\to\infty} k^{-\frac{n}{2}-1} d_{\mathcal{B}_k}(H_{k,1}, H_{k,2}) \le d_{\mathcal{H}}(\phi_1, \phi_2).$$

Proof. From Lemma 2.1, we know that for any $\epsilon > 0$, there exists a smooth ϵ-approximate geodesic $\phi(t)(t \in [0,1])$ in $\mathcal{H}$ connecting ϕ_1 and ϕ_2 such that the length

$$L_{\mathcal{H}}(\phi(t)) \le d_{\mathcal{H}}(\phi_1, \phi_2) + \frac{\epsilon}{2}.$$

For this path and k sufficiently large, by Theorem 3.2,

$$k^{-\frac{n}{2}-1} L_{\mathcal{B}_k}(\mathrm{Hilb}_k(\phi(t))) \le L_{\mathcal{H}}(\phi(t)) + \frac{\epsilon}{2}.$$

Then

$$k^{-\frac{n}{2}-1} d_{\mathcal{B}_k}(H_{k,1}, H_{k,2}) \le k^{-\frac{n}{2}-1} L_{\mathcal{B}_k}(\mathrm{Hilb}_k(\phi(t))) \le d_{\mathcal{H}}(\phi_1, \phi_2) + \epsilon. \qquad \square$$

To prove the reversed inequality, we need the following lemma.

Lemma 3.5. *Given a smooth path $\phi(t) \in \mathcal{H}(t \in [0,1])$, then*

$$\lim_{k\to\infty} k^{-n-2} \|\nabla_{\dot{H}_k} \dot{H}_k\|^2 = \|\nabla_{\dot{\phi}} \dot{\phi}\|^2 = \int_X |\nabla_{\dot{\phi}} \dot{\phi}|^2 d\mu_\phi,$$

for any $t \in [0,1]$.

Proof. First notice that

$$\nabla_{\dot{\phi}} \dot{\phi} = \ddot{\phi} - |\nabla \dot{\phi}|^2,$$

and

$$\nabla_{\dot{H}} \dot{H} = \frac{d}{dt}(H^{-1}\dot{H})H = \ddot{H} - \dot{H}H^{-1}\dot{H},$$

where again we drop the subscript k for convenience. As before, we only need to prove the theorem at $t = 0$. We can pick an orthonormal basis $\{s_i(t)\}$ with respect to $H(t)$ such that $\dot{H}(0)$ is diagonalized, i.e.,

$$\int (s_i(t), s_j(t))_\phi d\mu_\phi = \delta_{ij},$$

and

$$\dot{H}_{ij} = \int (s_i, s_j)(-k\dot{\phi} + \Delta\dot{\phi}) d\mu = \dot{H}_i \cdot \delta_{ij}, \tag{8}$$

holds at $t = 0$. Taking more derivatives, we obtain:

$$\ddot{H}_{ij} := \ddot{H}(s_i, s_j) = \int (s_i, s_j)[(-k\dot{\phi} + \Delta\dot{\phi})^2 - k\ddot{\phi} + \Delta\ddot{\phi} + \Psi(\dot{\phi})] d\mu, \tag{9}$$

where

$$\Psi(f) = -\sum_{i,j} f_{i\bar{j}} f_{j\bar{i}}.$$

$$
\begin{aligned}
\dddot{H}_{ij} &:= \dddot{H}(s_i, s_j)\\
&= \int (s_i, s_j)\Big[(-k\dot\phi + \Delta\dot\phi)^3 - k\ddot\phi(-k\dot\phi + \Delta\dot\phi) + (\Delta\ddot\phi + \Psi(\dot\phi))(-k\dot\phi + \Delta\dot\phi)\\
&\quad + 2k^2\dot\phi\ddot\phi - 2k\ddot\phi\Delta\dot\phi - 2k\dot\phi(\Delta\ddot\phi + \Psi(\dot\phi)) + \frac{d}{dt}(\Delta\dot\phi)^2 - k\dddot\phi + \frac{d^2}{dt^2}(\Delta\dot\phi)\Big]\,d\mu\\
&= \int (s_i, s_j)\Big[- k^3\dot\phi^3 + 3k^2(\dot\phi^2\Delta\dot\phi + \ddot\phi\dot\phi) - 3k\Big(\dot\phi(\Delta\dot\phi)^2 + \ddot\phi\Delta\dot\phi + \dot\phi\Delta\ddot\phi\\
&\quad + \dot\phi\Psi(\dot\phi) + \frac{1}{3}\dddot\phi\Big) + 3\Delta\dot\phi\frac{d}{dt}(\Delta\dot\phi) + \frac{d^2}{dt^2}(\Delta\dot\phi)\Big]\,d\mu. \qquad (10)
\end{aligned}
$$

Let $\dot s_i = \sum_j a_{ij}s_j$, then we can further assume that (a_{ij}) is Hermitian, i.e., $a_{ji} = \overline{a_{ij}}$. Then it is easy to see that

$$a_{ij} = -\frac{1}{2}\dot H_{ij}.$$

Let $\phi_k(t) = FS_k(H_k(t))$, i.e.,

$$\sum_i |s_i|^2_{h_0 e^{-k\phi_k}} = 1,$$

and

$$\dot{\phi_k}(t) = -\frac{1}{k}\sum_{i,j}\dot H_{ij}(s_j, s_i)_{\phi_k} = -\frac{1}{k\rho_k(\phi)}\sum_{i,j}\dot H_{ij}(s_j, s_i)_{\phi}. \qquad (11)$$

Taking time derivative, we get

$$\ddot{\phi_k}(t) = \frac{\dot\rho_k}{k\rho_k^2}\sum_{i,j}\dot H_{ij}(s_j, s_i)_\phi - \frac{1}{k\rho_k}\Big[\sum_{i,j}\ddot H_{ij}(s_j, s_i)_\phi - 2\sum_{i,j,l}\dot H_{il}\dot H_{lj}(s_j, s_i)_\phi\Big] - k\dot\phi\dot\phi_k, \qquad (12)$$

i.e.,

$$\sum_{i,j}\ddot H_{ij}(s_j, s_i)_\phi = -k\rho_k\ddot{\phi_k} - k\dot\rho_k\dot\phi_k - k^2\rho_k\dot\phi\dot\phi_k + 2\sum_{i,j,l}\dot H_{il}\dot H_{lj}(s_j, s_i)_\phi. \qquad (13)$$

Define:

$$
\begin{aligned}
g_k &= k^{-1}(-k+\Delta)^{-1}\Big[(-k\dot\phi + \Delta\dot\phi)^2 - k\ddot\phi + \Delta\ddot\phi + \Psi(\dot\phi)\Big]\\
&= -\dot\phi^2 - \frac{1}{k}(2|\nabla\dot\phi|^2 - \ddot\phi) - \frac{1}{k^2}\Big[(\Delta\dot\phi)^2 + \Psi(\dot\phi) - 2\Delta(\dot\phi\Delta\dot\phi) + \Delta^2(\dot\phi^2)\Big] + O\Big(\frac{1}{k^3}\Big).
\end{aligned}
$$

Now let $\psi_k = d\ \mathrm{FS}_k \circ d\ \mathrm{Hilb}_k(g_k)$, then

$$-\frac{1}{k^2\rho_k}\sum_{i,j}\ddot H_{ij}(s_j, s_i) = \psi_k = g_k - \frac{\Phi(\dot\phi^2)}{k^2} + O(k^{-3}),$$

where

$$\Phi(f) = \frac{1}{2}\delta_f S.$$

From (12) we then see that

$$
\begin{aligned}
&2\sum_{i,j,l} \dot{H}_{il}\dot{H}_{lj}(s_j, s_i)_\phi \\
&\quad = -k^2\rho_k g_k + k\rho_k\ddot{\phi_k} + k\dot{\rho_k}\dot{\phi_k} + k^2\rho_k\dot{\phi}\dot{\phi_k} + \Phi(\dot{\phi}^2)k^n + O(k^{n-1}) \\
&\quad = 2k^2\rho_k\dot{\phi}^2 + 2k\rho_k|\nabla\dot{\phi}|^2 + \Big[(\Delta\dot{\phi})^2 + \Psi(\dot{\phi}) - 2\Delta(\dot{\phi}\Delta\dot{\phi}) + 2\dot{\phi}\Phi(\dot{\phi}) \\
&\qquad + \Phi(\dot{\phi}^2) + \Delta^2(\dot{\phi}^2)\Big]k^n + O(k^{n-1}).
\end{aligned}
$$

Thus,

$$
\begin{aligned}
&\sum_{i,j}(\ddot{H}_{ij} - \dot{H}_i\dot{H}_{ij})(s_j, s_i) \\
&\quad = k\rho_k(|\nabla\dot{\phi}|^2 - \ddot{\phi}) + \frac{\rho_k}{2}\Big[\Phi(\dot{\phi}^2) + \Delta^2(\dot{\phi}^2) - 2\dot{\phi}\Phi(\dot{\phi}) + (\Delta\dot{\phi})^2 \\
&\qquad + \Psi(\dot{\phi}) - 2\Delta(\dot{\phi}\Delta\dot{\phi})\Big] + O(k^{n-1}).
\end{aligned}
\tag{14}
$$

By (10),

$$
\dddot{H} = k^2 d\ \mathrm{Hilb}_k(f_k),
$$

where

$$
\begin{aligned}
f_k &= k^{-2}(-k+\Delta)^{-1}\Big[-k^3\dot{\phi}^3 + 3k^2(\dot{\phi}^2\Delta\dot{\phi} + \ddot{\phi}\dot{\phi}) - 3k\Big(\dot{\phi}(\Delta\dot{\phi})^2 + \ddot{\phi}\Delta\dot{\phi} \\
&\quad + \dot{\phi}\Delta\ddot{\phi} + \dot{\phi}\Psi(\dot{\phi}) + \frac{1}{3}\dddot{\phi}\Big) + 3\Delta\dot{\phi}\frac{d}{dt}(\Delta\dot{\phi}) + \frac{d^2}{dt^2}(\Delta\dot{\phi})\Big] \\
&= -\frac{1}{k^3}\Big[-k^3\dot{\phi}^3 + 3k^2(\dot{\phi}^2\Delta\dot{\phi} + \ddot{\phi}\dot{\phi}) - 3k\Big[\dot{\phi}(\Delta\dot{\phi})^2 + \ddot{\phi}\Delta\dot{\phi} + \dot{\phi}\Delta\ddot{\phi} \\
&\quad + \dot{\phi}\Psi(\dot{\phi}) + \frac{1}{3}\dddot{\phi}\Big] - (3k^2\dot{\phi}^2\Delta\dot{\phi} + 6k^2\dot{\phi}|\nabla\dot{\phi}|^2) + 3k\Big[\Delta(\dot{\phi}^2\Delta\dot{\phi}) \\
&\quad + \Delta(\ddot{\phi}\dot{\phi}) - \Delta^2(\dot{\phi}^3)\Big] + O(1)\Big] \\
&= \dot{\phi}^3 - \frac{3}{k}\dot{\phi}(\ddot{\phi} - 2|\nabla\dot{\phi}|^2) + \frac{3}{k^2}\Big[\dot{\phi}(\Delta\dot{\phi})^2 + \ddot{\phi}\Delta\dot{\phi} + \dot{\phi}\Delta\ddot{\phi} + \dot{\phi}\Psi(\dot{\phi}) \\
&\quad + \frac{1}{3}\dddot{\phi} - \Delta(\dot{\phi}^2\Delta\dot{\phi}) - \Delta(\ddot{\phi}\dot{\phi}) - \Delta^2(\dot{\phi}^3)\Big] + O\Big(\frac{1}{k^3}\Big).
\end{aligned}
$$

Therefore,

$$
\begin{aligned}
\sum_{i,j}\dddot{H}_{ij}(s_j, s_i) &= -k^3\rho_k f_k - k^{n+1}\Phi(\dot{\phi}^3) + O(k^n) \\
&= -k^3\rho_k\dot{\phi}^3 + 3k^2\rho_k\dot{\phi}(\ddot{\phi} - 2|\nabla\dot{\phi}|^2) - k\rho_k\Big[\Phi(\dot{\phi}^3) + 3\dot{\phi}(\Delta\dot{\phi})^2 \\
&\quad + 3\ddot{\phi}\Delta\dot{\phi} + 3\dot{\phi}\Delta\ddot{\phi} + 3\dot{\phi}\Psi(\dot{\phi}) + \dddot{\phi} - 3\Delta(\dot{\phi}^2\Delta\dot{\phi}) \\
&\quad - 3\Delta(\ddot{\phi}\dot{\phi}) - \Delta^2(\dot{\phi}^3)\Big] + O(k^n).
\end{aligned}
\tag{15}
$$

Now differentiating (13), we obtain

$$
\begin{aligned}
&\sum_{i,j} \dddot{H}_{ij}(s_j, s_i) - \sum_{i,j} \ddot{H}_{i,j}(\dot{H}_i + \dot{H}_j)(s_j, s_i) - k\dot{\phi} \sum_{i,j} \ddot{H}_{ij}(s_j, s_i) \\
&\quad = -2k\dot{\rho}_k\ddot{\phi}_k - k\rho_k\dddot{\phi}_k - k\ddot{\rho}_k\dot{\phi}_k - k^2\dot{\rho}_k\dot{\phi}\dot{\phi}_k - k^2\rho_k\ddot{\phi}\dot{\phi}_k - k^2\rho_k\dot{\phi}\ddot{\phi}_k \\
&\qquad + 2\sum_{i,j} \ddot{H}_{i,j}(\dot{H}_i + \dot{H}_j)(s_j, s_i) - 6\sum_i \dot{H}_i^3|s_i|^2 - 2k\dot{\phi}\sum_i \dot{H}_i^2|s_i|^2. \qquad (16)
\end{aligned}
$$

Thus,

$$
\begin{aligned}
&3\sum_{i,j} \ddot{H}_{i,j}(\dot{H}_i + \dot{H}_j)(s_j, s_i) - 6\sum_i \dot{H}_i^3|s_i|^2 \\
&= \sum_{i,j} \dddot{H}_{i,j}(s_j, s_i) + k\dot{\phi}(k\rho_k\ddot{\phi}_k + k\dot{\rho}_k\dot{\phi}_k + k^2\rho_k\dot{\phi}\dot{\phi}_k) \\
&\quad + 2k\dot{\rho}_k\ddot{\phi}_k + k\rho_k\dddot{\phi}_k + k\ddot{\rho}_k\dot{\phi}_k + k^2\dot{\rho}_K\dot{\phi}\dot{\phi}_k + k^2\rho_k\ddot{\phi}\dot{\phi}_k + k^2\rho_k\dot{\phi}\ddot{\phi}_k \\
&= \sum_{i,j} \dddot{H}_{i,j}(s_j, s_i) + k^3\rho_k\dot{\phi}^2\dot{\phi}_k + 3k^2\rho_k\dot{\phi}\ddot{\phi} + k^{n+1}(2\Phi(\dot{\phi})\dot{\phi}^2 + \dddot{\phi}) + O(k^n) \\
&= 6k^2\rho_k\dot{\phi}(\ddot{\phi} - |\nabla\dot{\phi}|^2) + k^{n+1}\Big[3\Phi(\dot{\phi})\dot{\phi}^2 + \dddot{\phi} - \Phi(\dot{\phi}^3) - 3\dot{\phi}(\Delta\dot{\phi})^2 - 3\dot{\phi}\Delta\ddot{\phi} - 3\ddot{\phi}\Delta\dot{\phi} \\
&\quad - 3\dot{\phi}\Psi(\dot{\phi}) - \dddot{\phi} + 3\Delta(\dot{\phi}^2\Delta\dot{\phi}) + 3\Delta(\dot{\phi}\ddot{\phi}) - \Delta^2(\dot{\phi}^3)\Big] + O(k^n) \\
&= 6k^2\rho_k\dot{\phi}(\ddot{\phi} - |\nabla\dot{\phi}|^2) + k^{n+1}\Big[3\Phi(\dot{\phi})\dot{\phi}^2 - \Phi(\dot{\phi}^3) - \Delta^2(\dot{\phi}^3) - 3\dot{\phi}(\Delta\dot{\phi})^2 - 6\nabla\dot{\phi}\cdot\nabla\ddot{\phi} \\
&\quad - 3\dot{\phi}\Psi(\dot{\phi}) + 3\Delta(\dot{\phi}^2\Delta\dot{\phi})\Big] + O(k^n). \qquad (17)
\end{aligned}
$$

Putting (8), (9), (14), (17) together, we get

$$
\begin{aligned}
|\nabla_{\dot{H}}\dot{H}|^2 &= \sum_{i,j}(\ddot{H}_{ij} - \dot{H}_i\dot{H}_{ij}) \int (s_j, s_i)(k^2\dot{\phi}^2 - 2k\dot{\phi}\Delta\dot{\phi} - k\ddot{\phi} + O(1))d\mu \\
&\quad - \sum_{i,j}(\ddot{H}_{ij}\dot{H}_i - \dot{H}_i^3) \int (s_j, s_i)(-k\dot{\phi} + \Delta\dot{\phi})d\mu \\
&= \int \Big\{k\rho_k(|\nabla\dot{\phi}|^2 - \ddot{\phi}) + \frac{\rho_k}{2}\Big[\Phi(\dot{\phi}^2) - 2\dot{\phi}\Phi(\dot{\phi}) + \Delta^2(\dot{\phi}^2) + (\Delta\dot{\phi})^2 + \Psi(\dot{\phi}) \\
&\quad - 2\Delta(\dot{\phi}\Delta\dot{\phi})\Big] + O(k^{n-1})\Big\}(k^2\dot{\phi}^2 - 2k\dot{\phi}\Delta\dot{\phi} - k\ddot{\phi} + O(1))d\mu \\
&\quad + \int \Big\{k^2\rho_k\dot{\phi}(\ddot{\phi} - |\nabla\dot{\phi}|^2) + \frac{k\rho_k}{6}\Big[3\Phi(\dot{\phi})\dot{\phi}^2 - \Phi(\dot{\phi}^3) - \Delta^2(\dot{\phi}^3) \\
&\quad + 6\nabla\dot{\phi}\nabla\ddot{\phi} - 3\dot{\phi}\Psi(\dot{\phi}) - 3\dot{\phi}(\Delta\dot{\phi})^2 + 3\Delta(\dot{\phi}^2\Delta\dot{\phi}) + O(k^n)\Big]\Big\}(k\dot{\phi} - \Delta\dot{\phi})d\mu \\
&= \int -k^2\rho_k(|\nabla\dot{\phi}|^2 - \ddot{\phi})(2\dot{\phi}\Delta\dot{\phi} + \ddot{\phi}) + \frac{k^2}{2}\rho_k\dot{\phi}^2\Big[\Phi(\dot{\phi}^2) - 2\dot{\phi}\Phi(\dot{\phi}) + \Delta^2(\dot{\phi}^2) \\
&\quad + (\Delta\dot{\phi})^2 + \Psi(\dot{\phi}) - 2\Delta(\dot{\phi}\Delta\dot{\phi})\Big] - k^2\rho_k\dot{\phi}\Delta\dot{\phi}(\ddot{\phi} - |\nabla\dot{\phi}|^2)
\end{aligned}
$$

$$
\begin{aligned}
&+\frac{k^2\rho_k}{6}\dot\phi\Big[3\Phi(\dot\phi)\dot\phi^2-\Phi(\dot\phi^3)-\Delta^2(\dot\phi^3)+6\nabla\dot\phi\nabla\ddot\phi-3\dot\phi\Psi(\dot\phi)\\
&-3\dot\phi(\Delta\dot\phi)^2+3\Delta(\dot\phi^2\Delta\dot\phi)\Big]d\mu+O(k^{n+1})
\end{aligned}
$$

Further manipulating the above expression, we get

$$
\begin{aligned}
&|\nabla_{\dot H}\dot H|^2\\
&=-\frac{k^2}{2}\int\rho_k\Delta(\dot\phi^2)|\nabla\dot\phi|^2d\mu\\
&+\frac{k^2}{2}\int\rho_k\dot\phi^2[-R_{i\bar j}\dot\phi_{\bar i}\dot\phi_j+\Delta^2(\dot\phi^2)+\dot\phi\Delta^2\dot\phi+(\Delta\dot\phi)^2-2\Delta(\dot\phi\Delta\dot\phi)]d\mu\\
&+\frac{k^{n+2}}{6}\int\dot\phi\Big[3R_{i\bar j}\dot\phi\dot\phi_{\bar i}\dot\phi_j-\frac{3}{2}\dot\phi^2\Delta^2\dot\phi+\frac{1}{2}\Delta^2(\dot\phi^3)-\Delta^2(\dot\phi^3)\\
&-3\dot\phi(\Delta\dot\phi)^2+3\Delta(\dot\phi^2\Delta\dot\phi)\Big]d\mu+O(k^{n+1})\\
&=-k^2\int\rho_k(\dot\phi\Delta\dot\phi+|\nabla\dot\phi|^2)|\nabla\dot\phi|^2d\mu\\
&+k^{n+2}\int\frac{1}{2}\dot\phi^3\Delta^2\dot\phi+\frac{1}{4}\dot\phi^2\Delta^2(\dot\phi^2)+\frac{1}{2}\dot\phi^2(\Delta\dot\phi)^2-2\dot\phi^2(\Delta\dot\phi)^2-2\dot\phi\Delta\dot\phi|\nabla\dot\phi|^2d\mu\\
&-\frac{k^{n+2}}{3}\int\dot\phi^3\Delta^2\dot\phi d\mu+O(k^{n+1})\\
&=k^{n+2}\int(\ddot\phi-|\nabla\dot\phi|)^2d\mu+O(k^{n+1}).
\end{aligned}
$$

where we have used:

$$
\begin{aligned}
\Phi(\dot\phi)&=-\frac{1}{2}(\dot\phi_{i\bar j}R_{j\bar i}+\Delta^2\dot\phi),\\
\Phi(\dot\phi^2)&=-\dot\phi\dot\phi_{i\bar j}R_{j\bar i}-\dot\phi_i\dot\phi_{\bar j}R_{j\bar i}-\frac{1}{2}\Delta^2(\dot\phi^2),\\
\Phi(\dot\phi^3)&=-\frac{3}{2}\dot\phi^2\dot\phi_{i\bar j}R_{j\bar i}-3\dot\phi\dot\phi_i\dot\phi_{\bar j}R_{j\bar i}-\frac{1}{2}\Delta^2(\dot\phi^3).
\end{aligned}
$$

$$
\int\dot\phi^2\Delta^2(\dot\phi^2)d\mu=4\int\dot\phi^2(\Delta\dot\phi^2)d\mu+8\int\dot\phi\Delta\dot\phi|\nabla\dot\phi|^2d\mu+4\int|\nabla\dot\phi|^4d\mu.
$$

$$
\int\dot\phi^3\Delta^2(\dot\phi)d\mu=3\int\dot\phi^2(\Delta\dot\phi)^2d\mu+6\int\dot\phi\Delta\dot\phi|\nabla\dot\phi|^2d\mu. \qquad \square
$$

Proof of Theorem 1.1. In view of Corollary 3.4, it suffices to show that

$$
\liminf_{k\to\infty}k^{-\frac{n}{2}-1}d_{\mathcal{B}_k}(H_{k,0},H_{k,1})\ge d_{\mathcal{H}}(\phi_0,\phi_1).
$$

By Lemma 2.1, for any $\epsilon>0$ small enough, there exists a smooth family $\phi_\epsilon(\cdot):[0,1]\to\mathcal{H}$, such that $\phi_\epsilon(0)=\phi_0$, $\phi_\epsilon(1)=\phi_1$, and

$$
(\ddot\phi_\epsilon-|\nabla\dot\phi_\epsilon|_t^2)\omega_{\epsilon,t}^n=\epsilon\omega^n.
$$

Moreover, $\ddot{\phi}_\epsilon - |\nabla\dot{\phi}_\epsilon|_t^2 \leq C$, where C is a constant independent of t and ϵ. Let $H_\epsilon(t) = \mathrm{Hilb}_k(\phi_\epsilon(t))$, then by Lemma 3.5, for k large enough, we have

$$k^{-n-2}|\nabla_{\dot{H}_\epsilon}\dot{H}_\epsilon|^2 \leq C\epsilon.$$

By the following simple lemma, we know that

$$k^{-\frac{n}{2}-1}d_{\mathcal{B}_k}(H_{k,0}, H_{k,1}) \geq k^{-\frac{n}{2}-1}L(H_\epsilon) - \sqrt{C\epsilon} \to L(\phi_\epsilon) - \sqrt{C\epsilon},$$

as $k \to \infty$. Hence,

$$\liminf_{k\to\infty} k^{-\frac{n}{2}-1}d_{\mathcal{B}_k}(H_{k,0}, H_{k,1}) \geq d_{\mathcal{H}}(\phi_0, \phi_1) - \sqrt{C\epsilon}.$$

Let $\epsilon \to 0$, we obtain the desired result. □

Now we prove a simple lemma from Riemannian geometry.

Lemma 3.6. *Suppose (M, g) is a simply connected Riemmannian manifold with non-positive curvature. Let $\gamma : [0,1] \to M$ be a path in M such that $|\nabla_{\dot{\gamma}(t)}\dot{\gamma}(t)| \leq \epsilon$ for all $t \in [0,1]$. Then*

$$d(\gamma(0), \gamma(1)) \geq L(\gamma) - \epsilon.$$

Proof. Denote $p = \gamma(0)$ and $q = \gamma(1)$. By the theorem of Cartan-Hadamard, we know that the exponential map at p is a diffeomorphism. Now let $\gamma_t(s) (s \in [0,1])$ be the unique geodesic connecting p and $\gamma(t)$. By standard calculation of the second variation, we obtain:

$$\begin{aligned}\frac{d^2}{dt^2}L(\gamma_t) = & \frac{1}{L(\gamma_t)}\left\langle \nabla_{\dot{\gamma}(t)}\dot{\gamma}(t), \frac{\partial}{\partial s}\gamma_t(s)|_{s=1}\right\rangle + \frac{1}{L(\gamma_t)}\int_0^1 \left|\left(\frac{\partial^2}{\partial s\partial t}\gamma_t(s)\right)^{\perp}\right|^2 \\ & - R\left(\frac{\partial}{\partial s}\gamma_t(s), \frac{\partial}{\partial t}\gamma_t(s), \frac{\partial}{\partial s}\gamma_t(s), \frac{\partial}{\partial t}\gamma_t(s)\right) ds \\ \geq & -\epsilon,\end{aligned}$$

where $\perp$ denotes projection to the orthogonal complement of $\frac{\partial}{\partial s}\gamma_t(s)$. It is also easy to see that

$$L(\gamma_0) = 0,$$

and

$$\frac{d}{dt}L(t)|_{t=0} = |\dot{\gamma}(0)|.$$

Therefore,

$$L(t) \geq |\dot{\gamma}(0)|t - \frac{\epsilon}{2}t^2.$$

In particular, $d(p,q) = L(\gamma(1)) \geq |\dot{\gamma}(0)| - \frac{\epsilon}{2}$. On the other hand,

$$L(\gamma) = \int_0^1 |\dot{\gamma}(t)|dt \leq \int_0^1 (|\dot{\gamma}(0)| + \epsilon t)dt = |\dot{\gamma}(0)| + \frac{\epsilon}{2}.$$

Hence, $d(p,q) \geq L(\gamma) - \epsilon$. □

4. Weak convexity of K-energy

In this section, we shall prove Corollary 1.7 and 1.6. Before doing this, we want to show the functional I is quantized by I_k, which is an analogue of the fact that Z_k quantizes E.

Proposition 4.1. *There are constants c_k such that for $\phi \in \mathcal{H}$, and $H_k = \mathrm{Hilb}_k(\phi)$, we have*

$$\lim_{k\to\infty} k^{-n-1} I_k(H_k) + c_k = I(\phi).$$

The convergence is uniform when ϕ varies in a C^l bounded sets as before.

Proof. It suffices to show for any $\psi \in T_\phi \mathcal{H}$

$$d_\phi\ I(\psi) = d_{\mathrm{Hilb}_k(\phi)}\ I_k \circ d_\phi\ \mathrm{Hilb}_k(\psi). \tag{18}$$

By definition,

$$d_\phi\ I(\psi) = \int \psi d\mu_\phi.$$

On the other hand, by Lemma 3.1,

$$\begin{aligned} d_{\mathrm{Hilb}_k(\phi)}\ I_k \circ d_\phi\ \mathrm{Hilb}_k(\psi) &= \sum_i \int |s_i|_\phi^2(-k\psi + \Delta\psi) d\mu_\phi \\ &= \int \rho_k(\phi)(-k\psi + \Delta\psi) d\mu_\phi \end{aligned}$$

Then the result follows from Lemma 3.1. □

Proposition 4.2. *Let $\phi \in \mathcal{H}$, and $H_k = \mathrm{Hilb}_k(\phi)$, then*

$$\lim_{k\to\infty} k^{n+2} \|\nabla Z_k(H_k)\|^2 = \|\nabla E(\phi)\|^2.$$

Proof. Note

$$\|\nabla E(\phi)\|^2 = \int (S - \underline{S})^2 d\mu_\phi$$

is simply Calabi's functional. An inequality of this form is essentially proved in [13] for obtaining a lower bound of the Calabi functional. We can easily calculate the first variation of Z_k:

$$\delta Z_k = -\frac{d_k}{k^n \cdot V} \int_X \delta H_{ij} [(s_j, s_i)_{FS_k(H)}]_0 d\mu_{\omega_{FS_k(H)}},$$

where $[A]_0$ denote the trace-free part of a matrix A, and $\{s_i\}$ is an orthonormal basis with respect to H. Here

$$V = \int_X d\mu_\phi,$$

and

$$d_k = \dim H^0(X, L^k) = \int_X \rho_k(\phi) d\mu_\phi = k^n \cdot V + \frac{k^{n-1} \cdot V}{2} \underline{S} + O(k^{n-2}).$$

So

$$(\nabla Z_k)_{ij} = -\frac{d_k}{k^n \cdot V}\int_X [(s_i, s_j)_{FS_k(H)}]_0 d\mu_{\omega_{FS_k(H)}}.$$

Diagonalize the above matrix so that its diagonal entries are

$$\lambda_i = \frac{d_k}{k^n \cdot V}\int_X (|s_i|^2_{\psi_2} - \frac{1}{d_k}) d\mu_{\psi_2}.$$

Let $\phi_k = FS_k(H_k)$, then

$$\phi_k - \phi = \frac{1}{k}\log(\rho_k(\phi)) = \frac{1}{k}\log(k^n + \frac{S}{2}k^{n-1} + O(k^{n-2})).$$

So

$$|s_i|^2_{\phi_k} = k^{-n}|s_i|^2_{\phi}\left(1 - \frac{S}{2k} + O\left(\frac{1}{k^2}\right)\right),$$

and

$$\omega^n_{\phi_k} = \omega^n_{\phi}(1 + O(k^{-2})).$$

Since

$$\int_X |s_i|^2_\phi d\mu_\phi = 1,$$

we obtain

$$\lambda_i = -\frac{1}{k^{n+1}}\int_X |s_i|^2_\phi (S - \underline{S}) d\mu_\phi + O\left(\frac{1}{k^{n+2}}\right).$$

Hence by Remark 3.3,

$$\begin{aligned}
\|\nabla Z_k\|^2 &= \sum_i |\lambda_i|^2 \\
&= k^{-2n-2}\sum_i \left[\int_X |s_i|^2_\phi (S - \underline{S}) d\mu_\phi + O\left(\frac{1}{k^{n+2}}\right)\right]^2 \\
&= k^{-n-2}\left(Ca(\phi_2) + O\left(\frac{1}{k}\right)\right).
\end{aligned}$$

□

The proof of the following proposition is similar and we omit it.

Proposition 4.3. *Let $\phi(t)$ be a smooth path in $\mathcal{H}$, and $H_k(t) = \mathrm{Hilb}_k(\phi)$, then*

$$\lim_{k\to\infty} \frac{d}{dt} Z_k(H_k) = \frac{d}{dt} E(\phi).$$

Proof of Corollary 1.7. By Lemma 2.5, we already know that Z_k is a genuine convex function on $\mathcal{B}_k$. So it is easy to see that

$$Z_k(H_2) - Z_k(H_1) \le d_{\mathcal{B}_k}(H_1, H_2) \cdot \|\nabla Z_k(H_2)\|.$$

Using Theorem 1.1, Lemma 2.4 and Proposition 4.2, and let $k \to \infty$, we get the desired inequality for E. □

Proof of Corollary 1.6. As in the proof of Theorem 1.1, for any $\epsilon > 0$ sufficiently small, we choose ϵ-approximate geodesic $\phi_\epsilon : [0,1] \to \mathcal{H}$, such that $|\nabla_{\dot{\phi}_\epsilon} \dot{\phi}_\epsilon| \leq C\epsilon$. Denote $H_\epsilon(t) = \mathrm{Hilb}_k(\phi_\epsilon(t))$, and $H_i = \mathrm{Hilb}_k(\phi_i)$, $i = 0, 1$. Then by Lemma 3.5, we see that

$$k^{-\frac{n}{2}-1} |\nabla_{\dot{H}_\epsilon} \dot{H}_\epsilon| \leq C\epsilon,$$

for k sufficiently large and a different constant C. Denote by $\tilde{H} : [0,1] \to \mathcal{B}_k$ the geodesic connecting H_0 and H_1. By Lemma 4.5 which will be proved later, we know that

$$k^{-\frac{n}{2}-1} |\dot{H}_\epsilon(i) - \dot{\tilde{H}}(i)| \leq C\sqrt{\epsilon},$$

for $i = 0, 1$. Since $\tilde{H}(t)$ is a geodesic in $\mathcal{B}_k$, we have by Lemma 2.5 that

$$dZ_{k_{H_1}}(\dot{\tilde{H}}(1)) \geq dZ_{k_{H_0}}(\dot{\tilde{H}}(0)).$$

On the other hand, by Proposition 4.2,

$$|dZ_{k_{H_1}}(\dot{H}_\epsilon(1) - \dot{\tilde{H}}(1))| \leq |\nabla Z_k|_{H_1} \cdot |\dot{H}_\epsilon(1) - \dot{\tilde{H}}(1)| \leq C\sqrt{\epsilon} \left(\sqrt{Ca(\phi_1)} + O\left(\frac{1}{k}\right)\right).$$

Similarly,

$$|dZ_{k_{H_0}}(\dot{H}_\epsilon(0) - \dot{\tilde{H}}(0))| \leq C\sqrt{\epsilon}(\sqrt{Ca(\phi_0)} + O(\frac{1}{k})).$$

So,

$$dZ_{k_{H_1}}(\dot{H}_\epsilon(1)) \geq dZ_{k_{H_0}}(\dot{H}_\epsilon(0)) - C\sqrt{\epsilon}.$$

By Proposition 4.3, that as $k \to \infty$

$$dZ_{k_{H_1}}(\dot{H}_\epsilon(1)) \to dE_{\phi_1}(\dot{\phi}_\epsilon(1)),$$

and

$$dZ_{k_{H_0}}(\dot{H}_\epsilon(0)) \to dE_{\phi_0}(\dot{\phi}_\epsilon(0)).$$

Thus,

$$dE_{\phi_1}(\dot{\phi}_\epsilon(1)) \geq dE_{\phi_0}(\dot{\phi}_\epsilon(0)) - C\sqrt{\epsilon}.$$

Letting $\epsilon \to 0$, this proves Corollary 1.6. □

Remark 4.4. If we denote $\phi_{k,\epsilon}(t) = FS_k(\mathrm{Hilb}_k\, \phi_\epsilon(t))$ and $\tilde{\phi}_k(t) = FS_k(H(t))$, then by a similar estimate as in the proof of Proposition 4.2 we have

$$|\dot{\tilde{\phi}}_k(0) - \dot{\phi}_{k,\epsilon}(0)|_{L^2} \leq C\sqrt{\epsilon}.$$

Thus, the time derivative of $\tilde{\phi}_k$ at the end points converge in L^2 to the time derivative of the $C^{1,1}$ geodesic. This seems to be an interesting fact, compare [18].

Lemma 4.5. *Suppose (M, g) is a simply connected Riemannian manifold with non-positive curvature. Let $\gamma : [0,1] \to M$ be a path in M such that $|\nabla_{\dot{\gamma}(t)} \dot{\gamma}(t)| \leq \epsilon$ for all $t \in [0,1]$. Let $\tilde{\gamma} : [0,1]$ be the unique geodesic segment joining $\gamma(0)$ and $\gamma(1)$. Then*

$$|\dot{\gamma}(0) - \dot{\tilde{\gamma}}(0)|^2 \leq \frac{9}{4}\epsilon^2 + 4\epsilon|\dot{\gamma}(1)|,$$

and

$$|\dot{\gamma}(1) - \dot{\tilde{\gamma}}(1)|^2 \leq \frac{9}{4}\epsilon^2 + 4\epsilon|\dot{\gamma}(0)|,$$

Proof. It suffices to prove the second inequality. As before, let $\gamma_t(s) : [0,1] \to M$ be the smooth family of geodesic segments connecting $\gamma(0)$ and $\gamma(t)$. Then by the proof of Lemma 3.6 we see that

$$\frac{d}{dt}\left\langle \frac{d\gamma}{dt}, \frac{\dot{\gamma}_t(1)}{|\dot{\gamma}_t(1)|} \right\rangle \geq -\epsilon,$$

and so

$$\left\langle \frac{d\gamma}{dt}, \frac{\dot{\gamma}_t(1)}{|\dot{\gamma}_t(1)|} \right\rangle \geq \lim_{t\to 0}\left\langle \frac{d\gamma}{dt}, \frac{\dot{\gamma}_t(1)}{|\dot{\gamma}_t(1)|} \right\rangle - \epsilon t = \left|\frac{d\gamma}{dt}(0)\right| - \epsilon t.$$

Let

$$\frac{d\gamma}{dt} = A_t \cdot \frac{\dot{\gamma}_t(1)}{|\dot{\gamma}_t(1)|} + B_t$$

be the orthogonal decomposition. Then

$$A_t \geq \left|\frac{d\gamma}{dt}(0)\right| - \epsilon t.$$

It is clear that

$$\left|\frac{d}{dt}\left|\frac{d\gamma}{dt}\right|\right| \leq \left|\frac{d^2\gamma}{dt^2}\right| \leq \epsilon,$$

so

$$\left|\frac{d\gamma}{dt}(0)\right| - \epsilon t \leq \left|\frac{d\gamma}{dt}\right| \leq \left|\frac{d\gamma}{dt}(0)\right| + \epsilon t.$$

Hence,

$$|B_t|^2 \leq \left(\left|\frac{d\gamma}{dt}(0)\right| + \epsilon t\right)^2 - \left(\left|\frac{d\gamma}{dt}(0)\right| - \epsilon t\right)^2 = 4\epsilon t\left|\frac{d\gamma}{dt}(0)\right|.$$

From the proof of Lemma 3.6, we know that

$$\left|\frac{d\gamma}{dt}(0)\right| t - \frac{\epsilon}{2}t^2 \leq |\dot{\gamma}_t(1)| \leq \left|\frac{d\gamma}{dt}(0)\right| t + \frac{\epsilon}{2}t^2.$$

Finally,

$$\begin{aligned}\left|\frac{d\gamma}{dt} - \frac{1}{t}\dot{\gamma}_t(1)\right|^2 &= \left(A_t - \frac{1}{t}|\dot{\gamma}_t(1)|\right)^2 + B_t^2 \\ &\leq \frac{9}{4}\epsilon^2 t^2 + 4\epsilon t\left|\frac{d\gamma}{dt}(0)\right|,\end{aligned}$$

Let $t = 1$, we obtain the desired bound. □

5. Open problems

In this section, we list some open problems and speculations on which the results of this paper may help in the future.

Problem 5.1. Theorem 1.1 essentially says that $\mathcal{H}$ is a weak Gromov-Hausdorff limit of a sequence of finite-dimensional symmetric spaces and $\mathcal{H}$ can be viewed as a generalized $Cat(0)$ space which has been extensively studied in the literature (cf. [1]). It would be interesting to investigate this more carefully and develop a suitable notion for this type of convergence. From this convergence it follows that the negativity of curvature is inherited by the limit. A natural question would be what else properties for the finite-dimensional symmetric spaces would survive on $\mathcal{H}$. For example can we describe the algebraic structure of $\mathcal{H}$. In particular does it admit a local involution? In other words, for any $\phi \in \mathcal{H}$, there exists a small constant $\delta(\phi)$ such that there is an local involution

$$\sigma : B_\delta(\varphi) \bigcap \mathcal{H} \to B_\delta(\varphi) \bigcap \mathcal{H}$$

with $\sigma^2 = id$. This is not trivial since the initial value problem for the geodesic equation in $\mathcal{H}$ is not well posed.

We know that $\mathcal{H}$ is not complete, then the following is very interesting:

Problem 5.2. What is the structure of $\partial\mathcal{H}$?

It would be extremely interesting to understand the structure of the boundary. Note that for every k, the Bergman metric space $\mathcal{B}_k$ is complete and $\mathcal{H}$ is a limit of these nice symmetric spaces after appropriate scaling down (by a factor of $k^{-\frac{n+2}{2}}$). It is natural to hope that $\mathcal{H}$ should look like the limit of the tangent cone of $\mathcal{B}_k$ at infinity.

By the work of J. Fine [14], we can approximate Calabi flow over a bounded interval by the Balancing flow in $\mathcal{B}_k$. The next intriguing question is can we approximate the Calabi flow over $[0, \infty)$ if we know it converges (the complex structure might jump). This is related to

Problem 5.3. In an algebraic manifold, if there is a cscK metric, does the Calabi flow exist for long time and converge to a cscK metric?

This is only proved for metrics near a cscK metric. The corresponding problems for Kähler-Ricci flow was proved first by G. Perelman and a written version is provided by Tian-Zhu([25]).

Problem 5.4. The existence of cscK metrics implies that the K-energy is proper.

The corresponding results for Kähler-Einstein metrics is proved by Tian in [24]. Note the definition of properness could vary. In [24], properness means bounded below by another positive functional J. Since the K-energy is convex along smooth geodesics and its Hessian at a cscK metric is strictly positive (if the automorphism group of the manifold is discrete), we would expect it bounds the distance function. Note the corresponding statements in the finite-dimensional case is clear: a convex function with a non-degenerate critical point automatically bounds the distance function. Since both the distance function and the K-energy could be approximated by the corresponding quantities on $\mathcal{B}_k$, to prove the corresponding statement for $\mathcal{H}$, it is important to derive a uniform constant.

Acknowledgment

Both authors wish to thank Professor S. Zelditch for his interest in this work. The second author wishes to thank Professor S. Donaldson, Professor B. Lawson and Professor G. Tian for their helpful comments. He also would like to thank Garrett Alston for suggestions improving the exposition of this article.

References

[1] M. Bridson, A. Haefliger, *Metric spaces of non-positive curvature*, Springer, 1999.

[2] E. Calabi, X.-X. Chen. *Space of Kähler metrics and Calabi flow*, J. Differential Geom. 61 (2002), no. 2, 173–193.

[3] X.-X. Chen. *The space of Kähler metrics*, J. Differential Geom. 56 (2000), no. 2, 189–234.

[4] X.-X. Chen. *Space of Kähler metrics III – The greatest lower bound of the Calabi energy*, Invent. math. 175, 453–680 (2009).

[5] X.-X. Chen. *Space of Kähler metrics IV – On the lower bound of the K-energy*, arxiv: math/0809.4081.

[6] X.-X. Chen. *On the lower bound of the Mabuchi energy and its application*, Internet. Math. Res. Notices 2000, no. 12, 607–623.

[7] X.-X. Chen, G. Tian. *Geometry of Kähler metrics and foliations by holomorphic discs*, Publications Mathématiques de L'IHÉS, Vol 107, 1–107.

[8] S.K. Donaldson. *Remarks on gauge theory, complex geometry and 4-manifold topology*, Fields Medalists' Lectures, World Sci. Publ., Singapore, 1997, 384–403.

[9] S.K. Donaldson. *Symmetric spaces, Kähler geometry and Hamiltonian dynamics*, Northern California Symplectic Geometry Seminar, 13–33, Amer. Math. Soc. Transl. Ser. 2, 196, Amer. Math. Soc., Providence, RI, 1999.

[10] S.K. Donaldson. *Scalar curvature and projective embeddings*, I. J. Differential. Geom. 59 (2001), no. 3, 479–522.

[11] S.K. Donaldson. *Holomorphic discs and the complex Monge-Ampère equation*, J. Symplectic Geom. 1 (2002), no. 2, 171–196.

[12] S.K. Donaldson. *Scalar curvature and projective embeddings*, II. Q. J. Math. 56 (2005), no. 3, 345–356.

[13] S.K. Donaldson. *Lower bounds on the Calabi functional*, J. Differential Geom. 70(2005), 453–472.

[14] J. Fine. *Calabi flow and projective embeddings*, arxiv: math/0811.0155.

[15] Z. Lu. *On the lower order terms of the asymptotic expansion of Tian-Yau-Zelditch*, Amer. J. Math, 122 (2000), 235–273.

[16] Z. Lu, G. Tian. *The log term of the Szegö kernel*, Duke Math. J. 125 (2004), no. 2, 351–387.

[17] T. Mabuchi. *Some symplectic geometry on compact Kähler manifolds I*, Osaka, J. Math, no. 24, 227–252, 1987.
[18] D.H. Phong, J. Sturm. *The Monge-Ampère operator and geodesics in the space of Kähler potentials*, Invent. Math. 166 (2006), 125–149.
[19] D.H. Phong, J. Sturn. *Test Configurations for K-stability and Geodesic Rays*, J. Symplectic Geom. 5 (2007), no. 2, 221–247.
[20] W.-D. Ruan. *Canonical coordinates and Bergman metrics*, Comm. Anal. Geom. 6 (1998), no. 3, 589–631.
[21] Y.A. Rubinstein, *Geometric quantization and dynamical constructions on the space of Kähler metrics*, Ph.D. Thesis, MIT, 2008.
[22] Y.A. Rubinstein, S. Zelditch. *Bergman approximations of harmonic maps into the space of Kähler metrics on toric varieties*, arxiv:math/0803.1249.
[23] J. Song, S. Zelditch. *Bergman metrics and geodesics in the space of Kähler metrics on toric varieties*, arxiv:math/0707.3082.
[24] G. Tian. *On a set of polarized Kähler metrics on algebraic manifolds*, J. Differential Geometry. 32 (1990), 99–130.
[25] G. Tian, X-H. Zhu. *Convergence of Kähler-Ricci flow*, J. Amer. Math. Soc. 20 (2007), no. 3, 675–699.
[26] S. Zelditch. *Szegö kernel and a theorem of Tian*, Int. Math. Res. Notices 6(1998), 317–331.

Xiuxiong Chen
Department of Mathematics
University of Wisconsin-Madison
480 Lincoln Drive
Madison, WI 53706, USA

and

Department of Mathematics
Stony Brook University
Stony Brook, NY 11794, USA
e-mail: `xxchen@math.wisc.edu`
`xiu@math.sunysb.edu`

Song Sun
Department of Mathematics
University of Wisconsin-Madison
480 Lincoln Drive
Madison, WI 53706, USA

and

Department of Mathematics
Imperial College London
SW7 2AZ London, UK
e-mail: `ssun@math.wisc.edu`
`s.sun@imperial.ac.uk`

Split Special Lagrangian Geometry

F. Reese Harvey and H. Blaine Lawson, Jr.

Abstract. One purpose of this article is to draw attention to the seminal work of J. Mealy in 1989 on calibrations in semi-riemannian geometry where split SLAG geometry was first introduced. The natural setting is provided by doing geometry with the complex numbers **C** replaced by the double numbers **D**, where i with $i^2 = -1$ is replaced by τ with $\tau^2 = 1$. A rather surprising amount of complex geometry carries over, almost untouched, and this has been the subject of many papers. We briefly review this material and, in particular, we discuss Hermitian **D**-manifolds with trivial canonical bundle, which provide the background space for the geometry of split SLAG submanifolds.

A removable singularities result is proved for split SLAG subvarieties. It implies, in particular, that there exist no split SLAG cones, smooth outside the origin, other than planes. This is in sharp contrast to the complex case.

Parallel to the complex case, space-like Lagrangian submanifolds are stationary if and only if they are θ-split SLAG for some constant phase angle θ, and infinitesimal deformations of split SLAG submanifolds are characterized by harmonic 1-forms on the submanifold.

We also briefly review the recent work of Kim, McCann and Warren who have shown that split Special Lagrangian geometry is directly related to the Monge-Kantorovich mass transport problem.

Mathematics Subject Classification (2000). Primary 53C42; secondary 53C50, 35J96.

Keywords. Split special Lagrangian subvariety, calibration, Interior regularity, Monge-Ampère, mass transport, D-manifold, paracomplex manifold, bi-Lagrangian manifold.

1. Introduction

In an interesting recent series of papers [KM, W_1, KMW] Y.-H. Kim, R. McCann and M. Warren established a direct relationship between the classical Monge-Kantorovich mass transport problem and split special Lagrangian geometry. This geometry is one of the basic calibrated semi-riemannian geometries introduced by

The second author was partially supported by the N.S.F.

Jack Mealy in his seminal work [M$_1$, M$_2$] in 1989–90. The ambient spaces are analogues of complex Kähler manifolds for which the literature is vast and not unified[1] – they are sometimes called paraKähler and sometimes Bi-Lagrangian (we shall call them Kähler **D**-manifolds). Despite the widespread interest, Mealy's results about calibrations on such manifolds seem to have gone unnoticed. One purpose of this article is to give a brief self-contained discussion of **D**-manifolds so that we can discuss split special Lagrangian geometry and present the work of Mealy in this important case. We shall then explain how the work of Kim-McCann-Warren fits into the picture. Along the way we will also establish some new results concerning split special Lagrangian subvarieties. One is a removable-singularities result which implies, in particular, that any split SLAG cone, which is smooth and connected outside the origin, is a plane (in sharp contrast to the classical non-split case where such cones exist in abundance). Another can be interpreted as the strongest possible regularity result for d-closed split special Lagrangian rectifiable currents in dimension 2.

Throughout the exposition we emphasize the two distinct approaches to **D**-manifolds. The first is in strong analogy with complex geometry, where the double numbers **D** (see Section 2) are considered as a replacement for the complex numbers **C**, and a surprising number of definitions and results carry over. In the second viewpoint the double numbers are viewed as the algebra $\mathbf{D} = \mathbf{R} \oplus \mathbf{R}$. Here one focuses on the canonically defined pair of transverse n-dimensional foliations which give rise locally to null coordinates. In this setting many classical objects, such as holomorphic n-forms, which are written as usual in complex notation, appear quite differently in null coordinates. A striking example is the equation for the (split) special Lagrangian potential function. In "complex coordinates" the equation resembles that of the non-split case – all the odd elementary symmetric functions of the hessian must sum to zero. In null coordinates, it becomes the Monge-Ampère equation det Hess $u = 1$.

The algebra **D** mentioned in the paragraph above has at least four names: *the double numbers, the para-complex numbers, the Lorentz numbers, and the hyperbolic numbers.* It is the algebra over **R** generated by 1 and τ where $\tau^2 = 1$. In other words **D** is the Clifford algebra for the negative definite quadratic form on **R** (while **C** comes from the positive definite one). The algebras **C** and **D** are the only commutative normed algebras aside from **R**. They are like twin sisters living in parallel universes, one elliptic and the other hyperbolic. Writing elements in **D** as $z = x + \tau y$, and setting $\bar{z} = x - \tau y$ one can proceed formally to define **D** holomorphic functions (of one and several variables), and the analogues of complex manifolds (called **D**-manifolds here), holomorphic vector bundles, the Dolbeault complex, and much more. There are **D** projective spaces and grassmannians. In fact every real algebraic variety has an associated **D**-variety via base change. Even more striking are the metric analogues. There are hermitian **D**-manifolds, Kähler **D**-manifolds and Calabi-Yau **D**-manifolds. Many basic theorems carry over. This is discussed below.

[1] For example, see [AMT, CFG, EST, GM, GGM, IZ, SS] and references cited therein.

However, as the reader may have noticed, after setting $e = \frac{1}{2}(1+\tau)$ and $\bar{e} = \frac{1}{2}(1-\tau)$, $\mathbf{D}$ becomes the double algebra $\mathbf{R} \oplus \mathbf{R}$ of elements $ue + v\bar{e}$ where $u, v \in \mathbf{R}$ (and $e^2 = e, \bar{e}^2 = \bar{e}$ and $e\bar{e} = 0$). These are the null-coordinates mentioned above. The analogues of many fundamental objects look quite different in these coordinates. For example, $-\tau\partial\overline{\partial} = d_u d_v$ where d_u and d_v are the standard deRham differentials in the variables u and v.

The paper is organized as follows. After a brief introduction to analysis over $\mathbf{D}$ we examine the split special Lagrangian calibration $\Phi = \operatorname{Re} dz$ in $\mathbf{D}^n$ and prove Mealy's sharp inequality which states that Φ is a (reverse) calibration which is minimized on space-like special Lagrangian n-planes. By the fundamental lemma of calibrations this implies that split SLAG submanifolds are homologically volume-maximizing. This is done in Sections 1–5.

Recall that the natural exterior differential system (EDS) for SLAG submanifolds of $\mathbf{C}^n$ is generated by $\operatorname{Im} dz$ and the standard symplectic (Kähler) form ω. A submanifold is an integral submanifold for this EDS if and only if it is SLAG when correctly oriented. The parallel to this in the split case (Theorem 5.2) is almost identical. Again the differential system is generated by $\operatorname{Im} dz$ and ω (this time on $\mathbf{D}^n$), but an integral submanifold is split SLAG modulo orientation if and only if it is space-like. These more general integral submanifolds, which are not necessarily space-like, will be called *unconstrained* split SLAG submanifolds

In Section 6 we examine split SLAG submanifolds M which are graphs of mappings. If M is the oriented graph of $F : \Omega \to \operatorname{Im} \mathbf{D}^n$ where Ω is a simply-connected domain in $\operatorname{Re} \mathbf{D}^n \cong \mathbf{R}^n$, then following Mealy, $F = df$ for a potential function f for which

$$\operatorname{Im} \det_{\mathbf{D}}(I + \tau \operatorname{Hess} f) \;=\; 0 \qquad \text{and} \qquad -I \;<\; \operatorname{Hess} f \;<\; I. \tag{1.1}$$

If on the other hand, M similarly is expressed as the graph $v = G(u)$ over a domain in null coordinates, then $G = dg$ for a potential function g satisfying (cf. [Hi_1])

$$\det \operatorname{Hess} g \;=\; 1 \qquad \text{and} \qquad \operatorname{Hess} g \;>\; 0. \tag{1.2}$$

In Section 7 we prove the following removable-singularities result.

Theorem 7.1. *Let $\Omega \subset \mathbf{R}^n$ be a convex domain in $\mathbf{R}^n$ and consider the "tube" domain $\Omega \times \tau\mathbf{R}^n \subset \mathbf{R}^n \oplus \tau\mathbf{R}^n = \mathbf{D}^n$. Let $\Sigma \subset \Omega$ be a compact subset of Hausdorff $(n-2)$-measure zero. Then any closed, connected split SLAG submanifold*

$$M \subset (\Omega - \Sigma) \times \tau\mathbf{R}^n$$

has closure in $\Omega \times \tau\mathbf{R}^n$ which is the graph of a real analytic mapping $F : \Omega \to \mathbf{R}^n$. Furthermore, $F = df$ where $f : \Omega \to \mathbf{R}$ satisfies (1.1).

Corollary 7.2. (Absence of cones). *Suppose $C \subset \mathbf{D}^n$ is a split SLAG cone which is regular and connected outside the origin. Then C is an n-plane.*

This is in sharp contrast to the non-split case where SLAG-cones exist in all dimensions and there are many SLAG-varieties with singularities of high codimension (cf. [HL_1, J_1, J_2, H, HK_1, HK_2] and the references therein).

The proof of Theorem 7.1 relies on playing off the two different coordinate pictures for $z \in \mathbf{D}^n$. With $z = x + \tau y$ we show that the projection of M to $\Omega - \Sigma$ must be a covering map. This leads to a potential with a Lipschitz extension across Σ. We then rotate to null coordinates $z = ue + v\bar{e}$ (the Cayley transform) and apply deep results of Caffarelli [C_1, C_2, C_3] to get full regularity.

Corollary 7.2 would suggest that singularities of codimension ≥ 3 do not exist on split SLAG subvarieties. By contrast, singularities of codimension-2 certainly do exist in all dimensions. It turns out that when $n = 2$ the exterior differential system for (unconstrained) split SLAG submanifolds is equivalent to the Cauchy-Riemann system for complex curves. Specifically we have the following.

Theorem F.3. *There exists a permutation of real coordinates* $\mathbf{C}^2 \leftrightarrow \mathbf{D}^2$, *such that:*

(a) *holomorphic chains are transformed to unconstrained split SLAG currents and vice-versa.*
(b) *holomorphic chains satisfing a 45°-rule are transformed to split SLAG currents and vice-versa.*

Since complex curves have local uniformizing parameters, this can be interpreted as a very strong regularity result for split SLAG subvarieties when $n = 2$. (A split SLAG subvariety is a d-closed rectifiable current T whose tangent planes $\vec{T}$ are space-like special Lagrangian almost everywhere. See Appendices D and E for a fuller discussion.) Theorem F.3 follows from work of the first author and B. Shiffman [HS, S] which states that: a d-closed rectifiable 2-current defined in an open subset X of $\mathbf{C}^2$, whose (unoriented) tangent space $\vec{T}$ is a complex line $\|T\|$-a.e., is a T holomorphic chain in X. That is, T is a locally finite sum $T = \sum_j n_j[V_j]$ where for each j, $n_j \in \mathbf{Z}$ and V_j is a one-dimensional complex analytic subvariety of X.

This result also clearly implies that isolated singularities **do occur** on split SLAG subvarieties when $n = 2$. They are the singular points of the underlying holomorphic curves. Taking products of these with appropriately chosen linear subspaces gives codimension-2 singularities on split SLAG subvarieties in all dimensions n.

It is interesting to note that since locally irreducible holomorphic curves have well-defined tangent lines at singular points, this procedure does not yield interesting cones. So far, no non-planar irreducible cones are known.

In Sections 8 through 11 we discuss the geometry of $\mathbf{D}$-manifolds. In particular we examine hermitian $\mathbf{D}$-manifolds, Kähler $\mathbf{D}$-manifolds, and Ricci-Flat Kähler $\mathbf{D}$-manifolds. A number of examples are given. In Section 12 we discuss split SLAG submanifolds in this general setting. In Section 13 we consider the deformation theory for these submanifolds in the Ricci-flat Kähler case. It is shown that the basic results of McLean [Mc_1, Mc_2] for the classical (non-split) special Lagrangian submanifolds carry over to this case (cf. Warren [W_2]).

Then in Section 14 we discuss the work of Kim-McCann-Warren on the relevance of split special Lagrangian geometry to the mass transport problem.

In Section 15 we briefly survey the work of Mealy on other calibrations in geometries of indefinite signature.

In Section 16 we analyze the case of Lagrangian submanifolds of $\mathbf{D}^n$ (or more generally of Ricci-flat Kähler $\mathbf{D}$-manifolds) on which the restricted metric is everywhere non-degenerate of signature p, q $(p+q=n)$. We show that any such manifold M is stationary (mean curvature zero) if and only if the restriction of dz to M is of constant phase (see [HL$_1$, D] for parallel cases). When $p = n$ (respectively, 0), a compact oriented constant-phase submanifold maximizes volume among all oriented space-like (respectively, time-like) submanifolds with the same boundary. For $0 < p < n$, volume is not maximized in general, but it is maximized among oriented Lagrangian submanifolds of the same type. In all cases the volume inequality becomes an equality only when the competitor is also of the same constant phase.

Appendix A gives a canonical form for a space-like n-planes in $\mathbf{D}^n$, under the action of the $\mathbf{D}$-unitary group, which is used in proving Mealy's theorem.

Appendix B treats projective $\mathbf{D}$-subvarieties and their induced split Kähler geometry.

Appendix C shows that the result proved in [HL$_1$] characterizing special Lagrangian submanifolds with degenerate projections, carries over to the split case.

Appendix D presents the general theory of calibrations on semi-riemannian manifolds. This entails, among other things, the establishing of a canonical polar form for space-like currents which are representable by integration. The Fundamental Theorem for Semi-Riemannian Calibrations is then established for general ϕ-subvarieties (rectifiable ϕ-currents).

Appendix E examines the special case of split SLAG subvarieties which are shown to be volume maximizing (cf. [KMW]).

Appendix F is devoted to the case $n = 2$ which was discussed above.

We would like to thank Luis Caffarelli, Young-Heon Kim and Micah Warren for helpful comments.

2. Double numbers and double holomorphic functions

In this section we recall the basics of *double* or para-complex analysis. By the *double numbers* we mean the two-dimensional algebra $\mathbf{D}$ over $\mathbf{R}$ generated by 1 and τ with $\tau^2 = 1$. This is the only commutative normed algebra other than the real and complex numbers. In analogy with the complex numbers, each $z \in \mathbf{D}$ can be written as

$$z \;=\; x + \tau y \tag{2.1}$$

with $x, y \in \mathbf{R}$ defined to be the *real* and *imaginary parts*. However, choosing the basis $e = \frac{1}{2}(1-\tau)$ and $\bar{e} = \frac{1}{2}(1+\tau)$ each $z \in \mathbf{D}$ can be written as

$$z \;=\; ue + v\bar{e} \tag{2.2}$$

where (u, v) are called the *null-coordinates* of z. Note that $e^2 = e$, $\bar{e}^2 = \bar{e}$ and $e\bar{e} = 0$. Thus we see that $\mathbf{D}$ is just the algebra of diagonal 2×2-matrices (u, v), i.e., $\mathbf{D} = \mathbf{R} \oplus \mathbf{R}$ as algebras. Note the relations

$$\tau \cdot e \; = \; -e \qquad \text{and} \qquad \tau \cdot \bar{e} \; = \; \bar{e}. \tag{2.3}$$

The representation (2.1) leads to a vast development of analysis and geometry in parallel with the complex case. This development is frequently demystified by using the null coordinates (2.2).

Conjugation in $\mathbf{D}$ is defined exactly as in the complex case: $\bar{z} = x - \tau y = ve + u\bar{e}$. However

$$\langle z, z \rangle \; = \; z\bar{z} = x^2 - y^2 = uv$$

defines a quadratic form of signature $(1, 1)$. The algebra $\mathbf{D}$ is normed in the sense that $\langle zw, zw \rangle = \langle z, z \rangle \langle w, w \rangle$. The notion of vanishing or being zero in complex analysis is replaced by the notion of being *null*, i.e., $z\bar{z} = 0$, in which case having an inverse is impossible. Otherwise, $z^{-1} = \bar{z}/\langle z, z \rangle$ as in the complex case.

The multiplicative group $\mathbf{D}^*$ of non-null numbers has four connected components. Let $\mathbf{D}^+$ denote the component containing 1. Exponentiation gives an isomorphism:
$\exp : \mathbf{D} \xrightarrow{\cong} \mathbf{D}^+$ with inverse $\log : \mathbf{D}^+ \to \mathbf{D}$. Note that using (2.1) and (2.2)

$$\begin{aligned} \exp(z) &= \exp(x)\,(\cosh y + \tau \sinh y) \\ &= e \exp(u) + \bar{e} \exp(v). \end{aligned}$$

The *space-like unit sphere* has two components parameterized by $\pm \exp(\tau\theta)$.

As in complex analysis we now treat $\mathbf{D}$ as $\mathbf{R}^2$ and consider smooth $\mathbf{D}$-valued functions on open sets $U \subset \mathbf{D}$. This leads to several parallels. For example the $\mathbf{D}$-valued 1-forms

$$dz \; = \; dx + \tau dy \; = \; edu + \bar{e}dv \qquad \text{and} \qquad d\bar{z} \; = \; dx - \tau dy \; = \; edv + \bar{e}du$$

have duals

$$\frac{\partial}{\partial z} \; = \; \frac{1}{2}\left(\frac{\partial}{\partial x} + \tau \frac{\partial}{\partial y}\right) \; = \; e\frac{\partial}{\partial u} + \bar{e}\frac{\partial}{\partial v}$$

and

$$\frac{\partial}{\partial \bar{z}} \; = \; \frac{1}{2}\left(\frac{\partial}{\partial x} - \tau \frac{\partial}{\partial y}\right) \; = \; e\frac{\partial}{\partial v} + \bar{e}\frac{\partial}{\partial u}\,.$$

However,

$$4\frac{\partial^2}{\partial z \partial \bar{z}} \; = \; \frac{\partial^2}{\partial x^2} - \frac{\partial^2}{\partial y^2} \; = \; 4\frac{\partial^2}{\partial u \partial v} \tag{2.4}$$

is the wave equation, so that the regularity obtained in the complex case is lacking. Nevertheless, we define a smooth function $F : U \to \mathbf{D}$ to be $\mathbf{D}$-*holomorphic* if

$$\frac{\partial F}{\partial \bar{z}} \; = \; 0. \tag{2.5}$$

If we write $F = ef + \bar{e}g$ with f and g real-valued, then using null coordinates,

$$\overline{\partial} F \; \equiv \; \frac{\partial F}{\partial \bar{z}} d\bar{z} \quad \text{becomes} \quad \overline{\partial}(ef + \bar{e}g) \; = \; e\, d_v f + \bar{e}\, d_u g.$$

Also,

$$\partial F \equiv \frac{\partial F}{\partial z}dz \quad \text{becomes} \quad \partial(ef+\bar{e}g) = e\,d_u f + \bar{e}\,d_v g.$$

Here d_u and d_v are the usual exterior differentiation operators in the null coordinates u and v. Thus

$$\begin{aligned} F = ef+\bar{e}g \text{ is } \mathbf{D}\text{-holomorphic} \qquad \iff \\ f = f(u) \text{ and } g = g(v) \text{ are functions of the} \\ \text{single variable } u \text{ and } v \text{ respectively.} \end{aligned} \tag{2.6}$$

For any $F = ef + \bar{e}g$ we have $2\,\mathrm{Re}F = f + g$ and $2\,\mathrm{Im}F = f - g$. Consequently, we see that the real parts of $\mathbf{D}$-holomorphic functions are functions of the form $f(u) + g(v)$. Observe that just as in the complex case:

$$F \text{ is } \mathbf{D} \text{ holomorphic and } \mathrm{Re}\,F = 0 \quad \Rightarrow \quad F = \tau c \text{ is constant} \tag{2.7}$$

despite the hyperbolic character of (2.4). Finally note that

$$-\tau\partial\overline{\partial} = d_u d_v$$

is a real operator. One can show that any function in the kernel of this operator is locally the real part of a $\mathbf{D}$-holomorphic function.

3. Double n-space $\mathbf{D}^n$

The standard formulas in several complex variables have rather obvious analogues. Define

$$\partial = \sum_{k=1}^{n} dz_k \wedge \frac{\partial}{\partial z_k} \qquad \text{and} \qquad \overline{\partial} = \sum_{k=1}^{n} d\bar{z}_k \wedge \frac{\partial}{\partial \bar{z}_k} \tag{3.1}$$

operating on $\mathbf{D}$-valued differential forms (forms with coefficients in $\mathbf{D}$ at each point). These operators satisfy the usual relations: $\partial^2 = \overline{\partial}^2 = \partial\overline{\partial} + \overline{\partial}\partial = 0$ and $d = \partial + \overline{\partial}$. We can define the real operator $d^c = \tau(\partial - \overline{\partial}) = 2\mathrm{Im}(\partial)$ so that $\partial = \frac{1}{2}(d + \tau d^c)$. Then $-\frac{\tau}{2}\partial\overline{\partial} = \frac{1}{4}dd^c$ is a real operator. So far these are exactly the same (no sign changes) as in the complex case. Switching to null coordinates the operators are given by

$$\partial = e\,d_u + \bar{e}\,d_v \qquad \text{and} \qquad \overline{\partial} = e\,d_v + \bar{e}\,d_u. \tag{3.2}$$

The standard symplectic form ω on $\mathbf{R}^{2n}$ can be written in the two ways $\omega = \sum_j dx_j \wedge dy_j = \frac{1}{2}\sum_j du_j \wedge dv_j$ and has the $-\frac{1}{2}\tau\partial\overline{\partial} = \frac{1}{2}d_u d_v$ potential $\varphi = \sum_j z_j\bar{z}_j = \sum_j u_j v_j$.

Definition 3.1. A $\mathbf{D}$-valued function F on an open subset of $\mathbf{D}^n$ is $\mathbf{D}$-*holomorphic* if $\overline{\partial}F = 0$. Note that in null coordinates (u, v) if we write $F = e\,f + \bar{e}\,g$ with f and g real-valued, then

$$F \text{ is } \mathbf{D}\text{-holomorphic} \iff d_v f = 0 \text{ and } d_u g = 0 \iff f = f(u) \text{ and } g = g(v).$$

On $\mathbf{D}^n$ we have the standard $\mathbf{D}$-valued hermitian form and real inner product:

$$(z,w) = \sum_{j=1}^n z_j \bar{w}_j \qquad \text{and} \qquad \langle z,w\rangle = \mathrm{Re}(z,w) = \mathrm{Re}\sum_{j=1}^n z_j\bar{w}_j.$$

The associated real quadratic form $\langle z,z\rangle = \sum(x_j^2 - y_j^2)$ has split or neutral signature. Note that the standard $\mathbf{D}$-valued hermitian form has imaginary part $-\omega$. Hence, exactly as in the complex case

$$(\cdot,\cdot) = \langle\cdot,\cdot\rangle - \tau\omega(\cdot,\cdot). \tag{3.3}$$

The analogue of the complex structure operator J is the operator $Tz = \tau z$ of multiplication by τ (see (8.3)). Now (3.3) leads to an array of formulas such as: $\langle z', Tz\rangle = \omega(z',z)$ relating $\langle\cdot,\cdot\rangle$, ω and T (see (9.3)).

The $\mathbf{D}$-linear maps from $\mathbf{D}^n$ to $\mathbf{D}^m$ correspond to the set $M_{n,m}(\mathbf{D})$ of $n\times m$-matrices with entries in $\mathbf{D}$. By standard algebra each square matrix $A \in M_n(\mathbf{D})$ has a $\mathbf{D}$-determinant with the usual properties such as: $\det_{\mathbf{D}}(AB) = (\det_{\mathbf{D}} A)(\det_{\mathbf{D}} B)$, $\det_{\mathbf{D}} A^t = \det_{\mathbf{D}} A$, and $A\widetilde{A}^t = (\det_{\mathbf{D}} A)\, I$ where $\widetilde{A}$ is the cofactor matrix. Thus A has an inverse $A^{-1} \in M_n(\mathbf{D})$ if and only if $\det_{\mathbf{D}} A$ has in inverse, i.e., $\det_{\mathbf{D}} A \in \mathbf{D}^*$ is non-null. Let $\mathrm{GL}_n(\mathbf{D}) \subset M_n(\mathbf{D})$ denote the group of invertible elements. Setting $dz \equiv dz_1\wedge\cdots\wedge dz_n$ we have that if $z' = Az$ for $A \in M_n(\mathbf{D})$, then $dz' = \det_{\mathbf{D}} A\, dz$. Since $dz\wedge d\bar{z}$ is a non-null real multiple of $dx\wedge dy$, this proves that (as in the complex case)

$$(\det_{\mathbf{D}} A)\left(\overline{\det_{\mathbf{D}} A}\right) = \det_{\mathbf{R}} A.$$

Writing $A = eB + \bar{e}C \in M_n(\mathbf{D})$ with $B, C \in M_n(\mathbf{R})$ and noting that $dz = edu + \bar{e}dv$ yields the simple formula

$$\det_{\mathbf{D}}(eB + \bar{e}C) = e\,\det_{\mathbf{R}} B + \bar{e}\,\det_{\mathbf{R}} C$$

for the $\mathbf{D}$-determinant. In particular,

$$\det_{\mathbf{D}} A \in \mathbf{D}^+ \iff \det_{\mathbf{R}} B > 0 \text{ and } \det_{\mathbf{R}} C > 0.$$

Each real linear map from $\mathbf{D}^n$ to $\mathbf{D}^m$ decomposes into the sum of a $\mathbf{D}$-linear map and an anti-$\mathbf{D}$-linear map. Thus the Jacobian of a smooth mapping F from (an open subset of) $\mathbf{D}^n$ to $\mathbf{D}^m$ decomposes as

$$J_{\mathbf{R}}(F) = J_{\mathbf{D}}^{1,0}(F) + J_{\mathbf{D}}^{0,1}(F) \equiv \frac{\partial F}{\partial z} + \frac{\partial F}{\partial \bar{z}}$$

and F is said to be $\mathbf{D}$-*holomorphic* if the anti-linear part $J_{\mathbf{D}}^{0,1}(F) = 0$, i.e., if $J_{\mathbf{R}}(F)$ is $\mathbf{D}$-linear at each point.

In order for a function F between open subsets of $\mathbf{D}^n$ to be bi-holomorphic we must require that F be holomophic and that $\det_{\mathbf{D}} \frac{\partial F}{\partial z} \in \mathbf{D}^*$ be non-null, so that $\frac{\partial F}{\partial z}$ is invertible.

The $\mathbf{D}$-unitary group, denoted $U_n(\mathbf{D})$, can be defined, as in the complex case, in several equivalent ways. Given $A \in M_n(\mathbf{D})$, let $A^* = \overline{A}^t$, denote the conjugate transpose. Let $e_1,\dots,e_n, Te_1,\dots,Te_n$ denote the standard basis for

$\mathbf{D}^n = \mathbf{R}^n \oplus \tau\mathbf{R}^n$. A set of vectors $\epsilon_1, \ldots, \epsilon_n$ in $\mathbf{D}^n$ is a *space-like* $\mathbf{D}$*-unitary basis for* $\mathbf{D}^n$ if $\epsilon_1, \ldots, \epsilon_n, T\epsilon_1, \ldots, T\epsilon_n$ is a real orthonormal basis with $\langle \epsilon_j, \epsilon_j \rangle = 1$ $\langle T\epsilon_j, T\epsilon_j \rangle = -1$ for all j. A matrix $A \in M_n(\mathbf{D})$ is called $\mathbf{D}$*-unitary* if one of the following equivalent conditions holds:

$$\begin{aligned}
&(1)\ \ (Az, Az) = (z, z) \quad \text{for all } z \in \mathbf{D}^n \\
&(2)\ \ AA^* = I \ \text{ or } A^*A = I \\
&(3)\ \ Ae_1, \ldots, Ae_n \text{ is a space-like } \mathbf{D}\text{-unitary basis for } \mathbf{D}^n.
\end{aligned} \tag{3.4}$$

When $n = 1$, the unitary group $\mathrm{U}_1(\mathbf{D})$ is the space-like sphere $\{\pm e^{\tau\theta} : \theta \in \mathbf{R}\}$ with two components. In general, $\mathrm{U}_n(\mathbf{D})$ has two components determined by $\det_{\mathbf{D}} A = \pm e^{\tau\theta}$. Let $\mathrm{U}_n^+(\mathbf{D})$ denote the identity component. Computing in null coordinates, first express $A \in M_n(\mathbf{D})$ as $A = e\,B + \bar{e}\,C$ with $B, C \in M_n(\mathbf{R})$, then we see that: $A \in \mathrm{U}_n(\mathbf{D}) \iff A = e\,B + \bar{e}\,(B^t)^{-1}$ for some $B \in \mathrm{GL}_n(\mathbf{R})$, and that $A \in \mathrm{U}_n^+(\mathbf{D}) \iff B \in \mathrm{GL}_n^+(\mathbf{R})$. Thus $\mathrm{U}_n^+(\mathbf{D}) \cong \mathrm{GL}_n^+(\mathbf{R})$.

The *special unitary group* $\mathrm{SU}_n(\mathbf{D})$ is defined by the additional condition: $\det_{\mathbf{D}} A = 1$, i.e., $A = e\,B + \bar{e}\,(B^t)^{-1}$ with $B \in \mathrm{SL}_n(\mathbf{R})$. Thus $\mathrm{SU}_n(\mathbf{D}) \cong \mathrm{SL}_n(\mathbf{R})$. Note that $(B^t)^{-1} = (\det B)^{-1}\widetilde{B}$.

4. The Special Lagrangian Calibration

The Special Lagrangian Calibration, and its associated differential equation, which were introduced in [HL$_1$], have a complete analogue in $\mathbf{D}^n$. This work, due to Jack Mealy [M$_1$, M$_2$], will be presented in this and the following sections.

Some preliminaries. A real n-plane P in $\mathbf{D}^n$ is **Lagrangian** if $\omega\big|_P = 0$. Equivalently, $\langle z', Tz \rangle = \omega(z', z) = 0$ for all $z, z' \in P$, i.e., $z \in P \Rightarrow Tz \in P^\perp$. Let LAG denote the set of all Lagrangian n-planes. A real n-plane P in $\mathbf{D}^n$ is **space-like** if the inner product $\langle \cdot, \cdot \rangle$ on $\mathbf{D}^n$, when restricted to P, is positive definite. In particular, $P \cap \tau\mathbf{R}^n = \{0\}$ so that P can be graphed over $\mathbf{R}^n$ in $\mathbf{D}^n = \mathbf{R}^n \oplus \tau\mathbf{R}^n$ by a matrix $A \in M_n(\mathbf{R})$. Pulling back the inner product $\langle \cdot, \cdot \rangle$ on $\mathbf{D}^n$ to $\mathbf{R}^n$ by the graphing parameterization $x \mapsto x + \tau Ax$ yields $\langle x, x \rangle - \langle Ax, Ax \rangle$. Hence

$$P \text{ is space-like} \quad \iff \quad I - A^tA > 0. \tag{4.0}$$

Let $G_{\text{space}}(n, \mathbf{D}^n)$ denote the Grassmannian of oriented space-like n-planes in $\mathbf{D}^n$. There are two connected components $G^+_{\text{space}}(n, \mathbf{D}^n)$ and $G^-_{\text{space}}(n, \mathbf{D}^n)$. The graph of A, when given the orientation from $\mathbf{R}^n$, is defined to be in $G^+_{\text{space}}(n, \mathbf{D}^n)$, and when given the opposite orientation, is defined to be in $G^-_{\text{space}}(n, \mathbf{D}^n)$.

Recall that an oriented p-plane in a real vector space V does *not* uniquely determine a simple (or decomposable) vector $\xi \in \Lambda^p V$ by choosing an oriented basis $\epsilon_1, \ldots, \epsilon_p$ and defining $\xi = \epsilon_1 \wedge \cdots \wedge \epsilon_p$. However, ξ is determined up to a positive scale. If V is equipped with a positive definite inner product, then $\xi \in \Lambda^p V$ is uniquely determined by requiring $\epsilon_1, \ldots, \epsilon_p$ to be an oriented orthonormal basis, and the Grassmannian of oriented p-planes in V, denoted $G(p, V)$, can be identified with a subset of $\Lambda^p V$. This is not possible if V is equipped with an inner product such as the one on $\mathbf{D}^n$. However, if $\xi \in G_{\text{space}}(n, \mathbf{D}^n)$ is an oriented space-like

n-plane in $\mathbf{D}^n$, then again by requiring the basis $\epsilon_1, \ldots, \epsilon_n$ to be orthonormal with respect to the induced inner product, this subset $G_{\text{space}}(n, \mathbf{D}^n)$ of the full Grassmannian can (and will be) identified with a subset of $\Lambda^n_{\mathbf{R}}\mathbf{D}^n$.

Finally, denote by $\widetilde{\text{LAG}}$ the Lagrangian n-planes in $\mathbf{D}^n$ which are both oriented and space-like. The component $\widetilde{\text{LAG}}^+$ of $\widetilde{\text{LAG}}$ containing the standard oriented $\overrightarrow{\mathbf{R}}^n$ in $\mathbf{D}^n = \mathbf{R}^n \oplus \tau\mathbf{R}^n$ will be referred to as the set of *positive oriented space-like Lagrangian n-planes.* The sets $\widetilde{\text{LAG}}^+ \subset \widetilde{\text{LAG}} \subset G_{\text{space}}(n, \mathbf{D}^n)$ are all subsets of $\Lambda^n_{\mathbf{R}}\mathbf{D}^n$.

The special Lagrangian inequality. As before, let $dz = dz_1 \wedge \cdots \wedge dz_n$ denote the *standard* $(n,0)$-form on $\mathbf{D}^n$.

Lemma 4.1. *Suppose $\xi' \in \Lambda^n_{\mathbf{R}}\mathbf{D}^n \cong \Lambda^n\mathbf{R}^{2n}$ is a real n-vector and $A \in M_n(\mathbf{D})$. Then*

$$(dz)(A\xi') \;=\; (\det_{\mathbf{D}} A)\,(dz)(\xi').$$

Proof. Note that $(dz)(A\xi') = (A^t(dz))(\xi') = (\det_{\mathbf{D}} A^t)(dz)(\xi') = (\det_{\mathbf{D}} A)(dz)(\xi')$. □

In analogy with the complex case each $\xi \in \widetilde{\text{LAG}}^+$ has a *phase* $\theta \in \mathbf{R}$.

Corollary 4.2. *If $\xi \in \widetilde{\text{LAG}}^+$, then*

$$(dz)(\xi) \;=\; e^{\tau\theta} \quad \textit{for some } \theta \in \mathbf{R}.$$

Proof. Since $(dz)(\xi_0) = 1$ where $\xi_0 = \overrightarrow{\mathbf{R}}^n$, if $A \in \mathrm{U}^+_n(\mathbf{D})$, then

$$(dz)(A\xi_0) \;=\; \det_{\mathbf{D}} A. \tag{4.1}$$

By (3.4)(3), each $\xi \in \widetilde{\text{LAG}}^+$ is of the form $\xi = A\xi_0$ with $A \in \mathrm{U}^+_n(\mathbf{D})$. □

Since $A \in \mathrm{U}^+_n(\mathbf{D})$ fixes $\overrightarrow{\mathbf{R}}^n \iff A \in \mathrm{SO}_n(\mathbf{R}) \subset M_n(\mathbf{R})$, this proves that

$$\widetilde{\text{LAG}}^+ \;=\; \mathrm{U}^+_n(\mathbf{D})/\mathrm{SO}_n(\mathbf{R}). \tag{4.2}$$

Definition 4.3. A positively oriented space-like Lagrangian n-plane $\xi \in \widetilde{\text{LAG}}^+$ is said to be **(split) special Lagrangian** or to have **phase zero** if $\xi = A\xi_0$ for some $A \in \mathrm{SU}_n(\mathbf{D})$. The space of all (split) special Lagrangian n-planes in $\mathbf{D}^n$ is denoted by $\widetilde{\text{SLAG}}$.

Note that

$$\widetilde{\text{SLAG}} \;=\; \mathrm{SU}_n(\mathbf{D})/\mathrm{SO}_n(\mathbf{R}). \tag{4.3}$$

since $\mathrm{SO}_n(\mathbf{R})$ is contained in $\mathrm{SU}_n(\mathbf{D})$, and (4.2) holds.

Theorem 4.4. (Mealy).

$$(\mathrm{Re}\, dz)(\xi) \;\geq\; 1 \quad \textit{for all} \quad \xi \in G^+_{\text{space}}(n, \mathbf{D}^n)$$

with equality if and only if $\xi \in \widetilde{\text{SLAG}}$.

Proof. Let $\mathbf{e} = (1, 0)$ and $T\mathbf{e} = (0, 1)$ denote the standard basis for $\mathbf{D} \cong \mathbf{R}^2$. Then the dual basis for $(\mathbf{R}^2)^*$ is $\mathbf{e}^* = dx$ and $(T\mathbf{e})^* = dy$. Employing this same notation in $\mathbf{D}^n$, we have

$$dz \;=\; (\mathbf{e}_1^* + \tau(T\mathbf{e}_1)^*) \wedge \cdots \wedge (\mathbf{e}_n^* + \tau(T\mathbf{e}_n)^*)\,. \tag{4.4}$$

Expanding out this product expresses dz as a 2^n-fold sum. Each term α in the sum is the product of n of the $2n$ axis covectors $\mathbf{e}_1^*, \tau(T\mathbf{e}_1)^*, \ldots, \mathbf{e}_n^*, \tau(T\mathbf{e}_n)^*$. However, for any $j = 1, \ldots, n$,

$$\text{both } \; \mathbf{e}_j^* \; \text{ and } \; \tau(T\mathbf{e}_j)^* \; \text{ cannot be factors of } \; \alpha. \tag{4.5}$$

By Proposition A.3 each $\xi \in G^+_{\text{space}}(n, \mathbf{D}^n)$ is unitarily equivalent to

$$\xi' \;=\; \mathbf{e}_1 \wedge (\cosh\theta_1 \mathbf{e}_2^* + \sinh\theta_1 (T\mathbf{e}_1)^*) \wedge \mathbf{e}_3 \wedge (\cosh\theta_2 \mathbf{e}_4^* + \sinh\theta_2 (T\mathbf{e}_3)^*) \wedge \cdots\,, \tag{4.6}$$

that is, $\xi = A\xi'$ with $A \in \mathrm{U}_n^+(\mathbf{D})$. By (4.4), (4.5) and (4.6) we have

$$(dz)(\xi') \;=\; \cosh\theta_1 \cdots \cosh\theta_{[\frac{n}{2}]}.$$

Hence by Lemma 4.1

$$(dz)(\xi) \;=\; e^{\tau\theta} \cosh\theta_1 \cdots \cosh\theta_{[\frac{n}{2}]}, \tag{4.7}$$

where $e^{\tau\theta} = \det_{\mathbf{D}} A$. In particular

$$(\mathrm{Re}\, dz)(\xi) \;=\; \cosh\theta \cosh\theta_1 \cdots \cosh\theta_{[\frac{n}{2}]}. \tag{4.8}$$

Hence $(\mathrm{Re})(\xi) \geq 1$ and $= 1$ if and only if all the angles are zero, i.e., $\xi' = \mathbf{e}_1 \wedge \cdots \wedge \mathbf{e}_n = \xi_0$, $\xi = A\xi' = A\xi_0$, and $\det_{\mathbf{D}} A = 1$. $\square$

Corollary 4.5. *An oriented real n-plane ξ in $\mathbf{D}^n$ is (split) special Lagrangian if and only if $\xi \in \widetilde{\mathrm{LAG}}^+$ and $\mathrm{Im}\, dz\big|_\xi = 0$.*

The Null Viewpoint. We now revisit the material above in null coordinates. Several points are worth mentioning. Note first that (4.3) naturally becomes

$$\widetilde{\mathrm{SLAG}} \;=\; \mathrm{SL}_n(\mathbf{R})/\mathrm{SO}_n(\mathbf{R}). \tag{4.3$'$}$$

since $A \in \mathrm{SU}_n(\mathbf{D})$ if and only if $A = e\,B + \bar{e}\,(B^t)^{-1} = e\,B + \bar{e}\,\widetilde{B}$ for some $B \in \mathrm{SL}_n(\mathbf{R})$.

Note also that

$$dz \;=\; \frac{1}{2}(du + dv) + \frac{\tau}{2}(dv - du), \qquad \text{or} \tag{4.9}$$

$$\mathrm{Re}\, dz \;=\; \frac{1}{2}(du + dv) \qquad \text{and} \qquad \mathrm{Im}\, dz \;=\; \frac{1}{2}(dv - du). \tag{4.10}$$

To prove this note that $dz_j = e\,du_j + \bar{e}\,dv_j$ implies that $dz = e\,du + \bar{e}\,dv$.

Now Corollary 4.5 has the null form:

Corollary 4.5$'$. *An oriented real n-plane ξ in $\mathbf{D}^n$ is (split) special Lagrangian if and only if $\xi \in \widetilde{\mathrm{LAG}}^+$ and $du\big|_\xi = dv\big|_\xi$.*

5. Split SLAG submanifolds

The algebraic calculations of the previous sections apply to submanifolds. In what follows all submanifolds are assumed to have dimension n.

Definition 5.1. A closed oriented C^1-submanifold M of an an open subset of $\mathbf{D}^n$ is a **split SLAG** (or $\widetilde{\mathbf{SLAG}}$) **submanifold** if the oriented tangent space $\vec{T}_z M \in \widetilde{\mathrm{SLAG}}$ for all $z \in M$.

Theorem 5.2. *A closed oriented C^1-submanifold M of an an open subset of $\mathbf{D}^n$ is split SLAG if and only if*

(1) *M is Lagrangian, i.e., $\omega\big|_M = 0$,*
(2) *$\psi\big|_M = 0$ where $\psi \equiv \mathrm{Im}\, dz = \frac{1}{2}(dv - du)$, and*
(3) *M is positive space-like.*

Proof. Apply Corollary 4.5 and (4.10). □

If M satisfies only conditions (1) and (2), it will be called an *unconstrained* (or *not necessarily space-like*) split SLAG submanifold.

Each oriented space-like submanifold N inherits a volume form vol_N from $\mathbf{D}^n$.

Theorem 5.3. *Suppose that $(M, \partial M)$ is a compact oriented submanifold with boundary in $\mathbf{D}^n$. If M is split SLAG, then M is volume-maximizing, i.e.,*

$$\mathrm{vol}(M) \;\geq\; \mathrm{vol}(N) \tag{5.1}$$

for any other positive space-like compact submanifold N with $\partial N = \partial M$.

Equality holds in (5.1) if and only if N is also split SLAG.

Proof. We have

$$\mathrm{vol}(M) \;=\; \int_M \mathrm{Re}\, dz \;=\; \int_N \mathrm{Re}\, dz \;\geq\; \mathrm{vol}(N)$$

by Theorem 4.4, Stokes' Theorem, and Theorem 4.4 again. Equality in (5.1) implies $\mathrm{Re}\, dz\big|_N \equiv \mathrm{vol}(N)$ and therefore N is split SLAG by Theorem 4.4 once again. □

Remark 5.4. (Split SLAG$_\theta$-Calibrations). As in the complex case, each rotation of the calibration $\mathrm{Re}\, dz$ above gives a new calibration $\mathrm{Re}(e^{-\tau\theta}dz)$ on positively oriented space-like submanifolds. The corresponding calibrated submanifolds are the Lagrangian submanifolds with the property that the restriction

$$\mathrm{Im}(e^{-\tau\theta}dz)\big|_M \;=\; 0$$

or equivalently, that $\mathrm{Re}(e^{-\tau\theta}dz))\big|_M = d\mathrm{vol}_M$.

Note that Corollary 4.2 provides a fibration

$$\widetilde{\mathrm{LAG}}^+ \;\longrightarrow\; \mathbf{R}$$

whose fibre at $\theta \in \mathbf{R}$ is exactly the set $\widetilde{\mathrm{SLAG}}_\theta$ of n-planes in $G^+_{\mathrm{space}}(n, \mathbf{D}^n)$ calibrated by $\mathrm{Re}(e^{-\tau\theta}dz)$.

We shall not discuss these other cases since they are completely equivalent to the case $\theta = 0$. If $A \in \mathrm{U}_n^+(\mathbf{D})$ with $\det_{\mathbf{D}}(A) = e^{\tau\theta}$, then this isometry $A : \mathbf{D}^n \to \mathbf{D}^n$ carries $\widetilde{\mathrm{SLAG}}$-submanifolds to $\widetilde{\mathrm{SLAG}}_\theta$-submanifolds.

6. Split SLAG graphs in $\mathbf{D}^n$

Locally the graph of a smooth map $F : \mathbf{R}^n \to \mathbf{R}^n$ is Lagrangian in $\mathbf{R}^{2n}$ if and only if $F = df$ for some potential function f. This is because $\omega = \sum_j dx_j \wedge dy_j$ vanishes when restricted to M if and only if $\sum_j y_j dx_j\big|_M$ is d-closed. This is equivalent to requiring the Jacobian of F to be symmetric, in which case it equals the hessian of the potential f.

Any space-like submanifold M is locally the graph of a smooth map from $\mathbf{R}^n$ to $\tau\mathbf{R}^n$ since $T_zM \cap \tau\mathbf{R}^n = \{0\}$ for all $z \in M$. Furthermore, in Proposition 7.6 below we show that if M is a closed space-like Lagrangian submanifold of a tube $\Omega \times \tau\mathbf{R}^n \subset \mathbf{R}^n \times \tau\mathbf{R}^n = \mathbf{D}^n$, then the projection $\pi\big|_M : M \to \Omega$ is a covering map. In particular, if Ω is simply-connected, then each connected component of M is the graph of a gradient df for some potential function $f : \Omega \to \mathbf{R}$. This makes the following theorem particularly relevant.

Theorem 6.1. (Mealy). *Suppose that f is a smooth real-valued function defined on an open subset of $\mathbf{R}^n$. Let M denote the oriented graph of the gradient $F = df$ in $\mathbf{D}^n = \mathbf{R}^n \oplus \tau\mathbf{R}^n$. Then M is split special Lagrangian $\iff$ M is space-like and*

$$\sum_{k=0}^{[(n-1)/2]} \sigma_{2k+1}(\mathrm{Hess}\, f) \;=\; \mathrm{Im}\,\det\nolimits_{\mathbf{D}}(I + \tau\,\mathrm{Hess}\, f) \;=\; 0. \tag{6.1}$$

Furthermore M is space-like $\iff$

$$-I \;<\; \mathrm{Hess}\, f \;<\; I. \tag{6.2}$$

or, equivalently, the graphing map of df is uniformly Lipschitz with Lipschitz constant 1.

Proof. The tangent plane P to M at a point is the graph of $A \in M_n(\mathbf{R})$ in $\mathbf{R}^n \oplus \tau\mathbf{R}^n$, and is parameterized by $x \mapsto x + \tau Ax$, $x \in \mathbf{R}^n$. Pulling back the symplectic form ω from P to $\mathbf{R}^n$ gives $\omega((x, Ax), (x', Ax')) = \langle x, Ax'\rangle - \langle Ax, x'\rangle$ so that, as noted above, P is Lagrangian $\iff$ A is symmetric. Pulling back the quadratic form $\langle\cdot,\cdot\rangle$ yields $\langle x, x\rangle - \langle Ax, Ax\rangle$. Hence, as noted in (4.0),

$$P \;\equiv\; \mathrm{graph}\, A \text{ is space-like} \quad\iff\quad A^tA \;<\; I. \tag{6.2$'$}$$

It suffices to prove Theorem 6.1 when M is a plane. This case is handled explicitly in the following.

Proposition 6.2. *Suppose $\xi = \mathrm{graph}\, A$ is the oriented graph in $\mathbf{D}^n = \mathbf{R}^n \oplus \tau\mathbf{R}^n$ of a matrix $A \in M_n(\mathbf{R})$. Then $\xi \in \widetilde{\mathrm{SLAG}}$ $\iff$ A is symmetric with $-I < A < I$ and*

$$\mathrm{Im}\,\det\nolimits_{\mathbf{D}}(I + \tau A) \;=\; 0. \tag{6.3}$$

Proof. We have already shown that ξ is space-like and Lagrangian (i.e., $A \in \widetilde{\mathrm{LAG}}$) $\iff$ A is symmetric and $A^2 < I$, (which is equivalent to $-I < A < I$ for symmetric A). Since the space of such oriented graphs is connected and contains $\xi_0 = \overrightarrow{\mathbf{R}}^n$ (take $A = 0$), this proves that

$$\xi = \mathrm{graph}\, A \in \widetilde{\mathrm{LAG}}^+ \quad \iff \quad A \text{ is symmetric and } -I < A < I. \tag{6.4}$$

Now the pull-back of dz under the $\mathbf{D}$-linear map $I + \tau A : \mathbf{D}^n \to \mathbf{D}^n$ is equal to $\det_{\mathbf{D}}(I + \tau A)\, dz$. Hence, the pull-back of $dz\big|_\xi$ to $\mathbf{R}^n$ under the graphing map $x \mapsto x + \tau A x$ equals $\det_{\mathbf{D}}(I + \tau A)\, dx$. Hence (6.3) holds $\iff$ $\mathrm{Im}\, dz\big|_\xi = 0$.

Corollary 4.5 completes the proof of Proposition 6.2 and Theorem 6.1. □□

The null viewpoint. It is frequently more interesting to consider graphs over the null n-plane $e\,\mathbf{R}^n$ in $\mathbf{D}^n = e\,\mathbf{R}^n \oplus \bar{e}\,\mathbf{R}^n$.

Each space-like submanifold M in $\mathbf{D}^n$ is locally the graph of a smooth map from a domain $\Omega \subseteq \mathbf{R}^n$ to $\bar{e}\mathbf{R}^n$ since $T_z M \cap \bar{e} = \{0\}$ for all $z \in M$. Furthermore, since $\omega = \sum_j du_j \wedge dv_j$ we see that if M is also Lagrangian and Ω is simply-connected, then this graph is the gradient df of a potential function f on Ω.

Theorem 6.3. *Suppose that f is a smooth function defined on an open subset of $\mathbf{R}^n$. Let M denote the oriented graph of the gradient $F = df$ in $e\mathbf{R}^n \oplus \bar{e}\mathbf{R}^n \equiv \mathbf{D}^n$. Then*

$$M \text{ is split special Lagrangian} \iff f \text{ is convex and } \det_{\mathbf{R}} \mathrm{Hess}\, f \equiv 1.$$

Proof. The standard plane $\xi_0 = \overrightarrow{\mathbf{R}}^n$ is the graph of $A = I \in M_n(\mathbf{R})$. For any $A \in M_n(\mathbf{R})$, the previous discussion shows that $\xi = \mathrm{graph}\, A$ is Lagrangian $\iff$ A is symmetric. The quadratic form $\langle \cdot, \cdot \rangle$ restricted to the graph of A pulls back to $\langle eu + \bar{e}Au, eu + \bar{e}Au \rangle = 2\langle eu, \bar{e}Au \rangle = \langle u, Au \rangle$ since eu and $\bar{e}Au$ are null and $\langle e, \bar{e} \rangle = \frac{1}{2}$ in $\mathbf{D}$. Thus:

$$\mathrm{graph}\, A \text{ is space-like} \iff A + A^t > 0. \tag{6.5}$$

By a connectivity argument it follows that: for the oriented graph ξ of $A \in \mathbf{M}_n(\mathbf{R})$

$$\xi = \mathrm{graph}\, A \in \widetilde{\mathrm{LAG}}^+ \iff A \text{ is symmetric and } A > 0. \tag{6.6}$$

Proposition 6.4. *One has $\xi \in \widetilde{\mathrm{SLAG}}$ $\iff$ ξ is the oriented graph over $e\,\mathbf{R}^n$ in $\mathbf{D}^n = e\,\mathbf{R}^n \oplus \bar{e}\,\mathbf{R}^n$ of a positive definite symmetric matrix $A \in M_n(\mathbf{R})$ satisfying*

$$\det_{\mathbf{R}} A = 1. \tag{6.7}$$

Proof. As noted above, each space-like n-plane P can be graphed over $e\,\mathbf{R}^n$ since $P \cap e\mathbf{R}^n = \{0\}$. Under the graphing map $u \mapsto eu + \bar{e}Au$, the n-form du pulls back to du and the n-form dv pulls back to $(\det_{\mathbf{R}} A)du$. Thus by Corollary 4.5′ we have $\xi = \mathrm{graph}\, A \in \widetilde{\mathrm{SLAG}}$ $\iff$ $\xi \in \widetilde{\mathrm{LAG}}^+$ and $\det A = 1$. This completes the proof of both Proposition 6.4 and Theorem 6.3. □□

7. Removable singularities for split SLAG varieties

In this section we show that split SLAG subvarieties tend to be less singular than their elliptic cousins. For example, any connected split SLAG submanifold in a simply-connected tube domain $\Omega \times \tau\mathbf{R}^n$ is the graph of a mapping with potential function satisfying (6.1). Moreover, singularities in codimension > 2 tend to be removable. In particular, there are no irreducible split SLAG cones with isolated singularity in $\mathbf{D}^n$ for $n \geq 3$.

The proofs of these results, which follow, illustrate an interesting interplay between the z-coordinate and the null-coordinate viewpoints. This shift of coordinates, sometimes called the *Cayley transformation*, turns out to be quite useful in the analysis.

The first main theorem of this section is the following.

Theorem 7.1. (Removable singularities I). *Let $\Omega \subset \mathbf{R}^n$ be a convex domain in $\mathbf{R}^n$ and consider the "tube" domain $\Omega \times \tau\mathbf{R}^n \subset \mathbf{R}^n \oplus \tau\mathbf{R}^n = \mathbf{D}^n$. Let $\Sigma \subset \Omega$ be a compact subset of Hausdorff $(n-2)$-measure zero. Then any closed, connected split SLAG submanifold*

$$M \subset (\Omega - \Sigma) \times \tau\mathbf{R}^n$$

has closure in $\Omega \times \tau\mathbf{R}^n$ which is the graph of a real analytic mapping $F : \Omega \to \mathbf{R}^n$. Furthermore, $F = \nabla f$ where $f : \Omega \to \mathbf{R}$ satisfies the equation (6.1).

Corollary 7.2. (Absence of cones). *Suppose $C \subset \mathbf{D}^n$ is a split SLAG cone which is regular and connected outside the origin. Then C is a $\widetilde{\text{SLAG}}$ n-plane.*

Proof of Corollary 7.2. The proof is immediate for $n \geq 3$ – one takes $\Omega = \mathbf{R}^n$ and $\Sigma = \{0\}$. The case $n = 2$ follows from Theorem F.3 in Appendix F and the fact that complex cones of complex dimension 1 are planes. ☐

Remark 7.3. This is in sharp contrast to the non-split case where non-planar SLAG-cones exist in all dimensions > 2 and there are many SLAG-varieties with singularities of high codimension. (See [HL_1, J_1, J_2, H, HK_1, HK_2] and the references therein). Also in coassociative geometry there is a coassociative cone, smooth outside 0 in $\mathbf{R}^7$ which is the graph of a Lipschitz map [HL_1].

Remark 7.4. The codimension-2 hypothesis in Theorem 7.1 is close to best possible. There do exist split SLAG subvarieties in all dimensions $n \geq 2$ with codimension-2 singularities. When $n = 2$ they appear because split SLAG subvarieties are complex curves for a certain complex structure (see Appendix E). The higher-dimensional cases are just products of these with $\mathbf{R}^{n-2}$.

On the other hand we still do not know of any non-trivial (irreducible) split SLAG cones.

Corollary 7.5. (Removable singularities II). *Consider a domain $\Omega \subset \mathbf{R}^n$ with $n > 2$, and suppose $\Sigma \subset \Omega$ is any closed subset of linear-measure zero. Then any closed, split SLAG submanifold $M \subset (\Omega - \Sigma) \times \tau\mathbf{R}^n$ has closure in $\Omega \times \tau\mathbf{R}^n$ which is an immersed real analytic split SLAG submanifold of $\Omega \times \mathbf{R}^n$.*

Proof of Corollary 7.5. Fix $(x, y) \in \overline{M}$. Choose a ball $B = B_r(x) \subset \Omega$ of radius r such that $\partial B \cap \Sigma = \emptyset$. One now applies Theorem 7.1 to the tube domain $B \times \mathbf{R}^n$ to conclude that the closure of each connected component of $M \cap (B - \Sigma) \times \mathbf{R}^n$ is the graph of an analytic mapping $B \to \mathbf{R}^n$. □

Proof of Theorem 7.1. We begin with the following observation.

Proposition 7.6. *Let $\Omega \subset \mathbf{R}^n$ be a domain in $\mathbf{R}^n$ and consider the "tube" domain $T \equiv \Omega \times \tau\mathbf{R}^n \subset \mathbf{R}^n \oplus \mathbf{R}^n = \mathbf{D}^n$ with projection $\pi : T \to \Omega$. If $M \subset T$ is a closed (embedded) submanifold which is space-like and Lagrangian, then $\pi\big|_M : M \to \Omega$ is a covering map.*

If, in addition, Ω is simply-connected, then each component of M is the graph of ∇f for some potential function $f : \Omega \to \mathbf{R}$, and this graph mapping $F = \nabla f$ is 1-Lipschitz.

Proof. To begin we note that from (6.2) we have the following:

$$\|\pi_*(v)\| \;\geq\; \frac{1}{\sqrt{2}}\|v\| \text{ for all } v \in T(M). \tag{7.1}$$

Now fix a point $x \in \Omega$ such that $x = \pi(y)$ for some $y \in M$. Let $\gamma : [0,1] \to \Omega$ be a smooth curve with $\gamma(0) = x$. From (7.1) one easily concludes that there exists a lift $\widetilde{\gamma} : [0,1] \to M$ with $\widetilde{\gamma}(0) = y$ and $\pi \circ \widetilde{\gamma} = \gamma$. (The set of $t \in [0,1]$ such that $\gamma\big|_{[0,t]}$ can be lifted, is both open and closed.)

Fix a closed ball $B = B(x, r)$ about x which is contained in Ω. Then lifting radial lines from x (as above) gives a mapping $\varphi : B \to M$ with $\varphi(x) = y$ and $\pi \circ \varphi = \mathrm{Id}_B$. It now follows that for each connected component M_0 of M, the mapping $\pi_0 = \pi\big|_{M_0} : M_0 \to \Omega$ is a covering map, i.e., π_0 is surjective and every point x has a neighborhood B which is evenly covered by π_0. The first assertion is proved.

If Ω is simply-connected and M connected, then by elementary covering space theory $\pi = \pi\big|_M : M \to \Omega$ is a diffeomorphism. Hence M is the graph of a function $F : \Omega \to \mathbf{R}^n$, which, since M is Lagrangian and Ω is simply-connected, is the gradient $F = \nabla f$ of a scalar function $f : \Omega \to \mathbf{R}$.

Since M is also space-like, we have from (6.2) that

$$-I \;<\; \mathrm{Jac}(F) \;<\; I \qquad \text{on } \Omega. \tag{7.2}$$

Hence F is 1-Lipschitz and the proof is complete. □

Returning to Theorem 7.1, we apply Proposition 7.6 to the simply-connected domain $\Omega - \Sigma$ to conclude that M is the graph of a gradient $F = \nabla f : \Omega - \Sigma \to \mathbf{R}^n$ which satisfies (7.2). The map F extends uniquely to a 1-Lipschitz mapping $\hat{F} : \Omega \to \mathbf{R}^n$ which can be written as the gradient $\hat{F} = \nabla \hat{f}$ of a unique extension of f to Ω. It follows that the closure of M in Ω is

$$\overline{M} \;=\; M \cup \Sigma_M$$

where

$$\Sigma_M \equiv (I \times \hat{F})(\Sigma) \text{ is compact with Hausdorff } (n-2) \text{ measure zero.} \tag{7.3}$$

We now pass to null coordinates (u, v) on $\mathbf{D}^n$ and consider the projection of $\overline{M}$ onto the u-axis. Composing this projection with the graphing map F gives a mapping $G = F + I : \Omega \to \mathbf{R}^n$ (i.e., $G(x) = F(x) + x$ for $x \in \Omega$). Note that $G = \nabla g$ where $g(x) = f(x) + \frac{1}{2}|x|^2$. From (7.2) we see that this map satisfies

$$0 \ < \ \mathrm{Jac}\,\mathrm{G} \ < \ 2I \qquad \text{on } \Omega - \Sigma.$$

In particular, the potential g is strictly convex on $\Omega - \Sigma$. It follows that for any two points $x, y \in \Omega$, we have that

$$G(x) \ \neq \ G(y) \ \text{ unless the segment } \overline{xy} \subset \Sigma. \tag{7.4}$$

In particular, G is injective on $\Omega - \Sigma$. Thus $G\big|_{\Omega - \Sigma}$ is a diffeomorphism onto its image. Moreover we have the following.

Lemma 7.7. *The map* $G\big|_{\Omega-\Sigma} : \Omega - \Sigma \to \Omega' - \Sigma'$ *is a diffeomorphism, where* $\Omega' \equiv G(\Omega)$ *and* $\Sigma' \equiv G(\Sigma) \subset\subset \Omega'$.

Proof. By (7.4) and the strict convexity we know that if $x \in \Omega - \Sigma$ and $y \in \Sigma$, then $G(x) \neq G(y)$. Hence, $G(\Omega - \Sigma) \cap G(\Sigma) = \emptyset$. By shaving Ω slightly we may assume it has a smooth boundary $\partial\Omega$ which does not meet Σ and along which G is a diffeomorphism. Evidently $G(\Sigma)$ will be a compact subset of the bounded component of $\mathbf{R}^n - G(\partial\Omega)$. □

Since G is 2-Lipschitz, the Hausdorff 2-measure $\mathcal{H}(\Sigma') = 0$. It follows that $\Omega' - \Sigma'$ is simply-connected. Now note that $\overline{M}$ is the graph in null coordinates of a mapping $v = H(u)$ where $H : \Omega' \to \mathbf{R}^n$ is well defined and smooth except possibly over points of Σ'. (Points where it is not well defined correspond to points lying on non-trivial straight line segments in Σ). Hence, in $\Omega' - \Sigma'$ we know that $H = \nabla h$ where h is a smooth strictly convex function which satisfies the equation

$$\det \mathrm{Hess}(h) \ = \ 1 \tag{7.5}$$

on $\Omega' - \Sigma'$. Here we are using the assumption that M is $\widetilde{\mathrm{SLAG}}$ and Theorem 6.3.

Now any convex function defined on $\Omega' - S$ where $S \subset \Omega$ is closed with $\mathcal{H}^{n-2}(S) = 0$, extends as a convex function across S. Thus h is well defined and convex on Ω'

We claim that h is a weak solution of equation (7.5) (and therefore a viscosity solution by [C_3, Lemma 3]) on all of Ω'. To see this note that for any Borel set $\mathcal{O} \subset \Omega' - \Sigma'$ equation (7.5) implies that

$$\mathrm{meas}(\nabla h(\mathcal{O})) \ = \ \mathrm{meas}(\mathcal{O}).$$

However, by (7.3) this holds also for any Borel set $\mathcal{O} \subset \Omega'$.

We now apply deep results of Caffarelli (Theorem 7.8 below) to conclude that h is real analytic in Ω'. Hence, $M = \mathrm{graph}(\nabla h)$ is a real analytic submanifold of $\Omega \times \mathbf{R}^n$. Thus the graph of $F = \nabla f$ over Ω is real analytic. Since F is 1-Lipschitz, it is also real analytic on Ω. □

The following is well known but not explicitly stated in the literature.

Theorem 7.8. (Caffarelli [C$_2$, C$_3$]). *Let h be a viscosity solution of (7.5) on a bounded domain $\Omega \subset \mathbf{R}^n$ with smooth boundary $\partial\Omega$. Assume h is smooth in a neighborhood of $\partial\Omega$ in $\overline{\Omega}$. Then h is real analytic in Ω.*

Example 7.9. Take the union of the standard $\mathbf{R}^n \subset \mathbf{D}^n$ with the n-plane: $y_1 = \cdots = y_{n-2} = 0$, $y_{n-1} = \lambda x_{n-1}$, $y_n = -\lambda x_n$. Both are split SLAG. This union is singular (but not irreducible) along the intersection $(\mathbf{R}^{n-2} \times \{0\}) \times \{0\}$.

Example 7.10. The following classical example of Pogorelov is of interest in this context. Let (u, v) be null coordinates on $\mathbf{D}^n$, and write $u = (u_1, \mathbf{u})$ and $v = (v_1, \mathbf{v})$ for $\mathbf{R}^n = \mathbf{R} \times \mathbf{R}^{n-1}$. Consider the potential

$$\varphi(u_1, \mathbf{v}) \;=\; -k|\mathbf{v}|^{\frac{2(n-1)}{n-2}} f(u_1)$$

where $f(t)$ satisfies $f'' + f = 0$ and $f > 0$. The graph M of the gradient of φ is:

$$v_1 = -k|\mathbf{v}|^{\frac{2(n-1)}{n-2}} f'(u_1)$$

$$\mathbf{u} = -k\frac{2(n-1)}{n-2}|\mathbf{v}|^{\frac{2}{n-2}} f(u_1)\mathbf{v}.$$

Note that M is real analytic if $n = 3$, but just $C^{1,\alpha}$ along the u_1-axis if $n \geq 4$ (where $\alpha = ([\frac{n}{2}] - \frac{1}{2})^{-1}$ when n is odd, and $\alpha = (\frac{n}{2} - 1)^{-1}$ when n is even). The tangent plane at points on the u_1-axis is the $(u_1, \mathbf{v})$-plane, so M is being graphed over its tangent plane at these points.

With the right choice of k, M is an example of a split SLAG manifold outside of a singular line which is null. In particular, when $n = 3$, M is a real analytic unconstrained split SLAG submanifold. Outside the $u_1 = x_1 - y_1$ axis it is split SLAG, but at points on this axis it is not split SLAG since this axis is tangent to M and null.

Remark 7.10. Although the discussion in this section was carried out without defining the notion of a (singular) split SLAG variety, we call Appendix D to the reader's attention for this definition.

8. Double manifolds (D-manifolds)

Certain standard notions from complex analysis can now be carried over to $\mathbf{D}$-analysis.

Definition 8.1. A *double manifold* or $\mathbf{D}$-*manifold* is a smooth manifold X equipped with an atlas of charts $\{(U_\alpha, \psi_\alpha)\}_\alpha$, $\psi_\alpha : U_\alpha \to \mathbf{D}^n$, whose transition functions are $\mathbf{D}$-holomorphic and orientation-preserving, that is, the transition functions are of the form

$$z' \;=\; F(z) \qquad \text{with } \frac{\partial F}{\partial \bar{z}} = 0 \qquad \text{and} \qquad \det{}_{\mathbf{D}}\left(\frac{\partial F}{\partial z}\right) \in \mathbf{D}^+ \tag{8.1}$$

or equivalently

$$u' = f(u), \quad v' = g(v) \qquad \text{with } f, g \text{ orientation preserving.} \tag{8.2}$$

Let's examine these manifolds from the two points of view. In the complex picture, we have a Dolbeault decomposition

$$\left(\Lambda^k T^* X\right) \otimes_{\mathbf{R}} \mathbf{D} \;=\; \bigoplus_{p+q=k} \Lambda^{p,q}(X)$$

where, in local holomorphic coordinates $(z_1, \ldots, z_n)$, $\Lambda^{p,q}$ is spanned by forms of the type $dz_{i_1} \wedge \cdots \wedge dz_{i_p} \wedge d\bar{z}_{j_1} \wedge \cdots \wedge d\bar{z}_{j_q}$. Let $\mathcal{E}^{p,q}(X)$ denote the smooth sectons of $\Lambda^{p,q}(X)$. Then the operators

$$\partial : \mathcal{E}^{p,q}(X) \longrightarrow \mathcal{E}^{p+1,q}(X) \qquad \text{and} \qquad \overline{\partial} : \mathcal{E}^{p,q}(X) \longrightarrow \mathcal{E}^{p,q+1}(X),$$

already given in local holomorphic coordinates by (3.1), are well defined on the manifold X. A $\mathbf{D}$-valued function F defined on an open subset of X is holomorphic if and only if $\overline{\partial} F = 0$.

Suppose now that we rewrite our local charts on X in terms of null coordinates (u, v) in $\mathbf{D}^n = e \cdot \mathbf{R}^n \oplus \bar{e} \cdot \mathbf{R}^n$. Then from (8.2) it is clear that X is furnished with two transversal n-dimensional oriented foliations $\mathcal{F}^+$ and $\mathcal{F}^-$ given locally by the u and v coordinate planes respectively. Taking the tangent plane fields $N^{\pm} \equiv \mathcal{F}^{\pm}$ to be the ± 1-eigenspaces defines the **structure endomorphism**

$$\mathbf{T} : T(X) \to T(X) \qquad \text{with } \mathbf{T}^2 = Id. \tag{8.3}$$

which represents scalar multiplication by τ on the tangent spaces (cf. (2.3)). (This is the analogue of $J : T(X) \to T(X)$ with $J^2 = -Id$ in the complex case.) Indeed from this point of view a $\mathbf{D}$-manifold is simply a pair $(X, \mathbf{T})$, where $\mathbf{T}^2 = Id$ with eigenbundles $N^{\pm}$ of equal dimension both of which are integrable. (If one drops the integrability condition, this becomes an *almost* $\mathbf{D}$*-manifold.*) Even more simply, a $\mathbf{D}$-manifold is a $2n$-dimensional manifold with a pair of transversal n-dimensional oriented foliations. When $n = 1$ this is equivalent to a (fully oriented) conformal Lorentzian structure on the surface.

Remark 8.2. Any product $X = M_1 \times M_2$ of n manifolds is certainly a $\mathbf{D}$-manifold (with the foliations given by the factors). However, $\mathbf{D}$-manifolds can be much more complicated. Let $\mathcal{S}_k$ be a two-dimensional foliation on a compact 3-manifold M_k, $k = 1, 2$, and let $\mathcal{L}_k$ be a one-dimensional foliation determined by a line field transversal to $\mathcal{S}_k$. Then define the transversal three-dimensional foliations $\mathcal{F}_1 \equiv \mathcal{S}_1 \times \mathcal{L}_2$ and $\mathcal{F}_2 \equiv \mathcal{L}_1 \times \mathcal{S}_2$ on $M_1 \times M_2$. By a theorem of J. Wood [Wo] any 2-plane field on a 3-manifold is homotopic to a integrable one. Any such foliation can then be modified by introducing many Reeb components. The one-dimensional foliations can often be constructed to have dense orbits.

Remark 8.3. (D-submanifolds). A $\mathbf{D}$-submanifold is a submanifold whose tangent spaces are $\mathbf{T}$-invariant with ± 1-eigenspaces of the **same** dimension. This second condition is not automatic, but if it holds at a point, it holds on the connected component of that point. These subspaces are automatically integrable, with leaves given by intersection with the leaves of the ambient manifold. The graphs of a **D**-holomorphic mapping $f : X \to Y$ is a **D**-submanifold. So also is $f^{-1}(p)$ if p is a regular value of f.

We now address the question of **D**-line bundles and **D**-vector bundles. A **D**-*line bundle* on an arbitrary manifold is a family of free **D**-modules over X which is locally isomorphic to $U \times \mathbf{D}$ with transition functions which are smooth $\mathbf{D}^+$-valued functions.

Definition 8.4. A *holomorphic line bundle* on a **D**-manifold is a **D** line bundle whose transition functions can be chosen to be **D**-holomorphic.

Let $\mathcal{D}$ denote the sheaf of germs of **D**-holomorphic functions on a **D**-manifold X, and let $\mathcal{D}^+$ denote the sheaf of germs of holomorphic $\mathbf{D}^+$-valued functions on X. Then the Čech cohomology group

$$H^1(X, \mathcal{D}^+)$$

represents the isomorphism classes of holomorphic **D**-line bundles on X. Note that the sheaf sequence $0 \to \mathcal{D} \xrightarrow{\exp} \mathcal{D}^+ \to 1$ is exact, and therefore $H^1(X, \mathcal{D}) \cong H^1(X, \mathcal{D}^+)$. Now in local null coordinates we have $\mathcal{D} = e \cdot \mathcal{E}_u + \bar{e} \cdot \mathcal{E}_v \cong \mathcal{E}_u \oplus \mathcal{E}_v$ where $\mathcal{E}_u, \mathcal{E}_v$ are the sheaves of germs of functions of u and v respectively. Thus $H^1(X, \mathcal{D}^+) \cong H^1(X, \mathcal{E}_u) \oplus H^1(X, \mathcal{E}_v)$.

Example 8.5. (The canonical bundle). The bundle $\kappa \equiv \Lambda^{n,0}$ of holomorphic n-forms has transition functions $\det_{\mathbf{D}}\left(\frac{\partial z^\alpha}{\partial z^\beta}\right)$ which are holomorphic with values in $\mathbf{D}^+$. In local z-coordinates a holomorphic n-form can be written as $\Phi = F(z)dz = F(z)dz_1 \wedge \cdots \wedge dz_n$ where F is a **D**-holomorphic function. In null coordinates one has

$$\Phi = e\, f(u)du + \bar{e}\, g(v)dv. \tag{8.4}$$

From this we see that on a **D**-manifold which is a product $X = M_1 \times M_2$ as above, there always exists a global holomorphic n-form which is nowhere-null, i.e., $\Phi \wedge \bar{\Phi}$ is never zero. (For elements of $\Lambda^{n,0}\mathbf{D}^n$ the concepts of null and non-null make sense without a metric).

Remark 8.6. (Holomorphic D-vector bundles). There is also the notion of a holomorphic **D**-vector bundle of higher rank m where one has local trivializations $U_\alpha \times \mathbf{D}^m$ and transition functions which are holomophic maps to $\mathrm{GL}_m^+(\mathbf{D}) \equiv \{A \in \mathbf{M}_{m,m}(\mathbf{D}) : \det_{\mathbf{D}}(A) \in \mathbf{D}^+\}$. Note that one can twist the Dolbeault complex by such bundles.

9. Hermitian D-manifolds

Let X be a **D**-manifold and let $TX = N^+ \oplus N^-$ be the decomposition into the tangent spaces of the two foliations, i.e, the ± 1-eigenbundles of the endomorphism $\mathbf{T} : TX \to TX$ in (8.3).

Definition 9.1. By a *hermitian metric* on X we mean a non-degenerate $\mathbf{D}$-valued real-bilinear form $(\cdot,\cdot)$ on the fibres of TX such that

$$(V,W) \;=\; \overline{(W,V)} \qquad \text{and} \qquad (\mathbf{T}V,W) \;=\; \tau(V,W) \;=\; -(V,\mathbf{T}W) \tag{9.1}$$

for all $V,W \in T_xX$ at each point x.

Note that

$$(\mathbf{T}V,\mathbf{T}W) \;=\; -(V,W) \tag{9.2}$$

from which it follows that

$$N^+ \;\text{ and }\; N^- \;\text{ are null spaces for }\; (\cdot,\cdot).$$

Recall (3.3) which expresses $(\cdot,\cdot)$ in terms of its real and imaginary parts:

$$(V,W) \;=\; \langle V,W\rangle - \tau\,\omega(V,W)$$

and note that $\langle V,W\rangle \;=\; \langle W,V\rangle$, $\omega(V,W) \;=\; -\omega(W,V)$ and

$$\langle V,\mathbf{T}W\rangle \;=\; \omega(V,W). \tag{9.3}$$

Thus $\langle V,W\rangle$ is a semi-riemannian metric of type (n,n). The spaces $N^\pm$ are null for this metric and for its associated "Kähler" form ω. Both of these forms are characterized by the same bundle isomorphism:

$$A : N^+ \;\xrightarrow{\;\cong\;}\; (N^-)^*. \tag{9.4}$$

Writing $V = (V^+,V^-)$ with respect to the decomposition $TX = N^+ \oplus N^-$, we have $\langle V,W\rangle = A(V^+)(W^-) + A(W^+)(V^-)$ and $\omega(V,W) = A(V^+)(W^-) - A(W^+)(V^-)$. In terms of local z-coordinates and local null coordinates (u,v) one has

$$ds^2 = \sum_{j,k=1}^n a_{jk}dz_j \otimes d\bar z_k = \sum_{j,k=1}^n \tilde a_{jk}du_j \circ dv_k, \qquad \text{and}$$

$$\omega = \sum_{j,k=1}^n a_{jk}dz_j \wedge d\bar z_k \;=\; \sum_{j,k=1}^n \tilde a_{jk}du_j \wedge dv_k.$$

Note 9.2. The isomorphism (9.4) shows that the existence of a $\mathbf{D}$-hermitian metric on X puts further topological restrictions on the bundle TX (as opposed to the complex hermitian case). For example, any product $X = M_+ \times M_-$ of two n-manifolds is a $\mathbf{D}$-manifold, but (9.4) shows that a hermitian metric exists on X if and only if M_+ and M_- are both parallelizable. Also on any manifold X the condition $TX \cong N \oplus N$ is restrictive.

Note 9.3. A $\mathbf{D}$-submanifold $Y \subset X$ of a hermitian $\mathbf{D}$-manifold may not be hermitian in the induced metric. One needs the additional hypothesis that Y is a symplectic submanifold, i.e., that $\omega\big|_Y$ is non-degenerate on Y.

Note 9.4. On any hermitian $\mathbf{D}$-manifold there exists the canonical Levi-Civita connection for the semi-riemannian metric. There also exists a canonical hermitian connection, characterized by the fact that the metric and $\mathbf{T}$ are parallel and $\nabla^{0,1} = \overline{\partial}$.

Remark 9.5. In parallel with the complex case, canonical connections exist on *any* holomorphic **D**-bundle equipped with a hermitian metric (cf. Remark 8.6). The proof follows the complex case, where one computes the connection 1-form in a holomorphic frame as $\omega = h^{-1}\partial h$ where h is the matrix determined by the frame and the metric. The curvature 2-form $\Omega = d\omega - \omega \wedge \omega$ transforms to $A\Omega A^{-1}$ under a $\mathrm{GL}_n(\mathbf{D})$-valued frame change A, so that $\det_{\mathbf{D}}(I + \tau\Omega)$ is a globally defined form on X, called the *total* **D**-*Chern form of the bundle.* More generally, for any other $\mathrm{GL}_n(\mathbf{D})$-invariant polynomial ϕ, the ϕ-Chern form of the bundle, $\phi(\Omega)$, is a globally defined d-closed differential form. The standard transgression formula holds, and the cohomology class of $\phi(\Omega)$ in $H^*(X, \mathbf{D})$ is independent of the hermitian metric (or connection) on the bundle. Finally, if $h = e \cdot g + \bar{e} \cdot g^t$ defines a matrix g with values in $\mathrm{GL}_n(\mathbf{R})$, then one can show that in null coordinates

$$\tau\Omega = \tau\bar{\partial}(\partial h \cdot h^{-1}) = \bar{e} d_u \left((d_v g) \cdot g^{-1}\right) - e d_v \left((d_u g) \cdot g^{-1}\right).$$

Remark 9.6. Suppose X carries a nowhere-null holomorphic n-form Φ (see Example 8.4). Then changing the metric by the conformal factor $\|\Phi\|^{\frac{1}{n}}$ gives a new hermitian metric in which Φ has constant length. If Φ' is another holomorphic n-form with constant length in this new metric, then $\Phi' = \alpha\Phi$ for $\alpha \in \mathbf{D}$ with $\alpha\bar{\alpha} \neq 0$.

10. Kähler D-manifolds

We now consider the following natural class of hermitian double manifolds.

Definition 10.1. A hermitian **D**-manifold is said to be *Kähler* if the form ω is d-closed.

Interestingly, all the standard characterizations of complex Kähler manifolds carry over to this context. For example, we have the following.

Proposition 10.2. *A hermitian* **D**-*manifold* X *is Kähler if and only if the canonical hermitian connection on* TX *agrees with the Levi-Civita connection of the semi-riemannian metric*

Proposition 10.3. *Let* X *be a hermitian* **D**-*manifold with Kähler form* ω *and structure map* **T**. *Then* X *is Kähler if and only if* **T** *is parallel in the Levi-Civita connection.*

In the last assertion one does not need to assume *apriori* that the subbundles $N^{\pm}$ are integrable. The hypothesis $\nabla\mathbf{T} = 0$ implies that $N^{\pm}$ are integrable.

Note 10.4. In the null-coordinate approach, a Kähler **D**-manifold is simply a symplectic manifold with a pair of transverse Lagrangian foliations. For this reason Kähler **D**-manifolds are sometimes referred to in the literature as *bi-Lagrangian manifolds.*

Note 10.5. Any symplectic **D**-submanifold of a Kähler **D**-manifold, is Kähler in the induced metric.

Example 10.6. (Surfaces). When $n = 1$ a Kähler **D**-manifold is simply a surface equipped with a pair of 1-forms α and β such that $\alpha \wedge \beta$ is nowhere vanishing. Here the metric is $ds^2 = \alpha \circ \beta$ and the Kähler form is $\omega = \alpha \wedge \beta$. Equivalently it is a Lorentzian surface.

11. Ricci-flat Kähler D-manifolds

Let X be a hermitian **D**-manifold and suppose Φ is a nowhere-null holomorphic section of the canonical bundle $\kappa = \Lambda^{n,0}(X)$. Then the real 2-form

$$\Omega \equiv \tau\partial\overline{\partial}\log\|\Phi\| \tag{11.1}$$

is independent of Φ. To see this note first that in local holomorphic coordinates z we have $\Phi = a(z)dz_1 \wedge \cdots \wedge dz_n$, where $a(z)$ is a **D**-holomorphic function, and $\|\Phi\|^2 = \lambda(z)a(z)\bar{a}(z)$ where $\lambda > 0$. Hence we have $\tau\partial\overline{\partial}\log\|\Phi\|^2 = \tau\partial\overline{\partial}\log\lambda$. Changing coordinates to z' gives a new metric coefficient $\lambda' = \lambda F\overline{F}$ where F is a holomorphic function (the determinant of $\partial z/\partial z'$). Thus $\partial\overline{\partial}\log\lambda = \partial\overline{\partial}\log\lambda'$. This form Ω is the curvature of the canonical hermitian connection on κ. (This generalizes to any hermitian holomorphic line bundle.)

We observed in Remark 9.6 that given X and Φ as above, there is a conformally equivalent metric on X for which $\|\Phi\| \equiv 1$. Evidently, in this metric the curvature $\Omega \equiv 0$.

As remarked in (9.4), if one passes to null coordinates, then $\Phi = e\, f(u)du + \bar{e}\, g(v)dv$. For this reason **D** geometry plays an important role in the mass-transport problem.

When X is Kähler we have the following.

Proposition 11.1. *Let X be a Kähler* **D***-manifold, and Ω the curvature 2-form of the canonical line bundle as above. Then*

$$\Omega(V, W) = -\operatorname{Ric}(V, \mathbf{T}W) \tag{11.2}$$

where Ric *is the Ricci curvature tensor of the Levi-Civita connection on TX. In particular, the canonical bundle is flat if and only if the Ricci curvature of X is zero.*

Proof. The argument goes precisely as in the complex case, with i replaced by τ and the almost complex structure J replaced by the structure map $\mathbf{T}$.

Thus a Ricci-flat Kähler **D**-manifold is the exact analogue of a Calabi-Yau manifold. The holonomy will lie in the group $\mathrm{SU}_n(\mathbf{D})$ (cf. [B].)

These manifolds can be characterized, in analogy with Hitchin's [Hi$_1$] description of the complex case, as follows.

Proposition 11.2. *A Ricci-flat Kähler* **D***-manifold is equivalent to the data of a symplectic $2n$-dimensional manifold (X, ω) together with two d-closed real n-forms ϕ, ψ such that:*

(1) $\Phi = \phi + \tau\,\psi$ *is a simple (indecomposable)* **D***-valued n-form.*

(2) $\Phi \wedge \omega = 0$, *i.e.*, $\phi \wedge \omega = \psi \wedge \omega = 0$

(3) $\Phi \wedge \overline{\Phi} \;=\; \begin{cases} \omega^n & \textit{for } n \textit{ even} \\ -\tau\omega^n & \textit{for } n \textit{ odd.} \end{cases}$

This proposition can be restated as follows.

Proposition 11.2′. *A Ricci-flat Kähler* **D***-manifold is equivalent to the data of a symplectic* $2n$*-dimensional manifold* (X, ω) *together with two d-closed real* n*-forms* α, β *such that:*

(1) $\Phi = e\,\alpha + \bar{e}\,\beta$ *and the real* n*-forms* α *and* β *are simple.*

(2) $\alpha \wedge \omega = \beta \wedge \omega = 0$

(3) $\alpha \wedge \beta \;=\; \omega^n$.

To see the equivalence of these statements, set $\alpha = \phi - \psi$ and $\beta = \phi + \psi$ and note that $(\alpha_1 \wedge \cdots \wedge \alpha_n)\cdot e + (\beta_1 \wedge \cdots \wedge \beta_n)\cdot \bar{e} = (\alpha_1 \cdot e + \beta_1 \cdot \bar{e}) \wedge \cdots \wedge (\alpha_n \cdot e + \beta_n \cdot \bar{e})$

There are several interesting examples.

Example 11.3. By Proposition 11.2′ a Ricci-flat Kähler **D**-manifold of (real) dimension 2 is simply a surface Σ equipped with a pair of *closed* 1-forms α, β such that $\alpha \wedge \beta$ never vanishes. (The Kähler form is $\omega = \alpha \wedge \beta$.)

Note that given this data, $\xi \equiv \alpha + i\beta$ defines a conformal structure on Σ in which ξ is a (complex) holomorphic 1-form. Conversely if Σ is a Riemann surface with a nowhere vanishing holomorphic 1-form $\xi \;=\; \alpha + i\beta$, then (Σ, α, β) is a Ricci-flat Kähler **D**-manifold. This leads to the following.

Proposition 11.4. *Every non-compact connected surface carries a Ricci-flat Kähler* **D***-structure. The only compact connected surface to carry such a structure is the torus.*

Proof. Any surface Σ can be given a conformal structure. If Σ is a non-compact connected Riemann surface, then it carries a nowhere vanishing holomorphic 1-form since it is Stein. If Σ is compact, it carries such a 1-form if and only if its genus is zero. □

Since products of Ricci-flat Kähler **D**-manifolds are again of this type, we obtain examples with large fundamental groups in all dimensions.

Example 11.5. Let Σ, Σ' be Riemann surfaces endowed with nowhere-vanishing holomorphic 1-forms

$$\xi \;=\; a + ib \qquad \text{and} \qquad \xi' \;=\; a' + ib'.$$

On $X = \Sigma \times \Sigma'$ we have the symplectic form and metric given by

$$\omega \;=\; a \wedge a' + b \wedge b' \qquad \text{and} \qquad ds^2 \;=\; a \circ a' + b \circ b'.$$

The tensor $\mathbf{T} : T(\Sigma \times \Sigma') \to T(\Sigma \times \Sigma')$ is defined by $\mathbf{T} \;=\; \begin{pmatrix} I & 0 \\ 0 & -I \end{pmatrix}$. The holomorphic 1-form

$$\Phi \;=\; e \cdot a \wedge b + \bar{e} \cdot a' \wedge b' \;\equiv\; e \cdot \alpha + \bar{e} \cdot \beta$$

satisfies conditions (1), (2) and (3) above, and so X is a Ricci-flat Kähler **D**-manifold. One can also take products of such 4-manifolds.

Remark 11.6. (The Calabi question). Suppose (X, ω) is a Kähler **D**-manifold with trivial canonical bundle. It is natural to ask the *Calabi Question: Does there exist a Kähler form ω' on the* **D***-manifold X, which is cohomologous to ω and is Ricci-flat?* Interestingly, the answer is: *Not always.* Consider the **D**-structure on the torus $S^1 \times S^1$ given by the transverse foliations $\mathcal{F}_1$ and $\mathcal{F}_2$, where $\mathcal{F}_1$ consists of the circles $S^1 \times \{\theta\}$ for $\theta \in S^1$, and $\mathcal{F}_2$ has exactly one compact leaf $L = \{\theta_0\} \times S^1$ (whose complement is foliated by leaves which spiral barber-pole fashion from one side of L to the other). Any choice of nowhere vanishing 1-forms α and β so that α vanishes on $\mathcal{F}_1$ and β vanishes on $\mathcal{F}_2$ makes this a Kähler **D**-manifold. However, *it is not possible to choose α and β to be closed, that is, A Ricci-flat Kähler structure does not exist.* To see this, suppose β were closed, and consider a thin strip $S \equiv (\theta_0 - \epsilon, \theta_0 + \epsilon) \times S^1$ about the compact leaf L. Since $\beta\big|_L = 0$, we conclude that $\beta\big|_S$ is exact by the deRham Theorem. That is, $\beta = df$ where $f : S \to \mathbf{R}$ is a function whose level sets are the leaves of $\mathcal{F}_2$ in S. However, since $df = \beta \neq 0$ on S, one sees that the level sets of f near L must be compact (circles).

Taking products gives examples in all dimensions.

12. Split SLAG submanifolds in the general setting

Let X be a Hermitian **D**-manifold of **D**-dimension n with a nowhere null holomorphic n-form Φ. Let ω be the Kähler form of the hermitian metric. (We do not require $d\omega = 0$.)

Definition 12.1. An real oriented C^1 submanifold M of dimension n in X is said to be **split SLAG** (or $\widetilde{\textbf{SLAG}}$) if

(1) M is Lagrangian, i.e., $\omega\big|_M = 0$.
(2) $\operatorname{Im}\Phi\big|_M = 0$.
(3) M is positive space-like.

Theorem 12.2. *Suppose $\|\Phi\| \equiv 1$ on X. Then any compact split SLAG submanifold with boundary $(M, \partial M)$ in X is homologically volume-maximizing, i.e.,*

$$\mathrm{vol}(M) \ \geq \ \mathrm{vol}(N)$$

for any other positive space-like submanifold N such that $\partial N = \partial M$ and $M - N$ is homologous to zero in X.

Equality holds if and only if N is also split SLAG.

Proof. The same as the proof of Theorem 5.3. □

Remark 12.3. One can always change the metric on X by a conformal factor so that $\|\Phi\| \equiv 1$.

Remark 12.4. Theorem 12.2 carries over to split SLAG subvarieties. See Appendix D.

13. Deformations and the McLean theorem in the split case

In his thesis [Mc$_1$, Mc$_2$], R. McLean proved that the moduli space of Special Lagrangian submanifolds of a Ricci-flat Kähler manifold is smooth, and its tangent space at a point (submanifold) M is canonically identified with the space $\mathbf{H}^1 \cong H^1(M;\mathbf{R})$ of harmonic 1-forms on M. This result carries over to the split Special Lagrangian case (see Warren [W$_2$]). We present this result and sketch the proof based in part on the notes of Hitchin [Hi$_1$].

Theorem 13.1. *Let X, ω, Φ be as in Section 12 and assume $d\omega = 0$ (so X is a Ricci-flat Kähler $\mathbf{D}$-manifold). Let $M \subset X$ be a split SLAG submanifold, and suppose $M_t, -\epsilon < t < \epsilon$, is a variation through split SLAG submanifolds with $M = M_0$. Then the corresponding normal variational vector field ν canonically gives a harmonic 1-form θ on M by setting*

$$\theta = \nu \lrcorner \omega.$$

If M is compact, then, in a neighborhood of M, the moduli space $\mathfrak{M}$ of split SLAG submanifolds of X is a manifold of dimension $b_1(M)$ whose tangent space at $M' \in \mathfrak{M}$ is canonically identified with the space of harmonic 1-forms on M'. In particular, this endows $\mathfrak{M}$ with a natural riemannian metric.

Proof. Let $F : M \times (-\epsilon, \epsilon) \to X$ be a smooth variation of $M = M_0$ so that $M_t \equiv F(M \times \{t\})$ is positive space-like for all t. Then by Definition 12.1, this is a variation through split SLAG submanifolds if and only if

$$F^*\omega\big|_{M\times\{t\}} = 0 \qquad \text{and} \qquad F^*(\mathrm{Im}\Phi)\big|_{M\times\{t\}} = 0 \text{ for all } t,$$

or equivalently

$$F^*\omega = \widetilde{\theta} \wedge dt \qquad \text{and} \qquad F^*(\mathrm{Im}\Phi) = \widetilde{\varphi} \wedge dt$$

where $\widetilde{\theta}$ and $\widetilde{\varphi}$ are a 1-form and an $(n-1)$-form respectively on $M \times (-\epsilon, \epsilon)$. Note that the restrictions $\theta = \widetilde{\theta}\big|_{M\times\{t\}}$ and $\varphi = \widetilde{\varphi}\big|_{M\times\{t\}}$ are independent of the choice of $\widetilde{\theta}$ and $\widetilde{\varphi}$. Since $dF^*\omega = dF^*(\mathrm{Im}\Phi) = 0$, these restrictions satisfy $d\theta = d\varphi = 0$. Note also that θ and φ correspond to the restriction to M_t of the contraction of $\nu \equiv F_*(\frac{\partial}{\partial t})$ with ω and $\mathrm{Im}\Phi$ respectively. In fact, by 12.1 (1) and (2), only the normal part of $F_*(\frac{\partial}{\partial t})$ survives under restriction. The first part of the Theorem is now a consequence of the following algebraic assertion:

$$\varphi = *\theta \tag{13.1}$$

where $*$ denotes the Hodge star-operator with respect to the induced riemannian metric on M_t. The proof of (13.1) is straightforward and left to the reader.

The proof of the unobstructedness of the moduli space near M, when M is compact, follows exactly the argument given in [Mc$_2$]. □

14. Relation to the mass transport problem – work of Kim, McCann and Warren

We first recall the classical Monge-Kantorovich Optimal Mass Transport Problem in a smooth setting. Fix n-dimensional manifolds U and V with domains $\Omega_U \subset\subset U$, $\Omega_V \subset\subset V$ and smooth positive densities ρ_U, ρ_V on Ω_U and Ω_V respectively. Let $c : \Omega_U \times \Omega_V \to \mathbf{R}$ be a smooth "cost" function. Set

$$\mathcal{M} \equiv \{F : \Omega_U \to \Omega_V : F_*(\rho_U) = \rho_V\}.$$

Problem. Find a mapping in $\mathcal{M}$ which minimizes the functional

$$C(F) \equiv \int_{\Omega_U} c(u, F(u))\, \rho_U.$$

It turns out that this is equivalent to the following. Consider the space

$$\mathcal{L} \equiv \{(f, g) \in C(\Omega_U) \times C(\Omega_V) : f(u) + g(v) \le c(u, v)\ \forall\ u, v\}.$$

Dual problem. Find a pair $(f, g) \in \mathcal{L}$ which maximizes the functional

$$J(f, g) \equiv \int f(u)\, \rho_U + \int g(v)\, \rho_V.$$

Consider now the case $U = V = \mathbf{R}^n$. Under appropriate conditions on the cost function these equivalent problems have a unique solution. For the classical cost function $c(u, v) = \frac{1}{2}|u - v|^2$ the solutions are related by

$$F = df = (dg)^{-1} = F^{-1}.$$

Note that f and g are Legendre transforms of one-another. If we write $\rho_U = \rho(u)du$ and $\rho_V = \widetilde{\rho}(v)dv$ for smooth positive functions $\rho, \widetilde{\rho}$, then f satisfies the Monge-Ampère equation

$$\det \operatorname{Hess} f = \frac{\rho}{\widetilde{\rho}(df)}.$$

For a complete and detailed exposition of the modern state of knowledge pertaining to the optimal transport problem, the reader is referred to [V].

Even when one is working in the general case presented above, there are existence and uniqueness results. Kim, McCann and Warren have found that this problem and its solution fit beautifully into the framework of Kähler **D**-manifolds. Briefly it goes as follows.

Consider the **D**-manifold

$$U \times V.$$

The first assumption is that the given cost function gives a **global Kähler potential**, i.e.,

$$\omega \equiv d_u d_v c$$

is a symplectic (i.e., non-degenerate) 2-form on $U \times V$. (Otherwise said, the matrix $\frac{\partial^2 c}{\partial u_i \partial v_j}$ is everywhere non-singular,) This makes $U \times V$ into a **Kähler D-manifold**.

One also imposes a so-called "twist condition" on c which says that the maps $u \mapsto (d_v c)(u,v)$ and $v \mapsto (d_u c)(u,v)$ are injective, and which we can ignore here.

Observe now that our Kähler **D**-manifold is equipped with a **holomorphic** n**-form**

$$\Phi \;\equiv\; \rho(u)\, du \cdot e \;+\; \widetilde{\rho}(v)\, dv \cdot \bar{e}.$$

Theorem 14.1. [KMW]. *The graph of the unique solution to the Monge-Kantorovich optimal mass transport problem given above is a split SLAG submanifold of the Kähler* **D***-manifold* $U \times V$.

Note that in general $U \times V$ is not Ricci-flat. Hence the graph of F is only volume-maximizing for a conformally related metric (cf. Remark 12.3).

15. Further work of Mealy

In this section we give a brief overview of the primary work on semi-riemannian calibrated geometry [M_1, M_2]. Let $\mathbf{R}^{p,q}$ denote the standard semi-euclidean vector space of signature $p, q \geq 1$. Choose a connected component G of the grassmannian of oriented non-degenerate linear subspaces of $\mathbf{R}^{p,q}$ of dimension $k = r + s$ and signature r, s. There is a canonical embedding of G into $\Lambda^{r+s}\mathbf{R}^{p,q}$ via the wedge product of any oriented orthonormal basis for the subspace.

Definition 15.1. Given a subset $A \subset G$, a form $\phi \in \Lambda^{r+s}(\mathbf{R}^{p,q})^*$ is a **calibration of type** A if

$$\phi(\xi) \;\geq\; 1 \text{ for all } \xi \in A. \tag{15.1}$$

Moreover, the set

$$G(\phi) \;=\; \{\xi \in A : \phi(\xi) = 1\}$$

of calibrated planes is usually assumed to be non-empty. An oriented submanifold M is of **type** A if each tangent plane is in A. We say M is **calibrated** by ϕ if M is of type $G(\phi)$.

The cases where A can be taken to be a full connected component of G are very limited. Ignoring the two cases of dimension $k = 0$ and dimension $k = n$, where ϕ is $\pm$ the volume form on $\mathbf{R}^{p,q}$, and eliminating duplications from interchanging p and q, Mealy shows that only two cases (essentially one) remain. The Grassmannian $G_{\text{space}}(p, \mathbf{R}^{p,q})$ of maximally space-like oriented planes in $\mathbf{R}^{p,q}$, has two connected components. Let $G^+_{\text{space}}(p, \mathbf{R}^{p,q})$ denote the component containing $\overrightarrow{\mathbf{R}}^p = \mathbf{R}^p \times \{0\}$ with its standard orientation. The other component is $G^-_{\text{space}}(p, \mathbf{R}^{p,q}) = -G^+_{\text{space}}(p, \mathbf{R}^{p,q})$. Henceforth we only consider $G^+_{\text{space}}(p, \mathbf{R}^{p,q})$. Thus the problem of finding semi-riemannian calibrations ϕ of type A, where A is a connected component of the non-degenerate Grassmannian, is reduced to the case $A \equiv G^+_{\text{space}}(p, \mathbf{R}^{p,q})$ and the problem of finding a form

$$\phi \in \Lambda^{r+s}(\mathbf{R}^{p,q})^* \qquad \text{with } \phi(\xi) \geq 1 \quad \forall\, \xi \in \phi \in G^+_{\text{space}}(p, \mathbf{R}^{p,q}). \tag{15.2}$$

Remark 15.2. In the next section we will give an important example of "type A" where A is a proper subset of a component of the Grassmannian.

Theorem 15.3. *Suppose that ϕ is a calibration of type A and M is a compact submanifold of $\mathbf{R}^{p,q}$ calibrated by ϕ. If N is any other compact submanifold of type A with $\partial N = \partial M$, then*

$$\mathrm{vol}(M) \ \geq \ \mathrm{vol}(N)$$

with equality if and only if N is also calibrated by ϕ.

Proof. See the proof of Theorem 5.3. □

Example 1. (Point calibrations). The inner product $\langle \cdot, \cdot \rangle$ of signature p, q on $\mathbf{R}^{p,q}$ induces an isomorphism $\mathbf{R}^{p,q} \xrightarrow{\cong} (\mathbf{R}^{p,q})^*$ which extends to an equivalence between $\Lambda^k \mathbf{R}^{p,q}$ and $\Lambda^k(\mathbf{R}^{p,q})^*$. Given $\xi \in G^+_{\text{space}}(p, \mathbf{R}^{p,q})$, the equivalent form $\phi_\xi \in \Lambda^p(\mathbf{R}^{p,q})^*$ defined by $\phi_\xi(\eta) = \langle \xi, \eta \rangle$ is a calibration of type $G^+_{\text{space}}(p, \mathbf{R}^{p,q})$. More precisely one has:

$$\begin{aligned} \phi_\xi(\eta) \ &\geq \ 1 \quad \text{for} \quad \eta \in G^+_{\text{space}}(p, \mathbf{R}^{p,q}) \\ \text{with equality} \quad &\iff \quad \eta \ = \ \xi. \end{aligned} \tag{15.3}$$

With a conflict of terminology, the case $p = 1$ yields a calibration proof of the twin paradox for curves in $\mathbf{R}^{1,q}$ by taking the first axis to be time (see [H, Prop. 4.19]). The proof of the inequality above is immediate (see the bottom of page 797 in [M_2]) from a canonical form for η with respect to ξ under the action of the orthogonal group $\mathrm{O}(p, q)$ given on page 796 of [M_2].

Example 2. (Divided powers of the Kähler form). Now take $\mathbf{C}^{p,q}$ to be $\mathbf{C}^{p+q}$ equipped with the standard complex hermitian inner product of signature p, q, and let ω denote the corresponding "Kähler" form. In general, $\phi \equiv \frac{1}{k!}\omega^k$ is *not* a calibration. It is a calibration only when $k = p$.

Example 3. (Split associative, coassociative, etc.). Replacing the octonions $\mathbf{O}$ by their split companion $\widetilde{\mathbf{O}}$ (analogous to replacing $\mathbf{C}$ by its split companion $\widetilde{\mathbf{C}} = \mathbf{D}$), the associative, coassociative and Cayley calibrations have counterparts that are calibrations.

Finally it is worth noting that all possible calibrations of type $G^+_{\text{space}}(3, \mathbf{R}^{3,3})$ are classified in [M_1]. This is the lowest-dimensional non-trivial case.

16. Lagrangian submanifolds of constant phase and volume maximization

In this section we look at the general picture of all oriented Lagrangian submanifolds with non-degenerate metric. We show that such submanifolds are minimal (mean curvature zero) if and only if they are of *constant phase*. When and only when the submanifolds are also purely space-like or time-like, they are homologically volume-maximizing among purely space-like or time-like submanifolds of the same type. However, in all other signatures they are homologically volume-maximizing among Lagrangian submanifolds of the same type. For convenience we shall work in $\mathbf{D}^n$, however the results hold in any Ricci-flat $\mathbf{D}$-manifold.

The first result of this nature was noted in [HL$_1$, page 96] in the standard special Lagrangian setting. The phase θ of a Lagrangian submanifold M of $\mathbf{C}^n$ was defined using the fact that $dz(\overrightarrow{M}) = e^{i\theta}$ is of unit length (Proposition 1.14 and (2.18)). For any tangent vector V to M one has that $V(\theta) = \langle JV, H\rangle$ (equation (2.19)) where H is the mean curvature of M.

Dong [D] considered the case of non-degenerate Lagrangian submanifolds M of $\mathbf{C}^{p,q}$ ($\mathbf{C}^{p+q}$ with the standard hermitian inner product of signature p,q). Again the phase θ can be defined by $e^{i\theta} = (dz)(\overrightarrow{M})$ since it is of unit length. Despite the fact that $\mathrm{Re}\, dz$ is not a calibration, Dong noted that $V(\theta) = \langle JV, H\rangle$ still holds, so that M is of mean curvature zero if and only if the phase function is constant ([D, Lemma 2.1]). Each non-degenerate Lagrangian subspace in $\mathbf{C}^{p,q}$ must have signature p,q. We would like to make the further observation that $\mathrm{Re}\, dz$, although not a calibration in the pure sense, is a calibration of type $A = \mathrm{LAG}$, the set of oriented non-degenerate Lagrangian subspaces, because $(dz)(\xi) = e^{i\theta}(\xi)$ for $\xi \in \mathrm{LAG}$. By arguing as in the proof of Theorem 15.4, this proves that a constant phase Lagrangian submanifold is volume-minimizing among all other Lagrangian submanifolds with the same boundary.

We now turn attention back to $\mathbf{D}^n$. Let LAG denote the set of oriented non-degenerate Lagrangian n-planes in $\mathbf{D}^n$. The set LAG decomposes into $2n+2$ connected components

$$\mathrm{LAG} = \coprod_{p+q=n} \mathrm{LAG}^{\pm}_{p,q}$$

where $\mathrm{LAG}^{\pm}_{p,q}$ consists of planes for which the induced metric has signature (p,q) and orientation $+$ or $-$ when compared to an *a priori* chosen model. Each $\mathrm{LAG}^{+}_{p,q}$ and $\mathrm{LAG}^{-}_{p,q}$ is an orbit of the group $\mathrm{U}^{+}_{n}(\mathbf{D})$, and the pair $\mathrm{LAG}^{+}_{p,q} \cup \mathrm{LAG}^{-}_{p,q}$ is an orbit of $\mathrm{U}_n(\mathbf{D})$.

Note 16.1. The oriented model planes can be chosen so that

$$\pm\tau^q dz\big|_P = e^{\tau\theta_P} d\mathrm{vol}_P \qquad \text{for } P \in \mathrm{LAG}^{\pm}_{p,q}. \tag{16.1}$$

where $\theta_P \in \mathbf{R}$ and $d\mathrm{vol}_P$ is the unit (positive) volume form on P.

If M is an oriented connected non-degenerate Lagrangian submanifold of signature p,q, then all its tangent planes lie either in $\mathrm{LAG}^{+}_{p,q}$ or in $\mathrm{LAG}^{-}_{p,q}$ depending on the orientation of M and we say that M is of **type** $\mathrm{LAG}^{+}_{p,q}$ or of **type** $\mathrm{LAG}^{-}_{p,q}$. Therefore by (16.1) we have that

$$\pm\tau^q dz\big|_M = e^{\tau\theta} d\mathrm{vol}_M. \tag{16.2}$$

Definition 16.2. This smooth real-valued function θ on M will be called the **phase function** on M. If θ is constant, we say that M has **constant phase**.

Proposition 16.3. *Let $M \subset \mathbf{D}^n$ be an oriented non-degenerate Lagrangian submanifold which is connected. Then M is a minimal (mean curvature zero) submanifold if and only if M has constant phase.*

Proof. This follows immediately from the fact that for any tangent vector field V on M

$$V \cdot \theta \;=\; \langle \mathbf{T}V, H\rangle \tag{16.3}$$

where H is the mean curvature vector of M. By (9.3) this is equivalent to the statement that

$$d\theta \;=\; H \lrcorner \omega.$$

A proof of (16.3) is given at the end of the section. □

Proposition 16.4. *The form*

$$\phi_\theta \;=\; \pm \mathrm{Re}\left\{e^{-\tau\theta}\tau^q dz\right\} \tag{16.4}$$

is a calibration of type $A = \mathrm{LAG}^{\pm}_{p,q}$.

Proof. Apply (16.1) and (16.2). □

As noted in Theorem 5.3 and Remark 5.4, if M is a compact submanifold (with boundary) in $\mathbf{D}^n$ of type $\mathrm{LAG}^+_{n,0}$ which has constant phase θ, then for any submanifold N of type $G^+_{\mathrm{space}}(n, \mathbf{D}^n)$ with the same boundary as M we have $\mathrm{vol}(M) \geq \mathrm{vol}(N)$, and equality holds if and only if N is also of type $\mathrm{LAG}^+_{n,0}$ with the same constant phase. A similar result holds for $\mathrm{LAG}^+_{0,n}$ (and of course for $\mathrm{LAG}^-_{n,0}$), but not for $\mathrm{LAG}^+_{p,q}$ if $p, q \geq 1$ unless one restricts attention of a subset of the appropriate Grassmannian (See Remark 15.2).

Theorem 16.5. *Suppose that $(M, \partial M)$ is a compact submanifold of type $\mathrm{LAG}^{\pm}_{p,q}$ in $\mathbf{D}^n$ which is of constant phase θ. If $(N, \partial N)$ is any other compact submanifold of the same type $\mathrm{LAG}^{\pm}_{p,q}$ with $\partial N = \partial M$, then*

$$\mathrm{vol}(M) \;\geq\; \mathrm{vol}(N)$$

with equality if and only if N is also of constant phase θ.

Proof. Apply Theorem 15.3 and Proposition 16.4. □

Remark 16.6. This theorem carries over to any Ricci-flat Kähler $\mathbf{D}$-manifold. Here M is only homologically volume maximizing, i.e., maximizing among M' where $M - M' = \partial Y$ for some $(n+1)$-chain Y.

Proof of formula (16.3). Pick a local oriented orthonormal frame field $e_1, \ldots, e_n$ on M. Then

$$dz(e_1 \wedge \cdots \wedge e_n) \;=\; e^{\tau\theta}$$

defines the smooth phase function θ for M. Obviously we have that $V(\theta) = \tau e^{-\tau\theta} V(e^{\tau\theta}) = \tau e^{-\tau\theta} V(dz(e_1 \wedge \cdots \wedge e_n))$. Now with ∇ the riemannian connection on M, we have

$$\begin{aligned}
V\left(dz(e_1\wedge\cdots\wedge e_n)\right) &= dz\left(\sum_{k=1}^n e_1\wedge\cdots\wedge\nabla_V e_k\wedge\cdots\wedge e_n\right)\\
&= dz\left(\sum_{k=1}^n e_1\wedge\cdots\wedge\langle\nabla_V e_k,\mathbf{T}e_k\rangle\mathbf{T}e_k\wedge\cdots\wedge e_n\right)\\
&= \sum_{k=1}^n \langle\nabla_V e_k,\mathbf{T}e_k\rangle\tau e^{\tau\theta}
\end{aligned}$$

first because $\nabla dz=0$, second because $\langle\nabla e_k,e_k\rangle=0$ and $dz(e_\ell\wedge\mathbf{T}e_\ell\wedge\eta)=0$, with the third equality following because $dz(e_1\wedge\cdots\wedge\mathbf{T}e_k\wedge\cdots\wedge e_n)=\tau dz(e_1\wedge\cdots\wedge e_n)=\tau e^{\tau\theta}$. Combining gives

$$V(\theta) = \sum_{k=1}^n\langle\nabla_V e_k,\mathbf{T}e_k\rangle,$$

but

$$\langle\nabla_V e_k,\mathbf{T}e_k\rangle=\langle\nabla_{e_k}V,\mathbf{T}e_k\rangle=-\langle\nabla_{e_k}\mathbf{T}V,e_k\rangle=\langle\mathbf{T}V,\nabla_{e_k}e_k\rangle$$

and

$$H=\sum_k\left(\nabla_{e_k}e_k\right)^{\text{normal}}$$

is the mean curvature. □

Appendix A
A canonical form for space-like n-Planes in $\mathbf{D}^n$

In this appendix we give a canonical form for space-like n-planes under the unitary group. We first treat the case $n=2$ which is the key.

Lemma A.1. *Given a space-like 2-plane P in $\mathbf{D}^2$, there exist a space-like $\mathbf{D}$-unitary basis e_1,e_2 for $\mathbf{D}^2$ and an angle $\theta\geq 0$ such that e_1 and $\cosh\theta e_2+\sinh\theta\mathbf{T}e_1$ form an orthonormal basis for P.*

Proof. Pick any unit vector $e_1\in P$. Since P is space-like, there exists a unit time-like vector $\mathbf{T}e_2$ orthogonal to both P and $\mathbf{T}e_1$. Let this define the unit space-like vector e_2. Then the unit vector $\epsilon\in P$ orthogonal to e_1 belongs to the span of e_2 and $\mathbf{T}e_1$. Hence $\epsilon=\cosh\theta e_2+\sinh\theta\mathbf{T}e_1$, after possible sign changes of e_1 and e_2 to ensure $\theta\geq 0$. □

Note. Using the basis e_1,e_2 to define coordinates $z_1=x_1e_1+y_1\mathbf{T}e_1$ and $z_2=x_2e_2+y_2\mathbf{T}e_2$. The 2-plane P is defined by $y_1=(\tanh\theta)x_2$ and $y_2=0$ (where $-1<\tanh\theta<1$), i.e., P is the graph over $\mathbf{R}^2$ in $\mathbf{R}^2\oplus\tau\mathbf{R}^2$ of the nilpotent matrix $A=\begin{pmatrix}0&\tanh\theta\\0&0\end{pmatrix}$.

Proposition A.2. (Canonical form). *Each space-like n-plane P in $\mathbf{D}^n$ has an orthonormal basis*

$$\{e_1,\ \cosh\theta_1 e_2+\sinh\theta_1\mathbf{T}e_1,\ldots,e_{n-1},\ \cosh\theta_{\frac{n}{2}}e_n+\sinh\theta_{\frac{n}{2}}\mathbf{T}e_{n-1}\}\quad \textit{if } n \textit{ is even,}$$

and

$$\{e_1,\ \cosh\theta_1 e_2+\sinh\theta_1\mathbf{T}e_1,\ldots,e_{n-2},\ \cosh\theta_{\frac{n-1}{2}}e_{n-1}+\sinh\theta_{\frac{n-1}{2}}\mathbf{T}e_{n-2},\ e_n\}\quad \textit{if } n \textit{ is odd,}$$

where $e_1,\ldots,e_n$ is a space-like $\mathbf{D}$-unitary basis for $\mathbf{D}^n$ and $\theta_1\ge\cdots\ge\theta_{[\frac{n}{2}]}\ge 0$.

Proof. The case $n=2$ has already been proven. The 2-form $\omega\big|_P$ can be put in canonical form with respect to the positive definite inner product $\langle\cdot,\cdot\rangle$ on P. (This is the same as the canonical form for a skew $n\times n$-matrix under the conjugate action of the orthogonal group.) Namely, there exist an orthonormal basisi $\epsilon_1,\ldots,\epsilon_n$ for P and $\lambda_1\ge\cdots\ge\lambda_{[\frac{n}{2}]}\ge 0$ such that:

$$\omega\big|_P = \sum_{j=1}^{[\frac{n}{2}]}\lambda_j\epsilon_{2j-1}\wedge\epsilon_{2j}. \tag{A.1}$$

Note that $\operatorname{span}\{\epsilon_1,\epsilon_2,\mathbf{T}\epsilon_1,\mathbf{T}\epsilon_2\}\cong\mathbf{D}^2$ and $\operatorname{span}\{\epsilon_3,\epsilon_4,\mathbf{T}\epsilon_3,\mathbf{T}\epsilon_4\}\cong\mathbf{D}^2$ are orthogonal in $\mathbf{D}^n$ since $\langle\epsilon_i,\mathbf{T}\epsilon_j\rangle=\omega(\epsilon_i,\epsilon_j)$. Thus we can apply induction. □

Next we keep track of orientations. Let $\xi_0=\overrightarrow{\mathbf{R}}^n$ denote the space-like n-plane $\mathbf{R}^n\subset\mathbf{D}^n=\mathbf{R}^n\oplus\tau\mathbf{R}^n$ equipped with the standard orientation. Let $G^+_{\text{space}}(n,\mathbf{D}^n)$ denote the connected component of the Grassmannian of oriented space-like n-planes in $\mathbf{D}^n$ which contains ξ_0. Given $\xi\in G^+_{\text{space}}(n,\mathbf{D}^n)$ the orthonormal basis $\epsilon_1,\ldots,\epsilon_n$ for P, described in Proposition A.2, satisfies $\xi=\pm\epsilon_1\wedge\cdots\wedge\epsilon_n$. If n is odd and the minus sign occurs, we can replace $e_n=\epsilon_n$ by its negative and have a canonical oriented basis. If n is even and the minus sign occurs, replace e_n and $\theta_{[\frac{n}{2}]}$ by their negatives. This replaces $\epsilon_n=\cosh\theta_{\frac{n}{2}}e_n+\sinh\theta_{\frac{n}{2}}\mathbf{T}e_{n-1}$ by its negative, and we have proved the following oriented version of the canonical form.

Proposition A.3 (Canonical form – oriented version). *Each $\xi\in G^+_{\text{space}}(n,\mathbf{D}^n)$ can be put in the canonical form*

$$\begin{aligned}\xi &= e_1\wedge(\cosh\theta_1e_2+\sinh\theta_1\mathbf{T}e_1)\wedge\cdots\\ &\qquad\cdots\wedge e_{n-1}\wedge(\cosh\theta_{\frac{n}{2}}e_n+\sinh\theta_{\frac{n}{2}}\mathbf{T}e_{n-1})\quad \textit{if } n \textit{ is even, and}\\ \xi &= e_1\wedge(\cosh\theta_1e_2+\sinh\theta_1\mathbf{T}e_1)\wedge\cdots\\ &\qquad\cdots\wedge e_{n-2}\wedge(\cosh\theta_{\frac{n-1}{2}}e_{n-1}+\sinh\theta_{\frac{n-1}{2}}\mathbf{T}e_{n-2})\wedge e_n\quad \textit{if } n \textit{ odd,}\end{aligned}$$

where $e_1,\ldots,e_n$ is a space-like $\mathbf{D}$-unitary basis for $\mathbf{D}^n$ and $\theta_1\ge\cdots\ge\theta_{[\frac{n}{2}]}$; all ≥ 0 if n is odd, and all but one can be taken ≥ 0 if n is even. Moreover, $e_1\wedge\cdots\wedge e_n\in G^+_{\text{space}}(n,\mathbf{D}^n)$. Consequently, the map $A\in M_n(\mathbf{D})$ sending the *standard space-like unitary basis* $\mathbf{e}_1, \dots, \mathbf{e}_n$ *for* $\mathbf{D}^n$ *to the space-like unitary basis* $e_1, \dots, e_n$ *for* $\mathbf{D}^n$ *has* $\det_{\mathbf{D}} A = re^{\tau\theta}$, $r > 0$, *that is,* $A \in \mathrm{U}_n^+(\mathbf{D})$.

Proof. The only thing left to prove is that $e_1 \wedge \cdots \wedge e_n \in G^+_{\text{space}}(n, \mathbf{D}^n)$. The homothety connecting θ to 0 connects ξ to $e_1 \wedge \cdots \wedge e_n$. Since $\xi \in G^+_{\text{space}}(n, \mathbf{D}^n)$, this proves that $e_1 \wedge \cdots \wedge e_n \in G^+_{\text{space}}(n, \mathbf{D}^n)$. □

Appendix B
Projective and Hermitian projective varieties over D

Much of hermitian projective geometry carries over to the $\mathbf{D}$ setting, with some interesting twists. We begin with the following.

Definition B.1. The *n-dimensional projective space* $\mathbf{P}^n(\mathbf{D})$ over $\mathbf{D}$ is the set of all free rank-one submodules of $\mathbf{D}^{n+1}$. Otherwise said, $\mathbf{P}^n(\mathbf{D})$ is the set of all T-invariant real two-dimensional subspaces $\ell \subset \mathbf{D}^{n+1}$ for which the $+1$ and -1 eigenspaces $\ell_\pm$ are both one-dimensional.

Passing to null coordinates and writing $\mathbf{D}^{n+1} = e\,\mathbf{R}^{n+1} \oplus \bar{e}\,\mathbf{R}^{n+1}$ (the $\pm$-eigenspaces of T), we see that the decomposition $\ell = \ell_+ \oplus \ell_-$ gives

$$\mathbf{P}^n(\mathbf{D}) \;=\; \mathbf{P}^n(\mathbf{R}) \times \mathbf{P}^n(\mathbf{R}).$$

There is a corresponding decomposition for all the $\mathbf{D}$-grassmannians. In fact we have the following. For every real projective algebraic variety $V(\mathbf{R})$ there is a corresponding variety $V(\mathbf{D})$ over $\mathbf{D}$ defined by the same polynomials with arguments now in $\mathbf{D}^{n+1}$ (base change). Taking null coordinates $z = eu + \bar{e}v \in \mathbf{D}^{n+1} = e\,\mathbf{R}^{n+1} \oplus \bar{e}\,\mathbf{R}^{n+1}$, a polynomial $p(x)$ with real coefficients satisfies $p(z) = e\,p(u) + \bar{e}\,p(v)$ which shows that

$$V(\mathbf{D}) \;=\; V(\mathbf{R}) \times V(\mathbf{R}). \tag{B.1}$$

Note that in choosing *homogeneous coordinates* for $\mathbf{P}^n(\mathbf{D})$ one must restrict to elements $z \in \mathbf{D}^{n+1}$ which generate a *free* submodule. This means that $z = eu + \bar{e}v$ where $u \neq 0$ and $v \neq 0$. The space $\mathbf{P}^n(\mathbf{D})$ is then obtained by dividing by the group $\mathbf{D}^*$.

When we introduce the split Fubini-Study metric this all becomes more interesting. Recall the hermitian metric $(z, \zeta) = \sum_{k=0}^n z_k \bar{\zeta}_k$ with associated quadratic form

$$(z, z) \;=\; \sum_{k=0}^{n} u_k v_k \;\equiv\; u \bullet v.$$

In $\mathbf{P}^n(\mathbf{D})$ there is the *projective Stiefel variety*:

$$\begin{aligned} \mathrm{St} \;&\equiv\; \{[z] \in \mathbf{P}^n(\mathbf{D}) : (z, z) = 0\} \\ &=\; \{[e\,u + \bar{e}\,v] \in \mathbf{P}^n(\mathbf{D}) \;:\; u \bullet v = 0\} \\ &\cong\; \{(\ell_1, \ell_2) \in \mathbf{P}^n(\mathbf{R}) \times \mathbf{P}^n(\mathbf{R}) : \ell_1 \perp \ell_2\} \end{aligned}$$

which is a hypersurface (dimension $2n-1$) in $\mathbf{P}^n(\mathbf{R}) \times \mathbf{P}^n(\mathbf{R})$. It is not difficult to check that $(\mathbf{D}^{n+1})^* \equiv \mathbf{D}^{n+1} - \{(z,z)=0\} = \mathbf{D}^{n+1} - \{u \bullet v = 0\}$ has four connected components corresponding to the four components of the unit sphere $\$ \equiv \{u \bullet v = 1\}$. They are permuted by the group $\{\pm 1\} \times \{\pm \tau\}$. We define the **D** *hermitian projective space* to be the connected open set

$$\mathbf{P}^n_{\text{herm}}(\mathbf{D}) \equiv \mathbf{P}^n(\mathbf{D}) - \text{St}$$

and note the mapping $(\mathbf{D}^{n+1})^* \to (\mathbf{D}^{n+1})^*/\mathbf{D}^* = \mathbf{P}^n_{\text{herm}}(\mathbf{D})$ obtained by dividing by the group $\mathbf{D}^*$. Restricting to the unit sphere gives the *split Hopf mapping*

$$\$ \longrightarrow \mathbf{P}^n_{\text{herm}}(\mathbf{D})$$

with fibres which are orbits of the unit circle $S \equiv \{z \in \mathbf{D}^+ : \langle z,z\rangle = \pm 1\}$ acting by scalar multiplication.

We point out that the group $\text{GL}_{n+1}(\mathbf{D})$ acts transitively on $\mathbf{P}^n(\mathbf{D})$ and the group $\text{U}_{n+1}(\mathbf{D})$ acts transitively on the open subset $\mathbf{P}^n_{\text{herm}}(\mathbf{D})$. Thus,

$$\mathbf{P}^n(\mathbf{D}) = \text{GL}_{n+1}(\mathbf{D})/\text{GL}_{1,n}(\mathbf{D}) \supset \mathbf{P}^n_{\text{herm}}(\mathbf{D}) = \text{U}_{n+1}(\mathbf{D})/\text{U}_1(\mathbf{D}) \times \text{U}_n(\mathbf{D})$$

where $\text{GL}_{1,n}(\mathbf{D})$ is the subgroup preserving the initial line.

Now in hermitian homogeneous coordinates $(\mathbf{D}^{n+1})^*$ consider the real $\text{U}_{n+1}(\mathbf{D})$-invariant 2-form

$$\widetilde{\omega} \equiv -\tau\partial\overline{\partial}\log(z,z)$$

which in null coordinates can be written

$$\widetilde{\omega} = d_u d_v \log(u \bullet v) = \frac{1}{u \bullet v}\sum_j du_j \wedge dv_j - \frac{1}{(u \bullet v)^2}(v \bullet du) \wedge (u \bullet dv).$$

Fix a point $(u_0, v_0) \in \mathbf{D}^{n+1}$ with $u_0 \bullet v_0 = 1$. Then $\nu = (u_0, v_0)$, translated to (u_0, v_0) is the unit normal to the sphere $\$$ at this point, and the *horizontal subspace* $H_{(u_0,v_0)} \subset T_{(u_0,v_0)}\$$ is the **D**-orthogonal complement of ν. Note that $\nu \lrcorner \widetilde{\omega} = u_0 \bullet dv - v_0 \bullet du - (v_0 \bullet u_0)u_0 \bullet dv + (u_0 \bullet v_0)v_0 \bullet du = 0$. It follows that $(T\nu)\lrcorner\widetilde{\omega} = 0$, i.e., that $\widetilde{\omega}\big|_{\$}$ is a horizontal, $\text{U}_{n+1}(\mathbf{D})$-invariant 2-form. It is straightforward to check that $\widetilde{\omega}$ is non-degenerate on H_{u_0,v_0}. (It is particularly clear at points (u_0, u_0). Then use the transitivity of $\text{U}_{n+1}(\mathbf{D})$.) We have proved the following.

Lemma B.2. *The form $\widetilde{\omega}$ descends to a d-closed $\text{U}_{n+1}(\mathbf{D})$-invariant symplectic 2-form ω on $\mathbf{P}^n_{\text{herm}}(\mathbf{D})$. This form makes $\mathbf{P}^n_{\text{herm}}(\mathbf{D})$ a* **D**-*Kähler manifold.*

From Note 10.5 we have the following.

Corollary B.3. *Let $X \subset \mathbf{P}^n(\mathbf{D})$ be a* **D**-*submanifold and let $X_{\text{herm}} \subset \mathbf{P}^n_{\text{herm}}(\mathbf{D})$ denote the set of points in X where $\omega\big|_X$ is non-degenerate. Then X_{herm} is a* **D**-*Kähler manifold in the induced metric.*

Consider now a **D**-submanifold

$$V(\mathbf{D}) \subset \mathbf{P}^n(\mathbf{D}) \tag{B.2}$$

of $\mathbf{D}$-dimension k arising from an algebraic submanifold

$$V(\mathbf{R}) \subset \mathbf{P}^n(\mathbf{R}) \tag{B.3}$$

as above, and recall the canonical product structure (B.1). The embedding (B.2) induces a riemannian metric g on $V(\mathbf{R})$ from the standard spherical metric g_0 on $\mathbf{P}^n(\mathbf{R})$. The embedding (B.3) induces a Kähler metric with Kähler form ω on the subset $V(\mathbf{D})_{\text{herm}}$. There is also a nowhere null holomorphic k-form

$$\Phi = e \cdot \alpha + \bar{e} \cdot \beta$$

where α and β denote the volume form of $V(\mathbf{R})$, with metric g_0, pulled back from the first and second factors of (B.1) respectively. Note that

$$\text{Re}(\Phi) = \frac{1}{2}(\beta + \alpha) \qquad \text{and} \qquad \text{Im}(\Phi) = \frac{1}{2}(\beta - \alpha).$$

The following pretty geometry is straightforward to verify.

Proposition B.4. *Consider the diagonal embedding $\Delta : V(\mathbf{R}) \hookrightarrow V(\mathbf{D})$ given by $\Delta(u) = (u, u)$ with respect to* (B.1). *Set $V_\Delta = \Delta(V(\mathbf{R}))$. Then:*

(1) $V_\Delta \subset V(\mathbf{D})_{\text{herm}}$,
(2) *V_Δ is space-like Lagrangian, in particular*

$$\omega\big|_{V_\Delta} = 0,$$

(3) *V_Δ satisfies the "special" condition*

$$\text{Im}(\Phi)\big|_{V_\Delta} = 0,$$

(4) *the restriction of the (real part of) the Kähler metric to V_Δ coincides with $2g$,*
(5) *there is a constant $c = c(k)$ such that:*

$$\text{Re}(\Phi)\big|_{V_\Delta} = c\,\text{vol}(g).$$

The story is similar for the anti-diagonal embedding $\widetilde{\Delta} : V(\mathbf{R}) \hookrightarrow V(\mathbf{D})$ given by $\widetilde{\Delta}(u) = (u, -u)$ except that $V_{\widetilde{\Delta}}$ is purely time-like instead of space-like, and the induced metric is $-2g$.

Of course the Kähler metric on $V(\mathbf{D})_{\text{herm}}$ is not Ricci-flat. However, it is interesting to ask whether the metric and the holomorphic n-form Φ can be altered outside V_Δ so that so that Φ becomes parallel.

Appendix C
Degenerate projections

Many of the results in special Lagrangian geometry which appeared in [HL$_1$] carry over to the split case. One instance of this concerns submanifolds with degenerate projections. For what follows the reader is referred to [HL$_1$] for full details and proofs.

Decompose $\mathbf{R}^n$ as $\mathbf{R}^n = \mathbf{R}^p \oplus \mathbf{R}^{n-p}$ and consider a p-dimensional submanifold M in $\mathbf{R}^p \oplus \tau\mathbf{R}^{n-p}$ along with a smooth function h on M. Let $N_{\text{twist}}(M)$ denote the normal bundle to M twisted by the antipodal map on the fibres, defined precisely by (C.2) below. This is a Lagrangian submanifold of $\mathbf{D}^n$ as is its affine translate by dh,

$$X \equiv N_{\text{twist}}(M) + dh. \tag{C.1}$$

The case of interest is when M is space-like in $\mathbf{R}^p \oplus \tau\mathbf{R}^{n-p}$. Then $M \equiv \text{graph}(u)$ with $u : \mathbf{R}^p \to \mathbf{R}^{n-p}$, and we can assume $h : \mathbf{R}^p \to \mathbf{R}$.

Using the decomposition of $\mathbf{R}^n$ to decompose the coordinates $x = (x', x''), y = (y', y'')$ the manifold X is parameterized by sending $x \in \mathbf{R}^n$ to

$$\left(x',\, x'',\, x'' \cdot \frac{\partial u}{\partial x'}(x') + \frac{\partial h}{\partial x'}(x'),\, u(x')\right). \tag{C.2}$$

Therefore, X is the graph (over $\mathbf{R}^n$ in $\mathbf{D}^n = \mathbf{R}^n \oplus \tau\mathbf{R}^n$) of $\nabla\varphi$, the gradient of the potential function

$$\varphi(x) \equiv x'' \cdot u(x') + h(x'). \tag{C.3}$$

In the case $h \equiv 0$ we have:

Theorem C.1. *If M is a space-like austere submanifold of $\mathbf{R}^p \oplus \tau\mathbf{R}^{n-p}$, then $N_{\text{twist}}(M)$ is a special Lagrangian submanifold of $\mathbf{D}^n$.*

The case $n = 3$ and $p = 2$ is particularly nice even with $h \neq 0$. Note that $\varphi(x) = x_3 u(x_1, x_2) + h(x_1, x_2)$ is the potential function for X.

Theorem C.2. *The submanifold $X \equiv N_{\text{twist}}(M) + dh$ is special Lagrangian in $\mathbf{D}^3$ if and only if M^3 is space-like in $\mathbf{R}^2 \oplus \tau\mathbf{R}$ with vanishing mean curvature and h is a harmonic function on M.*

Proof. If $\varphi(x) = x_3 u(x_1, x_2) + h(x_1, x_2)$, then $\nabla\varphi = (x_3 u_1 + h_1, x_3 u_2 + h_2, u)$, and

$$\text{Hess}\,\varphi = \begin{pmatrix} x_3 u_{11} + h_{11} & x_3 u_{12} + h_{12} & u_1 \\ x_3 u_{21} + h_{21} & x_3 u_{22} + h_{22} & u_2 \\ u_1 & u_2 & 0 \end{pmatrix}.$$

Therefore

$$\begin{aligned} \Delta\varphi + \det \text{Hess}\,\varphi &= x_3 \left[(1 - u_2^2)u_{11} + 2u_1 u_2 u_{12} + (1 - u_1^2)u_{22}\right] \\ &\quad + \left[(1 - u_2^2)h_{11} + 2u_1 u_2 h_{12} + (1 - u_1^2)h_{22}\right]. \end{aligned}$$ □

Remark C.3. Each such special Lagrangian manifold $X \equiv N_{\text{twist}}(\text{graph}(u))$ has a potential $\bar{\varphi}(u_1, u_2, u_3)$ with $X \equiv \text{graph}\nabla\bar{\varphi}$ in $e\mathbf{R}^3 \oplus \bar{e}\mathbf{R}^3$. The function $\bar{\varphi}$ is a solution to:

$$\bar{\varphi} \text{ convex and } \det \text{Hess}\,\bar{\varphi} = 1. \tag{C.4}$$

It would be interesting to examine some explicit examples such as $y_3 = u(x_1, x_2) = \frac{1}{a}\arccos(ar + c),\ r = \sqrt{x_1^2 + x_2^2}$

Appendix D
Singularities and semi-riemannian calibrations

It is straightforward to extend the basic set-up of Section 15 to manifolds. Suppose that X is a semi-riemannian manifold of signature p, q with $p, q \geq 1$. We identify the Grassmannian $G_{\text{space}}(p)_x$ of oriented space-like p-planes in $T_x X$ with a subset of $\Lambda^p T_x X$ by choosing an oriented orthonormal basis for each such p-plane and taking the wedge product. Since $p, q \geq 1$, each $G_{\text{space}}(p)_x$ has two connected components. A **space-like orientation** on X is a continuous choice of one the components, denoted by $G^+_{\text{space}}(p)$, and referred to as the Grassmannian of **positive** (or "future") oriented space-like p-planes.

Definition D.1. The **mass ball** $B_x \subset \Lambda^p T_x X$ is the convex hull of $G^+_{\text{space}}(p)_x$. The **comass ball** $B^*_x \subset \Lambda^p T^*_x X$ is defined to be the polar:

$$B^*_x \equiv \{\phi \in \Lambda^p T^*_x X : \phi(\xi) \geq 1 \text{ for all } \xi \in G^+_{\text{space}}(p)_x\}$$

(or equivalently for all $\xi \in B_x$). The **mass cone** $\mathcal{C}_x$ is the (convex) cone on B_x with vertex at the origin in $\Lambda^p T_x X$, and the **comass cone** $\mathcal{C}^*_x$ is the (convex) cone on B^*_x with vertex at the origin in $\Lambda^p T^*_x X$.

Note that the closure $\overline{\mathcal{C}^*_x}$ is the polar cone of $\overline{\mathcal{C}_x}$, i.e., $\phi \in \overline{\mathcal{C}^*_x} \iff \phi(\xi) \geq 0$ for all $\xi \in \overline{\mathcal{C}_x}$. Given $\xi \in \mathcal{C}_x$, there exists a unique $\lambda > 0$ such that $\lambda\xi \in \partial B_x$. We define

$$\|\xi\| \equiv \frac{1}{\lambda} \quad \text{to be the } \textbf{mass norm} \text{ of } \xi. \tag{D.1}$$

Similarly, the **comass norm** of $\phi \in \mathcal{C}^*_x$, denoted by $\|\phi\|^*$, is defined so that $\|\phi\|^* = 1 \iff \phi \in \partial B^*_x$ and so that $\|\phi\|^*$ is homogeneous of degree 1 for positive scalars. (Note that in the analogous positive definite case the mass and comass cones are all of $\Lambda^p T_x X$ and $\Lambda^p T^*_x X$.) Both $\|\cdot\|$ and $\|\cdot\|^*$ are superadditive on $\mathcal{C}_x$ and $\mathcal{C}^*_x$ respectively.

Definition D.2. A d-closed p-form ϕ on X which takes its values in the comass cone $\mathcal{C}$ is a **calibration** if it has comass one at each point, i.e.,

$$\phi(\xi) \geq 1 \text{ for all } \xi \in G^+_{\text{space}}(p). \tag{D.2}$$

A calibration ϕ is said to **calibrate** the set $G(\phi) \equiv \{\phi = 1\} \cap G^+_{\text{space}}(p)$ of p-planes in $G^+_{\text{space}}(p)$ where ϕ attains its minimum. An oriented submanifold M of X is a **ϕ-submanifold** if $\overrightarrow{M} \in G(\phi)$ at each point, or equivalently, M is positive space-like and $\phi\big|_M$ is the unit volume form for M.

The fundamental theorem for semi-riemannian calibrations (see Theorem 15.3 for the case $X = \mathbf{R}^{p,q}$) can be stated for submanifolds as follows. See the generalization allowing singularities, Theorem D.10 below, for the proof.

Proposition D.3. *Suppose ϕ is a calibration of degree p on a semi-riemannian manifold with a space-like orientation an of signature p, q with $p, q \geq 1$. If M is a* *compact ϕ-submanifold (possibly with boundary), then*

$$\mathrm{vol}(M) \ \geq \ \mathrm{vol}(N) \tag{D.3}$$

for all positive space-like compact submanifolds N which are homologous to M. Moreover, if equality occurs, then N is also a ϕ-submanifold.

The purpose of this appendix is to generalize this result to allow singularities.

Currents of special type $\mathcal{S}$

A current T of dimension p on a manifold X which is representable by integration (see [F]) can be put in **polar form** in each coordinate system $U \subset \mathbf{R}^n$

$$T \ = \ \overrightarrow{T}\|T\| \tag{D.4}$$

where $\|T\|$ is a non-negative Radon measure called the *generalized volume measure for T*, and $\overrightarrow{T}$ is a $\|T\|$-measurable function taking values in the boundary of the unit mass ball in $\Lambda^p\mathbf{R}^n$, i.e., $|\overrightarrow{T}| = 1$ $\|T\|$-a.e. It is referred to as the *generalized tangent space* (since $\overrightarrow{T}(x)$ may not be simple). Of course

$$T(\varphi) \ = \ \int \varphi(\overrightarrow{T})\|T\| \tag{D.5}$$

for each test p-form φ on U.

Each riemannian metric on X determines a second notion of mass norm $|\cdot|'$ on $\Lambda^p T_x X$ (with $x \in U$). Defining

$$\|T\|' \ \equiv \ |\overrightarrow{T}|'\|T\| \qquad \text{and} \qquad \overrightarrow{T}'(x) \equiv \frac{1}{|\overrightarrow{T}(x)|'}\overrightarrow{T}(x) \tag{D.6}$$

yields the polar form

$$T \ = \ \overrightarrow{T}'\|T\|' \tag{D.7}$$

based on the riemannian metric. Here $\|T\|'$ is a Radon measure on X and $\overrightarrow{T}'(x)$ is a $\|T\|'$-measurable section of the bundle $\Lambda^p TX$ with unit mass norm almost everywhere.

Now given a riemannian metric we drop the prime from the notation and use $\overrightarrow{T}$ and $\|T\|$ based on the riemannian metric.

Suppose now that $\mathcal{S}$ is any cone set (i.e., $s \in \mathcal{S}_x \Rightarrow ts \in \mathcal{S}_x \ \forall t > 0$) in the total space of the bundle $\Lambda^p TX$. Given a current T of dimension p which is representable by integration, if

$$\overrightarrow{T}(x) \in \mathcal{S}_x \ \text{ for } \ \|T\| \text{ a.a. } x \tag{D.8}$$

is true for one riemannian metric, then it is true for all riemannian metrics. Consequently, the following concept is independent of the choice of riemannian metric on X

Definition D.4. Suppose $\mathcal{S}$ is a cone set in $\Lambda^p T_x X$. A current T of dimension p which is representable by integration is said to be of **type** $\mathcal{S}$ if

$$\vec{T}(x) \in \mathcal{S}_x \quad \text{for} \quad \|T\| \text{ a.a. } x.$$

The fundamental theorem for semi-riemannian calibrations

Suppose that X is a semi-riemannian manifold of signature $p, q \geq 1$ which is space-like oriented and with mass cone $\mathcal{C} \subset \Lambda^p TX$.

Definition D.5. A current T of dimension p which is representable by integration is said to be **positive space-like** if T is of type $\mathcal{C}$.

Proposition D.6. *Each positive space-like current T has a* semi-riemannian polar form

$$T \;=\; \vec{T}\|T\|$$

where $\|T\|$ is a non-negative Radon measure on X and $\vec{T}$ is a $\|T\|$-measurable section of $\Lambda^p TX$ with the semi-riemannian mass norm $|\vec{T}(x)| = 1$ for $\|T\|$-a.a. points x.

Proof. Let $T = \vec{T}'\|T\|'$ denote a local riemannian polar form. Since T is type $\mathcal{C}$, i.e., $\vec{T}'(x) \in \mathcal{C}_x$, there is a well-defined semi-riemannian mass norm $|\vec{T}'(x)|$. Set

$$\vec{T}(x) \equiv \frac{1}{|\vec{T}(x)|'}\vec{T}'(x) \qquad \text{and} \qquad \|T\| \;\equiv\; |\vec{T}|'\|T\|'. \tag{D.9}$$

Then $T = \vec{T}\|T\|$ and the semi-riemannian mass norm $|\vec{T}(x)|$ equals 1 for $\|T\|$-a.a. points x. ☐

Definition D.7. The **volume/mass** of a positive space-like current T is defined by

$$\mathrm{vol}_K(T) \;=\; \int_K \|T\| \qquad \text{for all compact } K \subset X$$

using the generalized volume measure given by Proposition D.6.

Remark D.8. We leave it to the reader to show that this definition of mass for positive space-like currents T agrees with the definition in [KMW]. Namely, $\int \|T\|$ equals the infimum of $T(\varphi)$ taken over all forms φ of type $\mathcal{C}^*$ with comass $\|\varphi\|_K^* \equiv \inf_{x\in K} \|\varphi\|^* \geq 1$. (Hint: Use (15.3).)

Now suppose that ϕ is a calibration of degree p on X.

Definition D.9. A **ϕ-subvariety** is a locally rectifiable current T of type $\mathcal{C}$, or equivalently of type $\mathbf{R}^+ \cdot G^+_{\text{space}}(p)$ (the cone on $G^+_{\text{space}}(p)$). The ϕ-subvariety is **without boundary** if $dT = 0$.

The fundamental theorem of semi-riemannian calibrations can be stated a follows.

Theorem D.10. *Suppose ϕ is a calibration of degree p on a space-like oriented, semi-riemannian manifold of signature $p, q \geq 1$. Suppose that T is a ϕ-subvariety with compact support. Then T is homologically volume maximizing. That is, for any other positive space-like rectifiable current S with compact support which is homologous to T ($S = T + dR$ with* $\mathrm{supp}(R)$ *compact), one has*

$$\mathrm{vol}(T) \geq \mathrm{vol}(S)$$

Moreover, equality occurs if and only if S is also a ϕ-subvariety.

Proof. Since $d\phi = 0$, $S = T + dR$, and ϕ is a calibration, we have

$$\begin{aligned}\mathrm{vol}(T) = \int \|T\| = \int \phi(\overrightarrow{T})\|T\| = T(\phi) = S(\phi) \\ = \int \phi(\overrightarrow{S})\|S\| \geq \int \|S\| = \mathrm{vol}(S)\end{aligned}$$

and equality occurs if and only if $\phi(\overrightarrow{S}) = 1$ for $\|S\|$-a.a. points. □

Remark D.11. We leave it to the reader to extend this theorem to currents representable by integration but not necessarily rectifiable, which are of type $\mathcal{C}(\phi)$ where $\mathcal{C}(\phi)$ is the convex cone on $G(\phi)$.

Remark D.12. (Null planes and the mass cone). Let $\overline{\mathcal{C}}$ be the closed convex cone on $G^+_{\text{space}}(p, \mathbf{R}^{p,q})$. Assume for simplicity that $p \leq q$. Let $\mathcal{I}(p)$ denote the set of isometries $A : \mathbf{R}^p \to \mathbf{R}^q$, i.e., $A^tA = I$, the identity on $\mathbf{R}^p$. Identify $\mathcal{I}(p)$ with a subset of $\Lambda^p\mathbf{R}^{p,q}$ by letting graph(A) denote the oriented graph of A and setting $\xi_A \equiv (e_1 + Ae_1) \wedge \cdots \wedge (e_p + Ae_p) \in \Lambda^p\mathbf{R}^{p,q}$. If $p = q = n$ we have the orthogonal group $\mathrm{O}(n) \subset \Lambda^n\mathbf{D}^n$.

Note that $\langle e_i + Ae_i, e_j + Ae_j\rangle = 0$ for all i, j since $A^tA = I$. That is ξ_A is totally null.

Proposition D.13. *The closed mass cone $\overline{\mathcal{C}} \subset \Lambda^p\mathbf{R}^{p,q}$ is the convex cone on $\mathcal{I}(p) \subset \Lambda^p\mathbf{R}^{p,q}$.*

Proof. If $A \in \mathcal{I}(p)$, note that for $0 \leq t < 1$, $\xi^t_A = (e_1 + tAe_1) \wedge \cdots \wedge (e_p + tAe_p)$ rescaled by $(1 - t^2)^{\frac{p}{2}}$ is an element of $G^+_{\text{space}}(p, \mathbf{R}^{p,q})$. Hence $\mathcal{I}(p) \subset \overline{\mathcal{C}}$.

Conversely, if $\xi \in G^+_{\text{space}}(p, \mathbf{R}^{p,q})$, then pick an oriented orthonormal basis $e_1, \ldots, e_p$ for ξ and an orthonormal set $\bar{e}_1, \ldots, \bar{e}_p$ in $\mathbf{R}^q$. Define $n^{\pm}_j = \frac{1}{2}(e_j \pm \bar{e}_j)$ so that $e_j = n^+_j + n^-_j$. Then $\xi = e_1 \wedge \cdots \wedge e_p = \sum_{\pm} n^{\pm}_1 \wedge \cdots \wedge n^{\pm}_p$ and each $2^p n^{\pm}_1 \wedge \cdots \wedge n^{\pm}_p \in \mathcal{I}(p)$. □

Appendix E
Singularities in split SLAG geometry

Recall the definition of type (Definition D.4) and the semi-riemannian polar form (Proposition D.6) for currents representable by integration.

Definition E.1. A locally rectifiable n-current T defined on an open subset of $\mathbf{D}^n$ is a **split SLAG subvariety** if

$$\overrightarrow{T} \in \widetilde{\mathrm{SLAG}} \quad \text{for } \|T\| \text{ a.a. points,}$$

and is **positive space-like** if

$$\overrightarrow{T} \in G^+_{\mathrm{space}}(n, \mathbf{D}^n) \quad \text{for } \|T\| \text{ a.a. points.}$$

The inequality in Theorem 4.4 reads

$$(\mathrm{Re}\, dz)(\overrightarrow{T}_x) \geq 1 \qquad \text{for all } \overrightarrow{T}_x \in G^+_{\mathrm{space}}(n, \mathbf{D}^n), \tag{D.2}$$

with equality $\iff$ $\overrightarrow{T}_x$ is split special Lagrangian, i.e., $\overrightarrow{T}_x \in \widetilde{\mathrm{SLAG}}$.

Theorem 5.3 generalizes to include "singularities".

Theorem E.2. *Suppose that T is a split SLAG subvariety with compact support in $\mathbf{D}^n$. Suppose S is a positive space-like compactly supported rectifiable current on $\mathbf{D}^n$ with $dS = dT$. Then*

$$\mathrm{vol}(T) \geq \mathrm{vol}(S)$$

and equality holds if and only if S is also a split SLAG subvariety.

Proof. This theorem is a special case of the fundamental Theorem D.10. □

Finally, we note that if the space-like requirement on $\overrightarrow{T}$ is dropped, the currents T are still of interest (see Appendix F on dimension two).

Definition E.3. A locally rectifiable n-current T defined on an open subset of $\mathbf{D}^n$ is **unconstrained split SLAG** (not necessarily space-like split SLAG) if

(1) $T \wedge \omega = 0$ (i.e., T is Lagrangian), and
(2) $T \wedge \psi = 0$ where $\psi = \mathrm{Im}\, dz = \frac{1}{2}(dv - du)$.

Appendix F
Split SLAG singularities in dimension two

This appendix is devoted to describing the structure in dimension $n = 2$ of d-closed currents which are unconstrained split SLAG (cf. Definition D.4) and currents that are split SLAG varieties. We use the corresponding results for complex curves which are collected together as a remark.

Remark F.1. (Complex curves in $\mathbf{C}^2$). Suppose V is an irreducible complex curve defined near a point $z \in \mathbf{C}^2$. Even if z is a singular point of V, the curve V has a well-defined tangent line (with multiplicity > 1). The current $[V]$ of integration over V is well defined and the singularities are isolated. The Gauss map is an open map which can be understood using a local uniformizing parameter. A **holomorphic chain** T in an open subset X of $\mathbf{C}^2$ is defined to be a locally finite sum of the form $\sum_j n_j[V_j]$ where each $n_j \in \mathbf{Z}$ and each V_j is a closed one-dimensional subvariety of X. Thus, holomorphic chains and divisors on X are equivalent concepts.

Locally rectifiable 2-currents T for which $\overrightarrow{T}$ is a field of complex lines, are completely understood because of the following result.

Theorem F.2. ([HS, S]). *Suppose that T is a d-closed rectifiable 2-current defined in an open subset X of $\mathbf{C}^2$. If the (unoriented) tangent space $\overrightarrow{T}$ to T is a complex line $\|T\|$-a.e., then T is a holomorphic chain in X.*

This result can be viewed as a very strong regularity result for "rectifiable" complex curves. Fortunately, it applies directly to split SLAG geometry.

Theorem F.3. *There exists a real orthogonal coordinate change $\mathbf{C}^2 \leftrightarrow \mathbf{D}^2$, such that holomorphic chains are transformed to unconstrained split SLAG currents and vice-versa.*

Proof. Let $z' \in \mathbf{C}^2$ denote complex coordinates and $z \in \mathbf{D}^2$ denote $\mathbf{D}$-coordinates with $z'_j = x'_j + iy'_j$ and $z_j = x_j + \tau y_j$. Define a coordinate change $z'_1 = x_1 + ix_2$ and $z'_2 = y_1 + iy_2$ from $\mathbf{D}^2$ to $\mathbf{C}^2$. Then

$$dz' \;=\; \omega_{\mathbf{D}} + i\mathrm{Im}\, dz. \tag{F.1}$$

Also we have

$$\omega_{\mathbf{C}} \;=\; dx_1\, dx_2 - dy_1\, dy_2 \qquad \text{and} \qquad \mathrm{Re}\, dz \;=\; dx'_1\, dx'_2 - dy'_1\, dy'_2.$$

Suppose that T is an unconstrained split SLAG current. Then by definition and (F.1) we have $T \wedge dz' = 0$ for $\|T\|$-a.a. points. This condition $T \wedge dz' = 0$ is equivalent to $\overrightarrow{T}$ being a complex line. Hence by the Structure Theorem F.2, T is a holomorphic chain with respect to the complex structure J on $\mathbf{D}^2$ defined by

$$J\left(\frac{\partial}{\partial x_1}\right) \;=\; \frac{\partial}{\partial x_2} \qquad \text{and} \qquad J\left(\frac{\partial}{\partial y_2}\right) \;=\; \frac{\partial}{\partial y_1}. \qquad \square$$

Note that the condition

$$\mathrm{Im}\, dz\big|_{\mathrm{graph}\, A} \;=\; 0 \qquad \text{for} \;\; A \in M_2(\mathbf{R})$$

is that $\mathrm{tr}\, A = 0$. Hence, $M = \mathrm{graph}\, A$ is unconstrained split SLAG $\iff$ A is symmetric with trace zero. Thus

$$A \;=\; \begin{pmatrix} a & b \\ b & -a \end{pmatrix}$$

and the graph $y = Ax$ becomes $z'_2 = (a + ib)z'_1$ in the complex coordinates z'_1, z'_2 above.

Now $M \equiv \mathrm{graph}\, A$, with A as above, is split SLAG $\iff$ in addition $A^t A = (a^2 + b^2) \cdot I < I$, i.e.,

$$|a + ib| \;<\; 1.$$

Definition F.4. A holomorphic chain $T = \sum_j n_j [V_j]$ in $\mathbf{C}^2$ satisfies the **45° rule** if each tangent plane is of the form $z'_2 = \alpha z'_1$ with $|\alpha| < 1$ (including tangent planes at singular points).

Corollary F.5. *Under the real orthogonal coordinate transformation given in Theorem* F.3*, a split SLAG variety T corresponds to a holomorphic chain T which satisfies the* 45° *rule.*

Proof. By the hypothesis on T that $\overrightarrow{T} \in \widetilde{\text{SLAG}}$ at $\|T\|$-a.a. points, we only know that T is a holomorphic chain whose tangent plane satisfies the 45° rule at $\|T\|$-a.a. points. However, since the Gauss map of each V_j is open (even in neighborhoods of singular points), it follows that then tangent planes satisfy the 45° rule at every point of the support of T. □

Remark F.6. These results can be restated in null coordinates on $\mathbf{D}^2$ as follows. Suppose T is a d-closed locally rectifiable current on an open subset X of $\mathbf{D}^n$, with the property that for $\|T\|$-a.a. points in X, the unoriented tangent space of T is the graph of some symmetric matrix $A > 0$ with $\det A = 1$ over $e\mathbf{R}^2$ in $\mathbf{D}^2 = e\mathbf{R}^2 \oplus \bar{e}\mathbf{R}^2$. Then T is a locally finite sum of currents of integration $\varphi_*([\Delta])$ where $\varphi : \Delta \to \mathbf{D}^2$ maps ζ to $z(\zeta)$ and

$$x_1(\zeta) + ix_2(\zeta) \qquad \text{and} \qquad y_2(\zeta) + iy_1(\zeta) \quad \text{are holomorphic.} \tag{F.2}$$

Conversely, if $T = \varphi_*([\Delta])$ where φ satisfies (F.2), and if

$$\left|\frac{\partial}{\partial\zeta}(x_1(\zeta) + ix_2(\zeta))\right| < \left|\frac{\partial}{\partial\zeta}(y_2(\zeta) + iy_1(\zeta))\right|,$$

then, except for isolated singularities, φ parameterizes a split SLAG submanifold of $X \subset \mathbf{D}^2$.

References

[A_1] A.D. Alexandrov, *Almost everywhere existence of the second differential of a convex function and properties of convex surfaces connected with it (in Russian)*, Lenningrad State Univ. Ann. Math. **37** (1939), 3–35.

[A_2] A.D. Alexandrov, *The Dirichlet problem for the equation* $\operatorname{Det}\|z_{i,j}\| = \psi(z_1, \dots, z_n, x_1, \dots, x_n)$, I. Vestnik, Leningrad Univ. **13** No. 1, (1958), 5–24.

[AMT] D.V. Alekseevsky, C. Medori and A. Tomassini, *Homogeneous para-Kähler Einstein manifolds*, Russian Math. Surveys 64, No. 1, (2009), 1–43. ArXiv:0806.2272.

[B] R. Bryant, *Classical, exceptional, and exotic holonomies: a status report*, Actes de la Table Ronde de Géométrie Différentielle (Luminy, 1992), 93–165, Sémin. Congr., 1, Soc. Math. France, Paris, 1996.

[C_1] L.A. Caffarelli, *Interior a priori estimates for solutions of fully non-linear equations*, Ann. of Math. **130** (1990), 189–213.

[C_2] L.A. Caffarelli, *A localization property of viscosity solutions to the Monge-Ampère equation and their strict convexity*, Ann. of Math. **131** (1990), 129–134.

[C_3] L.A. Caffarelli, *Interior* $W^{2,p}$ *estimates for solutions of the Monge-Ampère equation*, Ann. of Math. **131** (1990), 135–150.

[CNS_1] L. Caffarelli, L. Nirenberg and J. Spruck, *The Dirichlet problem for nonlinear second order elliptic equations. I: Monge-Ampère equation*, Comm. Pure Appl. Math. **37** (1984), 369–402.

[CNS_2] L. Caffarelli, L. Nirenberg and J. Spruck, *The Dirichlet problem for nonlinear second order elliptic equations, III: Functions of the eigenvalues of the Hessian*, Acta Math. **155** (1985), 261–301.

[CNS_3] L. Caffarelli, L. Nirenberg and J. Spruck, *The Dirichlet problem for the degenerate Monge-Ampère equation*, Rev. Mat. Iberoamericana **2** (1986), 19–27.

[CNS_4] L. Caffarelli, L. Nirenberg and J. Spruck, *Correction to: "The Dirichlet problem for nonlinear second order elliptic equations. I: Monge-Ampère equation"*, Comm. Pure Appl. Math. **40** (1987), 659–662.

[CIL] M.G. Crandall, H. Ishii and P.L. Lions *User's guide to viscosity solutions of second order partial differential equations*, Bull. Amer. Math. Soc. (N. S.) **27** (1992), 1–67.

[CFG] V. Cruceanu, P. Fortuny and P. Gadea, *A survey on paracomplex geometry*, Rocky Mountain J. Math. **26** (1996), 83–115.

[D] Y. Dong, *On indefinite special Lagrangian submanifolds in indefinite complex Euclidean spaces*, Journal of Geometry and Physics **59** (2009), 710–726.

[EST] F. Etayo, R. Santamaría amd U. Trías *The geometry of a bi-lagrangian manifold*, Diff. Geom. Appl. **24** (2006). 33–59.

[F] H. Federer, Geometric Measure Theory, Springer-Verlag, New York, 1969.

[Fu] L. Fu, *On the boundaries of Special Lagrangian submanifolds*, Duke Math. J. **79** no. 2 (1995), 405–422.

[GM] P. Gadea and A. Montesinos Amilibia, *Spaces of constant para-holomorphic sectional curvature*, Pacific J. Math. **136** (1989), 85–101.

[GGM] P. Gadea, J. Grifone and J. Muñoz Masqué, *Manifolds modelled over free modules over the double numbers*, Acta. Math. Hungar. **100** (2003). 187–203.

[H] F.R. Harvey, Spinors and Calibrations, Perspectives in Math. vol. 9, Academic Press, Boston, 1990.

[HL_1] F.R. Harvey and H.B. Lawson, Jr, *Calibrated geometries*, Acta Mathematica **148** (1982), 47–157.

[HL_2] F.R. Harvey and H.B. Lawson, Jr., *An introduction to potential theory in calibrated geometry*, Amer. J. Math. **131** no. 4 (2009), 893–944. ArXiv:math.DG/0710.3920.

[HL_3] F.R. Harvey and H.B. Lawson, Jr., *Duality of positive currents and plurisubharmonic functions in calibrated geometry*, Amer. J. Math. **131** no. 5 (2009), 1211–1240. ArXiv:math.DG/0710.3921.

[HL_4] F.R. Harvey and H.B. Lawson, Jr., *Dirichlet duality and the non-linear Dirichlet problem*, Comm. on Pure and Applied Math. **62** (2009), 396–443.

[HL_5] F.R. Harvey and H.B. Lawson, Jr., *Plurisubharmonicity in a general geometric context*, ArXiv 0804.1316.

[HL_6] F.R. Harvey and H.B. Lawson, Jr., *Lagrangian plurisubharmonicity and convexity*, Stony Brook Preprint (2007).

[HS] F.R. Harvey and B. Shiffman, *A characterization of holomorphic chains*, Ann. of Math., **99** (1974), 553–587.

[Ha_1] M. Haskins, *Constructing special lagrangian cones*, Ph.D. Thesis, Univ. of Texas at Austin, 2000.

[Ha_2] M. Haskins, *Special lagrangian cones*, Amer. J. Math **126** (2004), 845–871.

[Ha_3] M. Haskins, *The geometric complexity of Special lagrangian T^2-cones*, Invent. Math. **157** (2004), 11–70.

[HK_1] M. Haskins and N. Kapouleas, *Special lagrangian cones with higher genus links*, Invent. Math. **167** (2007), 223–294.

[HK_2] M. Haskins and N. Kapouleas, *Gluing constructions of special Lagrangian cones.* Handbook of geometric analysis. No. 1, 77145, Adv. Lect. Math. (ALM), 7, Int. Press, Somerville, MA, 2008.

[Hi_1] N. Hitchin, *The moduli space of special Lagrangian submanifolds*, Ann. Scuola Norm. Sup. Pisa Cl. Sci. (4) **25** (1997), 503–515.

[Hi_2] N. Hitchin, *Generalized Calabi-Yau manifolds*, Quart. J. Math. **54** (2003), 281–308.

[IZ] S. Ivanov and S. Zamkovoy, *Parahermitian and paraquaternionic manifolds*, Diff. Geom. and its Applications **23** (2005), 205–234.

[J_1] D. Joyce, pp. 163–198 in "Different Faces of Geometry", Int. Math. Ser. (N.Y.), 3, Kluwer/Plenum, New York, 2004.

[J_2] D. Joyce, *U(1)-Invariant special Lagrangian 3-folds. III. Properties of singular solutions*, Adv. Math. **192** (2005), 135–182.

[KM] Y.-H. Kim and R. J. McCann, *Continuity, curvature, and the general covariance of optimal transportation*, J. Eur. Math. Soc. (JEMS) 12 (2010), no. 4, 1009–1040.

[KMW] Y.-H. Kim, R.J. McCann and M. Warren, *Pseudo-Riemannian geometry calibrates optimal transportation.* Math. Res. Lett. 17 (2010), no. 6, 1183–1197.

[Mc_1] R.C. McLean, *Deformations of Calibrated Submanifolds*, Duke University Thesis, 1996.

[Mc_2] R.C. McLean, *Deformations of Calibrated Submanifolds*, Comm. Anal. Geom. **6** No. 4 (1998), 705–747.

[M_1] J. Mealy, Calibrations on semi-riemannian manifolds, Ph.D. Thesis, Rice University, 1989.

[M_2] J. Mealy, Volume maximization in semi-riemannian manifolds, Indiana Univ. Math. J. **40** (1991), 793–814.

[P_1] A.V. Pogorelov, *Monge-Ampère equations of elliptic type*, Noordhoff, Groningen, 1964.

[P_2] A.V. Pogorelov, *Extrinsic geometry of convex surfaces*, Translations AMS, Vol XXXV, 1973.

[RT] J.B. Rauch and B.A. Taylor, *The Dirichlet problem for the multidimensional Monge-Ampère equation*, Rocky Mountain J. Math **7** (1977), 345–364.

[SS] L. Schäfer and F. Schulte-Hengesbach, *Nearly pseudo-Kähler and nearly para-Kähler six-manifolds.* Handbook of pseudo-Riemannian geometry and supersymmetry, 425–453, IRMA Lect. Math. Theor. Phys., 16, Eur. Math. Soc., Zürich, 2010.

[S] B. Shiffman, *Complete characterization of holomorphic chains of codimension one*, Math. Ann. **274** (1986), 233–256.

[TU] N.S. Trudinger and J.n I.E. Urbas, *Second derivative estimates for equations of Monge-Ampère type*, Bull. Austral. Math. Soc. **30** (1984), 321–334.

[V] C. Villani, Optimal Transport: Old and New, Grundlehren der mathematischen Wissenschaften vol. 338, Springer, 2009.

[W$_1$] M. Warren, *Calibrations associated to Monge-Ampère equations*, Trans. Amer. Math. Soc. 362 (2010), no. 8, 3947–3962.

[W$_2$] M. Warren, *A McLean theorem for the moduli space of Lie solutions to mass transport equations*, ArXiv:1006.1334.

[Wo] J. Wood, *Foliations on 3-manifolds*, Annals of Math. **89** (1969), 336–358.

[Y] Yu Yuan, *A priori estimates for solutions of fully nonlinear special lagrangian equations*, Ann Inst. Henri Poincaré **18** (2001), 261–270.

F. Reese Harvey
Mathematics Department
Rice University
Houston, TX 77005-1892, USA
e-mail: `harvey@rice.edu`

H. Blaine Lawson, Jr.
Mathematics Department
Stony Brook University
Stony Brook, NY11794-3651, USA
e-mail: `blaine@math.sunysb.edu`

How Riemannian Manifolds Converge

Christina Sormani

Dedicated to Jeff Cheeger for his 65th birthday

Abstract. This is an intuitive survey of extrinsic and intrinsic notions of convergence of manifolds complete with pictures of key examples and a discussion of the properties associated with each notion. We begin with a description of three extrinsic notions which have been applied to study sequences of submanifolds in Euclidean space: Hausdorff convergence of sets, flat convergence of integral currents, and weak convergence of varifolds. We next describe a variety of intrinsic notions of convergence which have been applied to study sequences of compact Riemannian manifolds: Gromov-Hausdorff convergence of metric spaces, convergence of metric measure spaces, intrinsic Flat convergence of integral current spaces, and ultralimits of metric spaces. We close with a speculative section addressing possible notions of intrinsic varifold convergence, convergence of Lorentzian manifolds and area convergence.

Mathematics Subject Classification (2000). 53C23.

Keywords. Intrinsic flat, Gromov-Hausdorff, integral currents, varifolds.

1. Introduction

The strong notions of smoothly or Lipschitz converging manifolds have proven to be exceptionally useful when studying manifolds with curvature and volume bounds, Einstein manifolds, isospectral manifolds of low dimensions, conformally equivalent manifolds, Ricci flow and the Poincaré conjecture, and even some questions in general relativity. *However many open questions require weaker forms of convergence that do not produce limit spaces that are manifolds themselves.* Weaker notions of convergence and new notions of limits have proven necessary in the study of manifolds with no curvature bounds or only lower bounds on Ricci or scalar curvature, isospectral manifolds of higher dimension, Ricci flow through singularities, and general relativity. Here we survey a variety of weaker notions of convergence

Research funded in part by PSC CUNY and NSF DMS 10060059.

and the corresponding limit spaces covering both well-established concepts, newly discovered ones and speculations.

We begin with the convergence of submanifolds of Euclidean space as there is a wealth of different weak kinds of convergence that mathematicians have been applying for almost a century. Section 2.1 covers the *Hausdorff convergence of sets*, Section 2.2 covers Federer-Fleming's *flat convergence of integral currents*, and Section 2.3 covers Almgren's *weak convergence of varifolds*. Each has its own kind of limits and preserves different properties. Each is useful for exploring different kinds of problems. While Hausdorff convergence is most well known in the study of convex sets, flat convergence in the study minimal surfaces and varifold convergence in the study of mean curvature flow, each has appeared in diverse applications. Keep in mind that these are extrinsic notions of convergence which depend upon how a manifold is located within the extrinsic space.

It is natural to believe that each should have a corresponding intrinsic notion of convergence which should prove useful for studying corresponding intrinsic questions about Riemannian manifolds which do not lie in a common ambient space. The second section vaguely describes some of these intrinsic notions of convergence with pictures illustrating key examples. Section 3.1 covers *Gromov-Hausdorff convergence of metric spaces* which is an intrinsic Hausdorff distance, Section 3.2 covers various notions of *convergence of metric measure spaces*, Section 3.3 covers the *intrinsic flat convergence on integral current spaces*, and Section 3.4 covers the weakest notion of all: *ultralimits of metric spaces*. Illustrated definitions and key examples will be given for each kind of convergence. This survey is not meant to provide a thorough rigorous definition of any of these forms of convergence but rather to provide the flavor of each notion and direct the reader to further resources.

The survey closes with speculations on new notions of convergence: Section 4.1 describes difficulties arising when attempting to define an intrinsic varifold convergence. Section 4.2 describes the possible notion of area convergence and area spaces. Section 4.3 discusses the importance of weaker notions of convergence for Lorentzian manifolds. The author attempts to include all key citations of initial work in these directions.

The author apologizes for the necessarily incomplete bibliography and encourages the reader to consult mathscinet and the arxiv for the most recent results in each area. For some of the older forms of convergence entire textbooks have been written focussing on one application alone.

2. Converging submanifolds of Euclidean space

There are numerous textbooks written covering the three notions of extrinsic convergence we describe in this section. Textbooks covering all three notions include [48] and [45]. Morgan's text providing pictures and overviews of proofs with precise references to theorems and proofs in Federer's classic text [28]. The focus is on minimal surface theory. Another source with many pictures and intuition regarding Plateau's problem, is Almgren's classic textbook [6]. Lin-Yang's textbook [45] covers a wider variety of applications and provides a more modern perspective incorporating recent work of Ambrosio-Kirchheim. An excellent resource on varifolds is Brakke's book [11] which is freely available on his webpage. Simon's classic text [57] is another indispensable resource.

2.1. Hausdorff convergence of sets

The notion of Hausdorff distance dates back to the early 20th century. Here we define it on an arbitrary metric space, (Z, d_Z), although initially it was defined on Euclidean space. The Hausdorff distance between subsets $A_1, A_2 \subset Z$ is

$$d_H^Z(A_1, A_2) := \inf\Big\{ R : A_1 \subset T_R(A_2),\ A_2 \subset T_R(A_1)\Big\}, \tag{1}$$

where $T_r(A) := \{y : \exists x \in A \text{ s.t. } d_Z(x,y) < r\}$. We write $A_j \xrightarrow{\text{H}} A$ iff $d_H^Z(A_j, A) \to 0$.

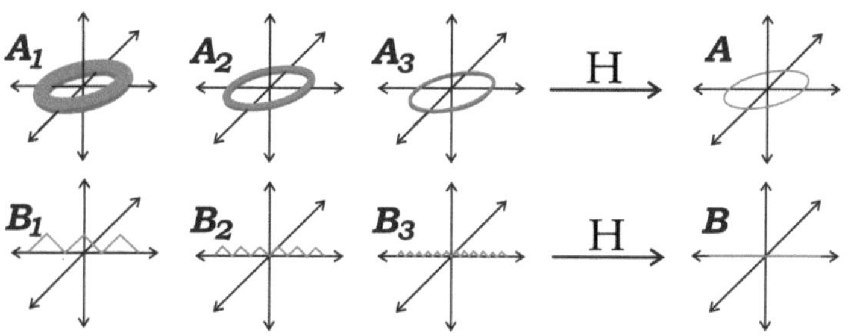

FIGURE 1. Hausdorff Convergence

In Figure 1, we see two famous sequences of submanifolds converging in the Hausdorff sense. The sequence A_i are tori which converge to a circle A, depicting how a sequence may lose both topology and dimension in the limit. Sequence B_i of jagged paths in the yz plane converges to a straight segment, B, in the y axis. Notice that $L(B_i) = \sqrt{2}$ while $L(B) = 1$. In addition to a sudden loss of length under a limit, we lose all information about the derivative (much like a C_0 limit). In fact one may construct a sequence of jagged curves that converge in the Hausdorff sense to the nowhere differentiable Weierstrass function.

One property which is preserved by Hausdorff convergence is convexity. For this reason, Hausdorff convergence has often been applied in the study of convex subsets of Euclidean space.

In 1916, Blaschke proved that if a sequence of nonempty compact sets K_i lie in a ball in Euclidean space, then a subsequence converges in the Hausdorff sense to a nonempty compact set [9]. If the sets are connected, then their limit is connected. Path connectedness, however, is not preserved as can be seen by a sequence approaching the well-known set:

$$\{(x, sin(1/x)) : x \in (0, 1]\} \cap \{(0, y) : y \in [-1, 1]\}. \tag{2}$$

Dimension and measure are also not well controlled. The k-dimensional Hausdorff measure of a set X is defined by covering X with countable collections of sets C_i of small diameter:

$$\mathcal{H}_k(X) := \lim_{r\to 0} \inf\left\{\sum_{i=1}^{\infty} \alpha(k)\left(\frac{\operatorname{diam}(C_i)}{2}\right)^k \, : \, X \subset \bigcup_{i=1}^{\infty} C_i, \ \operatorname{diam}(C_i) < r\right\}$$

where $\alpha(k)$ is the volume of a unit ball of dimension k in Euclidean space. For a beautiful exposition of this notion see [48]. When X is a submanifold of dimension k, then $\mathcal{H}_k(X)$ is just the k-dimensional Lebesgue measure: when $k = 1$ it is the length, when $k = 2$ it is the area and so on.

The Hausdorff dimension of a set, X, is

$$H_{\text{dim}}(X) := \inf\{k \in (0, \infty) : \mathcal{H}_k(X) = 0\}. \tag{3}$$

A compact k-dimensional submanifold, M^k, has $H_{\text{dim}}(M) = k$. Notice that in Figure 1, $H_{\text{dim}}(A_j) = 2$ while $H_{\text{dim}}(A) = 1$. It is also possible for the dimension to go up in the limit. In Figure 2, $H_{\text{dim}}(Y_j) = 2$ while $H_{\text{dim}}(Y) = 3$. It is also possible for a sequence of one-dimensional submanifolds, to converge to a space of fractional dimension like the von Koch curve. Such a sequence of curves must have length diverging to infinity.

In fact, if X_j have uniformly bounded length, $\mathcal{H}_1(X_j) < C$, and X_j converge to X in the Hausdorff sense, then

$$\mathcal{H}_1(X) \le \liminf_{j\to\infty} \mathcal{H}_1(X_j). \tag{4}$$

Note that the sequence B_j depicted in Figure 1 has $\mathcal{H}_1(B_j) = \sqrt{2}$ for all j, but $\mathcal{H}_1(B) = 1$, so lower semicontinuity is the best one can do. This is not true for higher dimensions as can be seen in Figure 2, where $\lim_{j\to\infty} \mathcal{H}_2(Y_j) = 1$ yet $\mathcal{H}_2(Y) = \infty$.

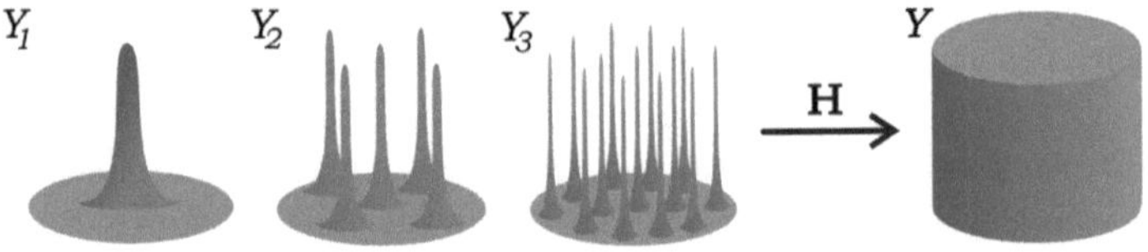

FIGURE 2. Disk of Many Splines

Figure 2 depicts a famous example of "a disk with splines" described in Almgren's text on Plateau's problem [6]. In Plateau's problem one is given a closed curve and is asked to find the surface of smallest area spanning that curve. In this example, the sequence of smooth surfaces, Y_j, have a common boundary, $\partial Y_j = \{(x, y, 0) : x^2 + y^2 = 1\}$ which is a given closed curve, and the area of the Y_j is approaching the minimal area filling that circle. However the sequence does not converge in the Hausdorff sense to the flat disk, D^2, which is the solution to the Plateau problem for a circle. Instead, due to the many splines, the sequence converges in the Hausdorff sense to the solid cylinder.

The disk with many splines example convinced mathematicians that the Hausdorff distance was not useful in the study of minimal surfaces. New notions of convergence had to be defined.

2.2. Flat convergence of integral currents

Federer and Fleming introduced the notion of an integral current and the flat convergence of integral currents to deal with examples like the disk with many splines depicted in Figure 2. When viewed as integral currents, the submanifolds depicted in that figure converge in the flat sense to the disk. All the splines disappear in the limit.

A current, T, is a linear functional on smooth k-forms. Any compact oriented submanifold, M, of dimension k with a smooth compact boundary, may be viewed as a k-dimensional current, T, defined by

$$T(\omega) := \int_M \omega. \tag{5}$$

Notice that in Figure 2, we have

$$\lim_{j\to\infty} \int_{Y_j} \omega = \int_{D^2} \omega \tag{6}$$

for any smooth differentiable 3-form, ω. So, viewed as currents, Y_j converge weakly to the flat disk D^2.

Federer-Fleming proved that any sequence of compact k-dimensional oriented submanifolds, M_j, in a disk in Euclidean space, with a uniform upper bound on $\mathcal{H}^k(M_j) \le V_0$ and a uniform upper bound on $\mathcal{H}^{k-1}(\partial M_j) \le A_0$, has a subsequence which converges when viewed as currents in the weak sense. The limit is an "integral current".

An integral current, T, is a current with a canonical set, R, and a multiplicity function θ. The canonical set R is countably $\mathcal{H}^k$ rectifiable, which means it is contained in the image of a countable collection of Lipschitz maps, $\varphi_i : E_i \to R$ from Borel subsets, E_i, of k-dimensional Euclidean space. The multiplicity function θ is an integer-valued Borel function. We define

$$T(\omega) := \int_R \theta\, \omega = \sum_{i=1}^{\infty} \int_{E_i} \theta \circ \varphi_i \;\; \varphi_i^* \omega. \tag{7}$$

It is further required that an integral current have finite mass

$$\mathrm{M}(T) := \int_R \theta \, d\mathcal{H}^k \tag{8}$$

and that the boundary, ∂T, defined by $\partial T(\omega) := T(d\omega)$ have finite mass, $\mathrm{M}(\partial T) < \infty$. Note that Federer-Fleming and Ambrosio-Kirchheim proved this implies that

∂T also has a rectifiable canonical set although it is one dimension lower than the dimension of T.[1]

When a submanifold M is viewed as an integral current T, then M itself is the canonical set, it has multiplicity 1, the boundary, ∂T, is just ∂M viewed as an integral current and the mass, $\mathrm{M}(T)$, is just the volume of M.

Also included as a k-dimensional integral current is the 0 current. In Figure 1, the sequence A_j may be viewed as integral currents. They converge in the weak sense as integral currents to the 0 current. More generally, if $\mathcal{H}^k(M_j)$ decreases to zero, then the weak limit of the k-dimensional submanifolds M_j viewed as integral currents is also the 0 current. In fact, whenever M_j have a uniform upper bound on $\mathcal{H}^k(M_j)$ and $M_j = \partial N_j$ where $\mathcal{H}^{k+1}(N_j) \to 0$, then M_j viewed as integral currents also converge weakly to the 0 current. The sequence A_j depicted in Figure 1 are the boundaries of solid tori whose volumes decrease to 0. It was not actually necessary that their areas decrease to 0. This idea of filling in the manifold to assess where it converges makes it much easier to see the limits of integral currents and leads naturally to the following definition.

The flat distance between two k-dimensional integral currents, T_1 and T_2, is defined by

$$d_{\mathcal{F}}(T_1, T_2) = \inf \left\{ \mathrm{M}(A) + \mathrm{M}(B) : \int \mathrm{curr}\, A, B \ s.t.\ A + \partial B = T_1 - T_2 \right\}, \tag{9}$$

where the infimum is taken over all k-dimensional integral currents, A, and all $k+1$-dimensional integral currents, B.[2]

In the figure to the right we see a choice of A with small area that looks like a catenoid, and then we fill in the space between the flat disk T_2 and the disk with many splines, T_1, to define B. If the surface with the splines actually shares the same boundary with the disk (as it does in Figure 2), then we can take A to be the 0 current and B just to be a filling (the region in between them). It is easy to see that the sequence of manifolds Y_j depicted in Figure 2 converges to the disk D^2 using the flat distance.

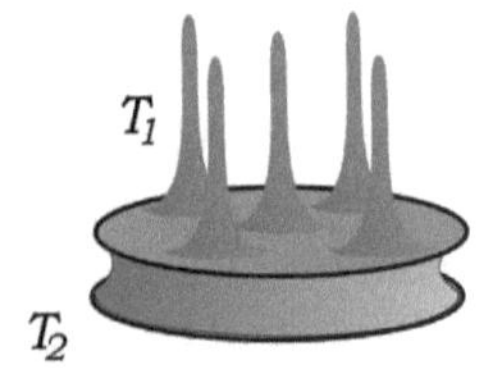

In Figure 1 the sequence B_j viewed as one-dimensional integral currents, T_j, converges to the Hausdorff limit B viewed as a one-dimensional current, T (as long as they are given the same orientation left to right). Here $T_j - T$ is again a cycle and we can find surfaces viewed as two-dimensional currents, S_j, such that $\partial S_j = T_j - T$ whose areas, $\mathrm{M}(S_j)$, converge to 0.

Federer-Fleming proved that when a sequence of integral currents has a uniform upper bound on mass and on the mass of their boundaries, then they converge weakly iff they converge with respect to the flat distance. Their compactness theorem can now be restated as follows: *if T_j is a sequence of k-dimensional integral*

[1] See [48] and [45] for more details. Note there are slight differences in the definition as [48] follows Federer-Fleming [29], while [45] follows the newer version introduced by Ambrosio-Kirchheim [7].
[2] See prior footnote.

currents supported in a compact subset with $M(T_j) \le V_0$ *and* $M(\partial T_j) \le A_0$ *then a subsequence converges in the flat sense to an integral current space.*

One of the beauties of this theorem is that the limit space is rectifiable with finite mass, and in fact the mass is lower semicontinuous. This makes flat convergence an ideal notion when studying Plateau's problem. The limit integral current has the same boundary as the sequence and minimal area.

In addition integral currents have a notion of an approximate tangent plane, which exists almost everywhere and is a subspace of the same dimension as the current. When M is a submanifold, this approximate tangent plane is the usual tangent plane and exists everywhere inside M. In 1966 Almgren was able to apply the strong control on the approximate tangent planes to obtain even stronger regularity results for the limits achieved when working on Plateau's problem [4]. The limit space ends up having multiplicity 1. In general the limit space may have higher multiplicity or regions with higher multiplicity [Figure 3], which we will discuss in more detail later.

In 1999, Ambrosio-Kirchheim extended the notion of an integral current from Euclidean space to arbitrary metric spaces, proving the existence of a solution to Plateau's problem on Banach spaces [7]. A key difficulty was that on a metric space, Z, there is no notion of a differential form. They applied a notion of DeGiorgi [27], replacing a k-dimensional differential form like $\omega = f dx_1 \wedge dx_2 \wedge \cdots \wedge dx_k$ with a $k+1$ tuple, $\omega = (f, x_1, x_2, \ldots, x_k)$, of Lipschitz functions satisfying a few rules, including $d\omega := (1, f, x_1, x_2, \ldots, x_k)$.

An integral current, T, is defined as a linear functional on $k+1$ tuples using a rectifiable set, R, and a multiplicity function, θ, so that as in (7), we have

$$T(\omega) := \sum_{i=1}^{\infty} \int_{E_i} \theta \circ \varphi_i \ \varphi_i^* \omega. \tag{10}$$

where $\varphi_i^* \omega := f \circ \varphi \, d(x_1 \circ \varphi) \wedge d(x_2 \circ \varphi) \wedge \cdots \wedge d(x_k \circ \varphi)$ is defined almost everywhere by Rademacher's Theorem. They can then define mass exactly as in (8) and boundary exactly as before as well. They require integral currents to have bounded mass and their boundaries to have bounded mass.[3]

Ambrosio-Kirchheim prove that if T_j are integral currents in a compact metric space and $\mathrm{M}(T_j) \le V_0$ and $\mathrm{M}(\partial T_j) \le A_0$ then a subsequence converges in the weak sense to an integral current. Furthermore the mass is lower semicontinuous. There is also a notion of an approximate tangent plane, however, here the approximate tangent plane is a normed space. The norm is defined using the metric differential described in earlier work by Korevaar-Schoen on harmonic maps and also by Kirchheim on rectifiable space [44], [42].

Wenger extended the class of metric spaces Z which have a solution to Plateau's problem and defined a flat distance exactly as in (9). We will use the notation $d_{\mathcal{F}}^Z$ for the flat distance in Z to be consistent with d_H^Z denoting the Hausdorff distance in Z. He proved that on this larger class of spaces, which includes

[3]See [7] or [62] for the precise definition.

Banach spaces, when $\mathbf{M}(T_j) \le V_0$ and $\mathbf{M}(\partial T_j) \le A_0$ then T_j converge weakly to T iff T_j converges in the flat sense to T [69], [70]. Recall that by Rademacher's Theorem, any separable metric space, Z, isometrically embeds into a Banach space, W, so we can always study the convergence of integral currents in Z using the flat distance of the push forwards of the currents in Banach space, $d_{\mathcal{F}}^{\mathrm{W}}$. If $\varphi : Z \to \mathrm{W}$, then the push forward of a current T in Z, is a current $\varphi_\# T$ in W defined by $\varphi_\# T\,(f, x_1, \dots, x_k) := T\,(f\circ\varphi, x_1\circ\varphi, \dots, x_k\circ\varphi)$.

Another extension of the notion of integral currents is that of the $\mathbb{Z}_n$ integral currents where the multiplicity function takes values in $\mathbb{Z}_n$. This was introduced by Fleming [32] and has been extended to arbitrary metric spaces by Ambrosio and Wenger [8]. One application for this notion is the study of minimal graphs (cf. [6]). To understand this better we will go over an example using Federer-Fleming's notation.

In Figure 3, the sequence of embedded curves C_i converges in C^1 to a limit curve C which is not embedded. Viewed as one-dimensional currents, T_i, they converge in the flat sense to T. It is perhaps easier to understand Figure 3 using weak convergence rather than flat convergence. Note that

$$T_i(\,f_1(x,y)dx + f_2(x,y)dy\,) = \int_0^1 (f_1\circ C_i)(t)\,d(x\circ C_i) + (f_2\circ C_i)(t)\,d(y\circ C_i)$$

converges to

$$T(\,f_1(x,y)dx + f_2(x,y)dy\,) \;=\; \int_0^1 (f_1\circ C)(t)\,d(x\circ C) + (f_2\circ C)(t)\,d(y\circ C).$$

Notice that T has multiplicity 1 where C does not overlap itself. It has multiplicity 2 on the segment C covers twice in the same direction. The segment depicted with dashes where C passes in opposite directions is not part of the canonical set of T because the integration cancels there. If one uses $\mathbb{Z}_n$-valued coefficients we get the same limit space for $n > 2$, but for $n = 2$ the doubled up segment disappears as well as the cancelled dashed segment.

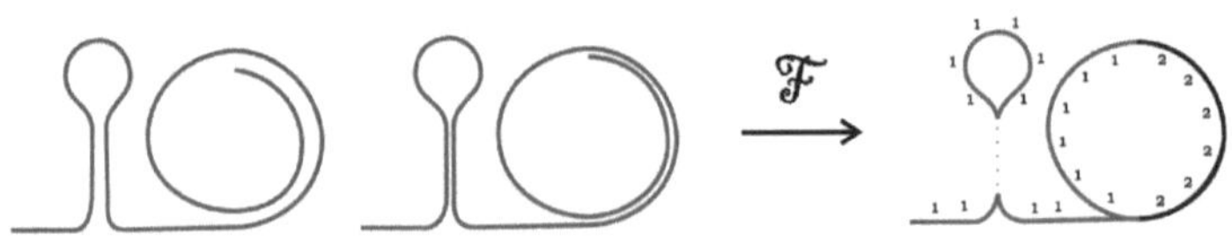

FIGURE 3. Doubling and Cancellation of Flat Limits

The cancellation depicted in Figure 3 is problematic for some applications. Note that the same effect can occur in higher dimensions, by extending the curve to a sheet in $\mathbb{R}^3$. While often it is useful for thin splines to disappear in the hopes of preserving dimension, cancelling sheets which overlap with opposing orientations is not necessary to obtain a rectifiable limit space. This is seen using the notion of a varifold.

2.3. Weak convergence of varifolds

Varifolds were introduced by Almgren in 1964 [5]. Significant further work was completed by Allard in 1972 [3]. Almgren's goal was to define a new notion of convergence for submanifolds which had rectifiable limit spaces and a notion of tangent planes for those limit spaces, but did not have the kind of cancellation that occurs when taking flat limits of integral currents. Under varifold convergence, the sequence of one-dimensional manifolds depicted in Figure 3, converges to the rectifiable set with weight 1 everywhere including the dashed segment. There is no notion of orientation on the limit, but there are tangent planes almost everywhere.

A k-dimensional varifold is a Radon measure on $\mathbb{R}^N \times \Gamma(k,N)$ where $\Gamma(k,N)$ is the space of k subspaces of Euclidean space, $\mathbb{R}^N$.[4] A submanifold $M^k \subset \mathbb{R}^n$, may be viewed as the varifold, V, defined on any $W \subset \mathbb{R}^N \times \Gamma(k,N)$ as follows:

$$V(W) := \mathcal{H}^k\left(W \cap \{(x, T_xM) : x \in M \subset \mathbb{R}^m\}\right), \tag{11}$$

where T_xM is the tangent space to x translated to the origin.

A sequence of varifolds, V_j, is said to converge weakly to a varifold, V,

$$\int f\, dV_j \to \int f\, dV \qquad \forall f \in C_0(\mathbb{R}^N \times \Gamma(k,N)). \tag{12}$$

For example, let us examine the sequence of curves C_j which converge C^1 to the curve C in Figure 3. They may be viewed as integral currents V_j,

$$V_j(W) = \mathcal{H}^1\left(W \cap \left\{\left(C_j(t), \pm C_j'(t)/|C_j'(t)|\right) : t \in [0,1]\right\}\right), \tag{13}$$

where we view points in $\Gamma(1,2)$ as $\pm v$ where v in a unit 2 vector. Then

$$\int f(x,v)dV_j = \int_0^1 f(C_j(t), \pm C_j'(t)/|C_j'(t)|)\, dt \tag{14}$$

converges to $\int_0^1 f(C(t), \pm C(t)/|C(t)|)dt = \int f(x.v)\, dV$ for the varifold V defined by

$$V(W) := \int_0^1 \chi_W(C(t), \pm C'(t)/|C'(t)|)\, dt \tag{15}$$

where χ_W is the indicator function of W. This varifold corresponds to viewing the limit curve C as having weight 1 on the segments where it doesn't overlap itself and weight 2 on the segments where it does overlap. There is no cancellation. The dashed segment has weight 2.

Not all limit varifolds end up with tangent planes that align well with the rectifiable set as they do in (15). If we examine instead the sequence B_j in Figure 1, as a sequence of curves C_j converging to a curve C, we see that the limit varifolds, V, has the form

$$V(W) := a\int_0^1 \chi_W(C(t), \pm a v_1)\, dt + a\int_0^1 \chi_W(C(t), \pm a v_2)\, dt \tag{16}$$

[4]In some books $\Gamma(k,N)$ is written as $GL(k,N)$

where $v_1 = (1,1)$, $v_2 = (-1,1)$ and $a = \sqrt{2}/2$ because $C_i'(t)$ is always in the direction of v_1 or v_2. This effect was observed by Young and is discussed at length in [45]. Varifolds like V are of importance but are not considered to be integral varifolds because $C'(t)$ is unrelated to v_1 and v_2.

An integral varifold, V, is defined as a positive integer weighted countable sum of varifolds, V_j, which are defined by embedded submanifolds M_j as in (11). While V is a measure on $\mathbb{R}^N \times \Gamma(N,k)$, it defines a natural measure $\|V\|$ on $\mathbb{R}^N$,

$$\|V\|(A) := V(A \times \Gamma(k,N)). \tag{17}$$

When V is defined by a submanifold, M, then $\|V\|(A) = \mathcal{H}^k(A \cap M)$.

A k-dimensional varifold V is said to have a tangent plane, $T \in \Gamma(k,N)$ with multiplicity $\theta \in (0,\infty)$ at a point $x \in \mathbb{R}^N$ if a sequence of rescalings of V about the point x converges to $\theta(x)T$.[5] An integral varifold, V, has a tangent space $\|V\|$ almost everywhere on $\mathbb{R}^N$.

Varifolds do not have a notion of boundary. Instead a varifold, V, has a notion of first variation, δV, which is a functional that maps functions $f \in C_0^\infty(\mathbb{R}^N, \mathbb{R}^N)$ to $\mathbb{R}$. There is no room here to set up the background for a precise definition[6], so we just describe δV when V is defined by a C^2 submanifold M with boundary ∂M. In this case, $\delta V(f)$ is the first variation in the area of M as it is flowed through a diffeomorphism defined by the vector field f:

$$\delta V(f) = -\int_M f(x)\cdot H(x)\, d\mathcal{H}^k(x) + \int_{\partial M} f(x)\cdot \eta(x)\, d\mathcal{H}^{k-1}(x) \tag{18}$$

where $H(x)$ is the mean curvature at $x \in M$ and $\eta(x)$ is the outward normal at $x \in \partial M$.

A varifold, V, is said to be stationary when $\delta V = 0$.

A varifold has bounded first variation if

$$\|\delta V\|(W) := \sup\{\delta V(f) : f \in C_0^\infty(\mathbb{R}^N, \mathbb{R}^n),\ |f| \le 1,\ \mathrm{spt}(f) \subset A\} < \infty. \tag{19}$$

When V corresponds to a submanifold, M, as above, then

$$\|\delta V\|(W) = \mathcal{H}^{k-1}(W \cap \partial M) + \int_{W\cap M} |H(x)|\, d\mathcal{H}^m(x). \tag{20}$$

There is an isoperimetric inequality for varifolds.

Allard proved that if a sequence of integral varifolds, V_j, with a uniform bound on $\|\delta V_j\|(W)$ depending only on W, converges weakly to V, then V is an integral varifold as well [3]. In particular a weakly converging sequence of minimal surfaces, M_j, with a uniform upper bound on the length of their boundaries, ∂M_j, converges to an integral varifold. For more general submanifolds, one needs only uniformly control the volumes of the boundaries and the L^1 norms of the mean curvatures.

---

[5] See [45] for a more precise definition.

[6] See [45] 6.2 for the full definition.

One key advantage of varifolds is that they have a notion of mean curvature defined using δV and an integral similar to the one in (18). This notion is used to define the Brakke flow, a mean curvature flow past singularities [11].

Recently White has set up a natural map, F, from integral varifolds to $\mathbb{Z}_2$ flat chains, which basically preserves the rectifiable set and takes the integer-valued weight to a $\mathbb{Z}_2$ weight. Each $\mathbb{Z}_2$ flat chain corresponds uniquely to a $\mathbb{Z}_2$ integral current. He has proven that if a sequence of submanifolds, M_j viewed as varifolds converge weakly to an integral varifold V and satisfy the conditions of Allard's compactness theorem, and if further ∂M_j converge as $\mathbb{Z}_2$ integral currents, then M_j viewed as $\mathbb{Z}_2$ integral currents converge to a $\mathbb{Z}_2$ integral current T corresponding to $F(V)$. This is easily seen to be the situation for the sequence depicted in Figure 3 [72].

3. Intrinsic convergence of Riemannian manifolds

When studying sequences of Riemannian manifolds, M_j, the strongest notions of convergence require that M_j be diffeomorphic to the limit space M with the metrics, g_j, converging smoothly:

$$\exists \varphi_j : M \to M_j \text{ such that } \varphi_j^* g_j \to g. \tag{21}$$

In this survey we are concerned with sequences of Riemannian manifolds which do not converge in such a strong sense. Here we describe a few weaker notions of convergence which allow us to better understand sequences which do not converge strongly. We begin with a pair of motivating examples.

The sequence of flat tori, $M_j = S^1_\pi \times S^1_{\pi/j}$, has volume converging to 0: $\mathrm{vol}(M_j) = 2\pi(2\pi/j) \to 0$. Sequences with this property are called collapsing sequences. They do not converge in a strong sense to a limit which is also a torus. Intuitively one would hope to define a weaker notion of convergence in which these tori converge to a circle. Examples like this lead to Gromov's notion of an intrinsic Hausdorff convergence.

If one views Figure 2 as a sequence of Riemannian disks with splines, M_j, with the induced Riemannian metrics, they do not converge in a strong sense to a flat Riemannian disk. While they are diffeomorphic to the disk and the volumes are converging, $\mathrm{vol}(M_j) \to \mathrm{vol}(D^2)$, the metrics do not converge smoothly. Examples like this lead to the notion of intrinsic flat convergence.

In this section we first present Gromov-Hausdorff convergence, then metric measure convergence, then the intrinsic flat convergence and finally, weakest of all, the notion of an ultralimit. We include a few key examples, applications and further resources for each notion.

3.1. Gromov-Hausdorff convergence of metric spaces

In 1981, Gromov introduced an intrinsic Hausdorff convergence for sequences of metric spaces [39]. A few excellent references are Gromov's book [40], the textbook of Burago-Burago-Ivanov[15], Fukaya's survey [36] and Bridson-Haefliger's book [12].

The Gromov-Hausdorff distance is defined between any pair of compact metric spaces,

$$d_{GH}(M_1, M_2) = \inf \left\{ d_H^Z\left(\varphi_1(M_1), \varphi_2(M_2)\right) : \text{ isom } \varphi_i : M_i \to Z \right\} \tag{22}$$

where the infimum is taken over all metric spaces, Z, and all isometric embeddings, $\varphi_i : M_i \to Z$. An isometric embedding, $\varphi : X \to Z$ satisfies

$$d_Z(\varphi(x_1), \varphi(x_2)) = d_X(x_1, x_2) \qquad \forall x_1, x_2 \in X. \tag{23}$$

We write $M_i \xrightarrow{\text{GH}} X$ iff $d_{GH}(M_j, X) \to 0$. See Figure 4.

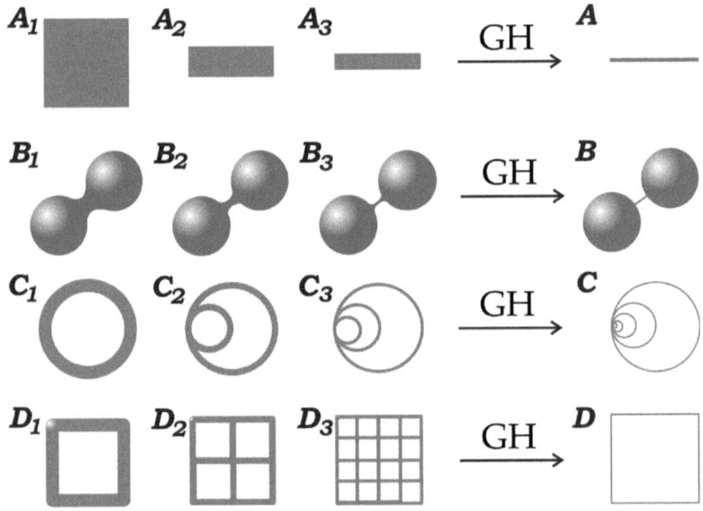

FIGURE 4. Gromov-Hausdorff Convergence

The sequences of Riemannian manifolds depicted in Figure 4 reveal a variety of properties that are not conserved under Gromov-Hausdorff convergence. The first sequence A_j are the flat tori $S^1_\pi \times S^1_{\pi/j}$ converging to a circle $A = S^1_\pi$. To see this one takes the common space $Z_j = A_j$ and isometrically embeds A into Z_j so that

$$d_{GH}(A_j, A) \leq d_H^{Z_j}(A_j, \varphi_j(A)) = \pi/(2j) \to 0. \tag{24}$$

Here we see the topology and Hausdorff dimension may decrease in the limit.

Note that if M_1 and M_2 are compact then $d_{GH}(M_1, M_2) = 0$ iff M_1 and M_2 are isometric. The Gromov-Hausdorff distance between M_1 and M_2 is almost 0 iff there is an almost isometry $f : M_1 \to M_2$ satisfying

$$|d_1(x, y) - d_2(f(x), f(y))| < \epsilon \text{ and } M_2 \subset T_\epsilon(f(M_1)). \tag{25}$$

Note that an almost isometry need not be continuous. In Figure 4, it is easy to construct ϵ_j almost isometries, $\varphi_j : B \to B_j$, such that $\epsilon_j \to 0$ and conclude that $B_j \stackrel{\text{GH}}{\longrightarrow} B$. This example reveals that the Gromov-Hausdorff limit of a sequence of Riemannian manifolds may not be a Riemannian manifold.

Gromov-Hausdorff limits of Riemannian manifolds are geodesic metric spaces. This means that the distance between any pair of points is equal to the length of the shortest curve between them. The shortest curve exists and is called a minimal geodesic. As in Riemannian geometry, a curve γ is called a geodesic if for every t, there is an $\epsilon > 0$ such that γ restricted to $[t-\epsilon, t+\epsilon]$ is a minimal geodesic.[7]

In the third sequence of Figure 4, C_j are also the boundaries of increasingly thin tubular neighborhoods. The limit, D, is the Hawaii Ring, a metric space of infinite topological type that has no universal cover. For more about the topology of Gromov-Hausdorff converging sequences of manifolds see [60] [61] and [54].

In the last sequence of Figure 4, D_j are the smoothed boundaries of increasing thin tubular neighborhoods of increasingly dense grids. They converge to a square $D = [0,1] \times [0,1]$. The metric on the square is the taxicab metric (also called the l_1 metric):

$$d_C((x_1, y_1), (x_2, y_2)) := |x_1 - x_2| + |y_1 - y_2|. \tag{26}$$

This is easiest to see by showing the D_j are Gromov-Hausdorff close to their grids and that the grids converge to the taxicab square.

Gromov proved that sequences of Riemannian manifolds, M_j, with uniform upper bounds on their diameter and on the number, $N(r)$, of disjoint balls of radius r have subsequences which converge in the Gromov-Hausdorff sense to a compact geodesic space, Y [39]. Conversely, when M_j converge to a compact Y, $N(r)$ is uniformly bounded. Consequently, if one views the sequence of disks with splines depicted in Figure 2 as Riemannian manifolds (with the intrinsic distance), the sequence does not converge in the Gromov-Hausdorff sense: a ball of radius $1/2$ about the tip of a spline does not intersect with the a ball of radius $1/2$ about the tip of another spline. As the number of splines approaches infinity so does the number of disjoint balls of radius $1/2$.

By the Bishop-Gromov volume comparison theorem, sequences of manifolds, M_j, with uniform lower bounds on Ricci curvature and upper bounds on diameter satisfy these compactness criteria [39]. This includes, for example, the sequence of flat tori, A_j, in Figure 4 as well as any other collapsing sequence of manifolds with bounded sectional curvature. This lead to a series of papers on the geometric properties of M_j with uniformly bounded sectional curvature and $\mathrm{vol}(M_j) \to 0$ by Cheeger-Gromov, Fukaya, Rong, Shioya-Yamaguchi and others [35], [24], [25], [55], [56] (cf. [53], [49]). Collapsing Riemannian manifolds with boundary have been studied by Alexander-Bishop and Cao-Ge [17], [1], [2]. This collapsing theory is an essential component of Perelman's proof of the Geometrization Conjecture using Hamilton's Ricci flow [49], [43], [16].

[7]Alexandrov space geometers use the term "geodesic" to refer to a "minimal geodesic" and "local geodesic" to refer to a geodesic.

In 1992, Greene-Petersen applied the notion of Gromov's filling volume to find a lower bound on the volume of a ball in a Riemannian manifold with a uniform geometric contractibility function. A geometric contractibility function is a function $\rho : (0, r_0] \to (0, \infty)$ with $\lim_{r\to 0} \rho(r) = 0$ such that any ball $B_p(r) \subset M$ is contractible in $B_p(\rho(r)) \subset M$. Applying Gromov's compactness theorem, one may conclude that a sequence M_j with a uniform geometric contractibility function and a uniform upper bound on volume, has a subsequence that converges in the Gromov-Hausdorff sense to a compact metric space. Note that in this setting, volume is uniformly bounded below so the sequence is not collapsing [37]. The limits of such sequences of spaces have been studied by Ferry, Ferry-Okun, Schul-Wenger and Sormani-Wenger [30], [31], [63] and [62].

Noncollapsing sequences of Riemannian manifolds with Ricci curvature bounded from below were studied by Colding and Cheeger-Colding in their work on almost rigidity in which they weakened the conditions of well-known rigidity theorems. One example of a rigidity theorem is the fact that if M^m has $\mathrm{Ricci} \geq (m-1)$ and $\mathrm{vol}(M^m) = \mathrm{vol}(S^m)$ then M^m is isometric to S^m. Colding proved a corresponding almost rigidity theorem which states that if M^m has $\mathrm{Ricci} \geq (m-1)$ then $\forall \epsilon > 0$, $\exists \delta_{m,\epsilon} > 0$ such that $\mathrm{vol}(M^m) \geq \mathrm{vol}(S^m) - \delta_{m,\epsilon}$ implies $d_{GH}(M^m, S^m) < \epsilon$. For a survey of such almost rigidity theorems see [26], [20] and Section 8 of [59]. The proofs generally involve an explicit construction of an almost isometry using distance functions and solutions to elliptic equations on the Riemannian manifolds.

When a sequence of M_j with uniform lower bounds on Ricci curvature does not collapse Cheeger-Colding proved the volume is continuous and the Laplace spectrum converges. Fukaya observed that when a sequence of Riemannian manifolds with uniform lower bounds on Ricci curvature is collapsing, then the Laplace spectrum does not converge. In Figure 5 there are two sequences of Riemannian manifolds A_i and B_i with nonnegative Ricci curvature converging to the same limit space $A = B = [0, 1]$ with the standard metric. The eigenvalues converge to real numbers,

$$\lambda_i^A := \lim_{j\to\infty} \lambda_i(A_j) \text{ and } \lambda_i^B := \lim_{j\to\infty} \lambda_i(B_j) \tag{27}$$

but $\lambda_i^A \neq \lambda_i^B$. While λ_i^B made sense as eigenvalues for $[0, 1]$, the other collection of numbers λ_i^A did not. The alternating sequence $\{A_1, B_1, A_2, B_2, \dots\}$ also converges in the Gromov-Hausdorff sense but the eigenvalues do not converge at all [33].

This example disturbed Fukaya in light of the successful isospectral compactness theorems of Osgood-Phillips-Sarnak, Brooks-Perry-Petersen and Chang-Yang [52], [13], [18] all of which imposed stronger conditions on the manifolds and involved smooth convergence or even conformal convergence of the sequences. This lead to Fukaya's notion of metric measure convergence.

3.2. Metric measure convergence of metric measure spaces

In 1987, Fukaya introduced the first notion of metric-measure convergence. A sequence of metric measure spaces (X_j, d_j, μ_j) converge in the metric measure sense to a metric measure space (X, d, μ) if there is a sequence of $1/j$ almost isometries, $f_j : X_j \to X$, such that push forwards of the measures, $f_{j*}\mu_j$, converge weakly to μ on X. Recall that $\varphi_{\#}\mu(A) := \mu(\varphi^{-1}(A))$.

In Figure 5, A_j and B_j are given probability measures, μ_{A_j} and μ_{B_j} proportional to $\mathcal{H}^2$. Then B_j converge in the metric measure sense to $[0,1]$ with the standard metric and $\mu_B(W) = \mathcal{H}^1(W)$. Meanwhile A_j converge in the metric measure sense to $[0,1]$ with the measure

$$\mu_A(W) := \int_W (2 - |2 - 4x|)\, dx. \tag{28}$$

Defining the Laplacian with respect to these measures, the spectrum for (A, μ_A) is $\{\lambda_i^A\}$ and the spectrum for (B, μ_B) is $\{\lambda_i^B\}$.[8]

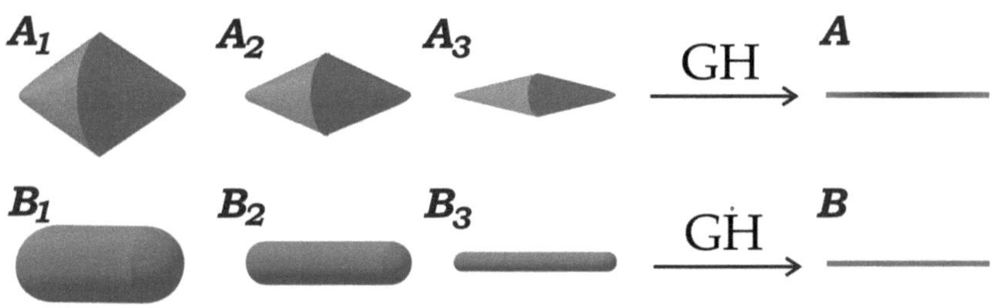

FIGURE 5. Metric Measure Convergence

Cheeger-Colding then proved that any sequence of compact Riemannian manifolds with uniform lower bounds on Ricci curvature endowed with a probability measure proportional to the Hausdorff measure, has a subsequence which converges in the metric measure sense. The measures on the limit space satisfy the Bishop-Gromov comparison theorem and are therefore doubling [21]. Sturm and Lott-Villani extended the notion of a Ricci curvature bound to general metric measure spaces using mass transport [64], [46] (cf. [68]). Recently Topping and others have been developing a notion of Ricci flow on this larger class of spaces in hopes of defining Ricci flow through a singularity [66].

Cheeger-Colding also prove a Poincaré inequality on these limit spaces and deduced that the eigenvalues converge [22], [23]. They proved that the limit spaces of a sequence of Riemannian manifolds with uniform lower bounds on Ricci curvature also have a notion of tangent plane almost everywhere. The tangent plane at a point is found by rescaling the space Y outward and taking a pointed Gromov-Hausdorff limit. At many points, one does not get a unique limit under rescaling and the limits are not necessarily planes or cones. However they do exist at every point and are called tangent cones. Cheeger-Colding proved tangent cones are Euclidean planes almost everywhere. In fact they prove Y is a countably $\mathcal{H}^m$ rectifiable space and is a $C^{1,\alpha}$ manifold away from the singularities. Key steps

[8] See [34] or [23] for more details.

in the proof involve the Splitting Theorem and the Poincaré inequality (cf. [19]). There is an example where the limit space has infinite topological type so it isn't a $C^{1,\alpha}$ manifold [47]. However, at least the manifold has a universal cover unlike the Hawaii Ring depicted in Figure 4 [60].

It is also natural to study metric measure convergence without Ricci curvature bounds on the sequence of manifolds. As long as the measure is doubling one can apply Gromov's compactness theorem. Sometimes the measure on the limit space is supported on a smaller set. This occurs for example for the sequence B_j of Figure 4. Viewed as a metric measure limit space, the limit space B has a measure which is supported on the two spheres.

In 1981, Gromov introduced $\square_\lambda$ convergence to handle this issue [40]. When $\lambda = 1$, the limit of the sequence of B_j is just the two spheres with the line segment removed. Basically, if (Y, d, μ) is the metric measure limit of a sequence, and if there is a ball such that $\mu(B_y(r)) = 0$ then y is removed from the space. This leaves us with a new, smaller limit space, $(\bar{Y}, d_Y, \mu)$, which may no longer be connected. This process may be called "reduction of measure". The spaces are then no longer close in the Gromov-Hausdorff topology.

In 2005, Sturm introduced a new distance between metric measure spaces which leads to such limits more naturally and also interacts well with mass transport notions mentioned above [65]. He uses the Wasserstein distance, W_p, to measure the distance between measures in a set Z and defines an intrinsic Wasserstein distance using an infimum over all metric spaces Z and all isometric embeddings $\varphi_i : X_i \to Z$:

$$d_{W_p}((X_1, d_1, \mu_1), (X_2, \mu_2, d_2)) := \inf\{d_{W_p}(\varphi_{1,\#}\mu_1, \varphi_{2\#}\mu_2)\}. \tag{29}$$

Sturm proved convergence with respect to this distance is equivalent to Gromov's $\square_1$ convergence [65].

Villani recently defined an intrinsic Prokhorov distance replacing the Wasserstein distance of order p in (29) with the Prokhorov distance between measures. As convergence with respect to the Wasserstein distance and with respect to the Prokhorov distance agree with weak convergence, the intrinsic Prokhorov and intrinsic Wasserstein limits agree as well. He refers to these kinds of convergence as "measure convergence" Two metric measure spaces are a zero distance apart with respect to these intrinsic measure distances iff there is a measure preserving isometry between them [68].

Villani also described "metric measure distances". The Gromov-Hausdorff Wasserstein distance is defined: $d_{W_p}((X_1, d_1, \mu_1), (X_2, d_2, \mu_2)) :=$

$$= \inf\{d_H(\varphi_1(X_1), \varphi_2(X_2)) + d_{W_p}(\varphi_{1\#}\mu_1, \varphi_{2\#}\mu_2)\}, \tag{30}$$

where the infimum is taken over all metric spaces Z and all isometric embeddings $\varphi_i : X_i \to Z$. The Gromov-Hausdorff Prokhorov distance has the same formula replacing the Wasserstein distance by the Prokhorov distance between measures. Convergence with respect to these metric measure distances agrees with Fukaya's metric measure convergence. The advantage of having multiple distances is that one may be easier estimate than another [68].

A sequence of metric measure spaces with doubling measures has a converging subsequence and the metric measure limit agrees with the measure limit of a sequence [68]. Any sequence of Riemannian manifolds with lower bounds on their Ricci curvature satisfies the doubling condition by the Bishop-Gromov volume comparison theorem. The sequence of B_j in Figure 4 does not, as the metric measure and measure limits do not agree.

Despite the immense success in applying these definitions of convergence to study manifolds with Ricci curvature bounds, there has been a need to introduce a weaker form of convergence to study sequences of manifolds which do not satisfy these strong conditions. Mathematicians studying manifolds with scalar curvature bounds and those interested only in sequences with an upper bound on volume and diameter without curvature bounds, need a weaker version of convergence. In particular, geometric analysis related to cosmology and the study of the spacelike universe requires a weaker form of convergence.

An important example introduced by Ilmanen may be called the 3-sphere of many splines. It is a sequence of three-dimensional spheres with positive scalar curvature whose volume converges to the volume of the standard three sphere, but has an increasingly dense set of thinner and thinner splines of "length" 1. Cosmologically, one may think of these splines as deep gravity wells. The sequence does not converge in the Gromov-Hausdorff sense because balls of radius $1/2$ centered on the tip of each spline are disjoint and the number of such disjoint balls approaches infinity. This example naturally lead Sormani and Wenger to develop a notion of intrinsic flat convergence.

3.3. Intrinsic flat convergence of integral current spaces

In 2008 Sormani and Wenger introduced the intrinsic flat distance between compact oriented Riemannian manifolds [63], [62]:

$$d_{\mathcal{F}}(M_1, M_2) := \inf \left\{ d_{\mathcal{F}}^Z(\varphi_{1\#}T_1, \varphi_{2\#}T_2) : \text{ isom } \varphi_i : M_i \to Z \right\} \tag{31}$$

where the flat distance in Z is defined as in (9), the $\varphi_i : M_i \to Z_i$ are isometric embeddings as in (23) and where T_i are defined by integration over M_i so that $\varphi_{i\#}T_i(f, x_1, \dots, x_k) =$

$$= T_i(f \circ \varphi_i, x_1 \circ \varphi_i, \dots, x_k \circ \varphi_i) = \int_{M_i} f \circ \varphi_i \; d(x_1 \circ \varphi_i) \wedge \cdots \wedge d(x_k \circ \varphi_i). \tag{32}$$

One may immediately note that $d_{\mathcal{F}}(M_1, M_2) \le \mathrm{vol}(M_1) + \mathrm{vol}(M_2) < \infty$ as we may always take the integral current A in (9) to be $\varphi_{1\#}T_1 - \varphi_{2\#}T_2$ and $B = 0$. The intrinsic flat distance is a distance between compact oriented Riemannian manifolds in the sense that $d_{\mathcal{F}}(M_1, M_2) = 0$ iff there is an orientation preserving isometry between M_1 and M_2 [62].

Note that in practice it is often possible to estimate the intrinsic flat distance using only notions from Riemannian geometry. That is, if M_1^k and M_2^k isometrically embed into a Riemannian manifold N^{k+1} such that $\partial N^{k+1} = \varphi_1(M_1^k) \cup \varphi_2(M_2^k)$ and the manifolds have been given an orientation consistent with Stoke's theorem on N^{k+1}, then $d_{\mathcal{F}}(M_1, M_2) \le \mathrm{vol}(N^{k+1})$ This viewpoint makes it quite easy to see that Ilmanen's 3-sphere of many splines example describes a sequence converging to the standard sphere. Note that one cannot just embed the sequence into four-dimensional Euclidean space and take N^4 to be the flat region lying between the two spheres because such φ_i would not be isometric embeddings. Instead, one rotates each spline into half a thin four-dimensional spline and glues it smoothly to a short $S^3 \times [0, \epsilon_j]$ where ϵ_j is "thinness" of the spline. This produces a four-dimensional manifold N^4 with S^3 isometrically embedded as one boundary and the 3-sphere of many splines isometrically embedded as the other boundary. So the intrinsic flat distance between these two spaces is $\le \mathrm{vol}(N^4)$ which is approximately $\epsilon_j \mathrm{vol}(S^3)$ plus the sum of the volumes of the four-dimensional splines each of which is approximately ϵ_j^3. See [62] for this example and many others.

The limit spaces are called integral current spaces, and are oriented weighted countably $\mathcal{H}^k$ rectifiable metric spaces. They have the same dimension as the sequence. They have a set of Lipschitz charts as well as a notion of an approximate tangent space. The approximate tangent space is a normed space whose norm is defined by the metric differential. The notion of the metric differential was developed in work of Korevaar-Schoen [44] and Kirchheim [42]. See [62] for a number of examples of limit spaces.

Integral current spaces are said to have a "current structure", T, defined by integration using the orientation and weight. One writes (X, d, T). The current structure defines a mass measure $\|T\|$. The points in x all have positive density with respect to this measure. These spaces have a notion of boundary coming from this current structure. The boundary is also an integral current space.[9]

An important integral current space is the 0 current space, and collapsing sequences of manifolds, $\mathrm{vol}(M_j) \to 0$, converge to the **0** space. Notice that in Figure 4, the sequences A_j, C_j and D_j all converge to 0 in the intrinsic flat sense. The sequence B_j converges to a pair of spheres in the intrinsic flat sense where the limit does not include the line segment.

Most results about integral current spaces and intrinsic flat convergence are proven by finding a way to isometrically embed the entire sequence into a common metric space, Z, and apply the theorems for integral currents in Z. The rectifiability, dimension and boundary properties of the limits follows from Ambrosio-Kirchheim's corresponding results on integral currents. So does the slicing theorem and the lower semicontinuity of mass. Combining Gromov's compactness theorem with Ambrosio-Kirchheim's compactness theorem, one sees that a sequence of manifolds which converge in the Gromov-Hausdorff sense that have a uniform upper bound on volume, also converge in the intrinsic flat sense [62].

In general the intrinsic flat and Gromov-Hausdorff limits do not agree: the intrinsic flat limit may be a strict subset of the Gromov-Hausdorff limit. If the

[9]See [62] for more details about the relationship between X and T and how to find the boundary.

Gromov Hausdorff limit is lower-dimensional than the sequence, then the intrinsic flat limit is the **0** space. Any regions in the Gromov-Hausdorff limit that are lower-dimensional disappear from the intrinsic flat limit [62]. Note that this is in contrast with metric measure limits, which do not lose regions of lower dimension that still have positive limit measures. Such a situation can occur, for example, if a thin region collapsing to lower dimension is very bumpy and has a uniform lower bound on volume.

Sequences of Riemannian manifolds may also disappear due to cancellation. Recall that in Euclidean space, a submanifold which curves in on itself as in Figure 3 will have cancellation in the limit due to the opposing orientation. An example described in [63] is a sequence of three-dimensional manifolds with positive scalar curvature created by taking a pair of standard three spheres and joining them by increasingly dense and increasingly thin and short tunnels. The Gromov-Hausdorff limit is a 3-sphere while the intrinsic flat limit is the 0 current space. If each tunnel is twisted, then the intrinsic flat limit is the standard three sphere with multiplicity two.

In some cases the intrinsic flat limits and Gromov-Hausdorff limits agree giving new insight into the rectifiability of the Gromov-Hausdorff limits. They agree for noncollapsing sequences of manifolds with nonnegative Ricci curvature and also for sequences of manifolds with uniform linear contractibility functions and uniform upper bounds on volume [63]. As a consequence the Gromov-Hausdorff limits are countably $\mathcal{H}^m$ rectifiable metric spaces. This was already shown by Cheeger-Colding for the sequences with bounded Ricci curvature but is a new result for the sequences with the linear geometric contractibility hypothesis. With only uniform geometric contractibility functions that are not linear, Schul and Wenger have shown the limits need not be so rectifiable [63] and when there is no upper bound on volume, Ferry has shown the limit spaces need not even be finite-dimensional [30]. Recall that Cheeger-Colding have proven limits of manifolds with nonnegative Ricci curvature have Euclidean tangent cones almost everywhere. In contrast, there is a sequence of Riemannian manifolds with uniform linear contractibility functions that converge to the taxicab space [62].

The key ingredient in the proofs of these noncancellation results is an estimate on the filling volumes of small spheres in the spaces. The filling volumes of the spheres are continuous with respect to the intrinsic flat distance and can be more easily applied to control sequence than the mass (or volume). It is conjectured that a sequence of three-dimensional Riemannian manifolds with positive scalar curvature, no interior minimal surfaces and a uniform upper bound on volume and on the area of the boundary converges in the intrinsic flat sense without cancellation. Such manifolds are important in the study of general relativity [63].

In fact, Wenger has proven that a sequence of oriented Riemannian manifolds with a uniform upper bound on diameter and on volume has a subsequence which converges in the intrinsic flat sense to an integral current space [71]. The proof involves an even weaker notion of convergence: the ultralimit.

3.4. Ultralimits

The notion of an ultralimit was introduced by van den Dries and Wilkie in 1984 [67] and developed by Gromov in [38]. An ultralimit of a sequence of metric spaces is defined using Cartan's 1937 notion of a nonprincipal ultrafilter: a finitely additive probability measure ω such that all subsets $S \subset \mathbb{N}$ are ω measurable, $\omega(S) \in \{0,1\}$ and $\omega(S) = 0$ whenever S is finite. Given an ultrafilter, ω, and a bounded sequence of $a_j \in \mathbb{R}$, there exists an ultralimit, $L = \omega - \lim_{j\to\infty} a_j$, such that $\forall \epsilon > 0$, $\omega\{j : |a_j - L| < \epsilon\} = 1$.

Given a sequence of compact metric spaces, (X_j, d_j), with a uniform bound on diameter, and given an ultrafilter, ω, there is an ultralimit (X, d) which is a metric space that may no longer be compact, but is at least complete. The space, X, is constructed as equivalence classes of sequences $\{x_j\}$ where $x_j \in X_j$ and the metric of X is defined by taking ultralimits:

$$d(\{x_j\},\{y_j\}) := \omega - \lim_{j\to\infty} d_j(x_j, y_j). \tag{33}$$

Two sequences are equivalent when the distance between them is zero. Notice that one does not need a subsequence to find a limit and that in general the ultralimit depends on ω.

If X_j has a Gromov-Hausdorff limit, then the ultralimit is the Gromov-Hausdorff limit and there is no dependence on ω. If one has two sequences $X_j \xrightarrow{\text{GH}} X$ and $Y_j \xrightarrow{\text{GH}} Y$, then the ultralimit of the alternating sequence $\{X_1, Y_1, X_2, Y_2, X_3, \dots\}$ is Y iff $\exists N$ such that $\forall n > N$ we have $\omega\{2n, 2n+2, 2n+4, 2n+6, \dots\} = 1$. Otherwise the ultralimit is X.

It is often useful to compute the ultralimits of sequences which do not have Gromov-Hausdorff limits. For example, the three sphere of many splines sequence, $\{M_j\}$, converges to a standard three sphere with countably many unit line segments attached at various points on the sphere. To see this, one may view each M_j as a union of regions $W_j \cup U_{1,j} \cup U_{2,j} \cup \cdots \cup U_{N_j,j}$, where each $U_{i,j}$ covers a spline and W_j covers the spherical portion. Then $W_j \xrightarrow{\text{GH}} S^3$, while $U_{i_j,j} \xrightarrow{\text{GH}} [0,1]$. Thus any ultralimit of these M_j is built by connecting countably many intervals to a three sphere. The locations where the intervals are attached to the sphere may depend on the ultrafilter, ω.

If X_j are geodesic spaces, then the ultralimit is a geodesic space. If the X_j are $CAT(0)$ spaces then the ultralimit is as well. Ultralimits have been applied extensively in the study of CAT(0) spaces and also Lie Groups. See for example [12] and [41] for more details as well as applications.

4. Speculation

While the methods of intrinsic convergence defined above have all proven to be useful in a variety of settings, and should in fact have more applications that have not yet been discovered or fully explored, it is clear that each has its disadvantages. Gromov-Hausdorff convergence, like Hausdorff convergence, provides no rectifiability for its limits. The intrinsic flat convergence, like Federer-Fleming's flat convergence, has difficulty with cancellation in the limit. Ultraconvergence has the same difficulties as Gromov-Hausdorff convergence and the additional concern that the spaces in the sequence are not particularly close to the limit space in a measurable way using a notion of distance.

There are important problems both related to General Relativity and Ricci flow which have not yet been addressed using the above methods. While intrinsic flat convergence may prove useful as it is further explored, other methods of convergence may be necessary to address all the problems that arise.

4.1. Intrinsic varifold convergence

For some time, mathematicians have been exploring possible methods of extending Ricci flow through singularities. Mean curvature flow behaves a lot like Ricci flow and the Brakke flow uses varifolds to extend this notion to the nonsmooth setting. The key missing ingredient that has kept people from directly extending the work on Brakke flow to better understand Ricci flow is that there has never been a notion of an intrinsic varifold space and intrinsic varifold convergence.

In the prior section we saw how mathematicians have repeatedly applied Gromov's trick of isometrically embedding metric spaces into a common space, measuring the distance between them in that common space, and then taking an infimum. Here, however, there is no metric measuring the distance between varifolds, just the natural notion of weak convergence against test functions. One might try to define a notion of distances between varifolds in Euclidean space which defines a convergence that agrees with the standard varifold convergence and then apply Gromov's trick.

Alternately, one might venture to say that a sequence of Riemannian manifolds, M_j, converges to a space, M, as varifolds if and only if there is a sequence of isometric embeddings from, M_j, into a common space, Z, such that the images of M_j converge as varifolds in Z to an image of M. However, there is as yet no notion of convergence as varifolds in a metric space. Recall that if Z is Euclidean space, convergence as varifolds, means weak convergence as measures on $\mathbb{R}^N \times \Gamma(N,k)$. We would need some sort of notion of a $\Gamma(Z,k)$ perhaps representing all possible "tangent spaces" of Z at a point. Perhaps this might be done by first isometrically embedding Z into a Banach space.

4.2. Area convergence

One may view Gromov-Hausdorff convergence as a notion of convergence defined by the fact that lengths of minimizing geodesics converge. Measure convergence is defined by the fact that volumes converge. It becomes natural to wonder, particularly in the case of three-dimensional manifolds, whether there is a notion of convergence defined by the fact that areas, or areas of minimal surfaces, converge.

Note that Ilmanen's example of the three sphere of many splines, $\{M_j\}$, converges in some area sense to the standard three sphere. There are radial projections

$f_j : M_j \to S^3$ which are almost area preserving, one-to-one and onto. Such a map seems perhaps to extend the notion of an almost isometry without involving a distance. However requiring a diffeomorphism to define area convergence seems too strong for any applications.

Burago, Ivanov and Sormani spent a few years investigating possible notions of area convergence and the area distance between spaces. They observed that two distinct metric spaces could easily have a surjection between them that is area preserving. One space could, for example, be the unit square, $X = [0,1] \times [0,1]$, with the Euclidean metric. The other space could be the unit square with a "pulled thread": $Y = [0,1] \times [0,1]/\sim$ where $(x_1, x_2) \sim (y_1, y_2)$ iff $x_1 = y_1 = 0$ with the metric:

$$d_Y((x_1,x_2),(y_1,y_2)) = \min\{\sqrt{(x_1-y_1)^2 + (x_2-y_2)^2}, |x_1| + |y_1|\}. \qquad (34)$$

This seemed to indicate the need for a notion of "area space" defined as an equivalence class of metric spaces. It was essential before even beginning the project to verify that each equivalence class would only include one Riemannian manifold.

Another difficulty arose in that one would not expect to always find an almost area preserving surjection between spaces which should intuitively be close in the area sense. In an almost isometry, there are two requirements: almost distance preserving and almost onto. Both requirements involve distance and one can prove that if there is an almost isometry from a metric space X to a metric space Y then one can find an almost isometry in the reverse direction (although the errors change slightly in reverse) [40]. To take advantage of Gromov's ideas, it was decided that one needs to convert the notion of area into a notion of distance.

Given a Riemannian manifold, M, one may examine the loop space, $\Omega(M)$. The flat distance between loops is defined using the notion of area. Recall that the flat distance defined in (9) involves both area and length, however, in the setting of closed loops one may choose $A = 0$ and only take an infimum over all possible fillings by surfaces B. Perhaps an area convergence of M_j to some sort of area space M could be defined by taking pointed Gromov-Hausdorff limits or ultralimits of the $\Omega(M_j)$.

Preliminary work in this direction was completed by Burago-Ivanov. They proved that if a pair of three-dimensional Riemannian manifolds, M^3 and N^3, have isometric loop spaces, then M^3 and N^3 are isometric [14]. This step alone was very difficult because M^3 is not isometrically embedded into its loop space, $\Omega(M)$, so one must localize points and planes in TM using sequences of loops without controlling the lengths of the loops.

Now one may move forward and investigate whether there is a natural compactness theorem for some notion of area convergence. It would not be surprising if positive scalar curvature on M_j control areas of minimal surfaces well enough to control $\Omega(M_j)$. Or perhaps one could skip compactness theorems, and apply ultralimits to the $\Omega(M_j)$ and see what properties are conserved under such a limit. Ultimately one does not just want to take a limit of $\Omega(M_j)$ but find an area space

X, such that $\Omega(X)$ is the limit of the $\Omega(M_j)$. Then one could say X is the area limit of M_j.

4.3. Convergence of Lorentzian manifolds

Another direction of research that is fundamental to General Relativity is the development of weak forms of convergence for Lorentzian manifolds. While Gromov-Hausdorff convergence has proven useful in the study of the stability of the space-like universe [58], one needs to extend Gromov-Hausdorff convergence to the Lorentzian setting to study the stability of the space-time universe. The techniques described above immediately fail for these spaces because they do not isometrically embed into metric spaces.

Noldus has developed a Gromov-Hausdorff distance between Lorentzian spaces [51]. Further work in this direction including a notion of the arising limit spaces appears in [50] and compactness theorems and an examination of causality in the limit spaces are proven with Bombelli in [10]. This work has not yet been applied to prove stability questions arising in general relativity.

Perhaps similar methods might be applied to extend the notion of the intrinsic flat distance to Lorentzian spaces.

Acknowledgement

The author is indebted to Jeff Cheeger for all the excellent advise he has given over the past fifteen years. His intuitive explanations of the beautiful geometry of converging Riemannian manifolds in his many excellent talks have been invaluable to the mathematics community. The author would also like to thank the editors for soliciting this article and William Wylie for discussions regarding the exposition.

References

[1] Stephanie B. Alexander and Richard L. Bishop. Thin Riemannian manifolds with boundary. *Math. Ann.*, 311(1):55–70, 1998.

[2] Stephanie B. Alexander and Richard L. Bishop. Spines and topology of thin Riemannian manifolds. *Trans. Amer. Math. Soc.*, 355(12):4933–4954 (electronic), 2003.

[3] William K. Allard. On the first variation of a varifold. *Ann. of Math.* (2), 95:417–491, 1972.

[4] F.J. Almgren, Jr. Some interior regularity theorems for minimal surfaces and an extension of Bernstein's theorem. *Ann. of Math.* (2), 84:277–292, 1966.

[5] Frederick J. Almgren, Jr. The theory of varifolds. 1964.

[6] Frederick J. Almgren, Jr. *Plateau's problem*, volume 13 of *Student Mathematical Library*. American Mathematical Society, Providence, RI, 2001. An invitation to varifold geometry, corrected reprint of the 1966 original, with forewords by Jean E. Taylor and Robert Gunning, and Hugo Rossi.

[7] Luigi Ambrosio and Bernd Kirchheim. Currents in metric spaces. *Acta Math.*, 185(1):1–80, 2000.

[8] Luigi Ambrosio and Stefan Wenger. Rectifiability of flat chains in Banach spaces with coefficients in z_p. *To appear in Mathematische Zeitschrift*, 2011.

[9] Wilhelm Blaschke. *Kreis und Kugel.* Chelsea Publishing Co., 1916.

[10] Luca Bombelli and Johan Noldus. The moduli space of isometry classes of globally hyperbolic spacetimes. *Classical Quantum Gravity*, 21(18):4429–4453, 2004.

[11] Kenneth A. Brakke. *The motion of a surface by its mean curvature*, volume 20 of *Mathematical Notes.* Princeton University Press, Princeton, N.J., 1978.

[12] Martin R. Bridson and André Haefliger. *Metric spaces of non-positive curvature*, volume 319 of *Grundlehren der Mathematischen Wissenschaften* [*Fundamental Principles of Mathematical Sciences*]. Springer-Verlag, Berlin, 1999.

[13] Robert Brooks, Peter Perry, and Peter Petersen, V. Compactness and finiteness theorems for isospectral manifolds. *J. Reine Angew. Math.*, 426:67–89, 1992.

[14] Dimitri Burago and Sergei Ivanov. Area spaces: First steps, with appendix by nigel higson. *Geometric and Functional Analysis*, 19(3):662–677, 2009.

[15] Dmitri Burago, Yuri Burago, and Sergei Ivanov. *A course in metric geometry*, volume 33 of *Graduate Studies in Mathematics.* American Mathematical Society, Providence, RI, 2001.

[16] Huai-Dong Cao and Xi-Ping Zhu. A complete proof of the Poincaré and geometrization conjectures – application of the Hamilton-Perelman theory of the Ricci flow. *Asian J. Math.*, 10(2):165–492, 2006.

[17] Jianguo Cao and Jian Ge. A simple proof of Perelman's collapsing theorem for three manifolds. 2009.

[18] Sun-Yung A. Chang and Paul C.-P. Yang. Isospectral conformal metrics on 3-manifolds. *J. Amer. Math. Soc.*, 3(1):117–145, 1990.

[19] Jeff Cheeger. *Degeneration of Riemannian metrics under Ricci curvature bounds.* Lezioni Fermiane. [Fermi Lectures]. Scuola Normale Superiore, Pisa, 2001.

[20] Jeff Cheeger and Tobias H. Colding. Lower bounds on Ricci curvature and the almost rigidity of warped products. *Ann. of Math.* (2), 144(1):189–237, 1996.

[21] Jeff Cheeger and Tobias H. Colding. On the structure of spaces with Ricci curvature bounded below. I. *J. Differential Geom.*, 46(3):406–480, 1997.

[22] Jeff Cheeger and Tobias H. Colding. On the structure of spaces with Ricci curvature bounded below. II. *J. Differential Geom.*, 54(1):13–35, 2000.

[23] Jeff Cheeger and Tobias H. Colding. On the structure of spaces with Ricci curvature bounded below. III. *J. Differential Geom.*, 54(1):37–74, 2000.

[24] Jeff Cheeger and Mikhael Gromov. Collapsing Riemannian manifolds while keeping their curvature bounded. I. *J. Differential Geom.*, 23(3):309–346, 1986.

[25] Jeff Cheeger and Mikhael Gromov. Collapsing Riemannian manifolds while keeping their curvature bounded. II. *J. Differential Geom.*, 32(1):269–298, 1990.

[26] Tobias H. Colding. Spaces with Ricci curvature bounds. In *Proceedings of the International Congress of Mathematicians, Vol. II* (*Berlin,* 1998), number Extra Vol. II, pages 299–308 (electronic), 1998.

[27] E. DeGiorgi. Problema di plateau generale e funzionali geodetici. *Atti Sem. Mat. Fis. Univ. Modena*, 43:285–292, 1995.

[28] Herbert Federer. *Geometric measure theory.* Die Grundlehren der mathematischen Wissenschaften, Band 153. Springer-Verlag New York Inc., New York, 1969.

[29] Herbert Federer and Wendell H. Fleming. Normal and integral currents. *Ann. of Math.* (2), 72:458–520, 1960.

[30] Steven C. Ferry. Topological finiteness theorems for manifolds in Gromov-Hausdorff space. *Duke Math. J.*, 74(1):95–106, 1994.

[31] Steven C. Ferry and Boris L. Okun. Approximating topological metrics by Riemannian metrics. *Proc. Amer. Math. Soc.*, 123(6):1865–1872, 1995.

[32] Wendell H. Fleming. Flat chains over a finite coefficient group. *Trans. Amer. Math. Soc.*, 121:160–186, 1966.

[33] Kenji Fukaya. Collapsing of Riemannian manifolds and eigenvalues of Laplace operator. *Invent. Math.*, 87(3):517–547, 1987.

[34] Kenji Fukaya. Collapsing of Riemannian manifolds and eigenvalues of Laplace operator. *Invent. Math.*, 87(3):517–547, 1987.

[35] Kenji Fukaya. Collapsing Riemannian manifolds to ones of lower dimensions. *J. Differential Geom.*, 25(1):139–156, 1987.

[36] Kenji Fukaya. Hausdorff convergence of Riemannian manifolds and its applications. In *Recent topics in differential and analytic geometry*, volume 18 of *Adv. Stud. Pure Math.*, pages 143–238. Academic Press, Boston, MA, 1990.

[37] Robert E. Greene and Peter Petersen V. Little topology, big volume. *Duke Math. J.*, 67(2):273–290, 1992.

[38] M. Gromov. Asymptotic invariants of infinite groups. In *Geometric group theory, Vol.* 2 (*Sussex,* 1991), volume 182 of *London Math. Soc. Lecture Note Ser.*, pages 1–295. Cambridge Univ. Press, Cambridge, 1993.

[39] Mikhael Gromov. *Structures métriques pour les variétés riemanniennes*, volume 1 of *Textes Mathématiques* [*Mathematical Texts*]. CEDIC, Paris, 1981. Edited by J. Lafontaine and P. Pansu.

[40] Misha Gromov. *Metric structures for Riemannian and non-Riemannian spaces*, volume 152 of *Progress in Mathematics*. Birkhäuser Boston Inc., Boston, MA, 1999. Based on the 1981 French original [MR0682063 (85e:53051)], with appendices by M. Katz, P. Pansu and S. Semmes, Translated from the French by Sean Michael Bates.

[41] Michael Kapovich. *Hyperbolic manifolds and discrete groups*, volume 183 of *Progress in Mathematics*. Birkhäuser Boston Inc., Boston, MA, 2001.

[42] Bernd Kirchheim. Rectifiable metric spaces: local structure and regularity of the Hausdorff measure. *Proc. Amer. Math. Soc.*, 121(1):113–123, 1994.

[43] Bruce Kleiner and John Lott. Notes on Perelman's papers. *Geom. Topol.*, 12(5):2587–2855, 2008.

[44] Nicholas J. Korevaar and Richard M. Schoen. Sobolev spaces and harmonic maps for metric space targets. *Comm. Anal. Geom.*, 1(3-4):561–659, 1993.

[45] Fanghua Lin and Xiaoping Yang. *Geometric measure theory – an introduction*, volume 1 of *Advanced Mathematics* (*Beijing/Boston*). Science Press, Beijing, 2002.

[46] John Lott and Cédric Villani. Ricci curvature for metric-measure spaces via optimal transport. *Ann. of Math.* (2), 169(3):903–991, 2009.

[47] Xavier Menguy. Noncollapsing examples with positive Ricci curvature and infinite topological type. *Geom. Funct. Anal.*, 10:600–627, 2000.

[48] Frank Morgan. *Geometric measure theory.* Elsevier/Academic Press, Amsterdam, fourth edition, 2009. A beginner's guide.

[49] John Morgan and Gang Tian. *Ricci flow and the Poincaré conjecture*, volume 3 of *Clay Mathematics Monographs.* American Mathematical Society, Providence, RI, 2007.

[50] Johan Noldus. The limit space of a Cauchy sequence of globally hyperbolic spacetimes. *Classical Quantum Gravity*, 21(4):851–874, 2004.

[51] Johan Noldus. A Lorentzian Gromov-Hausdorff notion of distance. *Classical Quantum Gravity*, 21(4):839–850, 2004.

[52] B. Osgood, R. Phillips, and P. Sarnak. Compact isospectral sets of surfaces. *J. Funct. Anal.*, 80(1):212–234, 1988.

[53] Xiaochun Rong. Collapsed Riemannian manifolds with bounded sectional curvature. In *Proceedings of the International Congress of Mathematicians, Vol. II (Beijing, 2002)*, pages 323–338, Beijing, 2002. Higher Ed. Press.

[54] Zhongmin Shen and Christina Sormani. The topology of open manifolds with nonnegative Ricci curvature. *Commun. Math. Anal.*, (Conference 1):20–34, 2008.

[55] Takashi Shioya and Takao Yamaguchi. Collapsing three-manifolds under a lower curvature bound. *J. Differential Geom.*, 56(1):1–66, 2000.

[56] Takashi Shioya and Takao Yamaguchi. Volume collapsed three-manifolds with a lower curvature bound. *Math. Ann.*, 333(1):131–155, 2005.

[57] Leon Simon. *Lectures on geometric measure theory*, volume 3 of *Proceedings of the Centre for Mathematical Analysis, Australian National University.* Australian National University Centre for Mathematical Analysis, Canberra, 1983.

[58] Christina Sormani. Friedmann cosmology and almost isotropy. *Geom. Funct. Anal.*, 14(4):853–912, 2004.

[59] Christina Sormani. Convergence and the length spectrum. *Adv. Math.*, 213(1):405–439, 2007.

[60] Christina Sormani and Guofang Wei. Hausdorff convergence and universal covers. *Trans. Amer. Math. Soc.*, 353(9):3585–3602 (electronic), 2001.

[61] Christina Sormani and Guofang Wei. The covering spectrum of a compact length space. *J. Differential Geom.*, 67(1):35–77, 2004.

[62] Christina Sormani and Stefan Wenger. Intrinsic flat convergence of manifolds and other current spaces. *To appear in the Journal of Differential Geometry*, 2010.

[63] Christina Sormani and Stefan Wenger. Weak convergence and cancellation, appendix by Raanan Schul and Stefan Wenger. *Calculus of Variations and Partial Differential Equations*, 38(1-2), 2010.

[64] Karl-Theodor Sturm. On the geometry of metric measure spaces. I. *Acta Math.*, 196(1):65–131, 2006.

[65] Karl-Theodor Sturm. On the geometry of metric measure spaces. I. *Acta Math.*, 196(1):65–131, 2006.

[66] Peter Topping. $\mathcal{L}$-optimal transportation for Ricci flow, on topping's webpage. *J. Reine Angew. Math.*, 636:93–122, 2009.

[67] L. van den Dries and A.J. Wilkie. Gromov's theorem on groups of polynomial growth and elementary logic. *J. Algebra*, 89(2):349–374, 1984.

[68] Cédric Villani. *Optimal transport*, volume 338 of *Grundlehren der Mathematischen Wissenschaften* [*Fundamental Principles of Mathematical Sciences*]. Springer-Verlag, Berlin, 2009. Old and new.

[69] S. Wenger. Isoperimetric inequalities of Euclidean type in metric spaces. *Geom. Funct. Anal.*, 15(2):534–554, 2005.

[70] Stefan Wenger. Flat convergence for integral currents in metric spaces. *Calc. Var. Partial Differential Equations*, 28(2):139–160, 2007.

[71] Stefan Wenger. Compactness for manifolds and integral currents with bounded diameter and volume. *Calc. Var. Partial Differential Equations*, 40(3-4):423–448, 2011.

[72] Brian White. Currents and flat chains associated to varifolds, with an application to mean curvature flow. *Duke Math. J.*, 148(1):41–62, 2009.

Christina Sormani
Mathematics Department
CUNY Graduate Center
365 Fifth Avenue
NY NY 10014, USA

and

Mathematics Department
Lehman College
Bronx, NY 10468, USA
e-mail: `sormanic@member.ams.org`

Existence of Einstein Metrics on Fano Manifolds

Gang Tian

Abstract. It is a long-standing problem to establish the existence of Kähler-Einstein metrics on Fano manifolds since Yau's solution for the Calabi conjecture in late 70s. It is also one of driving forces in today's study in Kähler geometry. In this paper, we discuss a program I started more than twenty years ago on this famous problem. It includes some of my results and speculations on the existence of Kähler-Einstein metrics on Fano manifolds, such as, holomorphic invariants, the K-stability, the compactness theorem for Kähler-Einstein manifolds, the partial C^0-estimates and their variations. I will also discuss some related problems as well as some recent advances.

Mathematics Subject Classification (2000). 14Jxx, 58Jxx.

Keywords. Kähler-Einstein metrics, Fano manifolds, holomorphic invariants, K-stability, compactness theorems.

This is largely an expository paper and dedicated to my friend J. Cheeger for his 65th birthday. The purpose of this paper is to discuss some of my works on the existence of Kähler-Einstein metrics on Fano manifolds and some related topics. I will describe a program I have been following for the last twenty years. It includes some of my results and speculations which were scattered in my previous publications or mentioned in my lectures. I also take this opportunity to clarify and make them more accessible. In the course of doing so, I will also discuss some recent advances and problems which arise from studying the existence problem.

A Fano manifold is a compact Kähler manifold with positive first Chern class. It has been one of the main problems in Kähler geometry to study if a Fano manifold admits a Kähler-Einstein metrics since the Aubin-Yau theorem on Kähler-Einstein metrics with negative scalar curvature and the Calabi-Yau theorem on Ricci-flat Kähler metrics in the 1970s. This problem is much more difficult because there are new obstructions to the existence. The classical one was given by Matsushima: If a Fano manifold M admits a Kähler-Einstein metric, then its Lie algebra of holomorphic vector fields must be reductive. In the early 80s,

Supported partially by a NSF grant; submitted in Feb. 2010.

A. Futaki introduced a new invariant, now referred to the Futaki invariant, whose vanishing is a necessary condition for M to have a Kähler-Einstein metric. Since late 80s, inspired by my works on Kähler-Einstein metrics on complex surfaces [Ti89], I have been developing methods of relating certain geometric stability of underlying manifolds to Kähler-Einstein metrics. In [Ti97], I introduced the K-stability for any Fano manifold and proved that a Fano manifold with trivial holomorphic vector fields and which admits a Kähler-Einstein metric is K-stable. An algebraic version of the K-stability was given by Donaldson in [Do02]. It was conjectured that the existence of Kähler-Einstein metrics on M is equivalent to the asymptotic K-stability.[1]

As said at the beginning, this is not intended to be a complete survey on Kähler-Einstein metrics with positive scalar curvature. Unfortunately, there are important works which I can not present here because of limited time and space. However, I do plan to write a much more complete survey and hope to include many of them there.

1. Preliminary

Let M be a Fano manifold of dimension n. A Kähler metric can be given by specifying its Kähler form ω, in local coordinates $z_1, \ldots, z_n$, it is of the form

$$\omega = \frac{\sqrt{-1}}{2} \sum_{i,j=1}^{n} g_{i\bar{j}} dz_i \wedge d\bar{z}_j,$$

where $\{g_{i\bar{j}}\}$ is a positive Hermitian matrix-valued function such that $d\omega = 0$. We will simply use ω to denote both a metric and its Kähler form.

Recall that the Kähler class of ω is the cohomology class $[\omega]$ in $H^2(M, \mathbb{R})$ represented by ω. It follows from Hodge theory that if ω' is another Kähler metric with $[\omega'] = [\omega]$, then there is a smooth function φ on M such that

$$\omega' = \omega + \frac{\sqrt{-1}}{2} \partial\bar{\partial}\varphi.$$

We will often denote the right side by ω_φ. Thus, the space $\mathcal{K}_{[\omega]}$ of Kähler metrics with the same Kähler class $[\omega]$ can be identified with

$$\left\{ \varphi \in C^\infty(M, \mathbb{R}) \mid \int_M \varphi\,\omega^n = 0,\ \omega + \frac{\sqrt{-1}}{2}\partial\bar{\partial}\varphi > 0 \right\}.$$

Since M is Fano, we can take ω such that $[\omega] = \pi\, c_1(M)$. We will assume this unless specified. In this paper, we call ω a Kähler-Einstein metric if $\mathrm{Ric}(\omega) = \omega$,

[1]The K-stability can be extended to any polarized projective algebraic manifolds and similar results can be proved for general Kähler metrics with constant scalar curvature. Also a conjecture, often referred as the Yau-Tian-Donaldson conjecture, can be made on existence of Kähler metrics with constant scalar curvature and the K-stability. I will refer the readers to [Ti97], [Do00], [Sto07] for more discussions and also to Section 4.1 (Conjecture 4.7) below.

where $\mathrm{Ric}(\omega)$ is the Ricci curvature form of ω, in local coordinates $z_1, \ldots, z_n$,

$$\mathrm{Ric}(\omega) = \frac{\sqrt{-1}}{2}\partial\bar{\partial}\log\det(g_{i\bar{j}}).$$

The following uniqueness theorem is due to Bando and Mabuchi [BM86].

Theorem 1.1. *Any given compact Fano manifold M admits at most one Kähler-Einstein metric up to automorphisms.*

The main concern of this paper is on the existence. First we recall an analytic obstruction introduced by Futaki in 1983. Let $\eta(M)$ be the space of holomorphic vector fields on M and ω be any fixed Kähler metric with $\pi c_1(M)$ as its Kähler class $\Omega \in H^2(M, \mathbb{R})$. We put

$$f_\Omega(v) = \int_M v(h_\omega)\omega^n, \; v \in \eta(M),$$

where h_ω is determined by the equations

$$\mathrm{Ric}(\omega) - \omega = \frac{\sqrt{-1}}{2}\partial\bar{\partial}h_\omega, \; \int_M \left(e^{h_\omega} - 1\right)\omega^n = 0,$$

where $\underline{s}_\omega$ denotes the average of $s(\omega)$. It was proved in [Fut83] that $f_\Omega(v)$ is actually independent of the choice of ω. Therefore, it is an invariant, referred as the Futaki invariant. Consequently, if M admits a Kähler-Einstein metric, then $f_\Omega \equiv 0$. On the other hand, there are examples of Fano manifolds with non-vanishing Futaki invariant, so there do not exist Kähler-Einstein metrics on such manifolds.

As usual, one can reduce the existence of Kähler-Einstein metrics to solving a complex Monge-Ampère equation. Let ω be a Kähler metric and h_ω be defined as above. Then ω_φ is a Kähler-Einstein metric if and only if modulo a constant, φ solves the following

$$\left(\omega + \frac{\sqrt{-1}}{2}\partial\bar{\partial}\varphi\right)^n = e^{h_\omega - \varphi}\omega^n. \tag{1.1}$$

There are two ways of solving this equation. One is to use the Kähler-Ricci flow. We will discuss this method in details in later sections. Another one is the continuity method. We will first discuss this continuity method. Consider

$$\left(\omega + \frac{\sqrt{-1}}{2}\partial\bar{\partial}\varphi\right)^n = e^{h_\omega - t\varphi}\omega^n. \tag{1.2}$$

Set $I = \{t \in [0,1] \mid (1.2)$ is solvable for $s \in [0,t]\ |$. Using Yau's solution for the Calabi conjecture [Ya76], there is a solution φ for (1.2) with $t = 0$, so $0 \in I$, i.e., I is non-empty. It was observed in [Au83] that I is open. Its proof can be outlined as follows: If φ is a solution of (1.2) for some $t_0 < 1$, then a direct computation shows that $\mathrm{Ric}(\omega_\varphi) > t_0\omega_\varphi$. So it follows from the Bochner identity that the Laplacian Δ_{t_0} of ω_φ has non-zero eigenvalue greater than t_0. On the other hand, the linearization of (1.2) is simply $\Delta_{t_0} + t_0$ which has non-zero eigenvalue, so it is invertible. Then the openness follows from the Implicit Function Theorem. To prove that I is closed, we need an a prior $C^{2,\alpha}$-estimate for any solutions of

(1.2). In view of the Aubin-Yau's second-order estimates (cf. [Au76], [Ya76]) and Calabi's 3rd estimate (cf. Appendix in [Ya76]), we only need to derive an a prior C^0-estimate for any solutions of (1.2).

Since there are analytic obstructions, such a C^0-estimate relies on geometry of underlying M.

2. Analytic criterion for existence

We have seen that in order to prove the existence of Kähler-Einstein metrics on M, we need some extra condition for M to get an a priori C^0-estimate. One analytic condition can be formulated in terms of the properness of the Lagrangian of (1.1).

The space of Kähler metrics with a fixed Kähler class $[\omega] = \pi\, c_1(M)$ is

$$\mathcal{K}_{[\omega]} = P(M,\omega)/\sim,\; P(M,\omega) = \left\{\varphi \in C^\infty(M,\mathbb{R}) \mid \omega + \frac{\sqrt{-1}}{2}\partial\overline{\partial}\varphi > 0\right\}.$$

Here $\varphi \sim \varphi'$ means $\varphi = \varphi' + c$ for some constant c.

Define

$$\mathbf{I}_\omega(\varphi) = \frac{1}{V}\int_M \varphi\left(\omega^n - \left(\omega + \frac{\sqrt{-1}}{2}\partial\overline{\partial}\varphi\right)^n\right), \tag{2.1}$$

where $V = \int_M \omega^n$. Also define

$$\mathbf{J}_\omega(\varphi) = \int_0^1 \frac{\mathbf{I}_\omega(t\varphi)}{t}dt = \frac{1}{V}\sum_{i=0}^{n-1}\frac{i+1}{n+1}\frac{\sqrt{-1}}{2}\int_M \partial\varphi\wedge\overline{\partial}\varphi\wedge\omega^i\wedge\omega_\varphi^{n-i-1}. \tag{2.2}$$

Then

$$\mathbf{I}_\omega(\varphi) - \mathbf{J}_\omega(\varphi) = \frac{1}{V}\sum_{i=0}^{n-1}\frac{n-i}{n+1}\frac{\sqrt{-1}}{2}\int_M \partial\varphi\wedge\overline{\partial}\varphi\wedge\omega^i\wedge\omega_\varphi^{n-i-1}.$$

Notice that for any $\varphi \in P(M,\omega)$,

$$\mathbf{I}_\omega(\varphi) \geq 0, \mathbf{J}_\omega(\varphi) \geq 0, \mathbf{I}_\omega(\varphi) - \mathbf{J}_\omega(\varphi) \geq 0.$$

Moreover, one can deduce from the definition

Lemma 2.1. *For any $\varphi \in P(M,\omega)$, we have*

$$\frac{1}{n}\mathbf{J}_\omega(\varphi) \leq \mathbf{I}_\omega(\varphi) - \mathbf{J}_\omega(\varphi) \leq n\,\mathbf{J}_\omega(\varphi).$$

The Lagrangian of (1.1) is given by

$$\mathbf{F}_\omega(\varphi) = \mathbf{J}_\omega(\varphi) - \frac{1}{V}\int_M \varphi\omega^n - \log\left(\frac{1}{V}\int_M e^{h_\omega - \varphi}\omega^n\right). \tag{2.3}$$

In this section, we give a sufficient condition for the existence of Kähler-Einstein metrics on a Fano manifold in terms of the properness of $\mathbf{F}$. We will also show that such a condition is necessary for Fano manifolds without non-trivial holomorphic fields.

2.1. Analytic stability

In this subsection, we introduce a notion of analytic stability and discuss its simple implications.

Definition 2.2. We say that a Fano manifold M is analytically stable if $\mathbf{F}_\omega$ is proper, that is, if there is an increasing function $f(t) \geq -c$ for some $c \geq 0$, where $t \in (-\infty, \infty)$, such that $\lim_{t\to\infty} f(t) = \infty$ and for any $\varphi \in P(M,\omega)$, we have[2]

$$\mathbf{F}_\omega(\varphi) \geq f(\mathbf{I}_\omega(\varphi) - \mathbf{J}_\omega(\varphi)). \tag{2.4}$$

We say that M is analytically semi-stable if $\mathbf{F}_\omega$ is bounded from below.

If $\omega' = \omega_\psi$ is another Kähler metric, then one can show

$$\mathbf{F}_\omega(\varphi) = \mathbf{F}_{\omega'}(\varphi - \psi) + \mathbf{F}_\omega(\psi). \tag{2.5}$$

Hence, the analytic stability is independent of the choice of the base metric ω. Clearly, the analytic stability also implies the semi-stability.

Denote by $\mathrm{Aut}(M)$ the group of all holomorphic automorphisms of M and by $\eta(M)$ its Lie algebra.

Remark 2.3. If G is a compact Lie group acting on M by automorphisms of $(M, [\omega])$ and ω is a G-invariant metric in $[\omega]$, then we say that $(M, [\omega])$ is analytically G-stable if (2.4) holds for all G-invariant φ in $P(M,\omega)$. Similarly, one can define analytic G-semi-stability.

For any $\sigma \in \mathrm{Aut}(M)$ and $\varphi \in P(M,\omega)$, there is a unique φ_σ such that

$$\sigma^*\omega_\varphi = \omega + \frac{\sqrt{-1}}{2}\partial\bar\partial\varphi_\sigma \text{ and } \int_M \left(e^{h_\omega - \varphi_\sigma} - 1\right)\omega^n = 0.$$

Let $X \in \eta(M)$ be a holomorphic vector field, then there is a unique θ_X satisfying:

$$i_X\omega = \frac{\sqrt{-1}}{2}\bar\partial\theta_X \text{ and } \int_M \theta_X\, e^{h_\omega}\,\omega^n = 0.$$

Since X is holomorphic, $\nabla^{0,1}\bar\partial\theta_X = 0$. It follows from this and the above that

$$\Delta_\omega\theta_X + \theta_X + X(h_\omega) = 0. \tag{2.6}$$

If $\sigma(t)$ is a one-parameter subgroup generated by the real part $\mathrm{Re}(X)$ of X, then $\dot\varphi_{\sigma(t)} = \theta_X(t,\cdot)$, where $\theta_X(t)$ is the corresponding potential θ_X of X when ω is replaced by ω_φ for $\varphi = \varphi_{\sigma(t)}$. It follows

$$\frac{d\mathbf{F}_\omega}{dt}(\varphi_{\sigma(t)}) = \frac{1}{V}\int_M \dot\varphi_{\sigma(t)}\,\sigma^*\omega_\varphi^n = -\,\mathrm{Re}\left(\frac{1}{V}\int_M X(h_{\sigma^*\omega_\varphi})\,\sigma^*\omega_\varphi^n\right). \tag{2.7}$$

The last integral is simply the Futaki invariant and is independent of ω_φ [Fut83]. Therefore, we have

Corollary 2.4. *If M is analytic semi-stable, then the Futaki invariant vanishes.*

---

[2] By Lemma 2.1, one can replace $\mathbf{I} - \mathbf{J}$ by $\mathbf{J}$ in the definition.

Let $\mathrm{Aut}_0(M)$ be the connected component of $\mathrm{Aut}(M)$ containing the identity. If $\mathrm{Aut}_0(M)$ is non-trivial, then by the above corollary, M can not be analytic stable. Hence, we need the following extension of the analytic stability.

Definition 2.5. We say that a Fano manifold M is weakly analytically stable if $\mathrm{Aut}_0(M)$ is reductive[3] and there is an increasing function $f(t) \geq -c$ for some $c \geq 0$, where $t \in (-\infty, \infty)$, such that $\lim_{t\to\infty} f(t) = \infty$ and for any $\varphi \in P(M,\omega)$, we have

$$\mathbf{F}_\omega(\varphi) \geq \inf_{\sigma \in \mathrm{Aut}_0(M)} f(\mathbf{I}_\omega(\varphi_\sigma) - \mathbf{J}_\omega(\varphi_\sigma)). \tag{2.8}$$

Of course, weak analytic stability implies the semi-stability and is independent of the choice of the base Kähler metric ω.

The above notion of analytic stability can be also defined by the K-energy in place of $\mathbf{F}$. The K-energy was introduced by T. Mabuchi in [Ma86]. It is defined as follows: For any $\varphi \in P(M,\omega)$, let φ_t be any path joining 0 to φ, then

$$\mathbf{T}_\omega(\varphi) = -\frac{1}{V}\int_0^1 \int_M \dot{\varphi}_t (s(\omega_{\varphi_t}) - \underline{s}_\omega)\, \omega_{\varphi_t}^n \, dt, \tag{2.9}$$

where $s(\cdot)$ denotes the scalar curvature and $\underline{s}$ denotes the average of scalar curvature. It is not hard to show that $\mathbf{T}_\omega(\varphi)$ is independent of the choice of the path. Also, as $\mathbf{F}$, we have the following cocycle condition for $\mathbf{T}$:

$$\mathbf{T}_\omega(\varphi) + \mathbf{T}_{\omega_\varphi}(\psi) = \mathbf{T}_\omega(\varphi + \psi).$$

The analytic stability using $\mathbf{T}$ is equivalent to the older one using $\mathbf{F}$. One direction follows easily from the following identity proved in [DT91]:

$$\mathbf{F}_\omega(\varphi) = \mathbf{T}_\omega(\varphi) + \frac{1}{V}\int_M h_{\omega_\varphi} \omega_\varphi^n - \frac{1}{V}\int_M h_\omega \omega^n.$$

Since $\int_M e^{h_{\omega_\varphi}} \omega_\varphi^n = V$, we have $\int_M h_{\omega_\varphi} \omega_\varphi^n \leq 0$ by the concavity of logarithm. The other direction is more tricky and follows from the results of [Ti97] by proving that the analytic stability defined by either $\mathbf{F}$ or $\mathbf{T}$ is equivalent to existence. The fact that the analytic semi-stability defined by $\mathbf{F}$ is equivalent to that defined by $\mathbf{T}$ was proved by Y. Rubinstein[Ru07].

One advantage of using $\mathbf{T}$ is that one can extend the notion of analytic stability to any polarized Kähler manifolds which are not necessarily Fano. A polarized Kähler manifold (M,Ω) is a compact Kähler manifold together with a Kähler class Ω. It was conjectured in [Ti98] that *a compact Kähler manifold M admits a Kähler metric of constant scalar curvature in the Kähler class Ω if the polarized manifold (M,Ω) is analytically stable. The converse is also true if $\mathrm{Aut}_0(M,\Omega)$ is trivial or in general cases, it holds in a suitable notion of analytic stability.* This conjecture was solved in the case of Fano manifolds [Ti97] as we will discuss in the following subsections.

[3]This may follow from the second condition imposed on $\mathbf{F}_\omega$.

2.2. From properness to existence

In this subsection, we prove the following

Theorem 2.6. *Let M be a weakly analytically stable Fano manifold. Then M admits a Kähler-Einstein metric.*

This theorem is a slight generalization of a known result (cf. [Ti97]): Any Fano manifold which is analytically stable admits a Kähler-Einstein metric. A converse to this result is proved in [Ti97]. In a later subsection, we will prove a converse to the above theorem.

The rest of this subsection is devoted to giving an outlined proof of Theorem 2.6. We will skip the arguments in the proof which have now become standard. We refer the readers to [Ti98] for those arguments. First we notice that we may assume ω is K-invariant for a maximal compact subgroup K of $\mathrm{Aut}_0(M)$.

As usual, we apply the continuity method to (1.2). By Yau's solution to the Calabi conjecture and the Implicit Function Theorem, we have a solution φ_t to (1.2) for each sufficiently small $t > 0$. By an observation of T. Aubin (cf. [Au83], also [Ti98]), the set S of t for which (1.2) admits a solution is open in [0,1]. So it suffices to show that S is closed. This amounts to establishing an a priori C^3-estimate for solutions φ_t of (1.2). On the other hand, by Aubin-Yau's C^2-estimate and Calabi's C^3-estimate, we only need to prove an a priori C^0-estimate. For this, we need to use the weak analytic stability.

Let φ_t be a solution of (1.2):

$$\left(\omega + \frac{\sqrt{-1}}{2}\partial\bar{\partial}\varphi_t\right)^n = e^{h_\omega - t\varphi_t}\omega^n.$$

We may assume that $t < 1$ and write $\omega_t = \omega + \frac{\sqrt{-1}}{2}\partial\bar{\partial}\varphi_t$. Taking $\partial\bar{\partial}$-derivative on both sides of the above, we obtain a known bound:

$$\mathrm{Ric}(\omega_t) = t\omega_t + (1-t)\omega > t\omega_t. \tag{2.10}$$

Hence, the Ricci curvature of ω_t is bounded from below by t. It follows that there is a uniform Sobolev inequality and the Poincaré inequality for ω_t. Then by using the Moser iteration (cf. [Ti87]), we get[4]

Lemma 2.7. *There is a uniform constant $c > 0$ such that for any solution φ_t of (1.2) ($t \in [0,1]$),*

$$\|\varphi_t\|_{C^0} \leq c(1 + \mathbf{J}_\omega(\varphi_t)).$$

So we only need to bound $\mathbf{J}_\omega(\varphi_t)$ or equivalently, $\mathbf{I}_\omega(\varphi_t) - \mathbf{J}_\omega(\varphi_t)$. It was shown in [Ti87]: If $\phi(s)$ be a smooth variation of $\phi \in P(M,\omega)$, i.e., $\phi(0) = \phi$, then

$$\frac{d}{ds}\left(\mathbf{I}_\omega(\phi(s)) - \mathbf{J}_\omega(\phi(s))\right)\Big|_{s=0} = -\frac{1}{V}\int_M \phi\,\Delta\dot{\phi}\,\omega_\phi^n, \tag{2.11}$$

where $\dot{\phi} = \frac{\partial\phi}{\partial s}$ at $s = 0$. Then we have

[4]One can also use the arguments as those in [BM86] by using the Green function.

Lemma 2.8. *Let $X \in \eta(M)$ and $\sigma(s)$ be an one-parameter subgroup in* $\mathrm{Aut}_0(M)$ *generated by the real part of X and such that $\sigma(0) = \mathrm{Id}$, then*

$$\frac{d}{ds}\left(\mathbf{I}_\omega(\phi_{\sigma(s)}) - \mathbf{J}_\omega(\phi_{\sigma(s)})\right)\big|_{s=0} = \mathrm{Re}\left(\frac{1}{V}\int_M X(\phi)\,\omega_\phi^n\right). \tag{2.12}$$

Proof. Differentiating $\sigma(s)^*\omega_\phi = \omega_{\phi(s)}$ on s, where $\phi(s) = \phi_{\sigma(s)}$, we obtain

$$d i_X \omega_\phi = \frac{\sqrt{-1}}{2}\partial\bar\partial\dot\phi.$$

On the other hand, as above, there is a unique θ satisfying:

$$i_X\omega_\phi = \frac{\sqrt{-1}}{2}\bar\partial\theta \text{ and } \int_M \theta\, e^{h_{\omega_\phi}}\,\omega_\phi^n = 0.$$

Then we have $\dot\phi = \mathrm{Re}(\theta + C)$. Plugging this into (2.11) and integrating by parts, we easily deduce (2.12). □

Lemma 2.9. *For any $\phi \in P(M,\omega)$, the function $\mathbf{I}_\omega(\phi_\sigma) - \mathbf{J}_\omega(\phi_\sigma)$ defined on the* $\mathrm{Aut}_0(M)$ *is K-invariant and convex.*

Proof. The K-invariance follows from

$$\mathbf{I}_\omega(\phi_\sigma) - \mathbf{J}_\omega(\phi_\sigma) = \mathbf{I}_{\sigma^*\omega}(\phi_\sigma) - \mathbf{J}_{\sigma*\omega}(\phi_\sigma) = \mathbf{I}_\omega(\phi) - \mathbf{J}_\omega(\phi),$$

where $\sigma \in K$. For the convexity, we first observe that being reductive, $\eta(M)$ is generated by holomorphic vector fields X satisfying: $i_X\omega = \frac{\sqrt{-1}}{2}\bar\partial\theta$ for some real-valued function θ. The condition means that $\mathrm{Im}(X)$ is a Killing field. Let $\sigma(s)$ be the one-parameter subgroup generated by $\mathrm{Re}(X)$. Write $\phi(s,x) = \phi_{\sigma(s)}(x)$, then one can easily show

$$\frac{\partial\phi}{\partial s}(s,x) = \frac{\partial\phi}{\partial s}(0,\sigma(s)(x)).$$

It follows that

$$\frac{\partial^2\phi}{\partial s^2} = d\left(\frac{\partial\phi}{\partial s}\right)(\mathrm{Re}(X)) = \left|\partial\left(\frac{\partial\phi}{\partial s}\right)\right|^2_{\omega_\phi}.$$

Here we need to use the fact that θ is real. Hence, $\phi_{\sigma(s)}$ is a geodesic with respect to the L^2-metric on the space of Kähler metrics. Then the convexity follows from an observation of X.X. Chen. For the readers' convenience, we sketch a proof as follows: As shown in [Ti87] (also see (2.12)), we have

$$\frac{d}{ds}\left(\mathbf{I}_\omega(\phi_{\sigma(s)}) - \mathbf{J}_\omega(\phi_{\sigma(s)})\right) = \frac{1}{V}\int_M \dot\phi\,(\omega - \omega_\phi)\wedge\omega_\phi^{n-1},$$

where $\dot\phi = \frac{\partial\phi}{\partial s}$. This implies

$$\begin{aligned}&\frac{d^2}{ds^2}\left(\mathbf{I}_\omega(\phi_{\sigma(s)}) - \mathbf{J}_\omega(\phi_{\sigma(s)})\right)\\ &= \frac{1}{V}\int_M |\partial\dot\phi|^2_{\omega_{\phi(s,\cdot)}}\,\omega\wedge\omega^{n-1}_{\phi(s,\cdot)} - (n-1)\frac{\sqrt{-1}}{2}\,\partial\dot\phi\wedge\bar\partial\dot\phi\wedge\omega\wedge\omega^{n-2}_{\phi(s,\cdot)}.\end{aligned}$$

It is easy to see that the integrand in the above integral is non-negative. The lemma is proved. □

Corollary 2.10. *For any solution $\phi = \varphi_t$ of (1.2) with $t < 1$, the minimum of $\mathbf{I}_\omega(\phi_\sigma) - \mathbf{J}_\omega(\phi_\sigma)$ is attained at $\sigma = \mathrm{Id}$.*

Proof. It follows from (2.10)

$$h_{\omega_t} = -(1-t)\varphi_t + a_t. \tag{2.13}$$

Combining this with (2.12), we get

$$\frac{d}{ds}\left(\mathbf{I}_\omega(\phi_{\sigma(s)}) - \mathbf{J}_\omega(\phi_{\sigma(s)})\right)\Big|_{s=0} = -\mathrm{Re}\left(\frac{1}{(1-t)V}\int_M X(h_{\omega_t})\,\omega_\phi^n\right). \tag{2.14}$$

The integral on the right is simply the Futaki invariant. Since M is analytically semi-stable, the Futaki invariant vanishes identically. Therefore, φ_t is a critical point of $\mathbf{I}_\omega - \mathbf{J}_\omega$ restricted to the orbit of φ_t by $\mathrm{Aut}_0(M)$, so by last lemma, it has to be the minimum. □

It follows from the above corollary and the definition of weak analytic stability, we obtain

$$\mathbf{F}_\omega(\varphi_t) \geq f(\mathbf{I}_\omega(\varphi_t) - \mathbf{J}_\omega(\varphi_t)).$$

But it is known that $\mathbf{F}_\omega$ is bounded from above, so $\mathbf{I}_\omega(\varphi_t) - \mathbf{J}_\omega(\varphi_t)$ is bounded. The theorem is proved.

2.3. From existence to properness

In this subsection, we give a converse to Theorem 2.6. This is amount to proving a differential inequality. For simplicity, we discuss only the case that $\eta(M) = \{0\}$. We will leave the general case to a future paper since the proof is more involved.

Theorem 2.11. ([Ti97]) *Let (M,ω) be a Kähler-Einstein manifold with $\mathrm{Ric}(\omega) = \omega$ and without any non-trivial holomorphic vector fields. Then there are $c > 0$ and $\epsilon \in (0,1)$[5] such that for any $\varphi \in P(M,\omega)$, we have*

$$\mathbf{F}_\omega(\varphi) \geq \frac{\mathbf{J}_\omega(\varphi)}{c(1+\|\varphi\|_{C^0})^{1-\epsilon}} - c. \tag{2.15}$$

This theorem was first given in [Ti97]. In the following, we will outline its proof. The readers can find all the details in [Ti97].

Let φ be given in the theorem and write $\omega' = \omega_\varphi$. The following lemma is well known and was first proved by Bando-Mabuchi ([BM86], also see [Ti97]).

Lemma 2.12. *There is a unique family $\{\varphi_t\} \subset P(M,\omega')$ such that $\varphi_1 = -\varphi$ and*

$$\left(\omega' + \frac{\sqrt{-1}}{2}\partial\bar\partial\varphi_t\right)^n = e^{h_{\omega'} - t\varphi_t}\,\omega'^n.$$

[5]This ϵ depends only on n. It is possible to give an explicit estimate on ϵ by examining the arguments in the proof. However, this is inessential.

Now we rewrite the equation in the above lemma as

$$\left(\omega + \frac{\sqrt{-1}}{2}\partial\bar{\partial}\psi_t\right)^n = e^{(1-t)\varphi_t - \psi_t}\omega^n,$$

where $\psi_t = \varphi_t - \varphi_1 = \varphi_t + \varphi$.

Put $\omega_t = \omega' + \frac{\sqrt{-1}}{2}\partial\bar{\partial}\varphi_t = \omega + \frac{\sqrt{-1}}{2}\partial\bar{\partial}\psi_t$, then $\omega_1 = \omega$ and

$$\mathrm{Ric}(\omega_t) = t\omega_t + (1-t)\omega' \geq t\omega_t, \; h_{\omega_t} = -(1-t)\varphi_t + a_t,$$

where a_t is chosen such that

$$\int_M \left(e^{-(1-t)\varphi + a_t} - 1\right)\omega_t^n = 0.$$

It follows that $|a_t| \leq (1-t)\|\varphi_t\|_{C^0}$ and $\Delta_{\omega_t} h_{\omega_t} + n\,(1-t) > 0$.

Applying Proposition 3.1 in [Ti97] to each ω_t, we obtain a new Kähler metric $\omega_t' = \omega_t + \partial\overline{\partial}u_t$ satisfying:

$$\|u_t\|_{C^0} \leq 3(1-t)\|\varphi_t\|_{C^0}, \tag{2.16}$$

$$\|h_{\omega_t'}\|_{C^{\frac{1}{2}}} \leq C(n,\omega_t')(1 + (1-t)^2\|\varphi_t\|^2_{C^0})^{n+1}(1-t)^{\beta}, \tag{2.17}$$

where $\beta = \beta(n) \in (0,1)$ and $C(n,\omega_t')$ depends only on the first non-zero eigenvalue and the Sobolev constant of (M,ω_t'). Note that ω_t' is obtained by evolving ω_t along the Kähler-Ricci flow in time 1 (see [Ti97], Section 3).

Choose μ_t such that

$$\int_M \left(e^{h_{\omega_t} - u_t + \mu_t} - 1\right)\omega_t^n = 0.$$

Then $|\mu_t| \leq 6(1-t)\|\varphi_t\|_{C^0}$.

An application of the maximum principle implies

$$\varphi_t = \varphi_1 - w_t - u_t + \mu_t + a_t, \tag{2.18}$$

where w_t is the unique solution of the following $\omega = \omega_t' + \partial\overline{\partial}w_t$ and

$$\omega^n = e^{h_{\omega_t'} - w_t}\omega_t'^n. \tag{2.19}$$

Hence, φ_t is uniformly equivalent to φ_1 as long as w_t is uniformly bounded.

Define

$$\Phi_t(w) = \log\left(\frac{(\omega_1 - \partial\overline{\partial}w)^n}{\omega_1^n}\right) + h_{\omega_t'} - w. \tag{2.20}$$

Clearly, $\Phi_t : C^{2,\frac{1}{2}}(M) \mapsto C^{0,\frac{1}{2}}(M)$. A direct computation shows that the linearization of Φ at $w = 0$ is $-\Delta - 1$. Since $\eta(M) = \{0\}$, using the standard Bochner identity, one can show that it is invertible. Then by the Inverse Function Theorem, there is a $\delta > 0$ such that $\Phi_t(w) = 0$ has a unique solution w whenever

$$\|h_{\omega_t'}\|_{C^{\frac{1}{2}}(\omega_1)} < \delta,$$

furthermore, we have

$$\|w\|_{C^{2,\frac{1}{2}}(\omega_1)} \leq C\delta.$$

Here C denotes a uniform constant.

Note that $C(n,\tilde{\omega})$ is uniformly bounded by a constant $\bar{c}$ for any metric $\tilde{\omega}$ satisfying:

$$\frac{1}{2}\omega \leq \tilde{\omega} \leq 2\omega.$$

Now we choose t_0 such that

$$\begin{aligned}\delta' &= (1-t_0)^\beta(1+(1-t_0)^2\|\varphi_{t_0}\|^2_{C^0})^{n+1}\\ &= \sup_{t_0\leq t\leq 1}(1-t)^\beta(1+(1-t)^2\|\varphi_t\|^2_{C^0})^{n+1},\end{aligned} \tag{2.21}$$

where $\bar{c}\delta' = \delta$. We may further assume that $C\delta < \frac{1}{4}$. Then for any $t \in [t_0, 1]$,

$$\|w_t\|_{C^{2,\frac{1}{2}}(\omega_1)} < \frac{1}{4}.$$

Summarizing the above estimates, we can deduce that for any $t \in [t_0, 1]$,

$$\|\varphi_t\|_{C^0} \geq (1-10(1-t))\|\varphi_1\|_{C^0} - 1. \tag{2.22}$$

Now we recall an identity due to Ding-Tian:

$$t\left(\mathbf{J}_{\omega'}(\varphi_t) - \frac{1}{V}\int_M \varphi_t\omega'^n\right) = -\int_0^t (\mathbf{I}_{\omega'}(\varphi_s) - \mathbf{J}_{\omega'}(\varphi_s))\,ds. \tag{2.23}$$

It follows

$$\mathbf{F}_{\omega'}(\varphi_1) = -\int_0^1 (\mathbf{I}_{\omega'}(\varphi_t) - \mathbf{J}_{\omega'}(\varphi_t))\,dt.$$

Since $I_\omega(\varphi_t) - J_\omega(\varphi_t)$ is nondecreasing, we deduce from this

$$\begin{aligned}\mathbf{F}_\omega(\varphi) &= -\mathbf{F}_{\omega'}(\varphi_1)\\ &\geq \min\left\{1-t_0, \frac{1}{12}\right\}(\mathbf{I}_{\omega'}(\varphi_{t_0}) - \mathbf{J}_{\omega'}(\varphi_{t_0}))\\ &\geq \min\left\{1-t_0, \frac{1}{12}\right\}(\mathbf{I}_{\omega'}(\varphi_1) - \mathbf{J}_{\omega'}(\varphi_1)) - 20(1-t_0)^2\|\varphi_{t_0}\|_{C^0} - 2.\end{aligned} \tag{2.24}$$

Then the theorem follows from this last inequality by some simple manipulation.

Since $J_\omega(\varphi)$ and $\|\varphi\|_{C^0}$ bound each other along (1.2), Theorem 2.11 can be regarded as a converse to Theorem 2.6. However, we can do better and get a genuine converse. In the following, we present a result of Tian-Zhu from [TZ97], which improves Theorem 2.11.

By (2.23) and the monotonicity of $\mathbf{I}_{\omega'} - \mathbf{J}_{\omega'}$, we get

$$\mathbf{F}_{\omega'}(\varphi_1) - t\left(\mathbf{J}_{\omega'}(\varphi_t) - \frac{1}{V}\int_M \varphi_t\omega'^n\right) \geq -(1-t)\left(\mathbf{I}_{\omega'}(\varphi_1) - \mathbf{J}_{\omega'}(\varphi_1)\right). \tag{2.25}$$

By the concavity of the logarithm, we have

$$-\log\left(\frac{1}{V}\int_M e^{h_{\omega'}-\varphi_t}\omega'^n\right) \leq \frac{1-t}{V}\int_M \varphi_t\omega_t^n.$$

Putting this and (2.25) together, we get

$$\begin{aligned}\mathbf{F}_\omega(\varphi_t-\varphi_1) &= \mathbf{F}_{\omega'}(\varphi_t)-\mathbf{F}_{\omega'}(\varphi_1) \\ &\le (1-t)\,(\mathbf{I}_{\omega'}(\varphi_1)-\mathbf{J}_{\omega'}(\varphi_1)-\mathbf{I}_{\omega'}(\varphi_t)+\mathbf{J}_{\omega'}(\varphi_t)) \\ &\le (1-t)(\mathbf{I}_{\omega'}(\varphi_1)-\mathbf{J}_{\omega'}(\varphi_1)) \;=\; (1-t)\,\mathbf{J}_\omega(\varphi).\end{aligned} \tag{2.26}$$

Next we recall the following lemma.

Lemma 2.13. *For any $t\in[\frac{1}{2},1]$,*

$$\|\varphi_t-\varphi_1\|_{C^0} \;\le\; c(1+\mathbf{J}_\omega(\varphi_t-\varphi_1)), \tag{2.27}$$

where c is a uniform constant.

Proof. Both ω and $\omega_t = \omega_{\varphi_t-\varphi_1}$ have positive Ricci curvature $\ge 1/2$. So there are uniform bounds on the Sobolev constants of both metrics. Then (2.27) follows from the standard Moser iteration. □

Using (2.27) and applying Theorem 2.11, we get

$$\mathbf{F}_\omega(\varphi_t-\varphi_1) \;\ge\; c'\mathbf{J}_\omega(\varphi_t-\varphi_1)^\epsilon - C. \tag{2.28}$$

Then we proceed as before

$$\begin{aligned}\mathbf{F}_\omega(\varphi) &= -\mathbf{F}_{\omega'}(\varphi_1) \\ &\ge (1-t)\,(\mathbf{I}_{\omega'}(\varphi_t)-\mathbf{J}_{\omega'}(\varphi_t)) \\ &\ge (1-t)\,(\mathbf{I}_{\omega'}(\varphi_1)-\mathbf{J}_{\omega'}(\varphi_1)) - c(1-t)\|\varphi_t-\varphi_1\|_{C^0} \\ &= (1-t)\mathbf{J}_\omega(\varphi) - c(1-t)\|\varphi_t-\varphi_1\|_{C^0} \\ &\ge (1-t)\mathbf{J}_\omega(\varphi) - c''(1-t)\mathbf{J}_\omega(\varphi_t-\varphi_1).\end{aligned} \tag{2.29}$$

If $\mathbf{J}_\omega(\varphi_t-\varphi_1)$ is uniformly bounded for $t\in[1/2,1]$, then we simply take $t=1/2$ in the above inequality. Otherwise, we first note that (2.26) and (2.28) imply

$$c'\mathbf{J}_\omega(\varphi_t-\varphi_1)^\epsilon - C \;\le\; (1-t)\,\mathbf{J}_\omega(\varphi).$$

Hence, we can choose $t\in[1/2,1]$ such that

$$(1-t)\mathbf{J}_\omega(\varphi)^{1-\epsilon}=1.$$

Then we have proved the following result due to Tian-Zhu:

Theorem 2.14. ([TZ97]) *Let (M,ω) be a Kähler-Einstein manifold with* $\mathrm{Ric}(\omega)=\omega$ *and without any non-trivial holomorphic vector fields. Then there are $C>0$ and $\epsilon\in(0,1)$ such that for any $\varphi\in P(M,\omega)$, we have*

$$\mathbf{F}_\omega(\varphi) \;\ge\; \mathbf{J}_\omega(\varphi)^\epsilon - C. \tag{2.30}$$

This gives an exact converse to Theorem 2.6 in the case that M does not have any non-trivial holomorphic fields. It is possible to remove this condition on holomorphic fields. More arguments are needed and will be discussed elsewhere.

Inspired by the above discussions, we can propose the following conjecture in [Ti97]. This is a natural extension of the famous Moser-Trudinger-Onofri inequality on S^2.

Conjecture 2.15. *Let (M, ω) be a Kähler-Einstein manifold with* $\mathrm{Ric}(\omega) = \omega$. *Then there are constants $\eta > 0$ and $C > 0$ such that for any $\varphi \in P(M, \omega)$ which is perpendicular to the kernel of $\Delta_\omega + 1$ (possibly trivial),*

$$\mathbf{F}_\omega(\varphi) \geq \eta \, \mathbf{J}_\omega(\varphi) - C. \tag{2.31}$$

When $\eta(M) = \{0\}$, this conjecture was verified by Phong et al. [PSSW06] following arguments in [Ti97] and [TZ97]. For the readers' convenience, we reproduce their arguments here. Recall that (2.26) and (2.27) yield for $t \geq 1/2$,

$$\mathbf{F}_\omega(\varphi_t - \varphi_1) \leq C'(1-t)\mathbf{J}_\omega(\varphi_t - \varphi_1).$$

Combining this with (2.28), we get

$$c'\mathbf{J}_\omega(\varphi_t - \varphi_1)^\epsilon - C \leq C''(1-t)\mathbf{J}_\omega(\varphi_t - \varphi_1).$$

If $C''\mathbf{J}_\omega(\varphi_{\frac{1}{2}} - \varphi_1)^{1-\epsilon} \leq c'$, then (2.29) with $t = 1/2$ yields (2.31) with $\eta = 1/2$ Otherwise, there is some $t' \in (1/2, 1)$ such that $C''(1-t')\mathbf{J}_\omega(\varphi_{t'} - \varphi_1)^{1-\epsilon} = c'/2$, then

$$c'\mathbf{J}_\omega(\varphi_{t'} - \varphi_1)^\epsilon \leq 2C.$$

Consequently, we have $1 - t' \geq \eta$ if η is sufficiently small, so we still deduce (2.31) from (2.29). therefore, the conjecture holds in the case that M has no non-trivial holomorphic fields.

3. Kähler-Einstein metrics on complex surfaces

This section concerns the existence of Kähler-Einstein metrics on complex surfaces with positive first Chern class, i.e., Del-Pezzo surfaces. This is completely solved in [Ti89]. Let us recall the main theorem in [Ti89].

Theorem 3.1. *Let M be a compact complex surface with positive first Chern class. Then M admits a Kähler-Einstein metric if and only if it has vanishing Futaki invariant.*

Of course, vanishing of the Futaki invariant is a necessary condition for the existence. In this section, we will first outline a proof of this theorem following arguments in [Ti89]. The main idea of my proof can be briefly described as follows: The main block of the proof is to prove the existence of Kähler-Einstein metrics on a Del-Pezzo surface M with vanishing Futaki invariant. First we establish the existence of Kähler-Einstein metrics on certain complex surface in the moduli of complex structures on M. This is done by computing the α-invariant I introduced. It turns out that the moduli is connected and the subset of those which admit Kähler-Einstein metrics is open. To prove that such a subset is also closed, we need an a prior C^0-estimate for solutions of certain complex Monge-Ampere equation. We achieve it in two steps:

1. We establish a partial C^0-estimate. This is done by proving a compactness theorem and using the L^2-estimate for the $\bar{\partial}$-operators;

2. With the partial C^0-estimate, one can reduce the properness of $\mathbf{F}_\omega$ to the properness of the restriction of $\mathbf{F}_\omega$ to certain finite-dimensional space of "algebraically" defined Kähler metrics.[6] We then develop a method of verifying this finite-dimensional properness.

3.1. The α-invariant and its applications

First let us introduce the α-invariant and state its basic properties for a general Fano manifold. As before, (M,ω) is a compact Kähler manifold with ω representing $\pi c_1(M)$. The α-invariant $\alpha(M)$, which is introduced in [Ti87], is defined as follows:

$$\alpha(M) := \sup\left\{\alpha \mid \int_M e^{-\alpha(\varphi - \sup_M \varphi)}\,\omega^n \leq C_\alpha, \forall\varphi \in P(M,\omega)\right\}, \tag{3.1}$$

where C_α denotes a constant depending only on α. If G is a maximal compact subgroup G in the automorphism group $\mathrm{Aut}(M)$, the we can choose ω to be G-invariant and define

$$\alpha_G(M) := \sup\left\{\alpha \mid \int_M e^{-\alpha(\varphi - \sup_M \varphi)}\,\omega^n \leq C_\alpha, \forall\varphi \in P(M,\omega)\right\}, \tag{3.2}$$

where $P_G(M,\omega)$ denotes the set of all $\varphi \in P(M,\omega)$ which is G-invariant. It is not hard to prove that both $\alpha(M)$ and $\alpha_G(M)$ are independent of the choices of ω and G, so they are holomorphic invariants of M. Moreover, we always have $\alpha_G(M) \geq \alpha(M) > 0$. The following theorem is proved in [Ti87] and provides a useful tool for establishing the existence of Kähler-Einstein metrics on Fano manifolds.

Theorem 3.2. *A Fano manifold M of dimension n admits a Kähler-Einstein metric if $\alpha(M) > \frac{n}{n+1}$ or more weakly, $\alpha_G(M) > \frac{n}{n+1}$.*

In fact, under the assumption on $\alpha(M)$ (resp. $\alpha_G(M)$), one can show (cf. [Ti98]) that $\mathbf{F}_\omega$ is proper on $P(M,\omega)$ (resp. $P_G(M,\omega)$), so this theorem follows from Theorem 2.6 or its variant for G-invariant functions.

Now we apply Theorem 3.2 to Del-Pezzo surfaces. By the classification theory of complex surfaces, M is either $\mathbb{C}P^2$ or $S^2 \times S^2$ or the blow-up Σ_m of $\mathbb{C}P^2$ at m points $(1 \leq m \leq 8)$ in general position. Here the general position means that no three points are collinear and no six points are on a common quadratic curve and there are no cubic curves which contain 7 points and the 8th point as a double point. This is equivalent to saying that $c_1(\Sigma_m) > 0$.

The first two surfaces are homogeneous and so have canonical Kähler-Einstein metrics. It is proved in [Fut83] that M has a Kähler-Einstein metrics only if its associated Futaki invariant vanishes. It was also known that the Futaki invariant of $M = \Sigma_m$ is nonzero if and only if $m = 1$ or 2. Therefore, it suffices to establish the existence of Kähler-Einstein metrics on Σ_m for $3 \leq m \leq 8$.

The following is the main result in [TY87] and proved by computing $\alpha_G(M)$ for a maximal compact subgroup G in $\mathrm{Aut}(M)$.

---

[6] In fact, this properness is equivalent to the K-stability condition.

Theorem 3.3. *For m between 3 and 8, there is at least one blow-up $M = \Sigma_m$ of $\mathbb{C}P^2$ with $c_1(M) > 0$ which admits a Kähler-Einstein metric.*

What we did in [TY87] is to prove $\alpha_G(\Sigma_m) > 2/3$ for each $m \in [3, 8]$ and certain configuration of blow-up points such that Σ_m has sufficiently many symmetries. Denote by $\mathcal{M}_m$ the moduli spaces of Σ_m. It consists of all possible m points in $\mathbb{C}P^2$ in general position. Here is a summary of our computation of $\alpha_G(M)$:

(1) There is only one complex surface $\Sigma_3 \in \mathcal{M}_3$ and $\alpha_G(\Sigma_3) \geq 1$;
(2) There is only one surface $\Sigma_4 \in \mathcal{M}_4$ and $\alpha_G(\Sigma_4) \geq 3/4$;
(3) Every surface $\Sigma_5 \in \mathcal{M}_5$ is a complete intersection of two quadrics in $\mathbb{C}P^4$. If two quadratic polynomials are given by $\sum_{i=0}^4 z_i^2 = 0$ and $\sum_{i=0}^4 \lambda_i z_i^2 = 0$, then $\alpha_G(\Sigma_5) \geq 1$. It was pointed out in [MaMu93] that this is always the case for smooth Σ_5;
(4) Every surface in $\mathcal{M}_6$ is a cubic surface in $\mathbb{C}P^3$. If Σ_6 is the Fermat surface, then $\alpha_G(\Sigma_6) \geq 1$;
(5) Every surface M in $\mathcal{M}_7$ is a double branch covering of $\mathbb{C}P^2$ along a quartic curve Q. Then $\alpha_G(M) \geq 3/4$ when Q is a quartic curve with certain finite symmetries[7];
(6) Certain Σ_8 in $\mathcal{M}_8$ with finite symmetries has $\alpha_G(\Sigma_8) \geq 5/6$.

Note that the complex structure on Σ_3 or Σ_4 is unique. When $m \geq 5$, there is a connected complex moduli of Σ_m of dimension $m - 4$.

Since [TY87], many advances have been made on computing $\alpha(\Sigma_m)$. Let us mention a few results which are directly related to Theorem 3.3. First we note that each Σ_5 is a complete intersection of two quadrics in $\mathbb{C}P^4$. As Mabuchi-Mukai stated in [MaMu93], one can diagonalize those two quadratics simultaneously. Then the arguments in [TY87] can be used to show that $\alpha_G(\Sigma_5) \geq 1$. In this way, Mabuchi-Mukai gave a simplified proof of Theorem 3.1 in the case Σ_5. More recently, I. Cheltsov et al. proved that $\alpha(\Sigma_m) \geq 2/3$ for all $m \geq 5$ and $\alpha(\Sigma_m) \geq 3/4$ for $m \geq 6$ except $m = 6$ and Σ_m has an Eckardt point [Ch07]. Hence, there is an alternative proof of Theorem 3.1 by using directly α-invariants for all Del-Pezzo surfaces except for those cubic surfaces Σ_6 with an Eckardt point. We refer the readers to the Appendix for more about cubic surfaces with an Eckardt point.

3.2. Compactness for Kähler-Einstein metrics

To prove Theorem 3.1, we may assume that $m = 5, 6, 7, 8$. One can show that $\mathcal{M}_m$ is connected. Now we can set up a new continuity method: Let $\mathcal{E}_m$ be the subset of all $M \in \mathcal{M}_m$ which admits a Kähler-Einstein metric. It follows from the last subsection that $\mathcal{E}_m$ is nonempty. Choose a smooth family of Kähler metrics ω_τ on $M_\tau \in \mathcal{M}_m$ with Kähler class $\pi c_1(M_\tau)$ and $M_0 \in \mathcal{E}_m$, where $\tau \in [0, 1]$. Then

---

[7] It was proved in [Ti98] that every surface in $\mathcal{M}_7$ has proper $\mathbf{F}_\omega$ on $P_G(M, \omega)$ for a maximal compact subgroup $G \subset \mathrm{Aut}(M)$, so it has a Kähler-Einstein metric. This gives a simpler proof of Theorem 3.1 for Σ_7.

M_τ admits a Kähler-Einstein metric if and only if the following Monge-Ampère equation is solvable

$$(\omega_\tau + \frac{\sqrt{-1}}{2}\partial\bar{\partial}\varphi)^2 = e^{h_\tau - \varphi}\omega_\tau^2, \ \omega_\tau + \frac{\sqrt{-1}}{2}\partial\bar{\partial}\varphi > 0 \text{ on } M_\tau, \tag{3.3}$$

where h_τ is determined by

$$\mathrm{Ric}(\omega_\tau) - \omega_\tau = \frac{\sqrt{-1}}{2}\partial\bar{\partial}h_\tau, \ \int_{M_\tau}(e^{h_\tau} - 1)\omega_\tau^2 = 0.$$

Since $m \geq 5$, any surface M_τ does not have any nontrivial holomorphic vector fields. Then by using the standard Bochner trick, one can show that for any solution φ of (3.3), the first non-zero eigenvalue of $\Delta_{\tau,\varphi}$ is strictly bigger than 1, where $\Delta_{\tau,\varphi}$ denotes the Laplacian of the metric $\omega_\tau + \frac{\sqrt{-1}}{2}\partial\bar{\partial}\varphi$. It follows that if (3.3) is solvable on M_τ, so is every $M_{\tau'}$ sufficiently close to M_τ. This is a simple application of the Implicit Function Theorem since the linearization of (3.3) is given by $\Delta_{\tau,\varphi} + 1$. It implies that the set I_m of $\tau \in [0,1]$ with $M_\tau \in \mathcal{E}_m$ is open, and consequently, $\mathcal{E}_m$ is open in $\mathcal{M}_m$. It remains to show that I_m is closed. For this, we need an a priori C^3-estimate on solutions of (3.3). As we explained in [Ti87] or at the end of Section 1, this C^3-estimate follows from an a priori C^0-estimate. In general, there does not exist such an estimate for an equation of the type like (3.3). The idea of [Ti89] is to derive a partial C^0-estimate and then use geometric information of underlying manifolds to check if the required C^0-estimate holds. Now let us recall the partial C^0-estimate.

Theorem 3.4. [Ti89] *There are two constants $c > 0$ and $l_0 > 0$ such that for any Kähler-Einstein surface (M,ω) with* $\mathrm{Ric}(\omega) = \omega$, *there is some $l \in [l_0, 2l_0]$ such that*

$$c^{-1} \geq \frac{1}{l}\sum_{i=0}^{N}\|S_i\|^2 \geq c > 0, \tag{3.4}$$

where $\{S_i\}_{0\leq i\leq N}$ is any orthonormal basis of $H^0(M, K_M^{-l})$ with respect to the inner product induced by ω.[8]

Let us first explain why it implies a partial C^0-estimate[9]: Let $M = \mathfrak{M}_\tau$ and φ be a solution of (3.3). Write ω for $\omega_\tau + \frac{\sqrt{-1}}{2}\partial\bar{\partial}\varphi$. Choose a Hermitian metric $\|\cdot\|_\tau$ on K_M^{-1} such that its curvature form is ω_τ. This Hermitian metric and ω_τ induces an inner product on $H^0(M, K_M^{-l})$. We may choose $\{S_i\}$ in (3.4) such that $\{\mu_i S_i\}$ is an orthonormal basis of this inner product associated to ω_τ for some positive constants μ_i $(i = 0, \ldots, N)$. It follows from the Maximum principle

$$\varphi - \frac{1}{l}\log\left(\frac{\sum_{i=0}^N \|S_i\|_\tau^2}{\sum_{i=0}^N \|S_i\|^2}\right) = c', \tag{3.5}$$

[8]In [Ti89], (3.4) is actually proved for any $l = 6k > 0$. It was also conjectured that (3.4) holds for any l sufficiently large.

[9]This derivation from (3.4) to the partial C^0-estimate, i.e., (3.6), for solutions of (3.3) holds for any dimensions.

where c' is some constant. Write $\sigma_i = \mu_i S_i$. We may arrange $\mu_0 \geq \mu_1 \geq \cdots \geq \mu_N$ and put $\lambda_i = \mu_N/\mu_1$. Then $\lambda_0 \leq \lambda_1 \leq \cdots \leq \lambda_N = 1$, also it follows from Theorem 3.4

$$\left\| \varphi - \sup_M \varphi - \frac{1}{l} \log \left(\sum_{i=0}^{N} \lambda_i^2 \|\sigma_i\|_\tau^2 \right) \right\|_{C^0} \leq C, \tag{3.6}$$

Note that C always denotes a uniform constant. In particular, $\varphi - \sup_M \varphi$ is bounded away from the zero locus of σ_N. So we have a partial C^0-estimate of $\varphi - \sup_M \varphi$.

Now we discuss how to prove Theorem 3.4. The key is the following compactness theorem for Kähler-Einstein metrics proved in [Ti89].

Theorem 3.5. *For any sequence of Kähler-Einstein surfaces (M_i, ω_i) with $M_i \in \mathcal{E}_m$ and $\mathrm{Ric}(\omega_i) = \omega_i$, there is a subsequence, for simplicity, still denoted by $\{(M_i, \omega_i)\}$, converging to a Kähler-Einstein orbifold $(M_\infty, \omega_\infty)$ in the Cheeger-Gromov topology satisfying:*

(1) *The singularities are of the form U/Γ and the number of them is uniformly bounded, where $U \subset \mathbb{C}^2$ and Γ is a finite group of $U(2)$ with uniformly bounded order;*
(2) *For each $\ell > 0$, $H^0(M_i, K_{M_i}^{-\ell})$ converge to $H^0(M_\infty, K_{M_\infty}^{-\ell})$ in the following sense[10]: There are orthonormal bases $\{S_0^i, \ldots, S_N^i\}$ of $H^0(M_i, K_{M_i}^{-\ell})$ converging to an orthonormal basis of $H^0(M_\infty, K_{M_\infty}^{-\ell})$.*

Remark 3.6. A real version of this theorem without (2) was given in [An90] with an earlier and weaker version given in [Na88]. We note that (2) is crucial in our deriving the partial C^0-estimate.

Remark 3.7. It was conjectured in [Ti89] that those quotient singularities are all rational double points, that is, the uniformization group Γ in (1) of Theorem 3.5 lies in $SU(2)$. Indeed, a result of Mabuchi-Mukai implies this for $m = 5$ [MaMu93]. In fact, we know more about Γ in [Ti89]: Choose $p_i \in M_i$ which converge to a singularity $p_\infty \in M_\infty$ which has a neighborhood of the form U/Γ, then there are open neighborhoods V_i of p_i in M_i which is a finite quotient of a smooth deformation of U/Γ_0 for some $\Gamma_0 \subset SU(2)$. It follows that either $\Gamma \subset SU(2)$ or Γ is cyclic. Using the volume comparison and the bound $c_1(M)^2([M]) = 9 - m$, one can show that the order $\Gamma/\Gamma_0 \leq 6$.

It follows that the moduli space $\mathcal{E}_m$ of Kähler-Einstein surfaces with positive scalar curvature can be compactified by adding Kähler-Einstein orbifolds with isolated singularities described in the above theorem. Clearly, Theorem 3.4 follows from this: Let ℓ is a product of orders of all finite groups which appear in the quotient singularities in the compactification, then for any orbifold $(M_\infty, \omega_\infty)$ in the boundary of $\mathcal{E}_m$ and any $p \in M_\infty$, there is at least one section $S \in H^0(M_\infty, K_{M_\infty}^{-\ell})$

[10]This can be regarded as a refined version of the flatness for a family of complex manifolds.

with its L^2-norm equal to 1 such that $\|S(p)\| \geq c$ for a uniform constant $c = c(\ell)$. Thus we can easily deduce (3.4).

To prove (1) in Theorem 3.5, we recall that for any Kähler-Einstein surface (M, ω) with $M \in \mathcal{E}_m$ and $\mathrm{Ric}(\omega) = \omega$, we have a fixed volume:

$$\mathrm{Vol}(M, \omega) := \int_M \omega^n = \pi^n c_1(M)^n([M]),$$

and by the Gauss-Bonnet-Chern formula,

$$\frac{1}{8\pi^2} \int_M \|Rm(\omega)\|_\omega^2 \omega^n = c_2(M)([M]),$$

where $Rm(\cdot)$ denotes the curvature tenor and $c_2(M)$ denotes the second Chern class. This implies that the L^2-norm of the curvature is uniformly bounded. Moreover, by the Meyer theorem, the diameter of (M, ω) is uniformly bounded, so we have the uniform Sobolev inequality. In [Ti89], using this Sobolev bound and adopting Uhlenbeck's estimate for Yang-Mills connections, I gave a curvature estimate in terms of the local L^2 norm of curvature. This enables us to bound curvature outside finitely many points of M. Then we can deduce (1) from this curvature estimate and a removable singularity theorem (cf. [Ti89]).

To prove (2) in Theorem 3.5, we use Hömander's L^2-estimate[11]. First we notice that using the Bochner identity and the Moser iteration, one can bound $\|S\|_{\omega_i}$ for any section $S \in H^0(M_i, K_{M_i}^{-\ell})$ with its L^2-norm equal to 1. Therefore, by taking a subsequence if necessary, we may assume that $\{S_a^i\}_{0 \leq a \leq N}$ converge to a sub-basis of $H^0(M_\infty, K_{M_\infty}^\ell)$. So it suffices to show that any section $S \in H^0(M_\infty, K_{M_\infty}^\ell)$ is the limit of a sequence of sections on M_i. This is done by using the fact that the singular set is of codimension at least 4 and the L^2-estimate.

3.3. Analytic stability in finite dimensions

We adopt the notations from the last subsection. In the final stage of proving Theorem 3.1, we will prove that for any solution φ of (3.3),

$$\mathbf{F}_{\omega_\tau}(\varphi) \geq \epsilon \mathbf{J}_{\omega_\tau}(\varphi) - C, \tag{3.7}$$

where $\epsilon > 0$ is independent of $\tau \in I_m$, and φ solves (3.3).[12] Note that C always denotes a uniform constant in this section. By (2.23), we have

$$\mathbf{F}_{\omega_\tau}(\varphi) \leq 0.$$

It follows from (3.7) that $\mathbf{J}_{\omega_\tau}(\varphi)$ is uniformly bounded and consequently, $\|\varphi\|_{C^{2,\frac{1}{2}}}$ is uniformly bounded. It follows that I_m is closed in [0,1]. Since $\{M_\tau\}$ can be any family in $\mathcal{M}_m$ with $M_0 \in \mathcal{E}_m$. we have $\mathcal{E}_m = \mathcal{M}_m$, thus Theorem 3.1 is proved.

[11]We should point out that arguments in proving (2) hold for all dimensions. More details will be given in the next section.
[12]This means that $\mathbf{F}_{\omega_\tau}$ is proper along the solutions of (3.3).

Now let us prove (3.7) for M_τ. Since φ is a solution of (3.3), we have

$$\begin{aligned}\mathbf{F}_{\omega_\tau}(\varphi) &= \frac{\sqrt{-1}}{3V}\int_M \partial\varphi\wedge\overline{\partial}\varphi\wedge\omega_\tau + \frac{\sqrt{-1}}{6V}\int_M \partial\varphi\wedge\overline{\partial}\varphi\wedge\omega - \frac{1}{V}\int_M \varphi\omega_\tau^2 \\ &= -\frac{1}{3V}\int_M \varphi\omega^2 - \frac{2}{3V}\int_M \varphi\omega_\tau^2 + \frac{\sqrt{-1}}{6V}\int_M \partial\varphi\wedge\overline{\partial}\varphi\wedge\omega_\tau. \end{aligned} \tag{3.8}$$

Since $\mathbf{F}_{\omega_\tau}(\varphi)\le 0$, we deduce from this

$$-\inf_M \varphi \le -\frac{1}{V}\int_M \varphi\omega^2 \le \frac{2}{V}\int \varphi\omega_\tau^2. \tag{3.9}$$

However, by using the Green function, we have (see [Ti87]):

$$\sup_M \varphi \le \frac{1}{V}\int_M \varphi\omega_\tau^2 + C. \tag{3.10}$$

Hence, we get

$$-\inf_M \varphi \le 2\sup_M \varphi + C. \tag{3.11}$$

We will use a certain finite-dimensional version of $\alpha(M_\tau)$ to derive (3.7). Recall the definition of $\alpha_{\ell,k}(M_\tau)$ $(1\le k\le N+1)$ (see the Appendix for more details): Choose any Hermitian metric h on $K_M^{-\ell}$ whose curvature form $\omega'>0$, where $M=M_\tau$. It induces a Hermitian inner product $(\cdot,\cdot)_h$ on $H^0(M,K_M^{-\ell})$. We define (cf. (5.1))

$$\alpha_{\ell,k}(M)=\sup\Big\{\alpha\mid \int_M \frac{\omega'^2}{\left(\sum_{i=1}^k \|S_i\|_h^2\right)^{\frac{\alpha}{\ell}}}\le C_\alpha,\ \forall \text{ orthonormal subbasis } \{S_i\}\Big\}.$$

Here C_α denotes a uniform constant depending only on α.

We will derive (3.7) by using $\alpha_{\ell,1}(M)$ and $\alpha_{\ell,2}(M)$. The following lemma is a refined version of a corresponding lemma proved in [Ti89]. We make the constants more precise, but the lemma in [Ti89] is sufficient for proving (3.7).

Lemma 3.8. *Let M be a smooth Del Pezzo surface obtained by blowing up $\mathbb{C}P^2$ at m points in general position $(5\le m\le 8)$. Assume that for some $\ell>0$, the partial C^0-estimate (3.4) holds and*

$$\frac{1}{\alpha_{\ell,1}(M)}+\frac{1}{\alpha_{\ell,2}(M)}<3,$$

then (3.7) holds.

Proof. Let us give an outline of its proof. First we recall that (3.4) implies

$$\left\|\varphi-\sup_M\varphi-\frac{1}{\ell}\log\left(\sum_{i=0}^N \lambda_i^2\|\sigma_i\|_\tau^2\right)\right\|_{C^0}\le C,$$

where $0<\lambda_0\le\cdots\le\lambda_{N-1}\le\lambda_N=1$ are given in (3.6).

Put

$$\psi = \frac{1}{\ell}\log\left(\sum_{i=0}^{N}\lambda_i^2\|\sigma_i\|_\tau^2\right).$$

Using the concavity of logarithm and (3.3), for any $\alpha < \alpha_{\ell,2}(M)$, we can have

$$\begin{aligned}
&\alpha\sup_M\varphi + \frac{1-\alpha}{V}\int_M \varphi\,\omega^2\\
&\leq \log\left(\frac{1}{V}\int_M e^{\alpha sup_M\varphi+(1-\alpha)\varphi}e^{h_\tau-\varphi}\omega_\tau^2\right)\\
&\leq \log\left(\frac{1}{V}\int_M e^{-\alpha(\varphi-\sup_M\varphi)}\omega_\tau^2\right) + \sup_M h_\tau\\
&\leq \log\left(\frac{1}{V}\int_M e^{-\frac{\alpha}{\ell}(\log\lambda_{N-1}^2+\log(\|\sigma_{N-1}\|_\tau^2+\|\sigma_N\|_\tau^2))}\,\omega_\tau^2\right) + C'\\
&\leq -\frac{\alpha}{\ell}\log\lambda_{N-1}^2 + C'_\alpha.
\end{aligned} \tag{3.12}$$

A direct computation shows that for any $\delta > 0$, there is a uniform C_δ such that

$$\frac{\sqrt{-1}}{2V}\int_M \partial\psi\wedge\overline{\partial}\psi\wedge\omega_\tau \geq -\frac{1-\delta}{\ell}\log\lambda_{N-1} - C_\delta. \tag{3.13}$$

Note that

$$\psi = \frac{1}{\ell}\log\left(\sum_{i=0}^{N}\lambda_i^2\|\sigma_i\|_\tau^2\right).$$

Plugging (3.13) into (3.8) and using (3.10), we have

$$\mathbf{F}_{\omega_\tau}(\varphi) \geq -\frac{1}{3V}\int_M \varphi\omega^2 - \frac{2}{3}\sup_M\varphi - \frac{1-\delta}{3\ell}\log\lambda_{N-1}^2 - C'_\delta.$$

Combining this with the estimate (3.12), we get

$$\mathbf{F}_{\omega_\tau}(\varphi) \geq -\left(\frac{1}{3} - \frac{(1-\delta)(1-\alpha)}{3\alpha}\right)\frac{1}{V}\int_M \varphi\omega^2 - \frac{1+\delta}{3}\sup_M\varphi - \tilde{C}_\delta. \tag{3.14}$$

On the other hand, for any $\beta < \alpha_{\ell,1}(M)$, using the arguments in deriving (3.12), we get

$$\beta\sup_M\varphi + \frac{1-\beta}{V}\int_M \varphi\,\omega^2 \leq C_\beta.$$

Combining this with (3.14), we have

$$\mathbf{F}_{\omega_\tau}(\varphi) \geq \frac{\beta}{3(1-\beta)}\left(3 - \frac{1-\delta}{\alpha} - \frac{1+\delta}{\beta}\right)\sup_M\varphi - C_1. \tag{3.15}$$

By our assumption, we can choose α, β and δ such that the coefficient of $\sup_M\varphi$ in (3.15) is positive. □

Remark 3.9. One can show that (3.7) is equivalent to the following finite-dimensional problem:

$$\mathbf{F}_{\omega_\tau}(\psi) \geq \epsilon \mathbf{J}_{\omega_\tau}(\psi) - C'. \tag{3.16}$$

Corollary 3.10. *For any $\tau \in \bar{I}_m$, if $\alpha_{\ell,1}(M_\tau) \geq 2/3$ and $\alpha_{\ell,2}(M_\tau) > 2/3$, then M_τ admits a Kähler-Einstein metric and consequently, I_m is closed.*

Proof. Let $\tau_i \in I_m$ converging to τ. Then we have the partial C^0-estimate (3.4) for $(M_{\tau_i}, \omega_{\tau_i})$. Using the above lemma and our assumptions, we see that (3.7) holds for $(M_{\tau_i}, \omega_{\tau_i})$. Because of Theorem 5.5, the constants in (3.7) can be made uniform on τ_i. Hence, M_τ admits a Kähler-Einstein metric. □

To finish the proof of Theorem 3.1, we only need to prove that $\alpha_{\ell,1}(M_\tau) \geq 2/3$ and $\alpha_{\ell,2}(M_\tau) > 2/3$ for some ℓ such that (3.4) holds for $\tau \in I_m$. In [Ti89], I prove (3.4) for $\ell = 6k$ and verify the lower bounds on $\alpha_{6,1}(M_\tau)$ and $\alpha_{6,2}(M_\tau)$. Then Theorem 3.1 follows.

Recently, in [Shi09], Yalong Shi proves that $\alpha_{\ell,2}(M) > 2/3$ for any smooth cubic surface $M \subset \mathbb{C}P^3$ (also see Theorem 5.2). Combining this with the results of I. Cheltsov ([Ch07] and [Ch08], also Theorem 5.1), he gives a simpler and elegant proof for Theorem 3.1.

Remark 3.11. It will be an interesting problem to study when there is a Kähler-Einstein orbifold metric on complex surfaces with isolated quotient singularities. Part of the proof of Theorem 3.1 can be adapted for the case of orbifolds, but there are substantial new difficulties due to the presence of singularities.

3.4. An approach by Kähler-Ricci flow

In this subsection, we show how to adapt the above arguments in last two subsections to give an alternative proof of Theorem 3.1 by using the Kähler-Ricci flow. The route for the proof is identical. A detailed proof has been recently produced by Chen-Wang [ChWa09].

First we collect some general facts for the Kähler-Ricci flow on any Fano manifold. Consider the Kähler-Ricci flow:

$$\frac{\partial \omega(t)}{\partial t} = -\mathrm{Ric}(\omega(t)) + \omega(t), \ \omega(0) = \omega_0. \tag{3.17}$$

There are φ such that $\omega = \omega_0 + \frac{\sqrt{-1}}{2}\partial\bar{\partial}\varphi$ and

$$\frac{\partial \varphi}{\partial t} = \log\left(\frac{(\omega_0 + \frac{\sqrt{-1}}{2}\partial\bar{\partial}\varphi)^n}{\omega_0^n}\right) + h_0 + \varphi, \ \varphi(0) = 0, \tag{3.18}$$

where $h_0 = h_{\omega_0}$. It is proved in [Cao86] that (3.18), and consequently (3.17), has a global solution $\varphi(t)$ for $t \geq 0$. Moreover, if $\|\varphi(t)\|_{C^0}$ is uniformly bounded, then there is a uniform bound on $\|\varphi(t)\|_{C^3}$ and $\omega(t)$ converges to a Kähler-Einstein metric on M. It implies Theorem 3.1. Therefore, as for the continuity method, we need an a prior C^0-estimate on any solution of (3.18).

Along the flow (3.18), we have

$$\frac{\partial}{\partial t}\mathbf{F}_{\omega_0}(\varphi) = -\int_M \frac{\partial \varphi}{\partial t}\left(e^{\frac{\partial\varphi}{\partial t}} - 1\right) e^{h_0-\varphi}\omega_0^n \leq 0. \tag{3.19}$$

It implies that $\mathbf{F}_{\omega_0}$ is non-increasing along (3.18). In particular, if $\mathbf{F}_{\omega_0}$ is proper, then we can bound $\mathbf{J}_{\omega_0}(\varphi)$ uniformly. If we have the uniform Sobolev inequality for $\omega(t)$, then $\|\varphi\|_{C^0}$ is uniformly bounded. The Sobolev inequality along the Kähler-Ricci flow has been established by Q. Zhang and R.G. Ye based on Perelman's fundamental works on Ricci flow (see [QZ07], [Ye07]). Hence, as before, the key is to establish the properness of $\mathbf{F}_{\omega_0}$.

Now we assume that $M = \Sigma_m$ is a Del-Pezzo surface. We may assume $m \in \{5, 6, 7, 8\}$, since for $m = 3, 4$ there is a unique Σ_m in the moduli space which is Kähler-Einstein by Theorems 3.2 and 3.3. Let ω_0 be a Kähler metric with $\pi c_1(M)$ as its Kähler class. In [Ch07], Cheltsov proved $\alpha(M) > 2/3$ unless $m = 5$ or $m = 6$ and M has an Eckardt point. If $m = 5$ and ω_0 is invariant under a maximal compact subgroup G, then $\alpha_G(M) > 2/3$. Actually, in [Ru08], Rubinstein proves that the Ricci flow (3.17) converges to a Kähler-Einstein metric if $\alpha(M) > 2/3$ or if ω_0 is G-invariant and $\alpha_G(M) > 2/3$.[13] Hence, we have a Ricci flow proof for Theorem 3.1 in those cases. An alternative approach to this is to use the properness of $\mathbf{F}_{\omega_0}$ (cf. [ChWa09]): $\mathbf{F}_{\omega_0}$ is proper in the case of $m > 6$ or $m = 6$ and M does not admit any Eckardt points, or $\mathbf{F}_{\omega_0}$ is proper on $P_G(M, \omega_0)$ if $m = 5$ and ω_0 is invariant under a maximal compact subgroup G.

But it remains to prove the case of $M = \Sigma_6$ with an Eckardt point. For this, we follow the route in [Ti89] or the arguments in last two subsections. But one needs to work out a new compactness theorem analogous to Theorem 3.5. The following is proved in [ChWa09].

Theorem 3.12. *For any sequence $\{t_i\}$ with $\lim t_i = \infty$, by taking a subsequence if necessary, $(M, \omega(t_i)$ converge to a Kähler-Einstein orbifold $(M_\infty, \omega_\infty)$ in the Cheeger-Gromov topology satisfying:*

(1) *The singularities are of the form U/Γ and the number of them is uniformly bounded, where $U \subset \mathbb{C}^2$ and Γ is a finite group of $U(2)$ with uniformly bounded order;*
(2) *For each $\ell > 0$, $H^0(M, K_M^{-\ell})$ converge to $H^0(M_\infty, K_{M_\infty}^{-\ell})$ in the following sense There are orthonormal bases $\{S_0^i, \ldots, S_N^i\}$ of $H^0(M, K_M^{-\ell})$ converging to an orthonormal basis of $H^0(M_\infty, K_{M_\infty}^{-\ell})$. Here the inner product is induced by a Hermitian metric on K_M^{-1} with $\omega(t_i)$ as its curvature form.*

To say a few words about its proof, we remark that in place of the Kähler-Einstein condition, one can use a result of Perelman: The scalar curvature of $\omega(t)$ is uniformly bounded. Moreover, the Sobolev constants are uniformly bounded.

[13]Rubinstein's result holds for any dimensions and is a Ricci flow version of the main theorem in [Ti87].

Hence, as before, one can develop the curvature estimate and use the L^2-estimate for the $\bar{\partial}$-operator.

Theorem 3.12 yields the partial C^0-estimate for $\varphi(t)$:

$$\left\| \varphi(t) - \sup_M \varphi(t) - \frac{1}{\ell} \log \left(\sum_{i=0}^{N} \lambda_i(t)^2 \|\sigma_i(t)\|_0^2 \right) \right\|_{C^0} \leq C.$$

Next one proves the properness of $\mathbf{F}_{\omega_0}$ for the above $\frac{1}{\ell}\log(\sum_{i=0}^N \lambda_i(t)^2 \|\sigma_i(t)\|_0^2)$. As before, it can be done by using $\alpha_{\ell,1}(M) \geq 2/3$ and $\alpha_{\ell,2}(M) > 2/3$ as we did in the last subsection. Hence a Ricci flow proof of Theorem 3.1 can be completed. We refer the readers to [ChWa09] for details.

4. What about higher dimensions

In this section, we discuss the existence of Kähler-Einstein metrics on Fano manifolds in higher dimensions. This problem faces new challenges:

1. The vanishing of Futaki invariants is no longer sufficient for the existence, in fact, there is a Fano 3-fold without any hon-trivial holomorphic vector fields and which does not admit any Kähler-Einstein metrics, either (cf. [Ti97]). Hence, a geometric condition needs to be found. It was already speculated in the late 80s that the correct condition would be in terms of a certain stability of the underlying manifold in the sense of the Geometric Invariant Theory. Now it seems to be apparent that the correct condition is the K-stability first introduced in [Ti97] and reformulated in [Do02] in a purely algebraic way;
2. It is much harder to establish the partial C^0-estimate (3.5) in higher dimensions. The difficulty arises from the fact that there may be more possible singularities in compactifying the moduli of Kähler-Einstein manifolds in higher dimensions, or more generally, in the Gromov-Hausdorff limits of Kähler spaces with Ricci curvature bounded from below by a positive number. Motivated by this, a compactness theorem was proved by Cheeger-Colding-Tian fifteen years ago. We will discuss this compactness theorem and how it can be applied to studying the existence problem and what remains to be done.

There are four subsections in this section. In the first, we discuss the K-stability. In the second, we discuss the partial C^0-estimate. In the third, we show how the K-stability implies the existence under the assumption of the partial C^0-estimate. In the last, we discuss what is known on the Kähler-Ricci flow.

4.1. The K-stability

In this subsection, we discuss the K-stability: First the definition from [Ti97] by using the generalized Futaki invariants, then a purely algebraic definition due to Donaldson [Do02]. We will also discuss how the K-stability is related to the existence of Kähler-Einstein metrics.

First let us recall the definition of the Futaki invariant [Fut83]: Let M be any Fano manifold and ω be a Kähler metric with $\pi c_1(M)$ as its Kähler class, for any holomorphic vector field X on M, Futaki defined

$$f_M(X) = \int_M X(h_\omega)\,\omega^n. \tag{4.1}$$

Note that $\mathrm{Ric}(\omega) - \omega = \frac{\sqrt{-1}}{2}\partial\bar{\partial}h_\omega$. Futaki proved in [Fut83] that $f_M(X)$ is independent of the choice of ω, so it is a holomorphic invariant. Later, Bando, Calabi and Futaki observed that the invariant can be defined for any polarized Kähler manifold (M,Ω): If ω is any Kähler metric with cohomology class Ω and h_ω satisfies $s(\omega) - \underline{s} = \Delta_\omega h_\omega$, where $s(\omega)$ denotes the scalar curvature and $\underline{s}$ is its average, then one can still show that the integral in (4.1) depends only on M, Ω and X. Hence, we have an invariant, denoted by $f_{M,\Omega}(X)$. The Futaki invariant is an obstruction to the existence of Kähler-Einstein metrics on Fano manifolds, but its vanishing does not assure the existence as shown in [Ti97]. It motivates us to introduce the K-stability. For this, we need to extend the Futaki invariant to singular varieties. In [DT92], it was done for normal Fano varieties. It turns out that the arguments also apply to more general cases. To be more convincing, we reformulate the definition of $f_{M,\Omega}(X)$. Note that

$$i_X\omega = \frac{\sqrt{-1}}{2}(H_X + \bar{\partial}\theta_X),$$

where H_X is a parallel (0,1)-form. One can show

$$f_{M,\Omega}(X) = -n\int_M \theta_X \left(\mathrm{Ric}(\omega) - \frac{\underline{s}}{n}\omega\right)\wedge\omega^{n-1}. \tag{4.2}$$

In particular, parallel (0,1)-forms have no effect in $f_{M,\Omega}(X)$. For simplicity, we will drop it and consider only those fields X with $H_X = 0$.

The formula (4.2) allows to extend the Futaki invariant to singular varieties. For our purpose, we only need to define the invariant for any polarized normal variety (M,L) and the following vector field: Assume that there is an embedding $M \subset \mathbb{C}P^N$ such that $L^\ell = \mathcal{O}(1)|_M$ and X is a holomorphic vector field of $\mathbb{C}P^N$ restricted to M, then there is a smooth function θ_X on $\mathbb{C}P^N$ such that

$$i_X\omega = \frac{\sqrt{-1}}{2}\bar{\partial}\theta_X.$$

where ω is the restriction of $\frac{1}{\ell}\omega_{FS}$ to the regular part of M.[14] It is well known that $\mathrm{Ric}(\omega)$ is bounded from above and its trace, the scalar curvature, is L^1-bounded. Therefore, the integral in (4.2) is still meaningful. Moreover, one can prove that it is independent of the choice of ω, so it defines the generalized Futaki invariant

[14]We will always denote ω_{FS} the Fubini-Study metric on $\mathbb{C}P^N$ in this paper.

$f_{M,L}(X)$ for polarized normal variety (M, L).[15] The proof is not very hard and the details will be skipped here.

The formula (4.2) can be used further as in [DT92]. Let (M, L) be a polarized manifold. By the Kodaira embedding theorem, for ℓ sufficiently large, a basis of $H^0(M, L^\ell)$ gives an embedding $\phi_\ell : M \mapsto \mathbb{C}P^N$, where $N = \dim_{\mathbb{C}} H^0(M, L^\ell) - 1$. Any other basis gives an embedding of the form $\sigma \cdot \phi_\ell$, where $\sigma \in G = \mathbf{SL}(N+1, \mathbb{C})$. We fix such an embedding and consider the action of G on M.

Fix a Hermitian metric $\|\cdot\|$ on L such that its curvature form ω is a Kähler metric. Then for any $\sigma \in G$, there is a unique function φ_σ such that

$$\phi_\ell^* \sigma^* (\|\cdot\|_{FS}^{\frac{2}{\ell}}) = e^{-\varphi_\sigma} \|\cdot\|^2, \tag{4.3}$$

where $\|\cdot\|_{FS}$ is a Hermitian metric on the hyperplane bundle over $\mathbb{C}P^N$ whose curvature form is ω_{FS}.

It is known that for any algebraic subgroup $G_0 = \{\sigma_t\}_{t \in \mathbb{C}^*}$ of $\mathbf{SL}(N+1, \mathbb{C})$, there is a unique limiting cycle

$$M_0 = \lim_{t \to 0} \sigma_t(M) \subset \mathbb{C}P^N.$$

Let X be the holomorphic vector field whose real part generates the action by $\sigma(e^{-s})$, that is,

$$\frac{d\sigma(e^{-s})}{ds} = \mathrm{Re}(X)(\sigma(e^{-s})).$$

The following lemma is proved in [DT92] for Fano manifolds. However, the proof for the general case is identical.

Lemma 4.1. *Assume that G_0 is as above and M_0 is a normal variety. Then we have*

$$\lim_{s \to \infty} \frac{d}{ds} \left(\mathbf{T}_\omega(\varphi_{\sigma(e^{-s})}) \right) = \mathrm{Re}(F_{M_0, \Omega}(X)), \tag{4.4}$$

where $\Omega = \frac{1}{\ell}[\omega_{FS}]|_{M_0}$.

Proof. We follow [DT92] to prove this lemma. Since $\mathrm{Im}(X)$ is a Killing field for ω_{FS}, there is a real function θ such that

$$\frac{1}{\ell} L_X \omega_{FS} = \frac{\sqrt{-1}}{2} \partial\bar{\partial}\theta.$$

Differentiating in s the identity

$$\frac{1}{\ell} \sigma(e^{-s})^* \omega_{FS} = \omega + \frac{\sqrt{-1}}{2} \partial\bar{\partial}\varphi_{\sigma(e^{-s})},$$

we get (possibly up to a constant)

$$\frac{d}{ds} \left(\varphi_{\sigma(e^{-s})} \right) (\sigma(e^s)(x)) = \theta(x), \ \forall x \in \sigma(e^{-s})(M).$$

[15] The integral (4.2) also makes sense for non-normal varieties, but the correct Futaki invariant has contributions from singularity of codimension 1 since we want to preserve certain natural continuity on any family of varieties.

Then by the definition (2.9), we can deduce

$$\frac{d\mathbf{T}_\omega(\varphi_{\sigma(e^{-s})})}{ds} = -\frac{1}{V}\int_{\sigma(e^{-s})(M)} \theta\,(s(\omega_{\varphi_{\sigma(e^{-s})}}) - \underline{s})\,\omega^n_{\varphi_{\sigma(e^{-s})}}.$$

There is a unique f_s on $\sigma(e^{-s})(M)$ such that

$$s(\omega_{\varphi_{\sigma(e^{-s})}}) - \underline{s} = \Delta_s f_s \text{ and } \inf f_s = 1,$$

where Δ_s is the Laplacian of the metric $\frac{1}{\ell}\omega_{FS}$ restricted to $\sigma(e^{-s})(M)$. Hence, we have

$$\frac{d\mathbf{T}_\omega(\varphi_{\sigma(e^{-s})})}{ds} = \frac{1}{V}\int_{\sigma(e^{-s})(M)} \nabla\theta\cdot\nabla f_s\,\omega^n_{\varphi_{\sigma(e^{-s})}}. \tag{4.5}$$

Since M_0 is a normal variety and $s(\omega_{\varphi_{\sigma(e^{-s})}})$ is uniformly bounded from above, we can prove that f_s converge to f_∞ outside the singular set of M_0 as $s\to\infty$. Moreover, we have uniformly bounded Sobolev constant C_s for $(\sigma(e^{-s})(M), \frac{1}{\ell}\omega_{FS})$, so we have

$$\begin{aligned}
&\left(\int_{\sigma(e^{-s})(M)} f_s^{\frac{2n}{2n-1}}\,\omega^n_{\varphi_{\sigma(e^{-s})}}\right)^{\frac{n-1}{2n}} \qquad (4.6)\\
&\le C_s\int_{\sigma(e^{-s})(M)} |\nabla f_s^{\frac{1}{2}(1-\frac{1}{2n-1})}|^2\,\omega^n_{\varphi_{\sigma(e^{-s})}}\\
&= \frac{n^2C_s}{2n-1}\int_{\sigma(e^{-s})(M)} f_s^{-\frac{1}{2n-1}}\Delta_s f_s\,\omega^n_{\varphi_{\sigma(e^{-s})}}\\
&= \frac{n^2C_s}{2n-1}\int_{\sigma(e^{-s})(M)} f_s^{-\frac{1}{2n-1}}(s(\omega_{\varphi_{\sigma(e^{-s})}}) - \underline{s})\,\omega^n_{\varphi_{\sigma(e^{-s})}}.
\end{aligned}$$

The last integral is uniformly bounded. It follows that

$$\begin{aligned}
&\left(\int_{\sigma(e^{-s})(M)} |\nabla f_s|\,\omega^n_{\varphi_{\sigma(e^{-s})}}\right)^2\\
&\le \int_{\sigma(e^{-s})(M)} f_s^{\frac{2n}{2n-1}}\,\omega^n_{\varphi_{\sigma(e^{-s})}}\int_{\sigma(e^{-s})(M)} f_s^{-\frac{2n}{2n-1}}|\nabla f_s|^2|^2\,\omega^n_{\varphi_{\sigma(e^{-s})}}.
\end{aligned}$$

Combining this with (4.5) and (4.6), we can easily deduce (4.4). Here we have used the fact that θ has bounded gradient. □

We will denote by $\mathbf{w}(M,L,G_0)$ the generalized Futaki invariant in (4.4) and call it the weight of G_0 associated to (M,L) (abbreviated as $\mathbf{w}(G_0)$ if no confusion). In fact, the limit in (4.4) exists without any assumption on M_0, so the weight $\mathbf{w}(M,L,G_0)$ for any G_0 can be defined. This is proved in [PT04] when M_0 has no components of multiplicity greater than one or in [Paul08] for the general case.

Now we are ready to define the K-stability introduced in [Ti97]. Because our main topic in this paper is on Kähler-Einstein metrics with positive scalar curvature, we first assume that $M\subset\mathbb{C}P^N$ is a Fano manifold and $L = K_M^{-1}$ even though the definition for the general cases is very similar. We always denote by G_0 an one-parameter subgroup of $\mathbf{SL}(N+1)$. Let M_0 be the corresponding limit cycle.

Definition 4.2. We say that M is K-semistable with respect to $K_M^{-\ell}$ if $\mathbf{w}(G_0) \geq 0$ for any $G_0 \subset \mathbf{SL}(N+1)$ such that the corresponding M_0 is a normal variety. We say that M is K-stable with respect to $K_M^{-\ell}$ if it is K-semistable and $\mathbf{w}(G_0) > 0$ for any $G_0 \subset \mathbf{SL}(N+1)$ such that the corresponding M_0 is a normal variety and not biholomorphic to M.[16]

For $G_0 \subset \mathbf{SL}(N+1)$, we can choose homogeneous coordinates $z_0, \ldots, z_N$ for $\mathbb{C}P^N$ such that $\sigma(t) \in G_0$ is represented by

$$\mathbf{diag}(t^{\alpha_0}, \ldots, t^{\alpha_N}),\ t \in \mathbb{C}^*,$$

where $\alpha_0 \leq \cdots \leq \alpha_N$ are integers. Define a height $\mathbf{h}(M, G_0)$ or simply $\mathbf{h}(G_0)$ to be the smallest $\alpha_N - \alpha_i$ such that $z_i|_M, \ldots, z_N|_M$ have no common zeroes. It is easy to show that M_0 is biholomorphic to M if M_0 is a normal variety and $\mathbf{h}(G_0) = 0$.

The following is proved in [Ti97]

Theorem 4.3. *Let M be a Fano manifold without non-trivial holomorphic vector fields and which admits a Kähler-Einstein metric. Then M is K-stable in the above sense with respect to any very ample $K_M^{-\ell}$.*

This follows from the properness of the K-energy and the computation of its derivative along any given one-parameter subgroup G_0.

For a general polarized Kähler manifold (M, L), we can not expect a uniform bound on diameter for Kähler metrics with constant scalar curvature and with Kähler class $c_1(L)$, consequently, we can not restrict the stability criterion to only those G_0 with normal limit variety M_0. However, we expect that any sequence of Kähler metrics with constant scalar curvature and with bounded Kähler classes contains a subsequence converging to a finite union of complete Kähler manifolds with constant scalar curvature and with finite volume. Because of this, it is reasonable to assume that M_0 has no multiple components.

Definition 4.4. Let (M, L) be a polarized Kähler manifold and $M \subset \mathbb{C}P^N$ by sections of L^ℓ. We say that M is K-semistable with respect to L^ℓ if $\mathbf{w}(G_0) \geq 0$ for $G_0 \subset \mathbf{SL}(N+1)$ such that the corresponding M_0 has no multiple components. We say that M is K-stable with respect to $L^{-\ell}$ if it is K-semistable and $\mathbf{w}(G_0) > 0$ for $G_0 \subset \mathbf{SL}(N+1)$ such that the corresponding M_0 has no multiple components and is not biholomorphic to M.

We say that (M, L) is asymptotically K-stable if M is K-stable with respect to L^ℓ for sufficiently large ℓ.

[16]We assume that M_0 is a normal variety because of the compactness we expect for the set of Kähler metrics with Ricci curvature bounded from below by a positive constant. For those Kähler metrics, the diameter is uniformly bounded and essential singularity should occur only along a closed subset of Hausdorff codimension at least 4.

Now we recall Donaldson's version of the K-stability [Do02]. It is more algebraic and neater. We say $(\mathcal{M},\mathcal{L}) \mapsto \mathbb{C}$ a test configuration of a polarized manifold (M,L) if it consists of a scheme $\mathcal{M}$ endowed with a $\mathbb{C}^*$-action that linearizes on a line bundle $\mathcal{L}$ over $\mathcal{M}$, and a flat $\mathbb{C}^*$-equivariant map $f:\mathcal{M}\mapsto\mathbb{C}$ (where $\mathbb{C}$ has the usual weight one $\mathbb{C}^*$- action) such that $\mathcal{L}|_{f^{-1}(0)}$ is ample on $f^{-1}(0)$ and we have $(f^{-1}(1),\mathcal{L}|_{f^{-1}(1)}) \cong (M;L^r)$ for some $r>0$. When (M,L) has a $\mathbb{C}^*$-action $\rho:\mathbb{C}^*\mapsto \mathrm{Aut}(M)$, a test configuration where $\mathcal{M} = M\times\mathbb{C}$ and $\mathbb{C}^*$ acts on $\mathcal{M}$ diagonally through ρ is called product configuration. This terminology is given in [Do02]. Also an analogous version of such a test configuration is given in [Ti97] under the name: *A special degeneration.*

Let (V,L) be an n-dimensional polarized variety or scheme and $\mathrm{Aut}(M,L)$ be the group of all automorphisms which can be lifted to L. Given a one parameter subgroup $\rho:\mathbb{C}^*\mapsto\mathrm{Aut}(V)$, we denote by $w(V,L)$ the weight of the $\mathbb{C}^*$-action induced on $\Lambda^{\mathrm{top}}H^0(V,L)$ and then we have the following asymptotic expansions as $\ell\gg 0$:

$$h^0(V,L^\ell) = a_0\ell^n + a_1\ell^{n-1} + O(\ell^{n-2}) \tag{4.7}$$

$$w(V,L^\ell) = b_0\ell^{n+1} + b_1\ell^n + O(\ell^{n-1}). \tag{4.8}$$

The Donaldson's version of the Futaki invariant of the action is defined as

$$F(V,L;\rho) = \frac{b_1a_0 - b_0a_1}{a_0^2}. \tag{4.9}$$

It is shown in [Do02] that $-F(V,L,\rho)$ is equal to the original Futaki invariant (possibly modulo multiplication by a universal positive constant) whenever V is smooth. It is also true for any reduced variety V as shown in [PT04] by using the so-called CM line bundles.

Now we can state Donaldson's version of the K-stability.

Definition 4.5. A polarized Kähler manifold (M,L) is K-semistable if for each test configuration $f:(\mathcal{M},\mathcal{L})\mapsto\mathbb{C}$ for (M,L) the Futaki invariant of the induced action on the central fiber $(f^{-1}(0),\mathcal{L}|_{f^{-1}(0)})$ is non-positive. We say (M,L) is K-stable if the Futaki invariant for any test configuration is strictly negative unless we have a product configuration.

In view of the identification between the original and Donaldson's definition of the Futaki invariant for reduced varieties in [PT04], it is expected that the above two versions of the K-stability are equivalent. Indeed, it follows from a recent work by Arezzo-La Nave-Della Vedova [ALV09] that the two versions of the K-semistability are equivalent.[17]

The importance of the K-stability is partly shown in the following result of J. Stoppa [Sto07].

[17]Note that signs in two definitions of the K-stability are opposite. This is because we used different orientations in the $\mathbb{C}^*$-actions. Also Arezzo-La Nave-Della Vedova has found a proof in the case of the K-stability.

Theorem 4.6. *Let (M, L) be a polarized Kähler manifold with trivial* $\mathrm{aut}(M, L)$. *Then M admits a Kähler metric with constant scalar curvature with Kähler class $c_1(L)$ only if (M, L) is K-stable.*

Its proof is based on Arezzo-Pacard's work [ArPa05] and the works on the K-semi-stability ([ChTi04], [Do05]). In [ChTi04] and [Do05], the authors prove independently that M admits a Kähler metric with constant scalar curvature with Kähler class $c_1(L)$ only if (M, L) is K-semistable. Recently, Stoppa's result has been extended by T. Mabuchi to the general case.

The following well-known conjecture also shows the importance of the K-stability.

Conjecture 4.7. *Let (M, L) be a polarized Kähler manifold. For simplicity, assume that* $\mathrm{Aut}(M, L)$ *is discrete. Then M admits a Kähler metric with constant scalar curvature and Kähler class $c_1(L)$ if and only if (M, L) is asymptotically K-stable.*[18]

4.2. Partial C^0-estimates

In this subsection, I present an approach to solving (1.1) under suitable geometric condition which I have been pursuing since the late 80s. My solution for complex surfaces in the last section can be regarded as a successful example of this approach. Now the geometric condition is much better understood and should be the K-stability.

First we recall that in order to establish the existence of Kähler-Einstein metrics on a Fano manifold M, we only need to establish the a priori C^0-estimate for the solutions of (1.2) for $t \geq t_0$ for some $t_0 > 0$ which may depend on (M, ω):

$$\left(\omega + \frac{\sqrt{-1}}{2}\partial\bar{\partial}\varphi\right)^n = e^{h_\omega - t\varphi}\omega^n. \tag{4.10}$$

As said before, there are two steps in this approach. First we need to prove a partial C^0-estimate which we have discussed for Del-Pezzo surfaces in a previous section.

Consider the set

$$\mathcal{K}(M, t_0) = \{\tilde{\omega} \mid [\tilde{\omega}] = c_1(M),\ \mathrm{Ric}(\tilde{\omega}) \geq t_0\tilde{\omega}\}.$$

For any $\tilde{\omega}$, choose a Hermitian metric $\tilde{h}$ with $\tilde{\omega}$ as its curvature form and any orthonormal basis $\{S_i\}_{0\leq i\leq N}$ of each $H^0(M, K_M^{-\ell})$ with respect to an induced inner product. Put

$$\rho_{\tilde{\omega},\ell}(x) = \sum_{i=0}^{N} \|S_i\|_{\tilde{h}}^2(x). \tag{4.11}$$

This is independent of the choice of $\tilde{h}$ and the orthonormal basis $\{S_i\}$.

---

[18] This conjecture is often referred to the Yau-Tian-Donaldson conjecture. In general, one can expect that (M, L) admits a constant scalar curvature metric if and only if M is asymptotically weakly K-stable. The weak K-stability means that M is semistable and $\mathbf{w}(G_0) > 0$ for any G_0 from Definition 4.4 which is transverse to the identity component of $\mathrm{Aut}(M, L)$ in a suitable sense. We refer the readers to recent works of Mabuchi for details.

Conjecture 4.8 ([Ti90]). *There are uniform constants $c_k = c(k,n) > 0$ for $k \geq 1$ and $\ell_i \to \infty$ with $i \geq 0$ and $\ell_0 = \ell_0(n)$ such that for any $\tilde{\omega} \in \mathcal{K}(M,t_0)$ and $\ell = \ell_i$ for each i,*

$$\rho_{\tilde{\omega},\ell} \geq c_\ell > 0. \tag{4.12}$$

Remark 4.9. There is a uniform upper bound on $\rho_{\tilde{\omega},\ell}$ which depends only on n and ℓ. This is proved in [Ti90]. The lower bound is more important and can be regarded as an effective version of the very ampleness in algebraic geometry.

Remark 4.10. A stronger version of Conjecture 4.8 may hold: There are uniform constants $c_k = c(k,n) > 0$ for $k \geq 1$ and $\ell_0 = \ell_0(n)$ such that for any $\tilde{\omega} \in \mathcal{K}(M,t_0)$, and $\ell \geq \ell_0$, $\rho_{\tilde{\omega},\ell} \geq c_\ell$.

It is known that if φ is a solution of (1.2) $(t \geq t_0)$, then $\omega_t := \omega_\varphi$ has its Ricci curvature greater than or equal to $t \geq t_0$. If Conjecture 4.8 is true, Then as we did for (3.6), we can deduce

$$\left\| \varphi - \sup_M \varphi - \frac{1}{\ell} \log \left(\sum_{i=0}^N \lambda_i^2 \|\sigma_i\|^2 \right) \right\|_{C^0} \leq C, \tag{4.13}$$

where $\|\cdot\|$ is a Hermitian metric whose curvature is ω and $\{\sigma_i\}$ is an orthonormal basis of $H^0(M, K_M^{-\ell})$ with respect to the inner product induced by $\|\cdot\|$.

For each orthonormal basis $\{S_i\}$ of $H^0(M, K_M^{-\ell})$, we have embedding Φ : $M \mapsto \mathbb{C}P^N$. Such an embedding is unique modulo the isometry group $U(N+1)$ of $\mathbb{C}P^N$. If ω_i is any sequence in $\mathcal{K}(M,t_0)$ and Φ_i is corresponding embedding, then by taking a subsequence of necessary, there is a limit cycle $\bar{M}_\infty \subset \mathbb{C}P^N$ of $\Phi_i(M)$. The partial C^0-estimate (4.12) is closely related to the irreducibility of this limit cycle $\bar{M}_\infty$. Under suitable regularity of the metric limits of (M,ω_i), they are equivalent.

On the other hand, since we have $\mathrm{Ric}(\omega_i) \geq t_0\omega_i$ and $[\omega_i] = c_1(M_i)$, it contains a subsequence converging to a length space (M_∞, d_∞) in the Gromov-Hausdorff topology. Note that the diameter of (M_∞, d_∞) is uniformly bounded. I expected the following[19]

Conjecture 4.11. *The above Gromov-Hausdorff limit M_∞ can be identified with the complex limit $\bar{M}_\infty$.*

We can say more about it in the case that ω_i is a sequence of Kähler-Einstein metrics with $\mathrm{Ric}(\omega_i) = \omega_i$ since we can apply certain works of Cheeger-Colding-Tian [CCT95]. We may assume that there is a Gromov-Hausdorff limit (M_∞, d_∞) of Kähler-Einstein manifolds (M_i, ω_i). The main theorem in [CCT95] gives partial regularity for this limit.

[19] It had come to my attention more than sixteen years ago. I have made many attempts to solving this for a sequence of Kähler-Einstein metrics.

Theorem 4.12. [CCT95] *Let* (M_i, ω_i) *be a sequence of Kähler-Einstein manifolds with* $\mathrm{Ric}(\omega_i) = \omega_i$ *and which converges to* (M_∞, d_∞) *in the Gromov-Hausdorff topology. Then there is a closed subset* $S \subset M_\infty$ *of Hausdorff codimension at least 4 such that* $M_\infty \backslash S$ *is a smooth Kähler manifold and* d_∞ *is induced by a Kähler-Einstein metric* ω_∞ *outside* S *with* $\mathrm{Ric}(\omega_\infty) = \omega_\infty$. *Moreover,* ω_i *converges to* ω_∞ *in the* C^∞*-topology outside* S.

In fact, even though $(M_\infty, \omega_\infty)$ may have singularity, one can still do integration as well as other analysis on M_∞. It was conjectured that the M_∞ *is a variety and* S *is a subvariety* (cf. [CCT95]). Indeed, J. Cheeger proved that S is rectifiable. The partial C^0-estimate follows from our understanding of the singular set.

Even though S is not completely understood, we can treat M_∞ as a "good" variety in many ways. By an element of $H^0(M_\infty, K_{M_\infty}^{-\ell})$, we mean a holomorphic section S of $K_{M_\infty}^{-\ell}$ on $M_\infty \backslash S$ with finite L^2-norm.[20] Then one can prove that the space $H^0(M_\infty, K_{M_\infty}^{-\ell})$ is of finite dimension. Choose a Hermitian metric h_∞ of $K_{M_\infty}^{-1}$ outside S with ω_∞ as its curvature. We may also choose h_i for $K_{M_i}^{-1}$ such that h_i converges to h_∞ outside S. The two-dimensional version of the following proposition was proved in [Ti89] and the same proof works for higher dimensions.

Proposition 4.13. *By taking a subsequence if necessary, for each* ℓ, *we have that* $H^0(M_i, K_{M_i}^{-\ell})$ *converges to* $H^0(M_\infty, K_{M_\infty}^{-\ell})$ *as* i *tends to* ∞ *in the sense: There are orthonormal bases* $\{S_a^i\}_{0\le a\le N}$ *of* $H^0(M_i, K_{M_i}^{-\ell})$ *with respect to* h_i *such that* S_a^i *converges to* S_a^∞ $(0 \le a \le N)$ *as* i *tends to* ∞ *and* $\{S_a^\infty\}$ *forms an orthonormal basis of* $H^0(M_\infty, K_{M_\infty}^{-\ell})$.

Proof. We outline its proof here. As in [Ti89], the proof uses the L^2-estimate for $\bar\partial$-operator and the theory for elliptic equations. First we observe: For each i,

$$\Delta_{\omega_i} \|S_a^i\|^2 = \|\nabla S_a^i\|^2 - \ell\, \|S_a^i\|^2.$$

Then we can apply the Moser iteration to deriving a uniform bound on $\|S_a^1\|$ as well as bounds on derivatives of S_a^i outside the singular set of M_∞. It follows that by taking a subsequence if necessary, we can assume that S_a^i converges to a S_a^∞ as i tends to ∞.

To complete the proof, we need to show that each section $\tilde S_\infty$ of $K_{M_\infty}^{-\ell}$ on M_∞ is the limit of a sequence of sections $\tilde S_i$ on M_i. This is done by using Hörmander's L^2-estimate. For any $\epsilon > 0$, choose a finite cover $\{B_{r_k}(x_k)\}$ of the singular set $S \subset M_\infty$ satisfying:

1. $r_k \le \epsilon$;
2. $\sum_k r_k^{2n-4} \le C < \infty$;[21]
3. For any $x \in M_\infty$, the number of balls $B_{r_k}(x_k)$ containing x is uniformly bounded, say C.

[20] This should be automatically true since S is of codimension at least 4.

[21] We will always use C to denote a uniform constant in this proof, though its actual value may vary in different places.

Let η be a smooth function on $\mathbb{R}$ such that $\eta(r) = 0$ for $r \leq 1$ and $\eta(r) = 1$ for $r \geq 2$. Put

$$S'_\infty = \left(\prod_k \eta(d(\cdot, x_k)/r_k)\right) \tilde{S}_\infty.$$

Then one can show

$$\int_{M_\infty} \|\bar{\partial} S'_\infty\|^2 \, \omega_\infty^n \leq C\epsilon^2.$$

For i sufficiently large, there is a diffeomorphism $\Phi_i : M_i \backslash K_i \mapsto M_\infty \backslash S$ satisfying:

(1) The measure of K_i is bounded by $C\epsilon^4$;
(2) The image of Φ_i contains the complement of $\cup_k B_{r_k}(x_k)$ in M_∞;
(3) $\|\bar{\partial}\Phi_i\| \leq \epsilon$.

Each Φ_i induces an isomorphism τ_i between the complex line bundles $K_{M_i}^{-\ell}$ and $K_{M_\infty}^{-\ell}$ such that $\|\bar{\partial}\tau_i\| \leq C_\ell \epsilon$, where C_ℓ is independent of i and ϵ. Put $S'_i = \tau^{-1}(S'_\infty)$. Then

$$\int_{M_i} \|\bar{\partial} S'_i\|^2 \, \omega_i^n \leq C_\ell \, \epsilon^2.$$

By applying Hörmander's L^2-estimate for $\bar{\partial}$-operators, we can have a C^∞-section u_i of $K_{M_i}^{-\ell}$ satisfying:

$$\bar{\partial} u_i = -\bar{\partial} S'_i, \quad \int_{M_i} \|u_i\|^2 \, \omega_i^n \leq C_\ell \, \epsilon^2.$$

It implies that $\tilde{S}_i = S'_i + u_i$ is a holomorphic section of $K_{M_i}^{-\ell}$ such that $\tilde{S}_i$ is within the $C\epsilon$-distance of $\tilde{S}_\infty$ for some uniform constant C. Then the proposition follows easily. □

Proposition 4.13 can be regarded as a metric version of the flatness for families of projective varieties.

Now we fix a large ℓ and consider embedding $\Phi_i(M) \subset \mathbb{C}P^N$ of M_i induced by $\{S_a^i\}$.

Proposition 4.14. *Assume that $\Phi_i(M_i)$ converges to an irreducible subvariety $\bar{M}_\infty$. Then there is a uniform constant c satisfying: $\rho_{\omega_i,\ell} \geq c$ for all i, i.e.,* (4.12) *holds for all i.*

Proof. We just outline the proof. First we observe

$$0 \leq \omega_i = \tilde{\omega}_i - \frac{\sqrt{-1}}{2\ell} \partial\bar{\partial} \log(\rho_{\omega_i,\ell}),$$

where $\tilde{\omega}_i$ is the restriction of ω_{FS} to $\Phi_i(M)$. In particular,

$$\Delta_{\tilde{\omega}_i} (-\log(\rho_{\omega_i,\ell})) \geq -n.$$

Since $\bar{M}_\infty$ is irreducible, we have uniform bounds on the Sobolev constant and the Poincaré constant for $(\Phi_i(M), \tilde{\omega}_i)$. Then the standard Moser iteration implies

$$-\inf_{M_i} \log(\rho_{\omega_i,\ell}) \le C\left(1 + \int_{M_i} |\log(\rho_{\omega_i,\ell})| \tilde{\omega}_i^n\right).$$

On the other hand, by the above proposition, we know that $\rho_{\omega_i,\ell}$ converges to $\rho_{\omega_\infty,\ell}$ in the C^∞-topology outside S, so they are uniformly bounded on any compact subsets away from S. Then one can easily deduce from the above $\rho_{\omega_i,\ell} \ge c$ for some uniform positive constant. □

Remark 4.15. The converse is true: If (4.12) holds, then $\bar{M}_\infty$ is irreducible.

Corollary 4.16. *Let M_i and $\bar{M}_\infty$ be as above. If $\bar{M}_\infty$ is irreducible, then $\bar{M}_\infty$ is the Gromov-Hausdorff limit of (M_i, ω_i). In particular, Conjecture* 4.11 *holds.*

It shows how important to have the irreducibility of the complex limits of $\Phi_i(M_i)$ under embedding given by $H^0(M_i, K_{M_i}^{-\ell})$. Also I would like to point out that the irreducibility can be derived if limits of the embeddings Φ_i stabilize for sufficiently large ℓ. This gives a strong evidence on the irreducibility. Let me elaborate a bit more: For all sufficiently large ℓ, we have embedding $\Phi_{i,\ell} : M_i \mapsto \mathbb{C}P^{N_\ell}$[22] and get a limit $\bar{M}_{\infty,\ell}$ of $\Phi_{i,\ell}(M_i)$ as i tends to ∞ (possibly after taking a subsequence). On the other hand, $\Phi_{i,\ell}$ converges to a limit map (possibly rational and after taking a subsequence):

$$\Phi_{\infty,\ell} : M_\infty \backslash S \mapsto \mathbb{C}P^{N_\ell}.$$

For any tubular neighborhood U of S, one can prove that for ℓ sufficiently large, $\Phi_{\infty,\ell}$ is an embedding on $M_\infty \backslash U$. Hence, for ℓ and ℓ' large, $\bar{M}_{\infty,\ell}$ can be identified with $\bar{M}_{\infty,\ell'}$ by a biholomorphic map outside a small closed subset. This means that $\Phi_{\infty,\ell}$ stabilize modulo a small subset. It may be sufficient for deducing the irreducibility and consequently, the partial C^0-estimate. I will give more analysis on this in a separate paper.

Remark 4.17. In view of my joint work with J. Viaclovsky [TV05], one can expect that Theorem 4.12 also holds for Kähler metrics with constant scalar curvature and bounded diameter. If so, following the same arguments as we did above, one can also derive a partial C^0-estimate for potential functions for Kähler metrics with constant scalar curvature and bounded diameter. This may be used in proving the convergence of K-stable Kähler manifolds with constant scalar curvature.

4.3. Relating K-stability to existence

In this subsection, we sketch how to deduce the existence of Kähler-Einstein metrics from the K-stability under the assumption of the partial C^0-estimate (4.8). These arguments were discussed before in many of my lectures. It may be useful

---

[22]This $\Phi_{i,\ell}$ is the Φ_i used above. Now we insert an extra subscript to emphasis dependence on ℓ.

to say a few words on them. More details can be provided later. I would like to point out that the arguments in this subsection work for general Kähler metrics with constant scalar curvature if there is a suitable version of partial C^0-estimate as we discussed at the end of the last subsection.

In order to establish the existence, we need a priori C^0-estimates on solutions for (4.10). Let φ_i be a solution for (4.10) for $t_i \geq t_0$. Write $\omega_i = \omega_{\varphi_i}$. We assume that the partial C^0-estimate holds for ω_i for a sufficiently large ℓ and M is K-stable with respect to the embedding given by $H^0(M, K_M^{-\ell})$. We want to show that φ_i are uniformly bounded.

First we observe that the K-energy $\mathbf{T}_\omega$ is monotonically decreasing along (4.10), so we have a uniform bound

$$\mathbf{T}_\omega(\varphi_i) \leq c = c(\omega). \tag{4.14}$$

We may assume that $M \subset \mathbb{C}P^N$ through an embedding given by an orthonormal basis of $H^0(M, K_M^{-\ell})$ with respect to ω. By our assumption on the partial C^0-estimate for ω_i, there is a $\sigma_i \in \mathbf{SL}(N+1)$ such that

$$\frac{1}{\ell}\sigma_i^* \omega_{FS} = \omega + \frac{\sqrt{-1}}{2}\partial\bar{\partial}\psi_i, \ \|\psi_\ell - \varphi_i\|_{C^0} \leq C. \tag{4.15}$$

Here C denotes a uniform constant. In fact,

$$\psi_\ell - \varphi_i = \frac{1}{\ell}\log\left(\sum_{a=0}^N \|S_a^i\|_i^2\right),$$

where $\|\cdot\|_i$ is a Hermitian norm whose curvature is ω_i and $\{S_a^i\}$ is an orthonormal basis with respect to ω_i. It follows

$$\mathbf{T}_\omega(\psi_i) \leq c' = c'(\omega). \tag{4.16}$$

For simplicity, we assume that M has no non-trivial holomorphic fields. If φ_i are not uniformly bounded, $\sigma_i(M)$ converges to a variety which is not biholomorphic to M.[23] For each i, join $\mathrm{I} \in \mathbf{SL}(N+1)$ to σ_i by the orbit O_i of a C^*-action, without loss of generality, we may assume that O_i converge to a C^*-orbit O_∞. Using appropriate compactification of $\mathbf{SL}(N+1)(M)$, one can show that if $\sigma(e^t)$ ($t \in \mathbb{C}$) is the limit C^*-action, $\sigma(e^t)(M)$ converge to the limit of $\sigma_i(M)$ as t tends to ∞. Then the K-stability assumption for M implies that $\mathbf{T}_\omega(\varphi_{\sigma(t)})$ diverges to ∞ as t tends to ∞. Next one shows that there are $t_i \to \infty$ satisfying:

$$\|\psi_i - \varphi_{\sigma(e^{t_i})}\|_{C^0} \leq C.$$

Then $\mathbf{T}_\omega(\psi_i)$ are unbounded, a contradiction. Hence, φ_i should stay bounded and there is a Kähler-Einstein metric on M.

---

[23]If M has non-trivial holomorphic fields, we can modify σ_i by automorphisms of M so that this still holds.

The same arguments also show the following: If $(M_i, \tilde{\omega}_i)$ is a sequence of Kähler manifolds with $c_1(M_i) = [\tilde{\omega}_i]$ converging to a Kähler manifold $(M_\infty, \omega_\infty)$ and M_∞ is asymptotically K-stable, we further assume that M_i admit Kähler-Einstein metrics ω_i with $c_1(M_i) = [\omega_i]$ for which the partial C^0-estimate holds, then ω_i converges to a Kähler-Einstein metric on M_∞. This is exactly what we did for complex surfaces in [Ti89].

4.4. Kähler-Ricci flow on Fano manifolds

In this subsection, we discuss some results on the Kähler-Ricci flow on a Fano manifold M. First we recall that the Kähler-Ricci flow (3.18) on M has a global solution $\varphi(t)$ for $t \geq 0$, i.e., $\varphi(t)$ satisfies:

$$\frac{\partial \varphi}{\partial t} = \log \left(\frac{(\omega_0 + \frac{\sqrt{-1}}{2\pi} \partial\bar{\partial}\varphi)^n}{\omega_0^n} \right) + h_0 + \varphi,\ \varphi(0) = 0, \tag{4.17}$$

where $h_0 = h_{\omega_0}$. Moreover, if $\|\varphi(t)\|_{C^0}$ is uniformly bounded, then $\varphi(t)$ converges to a smooth function φ_∞ such that ω_{φ_∞} is a Kähler-Einsetin metric. However, because of known obstructions, we can not expect a C^0-estimate for general Fano manifolds. What one can expect is to get the C^0-estimate for K-stable Fano manifolds.

Let $\varphi(t)$ be the global solution of (4.17) $(t \geq 0)$ and $\omega_t = \omega_{\varphi(t)}$. A folklore conjecture states: ω_t *converges to a Kähler-Ricci soliton (possibly with "mild" singularities of codimension at least* 4) *in the Cheeger-Gromov topology.* Partial progresses on this conjecture have been made in recent years. The following theorem is due to G. Perelman (cf. [SeTi07]):

Theorem 4.18. *We have*

(1) *The scalar curvature of* ω_t *is uniformly bounded;*
(2) *The diameter of* ω_t *is uniformly bounded;*
(3) *The time derivative* $\dot{\varphi}$ *of* $\varphi(t)$ *is uniformly bounded if we normalize* $\varphi(t)$ *by*

$$\int_M \left(e^{-\varphi(t)} - 1\right) \omega_t^n = 0.$$

Using this theorem, one can verify the above conjecture if one can further bound the Ricci curvature of ω_t (cf. [SeTi07]). If one can further bound the curvature, one can even prove that the limit soliton has no singularity, though its complex structure may differ from that of M. The above conjecture follows if one can extend the works of Cheeger-Colding on space of metrics with bounded Ricci curvature to the Ricci flow on a compact manifold with bounded diameter, scalar curvature as well as bounded Sobolev constants.

In the opposite direction, it is known that if the Fano manifold M admits a Kähler-Einstein metric, then the solution $\varphi(t)$ for (4.17) converges to a φ such that ω_φ is Kähler-Einstein (cf. [Pe03], [TZ07]). In [TZ07], this convergence result is extended to the case when M admits only a Kähler-Ricci soliton.

5. Appendix: Finite-dimensional α-invariants

In this appendix, we discuss a sequence of holomorphic invariants which were first introduced in my Harvard PhD thesis [Ti88] (also see [Ti89]). The motivation then was for solving a conjecture of Calabi for Del-Pezzo surfaces. There has been much progress on computing such invariants and understanding their relation to $\alpha(M)$ in [Ti87]. It was shown later that they have algebraic correspondences: log canonical thresholds in the study of projective manifolds and resolution of singularity.

Assume that (M, L) is a polarized projective manifold of dimension n.[24] Fix a Hermitian metric h on L such that its curvature ω gives a Kähler metric on M, then they induce a Hermitian inner product on $H^0(M, L)$. We can define a function $\rho_{L,h,k}$ on the Grassmannian manifold $M \times G(k, H^0(M, L))$ as follows: For any k-dimensional subspace $P \in G(k, H^0(M, L))$,

$$\rho_{L,h,k}(x, P) := \sum_{i=1}^{k} \|S_i\|_h^2(x),$$

where $\{S_i\}$ is any orthonormal basis of P with respect to the induced inner product. One can easily show that this function is well defined, i.e., it is independent of the choice of the orthonormal basis. Now we define

$$\alpha_{\ell,k}(M, L) := \sup \left\{ \alpha \mid \sup_P \int_M \left(\frac{1}{\rho_{L^\ell,h^\ell,k}(\cdot, P)} \right)^{\frac{\alpha}{\ell}} \omega^n < \infty \right\}. \tag{5.1}$$

One can prove that this is an invariant of (M, L), i.e., it is independent of the choice of h. Moreover, we have

$$0 < c \le \alpha_{\ell,1}(M, L) \le \alpha_{\ell,2}(M, L) \le \cdots \le \alpha_{\ell,k}(M, L) \le \cdots,$$

where c depends only on (M, L). When $L = K_M^{-1}$, we simply denote them by $\alpha_{\ell,k}(M)$. In Section 3, we have seen that they are closely related to the existence of Kähler-Einstein metrics.

There is an algebraic counterpart of $\alpha_{\ell,k}(M, L)$. For simplicity, we assume $k = 1$. Recall that we say a pair (M, D), where D is an effective divisor, is log canonical if there is a blow-up $\pi : M' \mapsto M$ such that $D' := \sum_i E_i + \pi_*^{-1} D$, where $\pi_*^{-1} D$ is the quadratic transformation of D, is a normal crossing and $K_{M'} + D' = \pi^*(K_M + D) + \sum_i a_i E_i$, where $a_i \ge -1$ and E_i are exceptional divisors of M'. The log canonical thresfold of any pair (M, D) is defined by

$$lct(M, D) = \sup\{ \lambda \in \mathbb{Q} \mid (M, \lambda D) \text{ is log canonical } \}. \tag{5.2}$$

The global log canonical threshold $lct(M, L)$ of (M, L) is defined as

$$lct(M, L) = \inf\{ lct(M, D) \mid D \in |L| \}. \tag{5.3}$$

[24]In [Ti88], M is assumed to be Fano and L is simply the anti-canonical bundle. It is not hard to see that this restriction is unnecessary (also see [Ch08]).

The log canonical threshold was introduced by V. Shokurov [Sho92] and has many applications in the classification of projective manifolds. It can be identified with $\alpha_{\ell,1}(M,L)$, more precisely, one can prove (cf. [Ch07])

$$\alpha_{\ell,1}(M,L) \;=\; \frac{1}{\ell}\, lct(M,L^\ell),\ \forall \ell \geq 1. \tag{5.4}$$

In particular, if M is Fano, we have

$$\alpha_{\ell,1}(M) \;:=\; \alpha_{\ell,1}(M,K_M^{-1}) \;=\; lct(M,K_M^{-\ell}))/\ell.$$

In [Ch07] and [Ch08], I. Cheltsov computes $lct(M,L)$ for Del-Pezzo surfaces and many Fano manifolds of higher dimensions. In particular, if M is a smooth Del-Pezzo surface with $K_M^2 \leq 4$, he shows

$$lct(M,K_M^{-\ell}) \;=\; \ell\, lct(M,K_M^{-1}).$$

He also estimates them from below. Identifying log canonical thresholds with $\alpha_{\ell,1}(M)$, we get

Theorem 5.1 (I. Cheltsov). *Let M be a smooth Del-Pezzo surface with $K_M^2 \leq 3$ and without any Eckardt points, then*

$$\alpha_{\ell,1}(M) \;=\; \alpha_{1,1}(M) \;\geq\; \frac{3}{4}.$$

There are fewer results on $\alpha_{\ell,k}(M)$. In [Ti89], I proved that $\alpha_{6,2}(M) > 2/3$ for any smooth cubic surface $M \subset \mathbb{C}P^3$. In [Shi09], Y.L. Shi proved the following:

Theorem 5.2 (Y.L. Shi). *Let M be a smooth cubic surface in $\mathbb{C}P^3$. Then*

$$\alpha_{\ell,2}(M) > 2/3.$$

This allows him to give an elegant and simpler proof of my theorem on the existence of Kähler-Einstein metrics on Del-Pezzo surfaces in [Ti89] (see [Shi09] and also Section 3).

In general, it is a difficult task to compute $\alpha_{\ell,k}(M,L)$ and understand relations among $\alpha_{\ell,k}$ for a fixed k. The following conjecture was proposed by myself a while ago:

Conjecture 5.3. *For any polarized manifold (M,L) and $k \geq 1$, there is an $\ell_0 > 0$ such that for all $\ell \geq \ell_0$,*

$$\alpha_{\ell,k}(M,L) \;=\; \alpha_{\ell_0,k}(M,L).$$

We can further speculate

Conjecture 5.4. *We can ask the stronger version of Conjecture 5.3: If the ring $R(M,L) := \oplus_{\ell\geq 0} H^0(M,L^\ell)$ is generated by $H^0(M,L)$, then for $k \in [1,k_0]$,*

$$\alpha_{\ell,k}(M,L) = \alpha_{1,k}(M,L),$$

where $k_0 \geq 1$ depends only on M and L. More generally, if $R(M,L)$ is generated by $\oplus_{\ell=0}^{\ell_0} H^0(M,L^\ell)$, then for $\ell \geq \ell_0$ and small k,

$$\alpha_{\ell,k}(M,L) = \alpha_{\ell_0,k}(M,L).$$

If G is a compact subgroup of the automorphism group $\mathrm{Aut}(M,L)$ of (M,L), then we can easily extend the above construction to have G-invariant versions $\alpha_{\ell,k,G}(M,L)$ of $\alpha_{\ell,k}(M,L)$. For this purpose, we only need to assume that h and P in (5.1) are G-invariant.

In applications of α-invariants, we often need a uniform bound on the integral in (5.1). More precisely, we have

Theorem 5.5 ([PhSt00, DeKo01]). *Assume that (M_j, L_j) is a sequence of polarized manifolds converging to a polarized manifold (M_∞, L_∞) and assume that $P_j \in G(k, H^0(M_j, L_j^\ell))$ converge to $P_\infty \in G(k, H^0(M_\infty, L_\infty))$. Let h_j be Hermitian metrics on L_j converging to a Hermitian metric h_∞ on L_∞ such that the curvature ω_∞ of h_∞ is a Kähler metric. Then for any $\alpha < \alpha_{\ell,k}(M_\infty, L_\infty)$, there is a uniform constant C such that*

$$\int_{M_j} \left(\frac{1}{\rho_{L_j^\ell, h_j^\ell, k}(\cdot, P_j)}\right)^\alpha \omega_j^n \le C, \tag{5.5}$$

where ω_j is the curvature form of h_j.

The special case of this theorem for $n = 2$ was first given and shown in the appendix of [Ti89]. The motivation is to derive the properness of $\mathbf{F}_{\omega_\tau}$ on the subset of functions in $P(M_\tau, \omega_\tau)$ induced by sections in $H^0(M_\tau, K_{M_\tau}^{-6})$ as shown in Section 3. Note that the proof of Theorem 5.5 is essentially local and can be reduced to a similar problem on a sequence of holomorphic functions.

Finally, we recall the definition of $\alpha(M,L)$: Choose any Kähler metric ω with Kähler class $\pi c_1(L)$. Then

$$\alpha(M,L) = \inf\left\{\alpha \mid \sup_{\varphi \in P(M,\omega)} \int_M e^{-\alpha(\varphi - \sup_M \varphi)}\,\omega^n < \infty\right\}. \tag{5.6}$$

If $G \subset \mathrm{Aut}(M,L)$ is a compact subgroup, then

$$\alpha_G(M,L) = \inf\left\{\alpha \mid \sup_{\varphi \in P_G(M,\omega)} \int_M e^{-\alpha(\varphi - \sup_M \varphi)}\,\omega^n < \infty\right\}. \tag{5.7}$$

Clearly, $\alpha_{\ell,1}(M,L) \ge \alpha(M,L)$ and $\alpha_{\ell,1,G}(M,L) \ge \alpha_G(M,L)$. The following theorem gives a closer relation among them.

Theorem 5.6. *For any polarized manifold (M,L), we have*

$$\alpha(M,L) = \lim_{\ell\to\infty} \alpha_{\ell,1}(M,L), \ \alpha_G(M,L) = \lim_{\ell\to\infty} \alpha_{\ell,1,G}(M,L).$$

This was shown by Demailly (see [Ch08], Appendix) and by Y.D. Shi, independently, in his PhD thesis (see [Shi09]).

References

[An90] Anderson, M.: Ricci curvature bounds and Einstein metrics on compact manifolds. J. Amer. Math. Soc., **3** (1990), 355–374.

[ArPa05] Arezzo, C. and Pacard, F.: Blowing up and desingularizing constant scalar curvature Kähler manifolds. Acta Math. **196** (2006), 179–228.

[ALV09] Arezzo, C., Lanave, G. and Vedova, A.: Singularity, test configurations and constant scalar curvature Kähler metrics. Preprint, 2009.

[Au76] Aubin, T.: Equations du type de Monge-Ampère sur les variétés Kähleriennes compactes. C. R. Acad. Sci. Paris, **283** (1976), 119–121.

[Au83] Aubin, T.: Réduction du cas positif de l'équation de Monge-Ampère sur les variétés Kähleriennes compactes à la demonstration d'une inégalité. J. Funct. Anal., **57** (1984), 143–153.

[BM86] Bando, S. and Mabuchi, T.: Uniqueness of Einstein Kähler metrics modulo connected group actions. Algebraic Geometry, Adv. Studies in Pure Math., **10** (1987).

[Cal82] Calabi, E.: Extremal Kähler metrics. Seminar on Diff. Geom., Ann. of math. Stud., 102, Princeton Univ. Press, 1982.

[Cao86] Cao, H.D.: Deformation of Kähler metrics to Kähler-Einstein metrics on compact Kähler manifolds. Inv. Math., **81** (1985), 359–372.

[CCT95] Cheeger, J., Colding, T. and Tian, G.: Constraints on singularities under Ricci curvature bounds. C. R. Acad. Sci. Paris Sér. I, Math. **324** (1997), 645–649.

[Ch07] Cheltsov, I.: Log canonical thresholds of Del Pezzo surfaces, Geometric Analysis and Functional Analysis, **11** (2008), 1118–1144.

[Ch08] Cheltsov, I. and Shramov, C.: Log canonical thresholds of smooth Fano threefolds. With an appendix by Jean-Pierre Demailly. Preprint, math.DG/0806.2107.

[ChTi04] Chen, X.X. and Tian, G.: Geometry of Kähler metrics and foliations by holomorphic discs. Publ. Math. Inst. Hautes Études Sci. **107** (2008), 1–107.

[ChWa09] Chen, X.X. and Wang, B.: Kähler Ricci flow on Fano manifolds (I). Preprint, math.DG/0909.2391.

[DeKo01] Demailly, P. and Kollar, J.: Semicontinuity of complex singularity exponents and Kähler-Einstein metrics on Fano orbifolds. Ann. Ec. Norm. Sup **34** (2001), 525–556.

[DT91] Ding, W. and Tian, G.: The generalized Moser-Trudinger inequality, in: Nonlinear Analysis and Microlocal Analysis: Proceedings of the International Conference at Nankai Institute of Mathematics (K.C. Chang et al., eds.), World Scientific, 1992, 57–70.

[DT92] Ding, W. and Tian, G.: Kähler-Einstein metrics and the generalized Futaki invariants. Invent. Math., **110** (1992), 315–335.

[Do00] Donaldson, S.: Scalar curvature and projective embeddings. I. J. Diff. Geom. **59** (2001), 479–522.

[Do02] Donaldson, S.: Scalar curvature and stability of toric varieties. J. Diff. Geom., **62** (2002), 289–349.

[Do05] Donaldson, S.: Lower bounds on the Calabi functional, J. Diff. Geom., **70** (2005), 453–472.

[Fut83] Futaki, A.: An obstruction to the existence of Einstein-Kähler metrics. Inv. Math., **73** (1983), 437–443.

[Fut90] Futaki, A.: Kähler-Einstein Metrics and Integral Invariants. Lecture Notes in Mathematics, 1314, Springer-Verlag.

[Ma86] Mabuchi, T.: Some symplectic geometry on compact Kähler manifolds. Osaka J. Math., bf 24 (1987), 227–252.

[MaMu93] Mabuchi, T. and Mukai, S.: Stability and Einstein-Kähler metric of a quartic del Pezzo surface. Einstein metrics and Yang-Mills connections (Sanda, 1990), 133–160, Lecture Notes in Pure and Appl. Math., 145, Dekker, New York, 1993.

[Na88] Nakajima, H.: Hausdorff convergence of Einstein 4-manifolds. J. Fac. Sci. Univ. **35** (1988), 411–424.

[Paul08] Paul, S.: Hyperdiscriminant polytopes, Chow polytopes, and Mabuchi energy asymptotics. Preprint math.DG/ 0811.2548.

[PT04] Paul, S. and Tian, G.: CM Stability and the Generalized Futaki Invariant I. math.DG/ 0605.278.

[Pe03] Perelman, G.: Private communication.

[PhSt00] Phong, D. and Sturm, J.: Algebraic estimates, stability of local zeta functions, and uniform estimates for distribution functions. Annals of Mathematics, **152** (2000), 277–329.

[PSSW06] Phong, D.; Song, J.; Sturm, J. and Weinkove, B.: The Moser-Trudinger inequality on Kähler-Einstein manifolds. Amer. J. Math. **130** (2008), 1067–1085.

[Ru07] Rubinstein, Y.: On energy functionals, Kähler-Einstein metrics, and the Moser-Trudinger-Onofri neighborhood, J. Funct. Anal. **255** (2008), 2641–2660.

[Ru08] Rubinstein, Y.: On the construction of Nadel multiplier ideal sheaves and the limiting behavior of the Ricci flow. Trans. Amer. Math. Soc. **361** (2009), 5839–5850.

[SeTi07] Sesum, N. and Tian, G.: Bounding scalar curvature and diameter along the Kähler Ricci flow (after Perelman). J. Inst. Math. Jussieu, **7** (2008), 575–587.

[Sho92] Shokurov, V.: Three-dimensional log perestroikas. Izvestiya: Mathematics, **56** (1992), 105-203.

[Shi09] Shi, Yalong: On the α-Invariants of Cubic Surfaces with Eckardt Points. Preprint, arXiv:0902.3203.

[Sto07] Stoppa, J.: K-stability of constant scalar curvature Kähler manifolds. Adv. Math. **221** (2009), 1397–1408.

[Ti87] Tian, G.: On Kähler-Einstein metrics on certain Kähler Manifolds with $C_1(M) > 0$. Invent. Math., **89** (1987), 225–246.

[Ti88] Tian, G.: Harvard PhD Thesis, May, 1988.

[Ti89] Tian, G.: On Calabi's conjecture for complex surfaces with positive first Chern class. Inv. Math. **101**, (1990), 101–172.

[Ti90] Tian, G.: Kähler-Einstein on algebraic manifolds. Proc. of ICM, Kyoto, 1990.

[Ti97] Tian, G.: Kähler-Einstein metrics with positive scalar curvature. Invent. Math., **130** (1997), 1–39.

[Ti98] Tian, G.: Canonical Metrics on Kähler Manifolds. Lectures in Mathematics ETH Zürich, Birkhäuser Verlag, 2000.

[TV05] Tian, G. and Viaclovsky, J.: Moduli spaces of critical Riemannian metrics in dimension four. Adv. Math. **196** (2005), 346–372.

[TY87] Tian, G. and Yau, S.T.: Kähler-Einstein metrics on complex surfaces with $C_1(M)$ positive. Comm. Math. Phys., **112** (1987).

[TZ97] Tian, G. and Zhu, X.H.: A nonlinear inequality of Moser-Trudinger type. Calc. Var. Partial Differential Equations, **10** (2000), 349–354.

[TZ00] Tian, G. and Zhu, X.H.: Uniqueness of Kähler-Ricci solitons. Acta Math. **184** (2000), no. 2, 271–305.

[TZ07] Tian, G. and Zhu, X.H.: Convergence of Kähler-Ricci flow. J. Amer. Math. Soc., **20** (2007), 675–699.

[Ya76] Yau, S.T.: On the Ricci curvature of a compact Kähler manifold and the complex Monge-Ampère equation, *I*. Comm. Pure Appl. Math., 31 (1978), 339–441.

[Ye07] Ye, R.G.: The logarithmic Sobolev inequality along the Ricci flow. Preprint, math.DG/0707.2424.

[QZ07] Zhang, Qi: A uniform Sobolev inequality under Ricci flow. Preprint Math. DG/ 0706.1594.

Gang Tian
Department of Mathematics
Beijing University and Princeton University

Part II

Metric Geometry

Analytic Properties of Quasiconformal Mappings Between Metric Spaces

P. Koskela and K. Wildrick

Dedicated to Jeff Cheeger for his 65th birthday

Abstract. We survey recent developments in the theory of quasiconformal mappings between metric spaces. We examine the various weak definitions of quasiconformality, and give conditions under which they are all equal and imply the strong classical properties of quasiconformal mappings in Euclidean spaces. We also discuss function spaces preserved by quasiconformal mappings.

Mathematics Subject Classification (2000). 30C65, 46E30, 46E35.

Keywords. Quasiconformal mapping, quasisymmetric mapping, Sobolev space, Triebel-Lizorkin space.

1. Introduction

A complete understanding of the behavior of quasiconformal mappings requires fluency in moving between the various definitions of quasiconformality. Of particular importance is that the Sobolev regularity and absolute continuity properties of quasiconformal mappings in fact follow from the easy to verify metric definition. In Euclidean spaces, these properties have been major theme in the literature from the initiation of the study of non-smooth quasiconformal mappings by Ahlfors in 1954 [1] to Gehring's seminal works in the early 1960s [8], [7]. By 1968, the celebrated work of Mostow demonstrated the need for a theory of quasiconformal mappings in the non-Riemannian setting [23]. This led to the study of rigidity and quasiconformal mappings in the Heisenberg group and other Carnot groups. In this setting, the techniques of Gehring, which are based on the foliation of Euclidean space by lines, become tenuous and delicate to employ, though with difficulty they still led to important results [24], [21].

An alternate approach to understanding the equivalence of the many different definitions of quasiconformal mappings was given by Heinonen and Koskela [13].

By considering quasiconformal mappings in the setting of arbitrary metric spaces, they were able to identify robust techniques that did not depend on the special structure of Euclidean spaces. The starting point is the simplest definition of a quasiconformal mapping, the *metric definition.*

For a homeomorphism $f\colon X \to Y$ of metric spaces, we define for all $x \in X$ and $r > 0$

$$L_f(x,r) := \sup\{d_Y(f(x), f(y)) : d_X(x,y) \le r\},$$
$$l_f(x,r) := \inf\{d_Y(f(x), f(y)) : d_X(x,y) \ge r\},$$
$$H_f(x) := \limsup_{r\to 0} \frac{L_f(x,r)}{l_f(x,r)} \quad \text{and} \quad h_f(x) := \liminf_{r\to 0} \frac{L_f(x,r)}{l_f(x,r)}.$$

The mapping f is H-*quasiconformal*, $H \ge 1$, if $H_f(x) \le H$ for all $x \in X$.

If the underlying space X lacks a useful infinitesimal structure, then we cannot expect any large-scale properties of quasiconformal mappings defined on X. On the other hand, if the underlying space X has enough structure, the infinitesimal definition given above in fact guarantees strong properties of quasiconformal mappings. Properties of particular importance are Sobolev regularity, absolute continuity on paths, and quasisymmetry, i.e., global rather than infinitesimal distortion bounds. In this survey, we examine the minimal assumptions on metric spaces X and Y and a homeomorphism $f\colon X \to Y$ that guarantee that f is quasiconformal and possesses these strong properties. We also discuss recent work regarding function spaces preserved by such mappings.

2. The metric space setting

A *metric measure space* is a triple (X, d, μ) where (X, d) is a metric space and μ is a measure on X. For our purposes, a *measure* is a non-negative countably subadditive set function defined on all subsets of X that gives the value 0 to the empty set. Moreover, we require that measures are Borel inner and outer regular.

A metric space is said to be *proper* if every closed and bounded set is compact. Unless otherwise mentioned, throughout this paper we let (X, d, μ) and (Y, d_Y, ν) be proper metric measure spaces.

Given a point $x \in X$ and a radius $r > 0$, we employ the following notation for balls:

$$B_{(X,d)}(x,r) = \{y \in X : d(x,y) < r\} \quad \text{and} \quad \bar{B}_{(X,d)}(x,r) = \{y \in X : d(x,y) \le r\}.$$

Where it will not cause confusion, we will replace $B_{(X,d)}(x,r)$ by $B_X(x,r)$, $B_d(x,r)$, or $B(x,r)$. A similar convention will be made for any other objects which depend on the ambient metric space. Given a ball $B = B(x,r)$ and a constant $\tau > 0$, we denote by τB the ball $B(x, \tau r)$.

A main theme in analysis on metric spaces is that the infinitesimal structure of a metric space can be understood via the paths that it contains. The reason for this is that rectifiable paths admit path-integration. We define a *path* in X to be a continuous, non-constant map $\gamma\colon I \to X$ where $I \subseteq \mathbb{R}$ is a compact interval. A path

$\gamma\colon I \to X$ is called *rectifiable* if it is of finite length. Any rectifiable path $\gamma\colon [a,b] \to X$ has a unique parameterization $\gamma_s\colon [0, \mathrm{length}(\gamma)] \to X$ such that for all $t \in [a,b]$,

$$\gamma(t) = \gamma_s(\mathrm{length}(\gamma|_{[a,t]})).$$

The path γ_s is called the *arc length parameterization of* γ, and it is 1-Lipschitz. Given a Borel function $\rho\colon X \to [0,\infty]$ and a rectifiable path γ in X, we define the integral of ρ over γ by

$$\int_\gamma \rho\, ds := \int_{[0,\mathrm{length}(\gamma)]} \rho \circ \gamma_s(t)\, dt.$$

A measurement of the size of a given collection of paths Γ in X is the *p-modulus of* Γ, $p \geq 1$, which is defined by

$$\mathrm{mod}_p(\Gamma) = \inf \int_X \rho^p\, d\mu,$$

where the infimum is taken over all Borel functions $\rho\colon X \to [0,\infty]$ such that for all locally rectifiable paths $\gamma \in \Gamma$,

$$\int_\gamma \rho\, ds \geq 1.$$

Such a function ρ is said to be *admissible* for the path family Γ. A condition is said to be true on *p-almost every path in X* if the collection of paths in X where the condition does not hold has p-modulus 0.

An upper gradient of f is a generalization of the norm of the gradient of f developed in connection with quasiconformal mappings in [13]. Philosophically, the more rectifiable curves a metric space contains, the more stringent the upper gradient condition becomes. Given an open set $U \subseteq X$ and a mapping $f\colon U \to Y$, we say that a Borel function $\rho\colon U \to [0,\infty]$ is an *upper gradient of f* in U if, for each rectifiable path $\gamma\colon [0,1] \to U$, we have

$$d_Y(f(\gamma(0)), f(\gamma(1))) \leq \int_\gamma \rho\, ds. \tag{2.1}$$

If (2.1) holds only for p-almost every path in U, then we say that ρ is a *p-weak upper gradient* of f in U.

Real-valued Sobolev spaces based on upper gradients were used to great success [6] and explored in-depth in [27]. They have been extended to the metric-valued setting in [14] and have seen numerous generalizations. A simple definition is as follows. Let $f\colon X \to Y$ be a continuous map. Then f is in the Sobolev space $W^{1,p}_{\mathrm{loc}}(X;Y)$, $1 \leq p \leq \infty$, if for each relatively compact open subset $U \subseteq X$, the map f has an upper gradient $g \in L^p(U)$ in U, and there is a point $x_0 \in U$ such that $u(x) := d_Y(f(x_0), f(x)) \in L^p(U)$. If the space Y is not specified, it assumed to be $\mathbb{R}$.

A continuous mapping $f\colon X \to Y$ is said to be *absolutely continuous on a rectifiable path γ in X* if the map $f \circ \gamma_s\colon [0, \mathrm{length}(\gamma)] \to Y$ is absolutely continuous in the usual sense. As in the Euclidean setting, Sobolev maps of metric spaces (which are defined to be continuous) have absolute continuity properties. Namely, if X is proper, then each $f \in W^{1,p}_{loc}(X;Y)$ is absolutely continuous on p-almost every rectifiable path in X [27, Prop. 3.1].

In the Euclidean setting, the dimension of the space obviously plays a key role in the Sobolev theory. Moreover, the dimension of the space is reflected in the uniform scaling of Lebesgue measure. The following condition is a relatively strong generalization of this phenomenon to the metric measure space setting. The metric measure space (X, d, μ) is called *Ahlfors Q-regular*, $Q \geq 0$, if there exists a constant $K \geq 1$ such that for all $a \in X$ and $0 < r \leq \operatorname{diam} X$, we have

$$\frac{r^Q}{K} \leq \mu(\bar{B}_d(a,r)) \leq K r^Q. \tag{2.2}$$

We say that (X, d, μ) is *locally Ahlfors Q-regular* if for every compact subset $V \subseteq X$, there is a constant $K \geq 1$ and a radius $r_0 > 0$ such that for each point $a \in V$ and radius $0 < r \leq r_0$, the inequalities in (2.2) are satisfied.

We also require the following non-standard definition. Let $E \subseteq X$. We say that (X, d, μ) is *locally Ahlfors Q-regular off E* if there is a constant $K \geq 1$ such that for each point $a \in X \backslash E$, there is a radius $r_a > 0$ such that for each $0 < r \leq r_a$, the inequalities in (2.2) are satisfied.

There is also a weaker notion of dimensionality for measures that is useful. The metric measure space (X, d, μ) is *doubling* if there is a constant $C \geq 1$ such that for every $x \in X$ and $r > 0$,

$$\mu(B(x, 2r)) \leq C\mu(B(x,r)).$$

Iterating this condition leads to the notion of *Assouad dimension*; for more information, see for example [12].

The space (X, d, μ) is said to support a *p-Poincaré inequality*, $1 \leq p < \infty$ if there are constants $C, \tau \geq 1$ such that if B is a ball in X, $u\colon \tau B \to \mathbb{R}$ is a bounded continuous function, and ρ is an upper gradient of u, then

$$\fint_B |u - u_B| \, d\mu \leq C \operatorname{diam}(B) \left(\fint_{\tau B} \rho^p \, d\mu \right)^{1/p}.$$

Here and throughout the paper we employ the notation

$$u_B = \fint_B u \, d\mu = \frac{1}{\mu(B)} \int_B u \, d\mu,$$

whenever u is a μ-measurable function on B.

Note that if (X, d, μ) supports a p-Poincaré inequality, $1 \leq p < \infty$, then it also supports a q-Poincaré inequality for all $q \geq p$. A deep theorem of Keith and Zhong states that if (X, d, μ) is doubling and supports a p-Poincaré inequality, $p > 1$, then it also supports a p'-Poincaré inequality for some $p' < p$ [16].

The p-Poincaré inequality can be thought of as a requirement that a space contains "many" curves, in terms of the p-modulus of curves in the space. See [13], [12], and [15] for more information. For a small bit of intuition, assume that

(X, d, μ) is Ahlfors Q-regular, $Q > 1$, and supports a p-Poincaré inequality. Regardless of the value of p, it follows that (X, d) is quasiconvex. However, if $p > Q$, then X may contain local cut-points. For more geometric implications of the Poincaré inequality, see [11] and [17].

3. Weak definitions of quasiconformality

In this section, we discuss weak versions of the metric, analytic, and geometric definitions of quasiconformality. Some relations between these conditions are valid even in the absence of a Poincaré inequality. Their value resides in the fact that in the presence of an appropriate Poincaré inequality, they are all equivalent to the usual strong forms of quasiconformality. However, they are potentially much easier to verify in practice.

We begin with a weak formulation of the metric definition. The following definition allows for an exceptional set and employs h_f rather than H_f.

Definition 3.1. Let $1 \leq p \leq Q$. We say that f satisfies the *weak* (Q, p)*-metric definition of quasiconformality* if one of the following two conditions holds:

- $p < Q$, and there exists a set $E \subseteq X$ of σ-finite $(Q-p)$-dimensional Hausdorff measure, and a number $0 \leq H < \infty$ such that $h_f(x) < \infty$ for all $x \in X \backslash E$ and $h_f(x) \leq H$ for μ-almost every point $x \in X$,
- $p = Q$, and there exists a countable set $E \subseteq X$ and a number $0 \leq H < \infty$ such that $h_f(x) \leq H$ for all $x \in X \backslash E$

The classical definition of a quasiconformal homeomorphism $f \colon \Omega \to \Omega'$ of domains in $\mathbb{R}^n$ consists of the requirements that $f \in W^{1,n}_{\mathrm{loc}}(\Omega; \mathbb{R}^n)$ and that there is a constant $K \geq 1$ such that $\|Df\|^n \leq KJ_f$ almost everywhere, where Df is the weak differential matrix of f and J_f is the determinant of Df. Given a homeomorphism $f \colon (X, d, \mu) \to (Y, d_Y, \nu)$, the role of J_f is played by the *volume derivative* $\mu_f \colon X \to [0, \infty]$ defined by

$$\mu_f(x) = \limsup_{r \to 0} \frac{\nu(f(B(x,r)))}{\mu(B(x,r))}.$$

Definition 3.2. We say that f satisfies the Q*-analytic definition of quasiconformality* if $f \in W^{1,Q}_{\mathrm{loc}}(X; Y)$ and there is a constant $H \geq 1$ such that $H\mu_f^{1/Q}$ is a Q-weak upper gradient of f.

Finally, we consider the geometric (or modulus) definition of quasiconformality, first introduced by Ahlfors in Euclidean space [1]. Many properties of quasiconformal mappings may be derived directly from this definition.

Definition 3.3. We say that f satisfies the *weak* Q*-geometric definition of quasiconformality* if there is a constant $H \geq 1$ such that for every path family Γ in X

$$\mathrm{mod}_Q(\Gamma) \leq H \,\mathrm{mod}_Q(f(\Gamma)).$$

If f and f^{-1} satisfy the weak Q-geometric definition of quasiconformality, then we say that f satisfies the Q*-geometric definition of quasiconformality.*

The key point in connecting these definitions is absolute continuity on paths. The following result from [2] establishes absolute continuity on p-almost every path for mappings satisfying the weak (Q,p)-metric definition of quasiconformality.

Theorem 3.4 (Balogh-Koskela-Rogovin). *Let $Q > 1$ and let $1 \leq p \leq Q$. Suppose f satisfies the weak (Q,p)-metric definition of quasiconformality, and assume that X is locally Ahlfors Q-regular and that Y is locally Ahlfors Q-regular off $f(E)$. Then $f \in W^{1,p}_{\mathrm{loc}}(X;Y)$.*

The relationship between the weak metric and analytic definitions of quasiconformality is now provided by the following results.

Theorem 3.5 (Balogh-Koskela-Rogovin). *Assume the hypotheses of Theorem 3.4. Then*

$$g_f(x) := \begin{cases} H(\mu_f(x))^{1/Q} & h_f(x) \leq H, \\ \infty & h_f(x) > H, \end{cases}$$

is a p-weak upper gradient of f.

Corollary 3.6. *Assume the hypotheses of Theorem 3.4 with $p = Q$. Then f satisfies the Q-analytic definition of quasiconformality.*

The Q-analytic definition and the weak Q-geometric definition are closely linked. Note that in the following result from [31], no form of Ahlfors Q-regularity is assumed.

Theorem 3.7 (Williams). *Assume that (X, d_X, μ) and (Y, d_Y, ν) are separable metric measure spaces of locally finite measure, and that (X, d, μ) is doubling. Then f satisfies the Q-analytic definition of quasiconformality if and only if it satisfies the weak Q-geometric definition.*

Many properties of quasiconformal mappings between Ahlfors Q-regular metric spaces follow directly from the (strong) Q-geometric definition. Thus it is of practical interest to understand the properties of the inverse of a mapping that satisfies a weak definition of quasiconformality.

Theorem 3.8. *Assume that (X, d_X, μ) and (Y, d_Y, ν) are locally Ahlfors Q-regular, $Q > 1$. If $f\colon X \to Y$ satisfies the weak (Q,Q)-metric definition of quasiconformality, then $f^{-1} \in W^{1,Q}_{\mathrm{loc}}(X;Y)$. Moreover, f^{-1} satisfies the Q-analytic definition of quasiconformality, and f satisfies the Q-geometric definition of quasiconformality.*

Proof. The proof is nearly identical to the proof of Theorem 3.4 given in [2, Theorem 4.1]; the main philosophical difference is that one replaces balls with ball-like objects, namely, the images of balls under f. One constructs the same cover as in that proof, but replace the control function ρ_ϵ defined there by the quantity

$$\rho_\epsilon(y) = \sum_i \frac{r_i}{L_f(x_i, r_i)} \chi_{B(f(x_i), 2L_f(x_i, r_i))}(y).$$

The remainder of the proof proceeds as in the original to show that $f^{-1} \in W^{1,Q}_{\mathrm{loc}}(X;Y)$. A similar trick applied to the proof of Theorem 3.5 given in [2,

Proposition 4.3] shows that f^{-1} satisfies the Q-analytic definition of quasiconformality. The final statement now follows from Corollary 3.6 and Theorem 3.7. □

4. Quasiconformality and quasisymmetry

In the Euclidean setting, the natural Sobolev regularity of a quasiconformal mapping corresponds to the dimension of the space. It turns out that the assumption of a suitable Poincaré inequality allows for similar results in the metric space setting, as the following modified version of Theorem 3.4 shows.

Theorem 4.1. *Let $Q > 1$ and $1 \leq p \leq Q$. Suppose f satisfies the weak (Q,p)-metric definition of quasiconformality, and assume that X is locally Ahlfors Q-regular and satisfies a p-Poincaré inequality, and that Y is locally Ahlfors Q-regular off $f(E)$. Then $f \in W^{1,Q}_{\mathrm{loc}}(X;Y)$.*

It has been shown that Theorem 3.4 is also sharp in the sense that in the absence of a Poincaré inequality, the absolute continuity and Sobolev regularity cannot be improved to the Euclidean analogue [18].

Theorem 4.2. *For each integer $m \geq 1$ and real number $\epsilon > 0$, there is a homeomorphism $f\colon X \to Y$ of metric measure spaces and a set $E \subseteq X$ such that*

(i) *X is compact, quasiconvex, and Ahlfors 2-regular,*
(ii) *Y is compact and locally Ahlfors 2-regular off $f(E)$,*
(iii) *$(\log_3 2)/m \leq \dim_H(E) \leq (2\log_3 2)/m$, and $0 < \mathcal{H}^{\dim_H(E)}(E) < \infty$,*
(iv) *$H_f(x) = 1$ for all $x \in X \backslash E$,*
(v) *$f \notin W^{1,q}_{\mathrm{loc}}(X;Y)$ for some $q < 2 - \dim_H(E) + \epsilon$.*

The abundance of curves provided by a Poincaré inequality also allows for much stronger *global* distortion estimates in the form of quasisymmetry, at least in the presence of necessary geometric conditions.

A homeomorphism $f\colon (X,d_X) \to (Y,d_Y)$ of metric spaces is called *quasisymmetric* if there exists a homeomorphism $\eta\colon [0,\infty) \to [0,\infty)$ such that for all triples $a,b,c \in X$ of distinct points, we have

$$\frac{d_Y(f(a),f(b))}{d_Y(f(a),f(c))} \leq \eta\left(\frac{d_X(a,b)}{d_X(a,c)}\right).$$

It is easy to see that such a homeomorphism satisfies $L_f(x,r) \leq \eta(1) l_f(x,r)$ for all $x \in X$ and $r \geq 0$. If f is a quasisymmetric homeomorphism, then f^{-1} is as well; indeed, most proofs that the inverse of a quasiconformal mapping is quasiconformal involve first showing that the map is quasisymmetric.

As opposed to quasiconformal mappings, quasisymmetric mappings preserve boundedness and prevent the formation of cusps in a space. The latter property can be formalized as follows.

Let $\lambda > 1$. A metric space (X,d) is λ-*linearly locally connected* (λ-*LLC*) if for all $a \in X$ and $r > 0$ the following conditions are satisfied:

(λ-LLC_1) For each pair of distinct points $x, y \in B(a,r)$, there is a continuum $E \subseteq B(a, \lambda r)$ such that $x, y \in E$,

(λ-LLC_2) For each pair of distinct points $x, y \in X \backslash B(a,r)$, there is a continuum $E \subseteq X \backslash B(a, r/\lambda)$ such that $x, y \in E$.

Recall that a continuum is a connected, compact set containing more than one point. If $f \colon X \to Y$ is an η-quasisymmetric homeomorphism, and X is λ-LLC, then Y is λ'-LLC where λ' depends only on η and λ. If a metric measure space (X, d, μ) is Ahlfors Q-regular and supports a Q-Poincaré inequality, then it is λ-LLC for some $\lambda \geq 1$ depending only on the data associated to the conditions on the space [13].

The following theorem may now be derived from [2, Theorem 5.1 and Remark 5.3] and the techniques of the proof of Theorem 3.8.

Theorem 4.3 (Balogh-Koskela-Rogovin). *Suppose that X and Y are Ahlfors Q-regular metric spaces that are simultaneously bounded or unbounded, and that one of X and Y is linearly locally connected and the other satisfies a Q-Poincaré inequality. If $f \colon X \to Y$ satisfies the weak (Q,Q)-metric definition of quasiconformality, then $f \in W^{1,Q}_{\mathrm{loc}}(X;Y)$ and f is quasisymmetric. Moreover, f satisfies the Q-analytic and Q-geometric definitions of quasiconformality, and is absolutely continuous in measure.*

It is not true that quasisymmetric mappings must be absolutely continuous in measure, in the absence of a Poincaré inequality.

Example 4.4. For any $Q \geq 1$, there is an Ahlfors Q-regular and LLC metric measure space (X, d_X, μ) and a quasisymmetric mapping $f \colon X \to X$ such that f maps a set of measure zero to a set of full measure and a set of full measure to a set of zero measure.

Proof. It was shown by Tukia [29] that there exists an η-quasisymmetric mapping $g \colon \mathbb{R} \to \mathbb{R}$ mapping a set of measure zero to a set of full measure and vice-versa. Let $X = \mathbb{R}$, $d_X = |\cdot|^{1/Q}$, and $\mu = \mathcal{H}^Q_{d_X}$. Then (X, d_X, μ) is Q-regular, and for any $E \subseteq \mathbb{R}$, the quantity $\mu(E)$ is equal to the one-dimensional Lebesgue measure of E. Moreover, g is also a quasisymmetric when considered as a mapping from (X, d_X) to itself, with distortion function $\widetilde{\eta}(t) = (\eta(t^Q))^{1/Q}$. □

The most general setting in which even very strong definitions of quasiconformality imply quasisymmetry is not clear. The question is particularly intriguing in infinite dimensions, even for very simple function spaces.

Theorem 4.5 (Naor). *The space L_p quasisymmetrically embeds in the space L_2 if and only if $p \leq 2$.*

Question 4.6 (Naor). *Does L_p quasiconformally embed in L_2 when $p > 2$?*

The proof of Theorem 4.5 relies on deep work of Mendel and Naor that gives a metric characterization of Rademacher cotype in Banach spaces [22]. In particular, it is shown that cotype is preserved by quasisymmetric mappings between K-

convex Banach spaces [25]. The study of cotype in the general non-linear setting is still nascent [30].

While quasiconformal mappings on $\mathbb{R}^n$ are defined to be in $W^{1,n}_{\text{loc}}(\mathbb{R}^n)$, each such mapping actually lies in a smaller Sobolev class $W^{1,p}_{\text{loc}}(\mathbb{R}^n)$, where $p > n$ depends n and the distortion of the individual mapping [5], [9]. The key fact behind this improved regularity is the following *reverse Hölder inequality*, which has a self-improving property. For $t \in [1, \infty]$, we say that a real-valued function $u\colon X \to \mathbb{R}$ is in the class $\mathcal{B}_t(X)$ if there is a quantity $C \geq 1$ such that for every $x \in X$ and $r > 0$,

$$\left(\fint_{B(x,r)} \mu_f^t \, d\mu\right)^{1/t} \leq C \fint_{B(x,r)} \mu_f \, d_\mu$$

when $t < \infty$, or an analogous condition when $t = \infty$. For each quasiconformal mapping $f\colon \mathbb{R}^n \to \mathbb{R}^n$, the Jacobian determinant J_f "naturally" lies in the class $\mathcal{B}_1(\mathbb{R}^n)$, but in fact lies in a smaller class $\mathcal{B}_t(\mathbb{R}^n)$ for some $t > 1$ that depends on n and the distortion of the individual mapping. The following result shows that this phenomenon persists in the presence of an appropriate Poincaré inequality (Theorem 4.3, [13], [16]).

Theorem 4.7. *Assume the hypotheses of Theorem* 4.3. *Then there is* $t > 1$ *such that* $\mu_f \in \mathcal{B}_t(X)$.

5. Function spaces preserved by quasisymmetric mappings

A fundamental problem in the theory of quasiconformal mappings between metric spaces is determining when a metric space that is topologically equivalent to a "model" space (such as $\mathbb{S}^n$) is actually quasisymmetrically equivalent to that model space. This problem is of particular importance in geometric group theory [3]. One approach to this problem is to find a function space associated to each metric space that is preserved under quasisymetric mappings.

The uniformization problem described above is of interest even when the given space is not known to possess any rectifiable curves. An approach to gradients on metric spaces that does not rely on path integration was explored in [11] and [10]. Given a mapping $f\colon (X, d_X, \mu) \to (Y, d_Y)$, a measurable function $g\colon X \to [0, \infty]$ is a *Hajłasz gradient* of f if for almost every $x, y \in X$,

$$d_Y(f(x), f(y)) \leq d_X(x, y)(g(x) + g(y)).$$

This definition is both local and global in nature. One should view a Hajłasz gradient, in the Euclidean setting or in the presence of a suitable Poincaré inequality, as the maximal function of the usual gradient.

More precise variants of the concept of a Hajłasz gradient have led to interesting function spaces that are invariant under quasisymmetric mappings [20].

Definition 5.1. Let $s \in (0, \infty)$ and u be a measurable function on X. A sequence of nonnegative measurable functions, $\vec{g} \equiv \{g_k\}_{k \in \mathbb{Z}}$, is called a *fractional* s*-Hajłasz* *gradient* of u if there exists $E \subset X$ with $\mu(E) = 0$ such that for all $k \in \mathbb{Z}$ and $x, y \in X \setminus E$ satisfying $2^{-k-1} \le d(x, y) < 2^{-k}$,

$$|u(x) - u(y)| \le [d(x, y)]^s [g_k(x) + g_k(y)]. \tag{5.1}$$

Denote by $\mathbb{D}^s(u)$ the *collection of all fractional s-Hajłasz gradients of u.*

Relying on this concept we now introduce counterparts of Triebel-Lizorkin spaces. Let $p \in (0, \infty)$. In what follows, for $q \in (0, \infty]$, we always write $\|\{g_j\}_{j\in\mathbb{Z}}\|_{\ell^q} \equiv \{\sum_{j\in\mathbb{Z}} |g_j|^q\}^{1/q}$ when $q < \infty$ and $\|\{g_j\}_{j\in\mathbb{Z}}\|_{\ell^\infty} \equiv \sup_{j\in\mathbb{Z}} |g_j|$,

$$\|\{g_j\}_{j\in\mathbb{Z}}\|_{L^p(X,\, \ell^q)} \equiv \| \|\{g_j\}_{j\in\mathbb{Z}}\|_{\ell^q} \|_{L^p(X)}.$$

Definition 5.2. Let $s, p \in (0, \infty)$ and $q \in (0, \infty]$. The *homogeneous Hajłasz-Triebel-Lizorkin space* $\dot{M}^s_{p,\,q}(X)$ is the space of all measurable functions u such that

$$\|u\|_{\dot{M}^s_{p,\,q}(X)} \equiv \inf_{\vec{g}\in\mathbb{D}^s(u)} \|\vec{g}\|_{L^p(X,\, \ell^q)} < \infty.$$

Theorem 5.3 (Bourdon-Pajot). *Let X_1 and X_2 be Ahlfors Q_1-regular and Q_2-regular spaces with $Q_1, Q_2 \in (0, \infty)$, respectively. Let f be a quasisymmetric mapping from X_1 onto X_2. For $s_i \in (0, Q_i)$ with $i = 1, 2$, if $Q_1/s_1 = Q_2/s_2$, then f induces an equivalence between $\dot{M}^{s_1}_{Q_1/s_1,\, Q_1/s_1}(X_1)$ and $\dot{M}^{s_2}_{Q_2/s_2,\, Q_2/s_2}(X_2)$.*

In [4], Bourdon and Pajot proved the above invariance for the Besov spaces $\dot{\mathcal{B}}^{s_i}_{Q_i/s_i}$, consisting of all measurable u with

$$\|u\|_{\dot{\mathcal{B}}^{s_i}_{Q_i/s_i}(X_i)} \equiv \left(\int_{X_i}\int_{X_i} \frac{|u(x) - u(y)|^{Q_i/s_i}}{[d(x, y)]^{2Q}}\, d\mu(y)\, d\mu(x)\right)^{s_i/Q_i} < \infty.$$

The fact that these spaces coincide with the above spaces $\dot{M}^s_{p,\,q}(X)$ for the indicated indices was established in [20] together with more general invariance properties described below.

It is not claimed in Theorem 5.3 that f acts as a composition operator, but merely that every $u \in \dot{\mathcal{B}}^{s_2}_{Q_2/s_2}(X_2)$ has a representative $\widetilde{u}$ so that $\widetilde{u}\circ f \in \dot{\mathcal{B}}^{s_1}_{Q_1/s_1}(X_1)$ with a norm bound, and similarly for f^{-1}. Indeed, $u\circ f$ need not even be measurable in this generality.

If both f and f^{-1} are absolutely continuous in measure and $\mu_f \in \mathcal{B}_s(X)$ for some $s \in (1, \infty]$, then the third index in Theorem 5.3 may be replaced by an arbitrary $q > 0$. In particular, these conditions are met under the assumptions of Theorem 4.3.

Theorem 5.4. *Let X_1 and X_2 be Ahlfors Q_1-regular and Q_2-regular spaces with $Q_1, Q_2 \in (0, \infty)$, respectively. Let f be a quasisymmetric mapping from X_1 onto X_2, and assume that f and f^{-1} are absolutely continuous and $\mu_f \in \mathcal{B}_t(X_1)$ for some $t \in (1, \infty]$. Let $s_i \in (0, Q_i)$ with $i = 1, 2$ satisfy $Q_1/s_1 = Q_2/s_2$, and $q \in (0, \infty]$. Then f induces an equivalence between $\dot{M}^{s_1}_{Q_1/s_1,\, q}(X_1)$ and $\dot{M}^{s_2}_{Q_2/s_2,\, q}(X_2)$.*

In Theorem 5.4, f and f^{-1} act as composition operators. Moreover, with the assumptions of Theorem 5.4, by Lebesgue-Radon-Nykodym Theorem and [28], we have that $\mu_{f^{-1}}(y) = [\mu_f(f^{-1}(y))]^{-1}$ for almost all $y \in X_2$, and hence $\mu_{f^{-1}} \in \mathcal{B}_{t'}(X_2)$ for some $t' \in (1, \infty]$

It is immediate from the definition that $u \in \dot{M}^s_{Q/s,\infty}(X)$ if and only if there is a set $E \subset X$ with $\mu(E) = 0$ and $g \in L^{Q/s}(X)$ such that for all $x, y \in X \setminus E$,

$$|u(x) - u(y)| \leq [d(x, y)]^s [g(x) + g(y)]. \tag{5.2}$$

That is, u belongs to the space $\dot{M}^s_{Q/s}(X)$ introduced by Hajłasz in [10]. Thus these spaces are invariant in the setting of Theorem 5.4. If we further assume that X satisfies a Q-Poincaré inequality, then $\dot{M}^1_Q(X) = \dot{W}^{1,Q}(X)$ (see [16]), and thus Theorem 5.4 includes the invariance of this space under the Poincaré inequality assumption, generalizing [19]. In fact, the invariance of this space holds already under the assumptions of Theorem 3.8.

We further point out that the class of functions of bounded mean oscillation is invariant under quasisymmetric mappings of $\mathbb{R}^n$, $n \geq 2$ [26]. This space is also invariant under the assumptions of Theorem 5.4.

References

[1] Lars V. Ahlfors. On quasiconformal mappings. *J. Analyse Math.*, 3:1–58; correction, 207–208, 1954.

[2] Z.M. Balogh, P. Koskela, and S. Rogovin. Absolute continuity of quasiconformal mappings on curves. *Geom. Funct. Anal.*, 17(3):645–664, 2007.

[3] Mario Bonk and Bruce Kleiner. Conformal dimension and Gromov hyperbolic groups with 2-sphere boundary. *Geom. Topol.*, 9:219–246, 2005.

[4] Marc Bourdon and Hervé Pajot. Cohomologie l_p et espaces de Besov. *J. Reine Angew. Math.*, 558:85–108, 2003.

[5] B.V. Boyarskiĭ. Homeomorphic solutions of Beltrami systems. *Dokl. Akad. Nauk SSSR (N.S.)*, 102:661–664, 1955.

[6] J. Cheeger. Differentiability of Lipschitz functions on metric measure spaces. *Geom. Funct. Anal.*, 9(3):428–517, 1999.

[7] F.W. Gehring. The definitions and exceptional sets for quasiconformal mappings. *Ann. Acad. Sci. Fenn. Ser. A I No.*, 281:28, 1960.

[8] F.W. Gehring. Rings and quasiconformal mappings in space. *Trans. Amer. Math. Soc.*, 103:353–393, 1962.

[9] F.W. Gehring and E. Reich. Area distortion under quasiconformal mappings. *Ann. Acad. Sci. Fenn. Ser. A I No.*, 388:15, 1966.

[10] Piotr Hajłasz. Sobolev spaces on an arbitrary metric space. *Potential Anal.*, 5(4):403–415, 1996.

[11] Piotr Hajłasz and Pekka Koskela. Sobolev met Poincaré. *Mem. Amer. Math. Soc.*, 145(688):x+101, 2000.

[12] Juha Heinonen. *Lectures on analysis on metric spaces.* Universitext. Springer-Verlag, New York, 2001.

[13] Juha Heinonen and Pekka Koskela. Quasiconformal maps in metric spaces with controlled geometry. *Acta Math.*, 181(1):1–61, 1998.

[14] Juha Heinonen, Pekka Koskela, Nageswari Shanmugalingam, and Jeremy T. Tyson. Sobolev classes of Banach space-valued functions and quasiconformal mappings. *J. Anal. Math.*, 85:87–139, 2001.

[15] Stephen Keith. Modulus and the Poincaré inequality on metric measure spaces. *Math. Z.*, 245(2):255–292, 2003.

[16] Stephen Keith and Xiao Zhong. The Poincaré inequality is an open ended condition. *Ann. of Math.* (2), 167(2):575–599, 2008.

[17] Riikka Korte. Geometric implications of the Poincaré inequality. *Results Math.*, 50(1–2):93–107, 2007.

[18] P. Koskela and K. Wildrick. Exceptional sets for quasiconformal mappings in general metric spaces. *Int. Math. Res. Not. IMRN*, (9):Art. ID rnn020, 32, 2008.

[19] Pekka Koskela and Paul MacManus. Quasiconformal mappings and Sobolev spaces. *Studia Math.*, 131(1):1–17, 1998.

[20] Pekka Koskela, Dachun Yang, and Yuan Zhou. Hajłasz characterizations of Besov and Triebel-Lizorkin spaces and quasiconformal mappings. *Adv. Math.*, 226(4):3579–3621, 2011.

[21] G.A. Margulis and G.D. Mostow. The differential of a quasi-conformal mapping of a Carnot-Carathéodory space. *Geom. Funct. Anal.*, 5(2):402–433, 1995.

[22] Manor Mendel and Assaf Naor. Metric cotype. *Ann. of Math.* (2), 168(1):247–298, 2008.

[23] G.D. Mostow. Quasi-conformal mappings in n-space and the rigidity of hyperbolic space forms. *Inst. Hautes Études Sci. Publ. Math.*, 34:53–104, 1968.

[24] G.D. Mostow. A remark on quasiconformal mappings on Carnot groups. *Michigan Math. J.*, 41(1):31–37, 1994.

[25] Assaf Naor. An application of metric cotype to quasisymmetric embeddings, in "Metric and Differential Geometry. The Jeff Cheeger Anniversary Volume", Birkäuser Basel, 175–178, 2012 (this volume).

[26] H.M. Reimann. Functions of bounded mean oscillation and quasiconformal mappings. *Comment. Math. Helv.*, 49:260–276, 1974.

[27] N. Shanmugalingam. Newtonian spaces: an extension of Sobolev spaces to metric measure spaces. *Rev. Mat. Iberoamericana*, 16(2):243–279, 2000.

[28] Jan-Olov Strömberg and Alberto Torchinsky. *Weighted Hardy spaces*, volume 1381 of *Lecture Notes in Mathematics*. Springer-Verlag, Berlin, 1989.

[29] Pekka Tukia. Hausdorff dimension and quasisymmetric mappings. *Math. Scand.*, 65(1):152–160, 1989.

[30] Kevin Wildrick and Ellen Veomett. Spaces of small metric cotype. *J. Topol. Anal.*, 2(4): 581–597, 2010.

[31] Marshall Williams. Geometric and analytic quasiconformality in metric measure spaces. To appear in *Proc. Amer. Math. Soc.*

P. Koskela and K. Wildrick
Department of Mathematics and Statistics
University of Jyväskylä, PL 35 MaD
40014 Jyväskylän yliopisto, Finland
e-mail: pekka.j.koskela@jyu.fi
kevin.wildrick@jyu.fi

An Application of Metric Cotype to Quasisymmetric Embeddings

Assaf Naor

Dedicated to Jeff Cheeger for his 65th birthday

Abstract. We apply the notion of metric cotype to show that L_p admits a quasisymmetric embedding into L_q if and only if $p \leqslant q$ or $q \leqslant p \leqslant 2$.

Mathematics Subject Classification (2000). 46B85 and 51F99.

Keywords. Quasisymmetric embeddings, metric cotype.

This note is a companion to [4]. After the final version of [4] was sent to the journal for publication I learned from Juha Heinonen and Leonid Kovalev of a long-standing open problem in the theory of quasisymmetric embeddings, and it turns out that this problem can be resolved using the methods of [4]. The argument is explained below. I thank Juha Heinonen and Leonid Kovalev for bringing this problem to my attention.

Let (X, d_X) and (Y, d_Y) be metric spaces. An embedding $f : X \to Y$ is said to be a quasisymmetric embedding with modulus $\eta : (0, \infty) \to (0, \infty)$ if η is increasing, $\lim_{t \to 0} \eta(t) = 0$, and for every distinct $x, y, z \in X$ we have

$$\frac{d_Y(f(x), f(y))}{d_Y(f(x), f(z))} \leqslant \eta\left(\frac{d_X(x, y)}{d_X(x, z)}\right).$$

We refer to [1] and the references therein for a discussion of this notion.

It was not known whether every two separable Banach spaces are quasisymmetrically equivalent. This is asked in [6] (see problem 8.3.1 there). We will show here that the answer to this question is negative. Moreover, it turns out that under mild assumptions the cotype of a Banach space is preserved under quasisymmetric embeddings. Thus, in particular, our results imply that L_p does not embed quasisymetrically into L_q if $p > 2$ and $q < p$. The question of determining when L_p is quasisymetrically equivalent to L_q was asked in [6] (see problem 8.3.3 there).

Research supported in part by NSF grants CCF-0635078 and CCF-0832795, BSF grant 2006009, and the Packard Foundation.

We also deduce, for example, that the separable space c_0 does not embed quasisymetrically into any Banach space which has an equivalent uniformly convex norm.

We recall some definitions. A Banach space X is said to have (Rademacher) type $p > 0$ if there exists a constant $T < \infty$ such that for every n and every $x_1, \dots, x_n \in X$,

$$\mathbb{E}_\varepsilon \left\| \sum_{j=1}^n \varepsilon_j x_j \right\|_X^p \leqslant T^p \sum_{j=1}^n \|x_j\|_X^p,$$

where above the expectation $\mathbb{E}_\varepsilon$ is with respect to a uniform choice of signs $\varepsilon = (\varepsilon_1, \dots, \varepsilon_n) \in \{-1, 1\}^n$. X is said to have (Rademacher) cotype $q > 0$ if there exists a constant $C < \infty$ such that for every n and every $x_1, \dots, x_n \in X$,

$$\mathbb{E}_\varepsilon \left\| \sum_{j=1}^n \varepsilon_j x_j \right\|_X^q \geqslant \frac{1}{C^q} \sum_{j=1}^n \|x_j\|_X^q.$$

We also write

$$p_X = \sup\{p \geqslant 1 : \ X \text{ has type } p\},$$

and

$$q_X = \inf\{q \geqslant 2 : \ X \text{ has cotype } q\}.$$

X is said to have non-trivial type if $p_X > 1$, and X is said to have non-trivial cotype if $q_X < \infty$. For example, L_p has type $\min\{p, 2\}$ and cotype $\max\{p, 2\}$ (see [5]).

Theorem 1. *Let X be a Banach space with non-trivial type. Assume that Y is a Banach space which embeds quasisymmetrically into X. Then $q_Y \leqslant q_X$.*

Proof. Let $f : Y \to X$ be a quasisymmetric embedding with modulus η. Assume for the sake of contradiction that X has cotype q and that $p := q_Y > q$. By the Maurey-Pisier Theorem [2] for every $n \in \mathbb{N}$ there is a linear operator $T : \ell_p^n \to Y$ such that for all $x \in \ell_p^n$ we have $\|x\|_p \leqslant \|T(x)\|_Y \leqslant 2\|x\|_p$. For every integer $m \in \mathbb{N}$ consider the mapping $g : \mathbb{Z}_m^n \to X$ given by

$$g(x_1, \dots, x_n) = f \circ T\left(e^{\frac{2\pi i x_1}{m}}, \dots, e^{\frac{2\pi i x_n}{m}} \right).$$

By Theorem 4.1 in [4] there exist constants $A, B > 0$ which depend only on the type and cotype constants of X such that for every integer $m \geqslant A n^{1/q}$ which is divisible by 4 and every $h : \mathbb{Z}_m^n \to X$ we have

$$\sum_{j=1}^n \mathbb{E}_x \left[\left\| h\left(x + \frac{m}{2} e_j\right) - h(x) \right\|_X^q \right] \leqslant B^q m^q \, \mathbb{E}_{\varepsilon, x} \left[\| h(x+\varepsilon) - h(x) \|_X^q \right], \tag{1}$$

where the expectations above are taken with respect to uniformly chosen $x \in \mathbb{Z}_m^n$ and $\varepsilon \in \{-1, 0, 1\}^n$. Here, and in the following, we denote by $\{e_j\}_{j=1}^n$ the standard basis of $\mathbb{R}^n$).

From now on we fix m to be be the smallest integer which is divisible by 4 and satisfies $m \geqslant An^{1/q}$. Thus $m \leqslant 8An^{1/q}$. For every $x \in \mathbb{Z}_m^n$, $j \in \{1,\dots,n\}$ and $\varepsilon \in \{-1,0,1\}^n$ we have

$$\begin{aligned}\frac{\|g(x+\varepsilon)-g(x)\|_X}{\left\|g\left(x+\frac{m}{2}e_j\right)-g(x)\right\|_X} &\leqslant \eta\left(\frac{\left\|T\left(\sum_{k=1}^n\left(e^{\frac{\pi i \varepsilon_k}{m}}-1\right)e_j\right)\right\|_Y}{\|T(2e_j)\|_Y}\right)\\ &\leqslant \eta\left(\frac{\pi n^{1/p}}{m}\right)\\ &\leqslant \eta\left(\frac{\pi}{A}n^{\frac{1}{p}-\frac{1}{q}}\right).\end{aligned}$$

Thus, using (1) for $g=h$ we see that

$$\begin{aligned}&n\,\mathbb{E}_{\varepsilon,x}\|g(x+\varepsilon)-g(x)\|_X^q\\ &\leqslant \eta\left(\frac{\pi}{A}n^{\frac{1}{p}-\frac{1}{q}}\right)^q\sum_{j=1}^n\mathbb{E}_x\left\|g\left(x+\frac{m}{2}e_j\right)-g(x)\right\|_X^q\\ &\leqslant \eta\left(\frac{\pi}{A}n^{\frac{1}{p}-\frac{1}{q}}\right)^q(8AB)^q n\,\mathbb{E}_{\varepsilon,x}\|g(x+\varepsilon)-g(x)\|_X^q.\end{aligned}$$

Cancelling the term $\mathbb{E}_{\varepsilon,x}\|g(x+\varepsilon)-g(x)\|_X^q$ we deduce that

$$\eta\left(\frac{\pi}{A}n^{\frac{1}{p}-\frac{1}{q}}\right)\geqslant\frac{1}{8AB}.$$

Since $p>q$ this contradicts the fact that $\lim_{t\to 0}\eta(t)=0$. □

Using the same argument as in [4, Corollary 7.3] (and noting that the snowflake embedding from [3] is a quasisymmetric embedding), we obtain the following complete answer to the question when L_p embeds quasisymmetrically into L_q.

Corollary 2. *For $p,q>0$, L_p embeds quasisymmetrically into L_q if and only if $p\leqslant q$ or $q\leqslant p\leqslant 2$.*

References

[1] J. Heinonen. *Lectures on analysis on metric spaces.* Universitext. Springer-Verlag, New York, 2001.

[2] B. Maurey and G. Pisier. Séries de variables aléatoires vectorielles indépendantes et propriétés géométriques des espaces de Banach. *Studia Math.*, 58(1):45–90, 1976.

[3] M. Mendel and A. Naor. Euclidean quotients of finite metric spaces. *Adv. Math.*, 189(2):451–494, 2004.

[4] M. Mendel and A. Naor. Metric cotype. *Ann. of Math.* (2), 168(1):247–298, 2008.

[5] V.D. Milman and G. Schechtman. *Asymptotic theory of finite-dimensional normed spaces*, volume 1200 of *Lecture Notes in Mathematics.* Springer-Verlag, Berlin, 1986. With an appendix by M. Gromov.

[6] J. Väisälä. The free quasiworld. Freely quasiconformal and related maps in Banach spaces. In *Quasiconformal geometry and dynamics* (*Lublin,* 1996), volume 48 of *Banach Center Publ.*, pages 55–118. Polish Acad. Sci., Warsaw, 1999.

Assaf Naor
Courant Institute, New York University
251 Mercer Street
New York, NY 10012, USA
e-mail: naor@cims.nyu.edu

Part III

Index Theory

Index Theory and the Hypoelliptic Laplacian

Jean-Michel Bismut

Dedicated to Jeff Cheeger for his 65*th birthday*

Abstract. We review various aspects of index theory, in connection with the hypoelliptic Laplacian and with orbital integrals.

Mathematics Subject Classification (2000). 35H10, 58J20.

Keywords. Hypoelliptic equations, Index theory and related fixed point theorems.

Ah ! non ! c'est un peu court,
jeune homme!

EDMOND ROSTAND
Cyrano de Bergerac

Contents

The author is indebted to Ma Xiaonan for his careful reading of the paper.

Introduction

The purpose of this paper is to review various aspects of index theory in connection with the hypoelliptic Laplacian [B05, BL08, B08a], and also with our book [B11d] on orbital integrals. We have excluded from this survey the relations of local index theory to localization formulas and functional integrals, which were reviewed in [B05, B06, B08b], and also the construction on the hypoelliptic Laplacian in de Rham theory which was explained in some detail in [B08c]. Still the present paper can be read independently of the above references. One reason is that many of the supporting facts are elementary, and can be verified by simple computations. This is in particular the case of the examples involving the real line and the circle.

Let us review a few themes which are covered in this survey.

0.1. Gaussians and the index theorem

Consider the Gaussian integral

$$1 = \int_{\mathbf{R}} \exp\left(-y^2/2\right) \frac{dy}{\sqrt{2\pi}}. \tag{0.1}$$

For several interrelated reasons, we will consider equation (0.1) as the prototype of an index formula:

1. A first reason is that an index problem can be set up so that (0.1) is an index formula. Indeed 1 is the Euler characteristic of $\mathbf{R}$, and (0.1) is exactly the formula which is produced via local index theory, when considering the smooth de Rham Witten complex [W82] of $\mathbf{R}$.
2. The McKean-Singer index formula [McS67] for the index of a Dirac operator D_+^X,
$$\operatorname{Ind} D_+^X = \operatorname{Tr_s}\left[\exp\left(-tD^{X,2}\right)\right], \tag{0.2}$$
also has an obvious Gaussian flavour. This equation was first considered as a tool to prove a local index theorem for Dirac operators [Gi74, ABoP73]. Here we will consider the full operator $\exp\left(-tD^{X,2}\right)$ as the genuine object to which index theory applies. One reason is that it verifies an obvious compatibility to products of manifolds, in the same way as the Gaussians distributions on Euclidean vector spaces are compatible to direct sums, a consequence of the Pythagorean theorem.
3. Quillen's theory of superconnections [Q85] gives a formalism putting the McKean-Singer formula and the Chern-Weil theory for the Chern character on the same footing, by exploiting the common Gaussian outlook of the McKean-Singer formula and of the Chern-Weil formula for the Chern character. Superconnections have been used in [B86] to prove a local version of the families index theorem of Atiyah-Singer [AS71].
4. The Atiyah-Singer index formula for the index of a Dirac operator says that
$$\operatorname{Ind} D_+^X = \int_X \widehat{A}\left(TX, \nabla^{TX}\right) \operatorname{ch}\left(E, \nabla^E\right), \tag{0.3}$$
with the right-hand side written in Chern-Weil form. Formula (0.3) can be obtained by making $t \to 0$ in (0.2), while using the local 'fantastic cancellations' in the supertrace of the heat kernel associated with $\exp\left(-tD^{X,2}\right)$, which were anticipated by McKean-Singer [McS67]. We claim that (0.3) is also a Gaussian formula. This is clear for the Chern character form in the right-hand side of (0.3). The remarkable thing is that it is also true for $\widehat{A}\left(TX, \nabla^{TX}\right)$. Indeed superconnections can be used to reexpress the right-hand side as a Quillen like Chern character form, so that (0.2), (0.3) can be written as the doubly Gaussian formula
$$\operatorname{Ind} D_+^X = \operatorname{Tr_s}\left[\exp\left(-tD^{X,2}\right)\right] = \int_X \varphi \operatorname{Tr_s}\left[\exp\left(-A^{S^{TX}\otimes E,2}\right)\right]. \tag{0.4}$$

In (0.4), $A^{S^{TX}\otimes E}$ is the Levi-Civita superconnection along the fibres $\widehat{TX}$ of the total space of $\mathcal{X}$ of TX, with fibre now called $\widehat{TX}$, which is precisely the superconnection one would need if we were to establish a local version of the families index theorem for the projection $\pi : \mathcal{X} \to X$ [B86]. A computational explanation for (0.4) is that $A^{S^{TX}\otimes E,2}$ is precisely the Getzler operator [Ge86] in local index theory. Also note that equation (0.3), which expresses a global quantity in local terms, looks like a kind of Fourier transform.

0.2. The hypoelliptic Laplacian

Equation (0.4) raises a number of natural questions.

1. First, one can ask whether the doubly Gaussian character of equation (0.4) can be viewed as an inspiring tautology.
2. The right-hand side of (0.4) has been obtained by making $t \to 0$ in (0.2). If (0.4) is indeed tautologous, is it possible to deform $A^{S^{TX}\otimes E}$ back to D^X?
3. Can this possibility unify the theory of operators and the theory of characteristic forms?

The construction of the hypoelliptic Laplacian gives a positive answer to the above questions. It is a family of hypoelliptic operators acting over $\mathcal{X}$, which, up to lower-order terms, is a weighted sum of the harmonic oscillator along the fibre $\widehat{TX}$ and of the generator of the geodesic flow.

0.3. Bargmann isomorphism and the harmonic oscillator

The harmonic oscillator appears naturally in the context of index theory. Indeed the Getzler operator [Ge86] is a harmonic oscillator along the fibre $\widehat{TX}$. When localizing analytically index theoretic data, like in the context of Morse inequalities [W82], or in questions related to analytic torsion [BZ92], the harmonic oscillator provides a universal model of localization.

Also, if V is an Euclidean vector space, the Bargmann isomorphism identifies the Hilbert completion of the symmetric algebra $S^{\cdot}(V^*)$ with L_2^V. The Bargmann isomorphism also identifies the algebraic de Rham complex of V with the de Rham Witten complex of V. In real or complex geometry, it expresses various aspects of the embedding of $\{0\} \to V$. The fundamental elementary fact is that via the Bargmann isomorphism, the operator which defines the $\mathbf{Z}$-grading of $S^{\cdot}(V^*)$ is just the harmonic oscillator of V.

If X is a Riemannian manifold, one idea in the construction of the hypoelliptic Laplacian is to consider along each fibre $\widehat{TX}$ of $\pi : \mathcal{X} \to X$ the various forms of the Bargmann isomorphism, which provides different ways of expressing analytically the embedding $X \to \mathcal{X}$. This way, the fibres $\widehat{TX}$ carry universal infinite-dimensional algebraic and analytic objects. The question is then how to link geometrically the fibre $\widehat{TX}$ and the base manifold X.

0.4. Geodesic flow and the Fourier transform

The second key ingredient in the construction of the hypoelliptic Laplacian is the geodesic flow.

1. From an algebraic point of view, the geodesic flow can be interpreted as a kind of bosonic Hodge de Rham operator, in which the exterior algebra $\Lambda^{\cdot}(T^*X)$ is replaced by the symmetric algebra $S^{\cdot}(T^*X)$.
2. Geometrically, it provides the critical link between the fibre $\widehat{TX}$ and the physical tangent space TX. This way, the vector bundle $\widehat{TX}$ incorporates dynamic data.
3. Analytically, it is very much related to Fourier transform.

0.5. Operators and characteristic classes

Originally $\widehat{TX}$ is a vector bundle on X. The introduction of the geodesic flow treats $\widehat{TX}$ as a special vector bundle, since it is a copy of the tangent bundle TX, which can be moved around along geodesics by parallel transport. When considering the heat flow associated with the hypoelliptic Laplacian, the fibrewise harmonic oscillator makes this motion more erratic, until a final point is reached where in the fibre direction, only the kernel of the harmonic oscillator becomes relevant, the hypoelliptic Laplacian collapses to the Laplacian of the base, and the motion on X becomes Brownian motion. The associated analytic data, which were the characteristic forms of $\widehat{TX}$, or superconnections associated with the projection $\pi : \mathcal{X} \to X$, ultimately become operators on X. The distinction between characteristic forms on X and differential operators is now blurred.

0.6. The evaluation of orbital integrals

A significant application of the hypoelliptic Laplacian is the evaluation in [B11d] of semisimple orbital integrals associated with reductive groups, these integrals being a key ingredient in Selberg's trace formula. Here, we cover in detail the case of the real line and the case of the circle, for which the relevant Selberg formula is just the Poisson formula. In both cases, the computations can be done explicitly, and their relation to Fourier transform is obvious.

0.7. Heat and waves

A most remarkable aspect of the hypoelliptic Laplacian is its wave-like character. Indeed, observe that like the wave equation, the geodesic flow exhibits finite propagation speed. It turns out that, after projection on X, the heat equation for the hypoelliptic Laplacian exhibits a version of finite propagation speed, which plays an important role in its analysis. This wave-like aspect of the projected heat flow for the hypoelliptic Laplacian is a quantized version of the Hamiltonian-Lagrangian correspondence in the classical calculus of variations.

0.8. The organization of the paper

This paper is organized as follows. In Section 1, we describe the algebraic de Rham complex of a vector space, and the associated Bargmann isomorphism.

In Section 2, we review the Gaussian aspects of index theory.

In Section 3, we give an elementary construction of the degree 0 part of the hypoelliptic Laplacian in de Rham theory.

In Section 4, when X is a complex manifold, the construction of the hypoelliptic Dirac operator is described in detail, in particular with regard to the deformation of the underlying characteristic forms of the tangent bundle. For other aspects of the construction, we refer to [B08c, Section 4].

Finally, in Section 5, we give an introduction to the role of the hypoelliptic Laplacian in the evaluation of orbital integrals, and to the wave-like character of this operator.

1. The algebraic de Rham complex and the Bargmann isomorphism

The purpose of this section is to review well-known facts on the algebraic de Rham complex and the Bargmann isomorphism. Connections with index theory are also emphasized, as well as the formal role of Gaussian formulas.

This section is organized as follows. In Subsection 1.1, we construct the algebraic de Rham complex of a real vector space.

In Subsection 1.2, the Hodge theory of the algebraic de Rham complex is explained.

In Subsection 1.3, the dual algebraic complex is constructed.

In Subsection 1.4, given an Euclidean vector space, we describe the Bargmann isomorphism of the Hilbert completion of the symmetric algebra of the dual of a vector space with the corresponding L_2 space.

In Subsection 1.5, the Bargmann isomorphism is shown to interchange Dirac masses and Gaussian distributions.

In Subsection 1.6, the Bargmann isomorphism is applied to the algebraic de Rham complex of a vector space.

In Subsection 1.7, various formulas are given for the Euler characteristic $\chi = 1$ of the algebraic de Rham complex. Their Gaussian character is emphasized.

Finally, in Subsection 1.8, we give various applications of the Bargmann isomorphism in differential geometry.

1.1. The algebraic de Rham complex

Let $\mathcal{C}$ be the category of real finite-dimensional vector spaces. If $V \in \mathcal{C}$, let $S^{\cdot}(V^*), \Lambda^{\cdot}(V^*)$ be the symmetric and exterior algebras of V^*. These are commutative and supercommutative $\mathbf{Z}$-graded algebras. Let $N^{S^{\cdot}(V^*)}, N^{\Lambda^{\cdot}(V^*)}$ be the number operators which define their $\mathbf{Z}$-grading. Let $Y \in V$ be the tautological section of V. Then $S^{\cdot}(V^*)$ can be identified with the polynomial algebra $\mathbf{R}[V]$, via the map $P \to P(Y)$.

Note that if $e \in V, f \in V^*$, the derivative ∇_e and the multiplication operator $P(Y) \to \langle f, Y \rangle P(Y)$ act on $S^{\cdot}(V^*)$, and respectively decrease and increase $N^{S^{\cdot}(V^*)}$ by 1. Similarly $i_e, f\wedge$ act on $\Lambda^{\cdot}(V^*)$, and decrease and increase $N^{\Lambda^{\cdot}(V^*)}$ by 1. Finally, we have the commutation relations,

$$[\nabla_e, \langle f, Y \rangle] = \langle f, e \rangle, \qquad\qquad [i_e, f\wedge] = \langle f, e \rangle. \tag{1.1}$$

In (1.1), the first bracket is an ordinary commutator, and the second bracket is a supercommutator. The operators ∇_e, i_e are called annihilation operators, and the operators $\langle f, Y \rangle, f\wedge$ are called creation operators. Also we have the identity of operators acting on $S^{\cdot}(V^*)$,

$$\nabla_Y = N^{S^{\cdot}(V^*)}. \tag{1.2}$$

Let $\mathcal{E}$ be automorphism of $S^{\cdot}(V^*) \otimes_{\mathbf{R}} \mathbf{C}$,

$$\mathcal{E} = \exp\left(-i\frac{\pi}{2} N^{S^{\cdot}(V^*)}\right) = -i^{N^{S^{\cdot}(V^*)}}. \tag{1.3}$$

Equivalently, if $P \in S^{\cdot}(V^*) \otimes_{\mathbf{R}} \mathbf{C}$,

$$\mathcal{E} P(Y) = P(-iY). \tag{1.4}$$

Set

$$\mathcal{A}^{\cdot}(V^*) = S^{\cdot}(V^*) \otimes \Lambda^{\cdot}(V^*). \tag{1.5}$$

We define $N^{\mathcal{A}(V^*)}$ by the formula

$$N^{\mathcal{A}(V^*)} = N^{S^{\cdot}(V^*)} \otimes 1 + 1 \otimes N^{\Lambda^{\cdot}(V^*)}. \tag{1.6}$$

In the sequel, we will write (1.5) in the form

$$N^{\mathcal{A}(V^*)} = N^{S^{\cdot}(V^*)} + N^{\Lambda^{\cdot}(V^*)}. \tag{1.7}$$

The algebra $\mathcal{A}^{\cdot}(V^*)$ is $\mathbf{Z}$-graded by $N^{\Lambda^{\cdot}(V^*)}$, and $N^{\mathcal{A}(V^*)}$ defines an increasing filtration on $\mathcal{A}^{\cdot}(V^*)$, so that $\mathcal{A}^{\cdot}(V^*)$ is a $\mathbf{Z}$-graded filtered supercommutative algebra. Also the operators in (1.1) act on $\mathcal{A}^{\cdot}(V^*)$ and verify the same commutation relations.

Observe that $V \to S^{\cdot}(V^*), V \to \Lambda^{\cdot}(V^*), V \to \mathcal{A}^{\cdot}(V^*)$ define multiplicative functors, so that if $V, V' \in \mathcal{C}$,

$$\begin{aligned} S^{\cdot}\left((V \oplus V')^*\right) &= S^{\cdot}(V^*) \otimes S^{\cdot}(V'^*), \\ \Lambda^{\cdot}\left((V \oplus V')^*\right) &= \Lambda^{\cdot}(V^*) \widehat{\otimes} \Lambda^{\cdot}(V'^*), \\ \mathcal{A}^{\cdot}\left((V \oplus V')^*\right) &= \mathcal{A}^{\cdot}(V^*) \widehat{\otimes} \mathcal{A}^{\cdot}(V'^*). \end{aligned} \tag{1.8}$$

In (1.8), $\otimes$ is the ordinary tensor product, and $\widehat{\otimes}$ is the $\mathbf{Z}$-graded tensor product of filtered algebras. In particular,

$$N^{\mathcal{A}^{\cdot}((V \oplus V')^*)} = N^{\mathcal{A}^{\cdot}(V^*)} \otimes 1 + 1 \otimes N^{\mathcal{A}^{\cdot}(V'^*)}. \tag{1.9}$$

As in (1.7), we rewrite (1.9) in the simpler form

$$N^{\mathcal{A}^{\cdot}((V\oplus V')^{*})} = N^{\mathcal{A}^{\cdot}(V^{*})} + N^{\mathcal{A}^{\cdot}(V'^{*})}. \tag{1.10}$$

Let d be the de Rham operator acting on $\mathcal{A}^{\cdot}(V^{*})$. If $e_1, \dots, e_n$ is a basis of V, and if $e^1, \dots, e^n$ is the dual basis of V^*, then

$$d = \sum_{i=1}^{n} e^i \nabla_{e_i}. \tag{1.11}$$

Then $(\mathcal{A}^{\cdot}(V^{*}), d)$ is the algebraic de Rham complex of V. The de Rham operator d increases the degree in $\mathcal{A}^{\cdot}(V^{*})$ by 1, and it commutes with $N^{\mathcal{A}(V^{*})}$, so that $(\mathcal{A}^{\cdot}(V^{*}), d)$ is also a filtered complex. We identify Y with the tautological radial vector field. Let i_Y denote the contraction by Y. Then i_Y decreases the degree by 1, and it commutes with $N^{\mathcal{A}^{\cdot}(V^{*})}$. Let L_Y be the corresponding Lie derivative operator. Cartan's formula asserts that

$$L_Y = [d, i_Y], \tag{1.12}$$

where $[\ ]$ still denotes a supercommutator. We have the trivial identity

$$L_Y = N^{\mathcal{A}^{\cdot}(V^{*})}. \tag{1.13}$$

The fact that d, i_Y commute with $N^{\mathcal{A}^{\cdot}(V^{*})}$ can be viewed as a consequence of (1.13).

For $k \in \mathbf{N}$, let $\mathcal{A}^{\cdot}_k(V^{*})$ be the subcomplex of elements of $\mathcal{A}^{\cdot}(V^{*})$ which are of degree k with respect to $N^{\mathcal{A}^{\cdot}(V^{*})}$. Then $(\mathcal{A}^{\cdot}(V^{*}), d)$ splits as a sum of finite-dimensional complexes

$$(\mathcal{A}^{\cdot}(V^{*}), d) = \bigoplus_{k\in\mathbf{N}} (\mathcal{A}^{\cdot}_k(V^{*}), d), \tag{1.14}$$

and i_Y also acts on the splitting. By (1.12), (1.13), except for $k = 0$, the complexes $(\mathcal{A}^{\cdot}_k(V^{*}), d)$ are exact. This shows that the cohomology of $(\mathcal{A}^{\cdot}(V^{*}), d)$ is concentrated in degree 0 and is one-dimensional. In the above, we have used implicitly the homotopy on $(\mathcal{A}^{\cdot}_{>0}(V^{*}), d)$,

$$1 = \left[N^{\mathcal{A}^{\cdot}(V^{*})}\right]^{-1} [d, i_Y], \tag{1.15}$$

which is the algebraic version of Poincaré's lemma.

The argument is exactly the same when proving Poincaré's lemma on the smooth de Rham complex $(\Omega^{\cdot}(V), d)$. Indeed, let $\varphi_t(Y) = e^t Y$ be the group of diffeomorphisms associated with the radial vector field. If $\omega \in \Omega^{\cdot}(V)$, then

$$\frac{\partial}{\partial t} \varphi_t^* \omega = \varphi_t^* L_Y \omega, \tag{1.16}$$

so that

$$\omega - \omega^{(0)}(0) = \int_{-\infty}^{0} \varphi_t^* L_Y \omega dt. \tag{1.17}$$

One verifies easily that (1.17) is just the formal analogue of (1.15) in smooth de Rham theory. When ω is a polynomial form, these two equations are equivalent.

1.2. Algebraic Hodge theory

Let h^V be a nondegenerate symmetric bilinear form on V. Then h^V induces nondegenerate symmetric bilinear forms on $S^{\cdot}(V^*)$ and $\Lambda^{\cdot}(V^*)$. Let $h^{\mathcal{A}^{\cdot}(V^*)}$ be the corresponding symmetric bilinear form on $\mathcal{A}^{\cdot}(V^*)$.

Let d^* denote the adjoint of d with respect to $h^{\mathcal{A}^{\cdot}(V^*)}$.

Proposition 1.1. *The following identity holds:*

$$d^* = i_Y. \tag{1.18}$$

In particular d^* does not depend on h^V. Moreover,

$$[d, d^*] = N^{\mathcal{A}^{\cdot}(V^*)}. \tag{1.19}$$

Proof. Equation (1.18) follows from an obvious computation. By (1.13), we get (1.19). □

That d^* does not depend on h^V is strange. This is because h^V is used in both factors $S^{\cdot}(V^*), \Lambda^{\cdot}(V^*)$. If h^V is a scalar product, then $[d, d^*]$ is a Hodge Laplacian. It follows from the above that L_Y, a differential operator of order 1, is a Hodge Laplacian. The fact that $\mathcal{N}^{\mathcal{A}^{\cdot}(V^*)}$ is nonnegative appears as a consequence of Hodge theory. Equation (1.15) is then a version of the standard assertion in Hodge theory that on the direct sum of eigenspaces associated with positive eigenvalues of the Hodge Laplacian, the corresponding complex is exact.

The above tends to indicate that the classical proof of Poincaré's lemma is nothing else than a standard application of Hodge theory.

1.3. Duality

The algebraic dual of $S^{\cdot}(V^*)$ is given by $S^{\cdot}_{\infty}(V) = \prod_{p=0}^{+\infty} S^p(V)$. Then $S^{\cdot}(V)$ embeds in the algebraic dual to $S^{\cdot}_{\infty}(V)$, $\Lambda^{\cdot}(V)$ is the algebraic dual to $\Lambda^{\cdot}(V^*)$. Also $S^{\cdot}(V)$ can be identified with the algebra of polynomials $\mathbf{R}[V^*]$ on V^*. Finally, $\mathcal{A}^{\cdot}(V)$ embeds in the algebraic dual $\mathcal{A}^{\cdot}_{\infty}(V)$ to $\mathcal{A}(V^*)$.

Let Z be the tautological section of V^*. The complex $(\mathcal{A}^{\cdot}(V), i_Z)$ is dual to the algebraic de Rham complex $(\mathcal{A}(V^*), d)$, and the algebraic de Rham complex $(\mathcal{A}^{\cdot}(V), d)$ is dual to $(\mathcal{A}^{\cdot}(V^*), i_Y)$. The complex $(\mathcal{A}^{\cdot}(V), i_Z)$ is $\mathbf{Z}$-graded by $n - N^{\Lambda^{\cdot}(V)}$, and is filtered by

$$N^{\mathcal{A}^{\cdot}(V)} = N^{S^{\cdot}(V)} + N^{\Lambda^{\cdot}(V)}. \tag{1.20}$$

Clearly,

$$\Lambda^{\cdot}(V) = \Lambda^{n-\cdot}(V^*) \otimes \Lambda^n(V). \tag{1.21}$$

Set

$$\mathcal{B}^{\cdot}(V) = S^{\cdot}(V) \otimes \Lambda^{\cdot}(V^*) \otimes \Lambda^n(V). \tag{1.22}$$

Then $(\mathcal{B}^{\cdot}(V), Z\wedge)$ is a complex which is $\mathbf{Z}$-graded by $N^{\Lambda^{\cdot}(V^*)}$, and filtered by

$$N^{\mathcal{B}^{\cdot}(V)} = N^{S^{\cdot}(V)} + n - N^{\Lambda^{\cdot}(V^*)}. \tag{1.23}$$

Set

$$d_* = -\sum_{i=1}^{n} i_{e_i} \nabla_{e^i}. \tag{1.24}$$

From the above, we have the identification of **Z**-graded filtered complexes

$$\begin{aligned}(\mathcal{B}^{\cdot}(V), Z\wedge) &= \left(\mathcal{A}^{n-\cdot}(V), i_Z\right),\\ (\mathcal{B}(V), -d_*) &= (\mathcal{A}^{\cdot}(V), d).\end{aligned} \tag{1.25}$$

The Lie derivative operator L_Z acts on $\mathcal{A}^{\cdot}(V)$, and moreover,

$$L_Z = [d, i_Z] = N^{\mathcal{A}^{\cdot}(V)}. \tag{1.26}$$

The action of L_Z on $\mathcal{B}(V)$ is given by

$$L_Z = [Z\wedge, -d_*] = \nabla_Z + n - N^{\Lambda^{\cdot}(V^*)} = N^{\mathcal{B}^{\cdot}(V)}. \tag{1.27}$$

By proceeding as before, we find that the cohomology of $(\mathcal{B}(V), Z\wedge)$ is concentrated in degree n and is one-dimensional, and $1 \in \Lambda^n(V^*) \otimes \Lambda^n(V)$ is a canonical section of the cohomology.

1.4. The Bargmann isomorphism

Let $\mathcal{D}$ be the category of real Euclidean vector spaces.

Let $\left(V, g^V\right) \in \mathcal{D}$. Set $n = \dim V$. Let dY be the corresponding Lebesgue measure. Let m^V be the Gaussian measure on V,

$$dm^V(Y) = \exp\left(-|Y|^2/2\right) \frac{dY}{(2\pi)^{n/2}}. \tag{1.28}$$

Then

$$\int_V dm^V(Y) = 1. \tag{1.29}$$

In the sequel we will often omit g^V, so that V comes with its Euclidean product g^V. Then $V \to m^V$ is multiplicative, in the sense that if $V, V' \in \mathcal{D}$,

$$m^{V\oplus V'} = m^V \otimes m^{V'}. \tag{1.30}$$

Conversely, if $V \to n^V$ is a map into the set of $\mathrm{O}(V)$-invariant nonnegative measures on V such that (1.29), (1.30) hold, then n^V can be deduced from m^V by the action of a universal dilation $\delta_s : s \geq 0$. In particular the Dirac mass δ_0^V is also such a measure.

Let $L_2^V\left(m^V\right)$ be the real L_2-space associated with the measure m^V. Again,

$$L_2^{V\oplus V'}\left(m^{V\oplus V'}\right) = L_2^V\left(m^V\right) \otimes L_2^{V'}\left(m^{V'}\right). \tag{1.31}$$

Let Δ^V be the Laplacian on V which is associated with g^V. Here Δ^V is a sum of second derivatives. Note that even though the operator $\exp\left(-\Delta^V/2\right)$ is not well defined, it acts on $S^{\cdot}(V^*)$ by expanding $\exp\left(-\Delta^V/2\right)$ as a power series. Using the operator $\mathcal{E}$ in (1.3), (1.4), we get

$$\exp\left(-\Delta^V/2\right) = \mathcal{E} \exp\left(\Delta^V/2\right) \mathcal{E}^{-1}. \tag{1.32}$$

Equivalently, if $P \in S^{\cdot}(V^*)$,

$$\begin{aligned}\exp\left(-\Delta^V/2\right) P(Y_0) &= (2\pi)^{-n/2}\int_V P(Y_0 - iY)\exp\left(-|Y|^2/2\right)dY \\ &= (2\pi)^{-n/2}\int_V \exp\left(\frac{1}{2}|Y_0|^2 - \frac{1}{2}|Y|^2 + i\langle Y_0, Y\rangle\right)P(-iY)\,dY. \end{aligned} \tag{1.33}$$

Let $F^{\cdot}S^{\cdot}(V^*)$ be the increasing filtration on $S^{\cdot}(V^*)$ which is associated with $N^{S^{\cdot}(V^*)}$, and let $\mathrm{Gr}^{\cdot}(V^*) = F^{\cdot}S^{\cdot}(V^*)/F^{\cdot -1}S^{\cdot}(V^*)$ be the corresponding $\mathrm{Gr}^{\cdot}$. We have the canonical isomorphism of algebras,

$$\mathrm{Gr}^{\cdot}S^{\cdot}(V^*) \simeq S^{\cdot}(V^*). \tag{1.34}$$

Then $\exp\left(-\Delta^V/2\right) : S^{\cdot}(V^*) \to S^{\cdot}(V^*)$ preserves the filtration of $S^{\cdot}(V^*)$, and induces the identity map on $\mathrm{Gr}^{\cdot}S^{\cdot}(V^*)$. Let $g^{S^{\cdot}(V^*)}$ be the scalar product on $S^{\cdot}(V^*)$ which is induced by g^V. Then one verifies easily that if $P \in S^{\cdot}(V^*)$,

$$\left\|\exp\left(-\Delta^V/2\right)P\right\|_{L_2^V(m^V)} = \|P\|_{g^{S^{\cdot}(V^*)}}. \tag{1.35}$$

The map $\exp\left(-\Delta^V/2\right)$ does not preserve the product on $S^{\cdot}(V^*)$. The pull-back of the product on $S^{\cdot}(V^*)$ by the map $\exp\left(-\Delta^V/2\right)$ can be computed using the Wick rules.

Let $\overline{S^{\cdot}(V^*)}$ be the Hilbert completion of $S^{\cdot}(V^*)$ with respect to $g^{S^{\cdot}(V^*)}$. A basic result is that the map $\exp\left(-\Delta^V/2\right)$ extends to an isometry

$$\overline{S^{\cdot}(V^*)} \simeq L_2^V\left(m^V\right). \tag{1.36}$$

If $V = \mathbf{R}$ is the canonical Euclidean vector space of dimension 1, for $m \in \mathbf{N}$,

$$P_m(x) = \exp\left(-\frac{1}{2}\frac{\partial^2}{\partial x^2}\right)x^m \tag{1.37}$$

is called the Hermite polynomial of order m. Also

$$\|P_m\|^2_{L_2^V(m^V)} = m!, \tag{1.38}$$

and the $P_m, m \in \mathbf{N}$ form an orthogonal basis of $L_2^V\left(m^V\right)$. Of course, the above extends to the case of arbitrary dimension.

If $f \in V$, let $f^* \in V^*$ correspond to f by g^V. Then

$$\begin{aligned}&\exp\left(-\Delta^V/2\right)\nabla_e \exp\left(\Delta^V/2\right) = \nabla_e, \\ &\exp\left(-\Delta^V/2\right)\langle f^*, Y\rangle \exp\left(\Delta^V/2\right) = -\nabla_f + \langle f^*, Y\rangle, \\ &\exp\left(-\Delta^V/2\right)\nabla_Y \exp\left(\Delta^V/2\right) = -\Delta^V + \nabla_Y.\end{aligned} \tag{1.39}$$

Clearly ∇_e maps $F^{\cdot}S^{\cdot}(V^*)$ into $F^{\cdot -1}S^{\cdot}(V^*)$, $-\nabla_f + \langle f^*, Y\rangle$ maps $F^{\cdot}S^{\cdot}(V^*)$ into $F^{\cdot +1}S^{\cdot}(V^*)$, and $-\Delta^V + \nabla_Y$ preserves the filtration $F^{\cdot}S(V^*)$.

Let L_2^V be the ordinary L_2 space of V with respect to dY. Let T denote the isometry from $L_2^V\left(m^V\right) \to L_2^V$,

$$Tf(Y) = \frac{1}{\pi^{n/4}}\exp\left(-|Y|^2/2\right)f\left(\sqrt{2}Y\right). \tag{1.40}$$

Definition 1.2. Let $B : \overline{S^{\cdot}(V^*)} \to L_2^V$ be given by

$$B = T\exp\left(-\Delta^V/2\right). \tag{1.41}$$

The isometry B is called the Bargmann isomorphism. An easy computation shows that if $P \in S^{\cdot}(V^*)$, $f \in L_2^V$,

$$BP(Y_0) = \int_V \left(2^{-2}\pi^{-3}\right)^{n/4} \exp\left(\frac{1}{2}|Y_0|^2 - \frac{1}{2}|Y|^2 + i\sqrt{2}\langle Y_0, Y\rangle\right) P(-iY)\,dY, \tag{1.42}$$

$$B^{-1}f(Y_0) = \int_V \pi^{-n/4}\exp\left(-\frac{1}{2}|Y_0|^2 - \frac{1}{2}|Y|^2 + \sqrt{2}\langle Y_0, Y\rangle\right) f(Y)\,dY.$$

As in (1.39), we get

$$\begin{aligned} B\nabla_e B^{-1} &= \frac{1}{\sqrt{2}}\left(\nabla_e + \langle e^*, Y\rangle\right), \\ B\langle f^*, Y\rangle B^{-1} &= \frac{1}{\sqrt{2}}\left(-\nabla_f + \langle f^*, Y\rangle\right), \\ B\nabla_Y B^{-1} &= \frac{1}{2}\left(-\Delta^V + |Y|^2 - n\right). \end{aligned} \tag{1.43}$$

In the sequel, we will use the notation

$$\mathcal{H}^V = \frac{1}{2}\left(-\Delta^V + |Y|^2 - n\right), \tag{1.44}$$

so that the last identity in (1.43) takes the form

$$B\nabla_Y B^{-1} = \mathcal{H}^V. \tag{1.45}$$

The operator $\mathcal{H}^V$ in the right-hand side of (1.43) is called the harmonic oscillator. By the above, it is self-adjoint, and its spectrum is $\mathbf{N}$.

Using $\mathcal{H}^V$, one can define corresponding abstract Sobolev spaces H^s. It is well known that H^∞ coincides with the Schwartz space $\mathcal{S}(V)$. Of course, $B^{-1}H^\infty$ is a vector space of formal power series with rapidly decreasing coefficients. When $V = \mathbf{R}$, an orthonormal basis of $S^{\cdot}(V^*)$ is given by the $x^m/\sqrt{m!}$, and the coefficients of the formal power series are evaluated with respect to this basis.

By the above, tempered distributions on V can be expressed as formal power series on V. Equivalently, tempered distributions can be written as linear combinations of Hermite polynomials, and from the coefficients of the expansion, we obtain a corresponding power series in a suitable completion of $S^{\cdot}(V^*)$.

Let $\mathcal{F}$ be the Fourier transform acting on L_2^V,

$$\mathcal{F}f(Y) = (2\pi)^{-n/2}\int_V \exp\left(-i\langle Y, Y'\rangle\right) f(Y')\,dY'. \tag{1.46}$$

By (1.42), one gets easily

$$\mathcal{F} = B\mathcal{E}B^{-1}. \tag{1.47}$$

By (1.2), (1.3), (1.45), and (1.47), we obtain

$$\mathcal{F} = \exp\left(-i\frac{\pi}{2}\mathcal{H}^V\right) = (-i)^{\mathcal{H}^V}. \tag{1.48}$$

Equation (1.48) explains the invariance of $\mathcal{H}^V$ under Fourier transform. It can also be obtained via Mehler's formula.

1.5. Gaussians and Dirac masses

Take $P \in S^{\cdot}(V^*)$. If $Y_0 \in V$, one verifies easily that

$$\begin{aligned} P(Y_0) &= \langle \exp(\langle Y_0, Y\rangle), P\rangle_{S^{\cdot}(V^*)} \\ &= \left\langle \exp\left(\langle Y_0, Y\rangle - \frac{1}{2}|Y_0|^2\right), \exp\left(-\Delta^V/2\right)P\right\rangle_{L_2^V(m^V)} \\ &= \left\langle \pi^{-n/4}\exp\left(-\frac{1}{2}|Y_0|^2 - \frac{1}{2}|Y|^2 + \sqrt{2}\langle Y_0, Y\rangle\right), BP\right\rangle_{L_2^V}. \end{aligned} \tag{1.49}$$

Similarly, if $P \in S^{\cdot}(V^*)$, then $P \in L_2^V(m^V)$, and moreover,

$$\begin{aligned} P(Y_0) &= (2\pi)^{n/2}\exp\left(|Y_0|^2/2\right)\langle \delta_{Y_0}, P\rangle_{L_2^V(m^V)} \\ &= \left\langle \exp\left(\langle Y, Y_0\rangle - \frac{1}{2}|Y|^2\right), \exp\left(\Delta^V/2\right)P\right\rangle_{\overline{S^{\cdot}(V^*)}} \\ &= \left\langle \pi^{-n/4}\exp\left(-\frac{1}{2}|Y_0|^2 - \frac{1}{2}|Y|^2 + \sqrt{2}\langle Y_0, Y\rangle\right), B^{-1}P\right\rangle_{\overline{S^{\cdot}(V^*)}} \end{aligned} \tag{1.50}$$

The last identity in (1.42) is equivalent to the last identity in (1.49). By the above, the Bargmann isomorphism exchanges Dirac masses and Gaussians. It is similar to a Fourier transform, in the sense that localization in one variable is necessarily delocalized in the other variable.

Finally, observe that

$$B1 = \pi^{-n/4}\exp\left(-|Y|^2/2\right), \qquad B^{-1}1 = 2^{n/2}\pi^{n/4}\exp\left(\frac{1}{2}|Y|^2\right). \tag{1.51}$$

1.6. Algebraic de Rham complex and Bargmann isomorphism

We use the same notation as in Subsection 1.4. The map $\exp\left(-\Delta^V/2\right) : S^{\cdot}(V^*) \to S^{\cdot}(V^*)$ extends to an automorphism of the $\mathbf{Z}$-graded filtered complex $(\mathcal{A}^{\cdot}(V^*), d)$. This automorphism induces the identity on $\mathrm{Gr}^{\cdot} S^{\cdot}(V^*)$.

Let $e_1, \ldots, e_n$ be an orthonormal basis of V. Set

$$d^* = -\sum_{i=1}^{n} i_{e_i}\nabla_{e_i}. \tag{1.52}$$

By (1.39),

$$\begin{aligned}\exp\left(-\Delta^V/2\right) d \exp\left(\Delta^V/2\right) &= d,\\ \exp\left(-\Delta^V/2\right) i_Y \exp\left(\Delta^V/2\right) &= d^* + i_Y,\\ \exp\left(-\Delta^V/2\right) L_Y \exp\left(\Delta^V/2\right) &= -\Delta^V + L_Y.\end{aligned} \tag{1.53}$$

Also by (1.43),

$$\begin{aligned}BdB^{-1} &= \frac{1}{\sqrt{2}}\left(d + Y^*\wedge\right),\\ Bi_YB^{-1} &= \frac{1}{\sqrt{2}}\left(d^* + i_Y\right),\\ BL_YB^{-1} &= \mathcal{H}^V + N^{\Lambda^\cdot(V^*)}.\end{aligned} \tag{1.54}$$

The fact that up to a constant, in the right-hand side of (1.54), we recover the Witten twist [W82] of d, d^* associated with the function $|Y|^2/2$ is not a surprise in view of (1.40), (1.41).

The strangest aspect of (1.53) is that with the proper proviso, the algebraic de Rham complex $(\mathcal{A}^\cdot(V^*), d)$ is isomorphic to a version of the smooth de Rham complex $(\Omega^\cdot(V), d)$, but this isomorphism does not preserve the product.

The kernel of L_Y in $\overline{S^\cdot(V^*)}$ is spanned by 1, and the kernel of $\mathcal{H}^V + N^{\Lambda^\cdot(V^*)}$ in $L_2^V \otimes \Lambda^\cdot(V^*)$ is spanned by $\pi^{-n/4}\exp\left(-|Y|^2/2\right)$. This is compatible with the first identity in (1.51).

Now we briefly consider the Bargmann transform of the algebraic dual complex $(\mathcal{B}^\cdot(V), Z\wedge)$. First of all, we identify V and V^* by the scalar product, so that

$$\mathcal{B}^\cdot(V) \simeq S^\cdot(V^*) \otimes \Lambda^\cdot(V^*) \otimes \Lambda^n(V). \tag{1.55}$$

By (1.55), we get

$$\mathcal{B}^\cdot(V) \simeq \mathcal{A}^\cdot(V^*) \otimes \Lambda^n(V). \tag{1.56}$$

Also the scalar product trivializes $\Lambda^n(V)$ up to a $\mathbf{Z}_2$ ambiguity. The canonical sections Y and Z are also identified. By (1.56), we have the identification of complexes,

$$\left(\mathcal{B}^\cdot(V), Z\wedge\right) \simeq \left(\mathcal{A}^\cdot(V^*) \otimes \Lambda^n(V), Y\wedge\right). \tag{1.57}$$

In what follows, we use the notation L_Z instead of L_Y to point out that here L_Z is given by (1.27). As in (1.53), we get

$$\begin{aligned}\exp\left(-\Delta^V/2\right) Y^*\wedge \exp\left(\Delta^V/2\right) &= -d + Y^*\wedge,\\ \exp\left(-\Delta^V/2\right)(-d_*) \exp\left(\Delta^V/2\right) &= -d^*,\\ \exp\left(-\Delta^V/2\right) L_Z \exp\left(\Delta^V/2\right) &= -\Delta^V + \nabla_Y + n - N^{\Lambda^\cdot(V^*)}.\end{aligned} \tag{1.58}$$

Moreover,

$$BY^* \wedge B^{-1} = \frac{1}{\sqrt{2}}\left(-d + Y^*\wedge\right), \qquad B\left(-d_*\right)B^{-1} = -\frac{1}{\sqrt{2}}\left(d^* - i_Y\right),$$
$$BL_ZB^{-1} = \mathcal{H}^V + n - N^{\Lambda^\cdot(V^*)}. \tag{1.59}$$

The operators in the right-hand side of (1.59) come just from the Witten twist associated with the function $-|Y|^2/2$. Incidentally, because the L_2 norm involves the choice of the volume form associated with the metric of V, it is better to view the Bargmann transform of $\overline{\mathcal{B}^\cdot(V)}$ as being the vector space of L_2 sections of $\Lambda^\cdot(V^*) \otimes o(V)$, where $o(V)$ is the orientation line of V.

As we saw in Subsection 1.3, the kernel of L_Z in $\mathcal{B}^\cdot(V)$ is concentrated in degree n and spanned by $1 \in \Lambda^n(V^*) \otimes \Lambda^n(V)$.

Similarly the kernel of $\mathcal{H}^V + n - N^{\Lambda^\cdot(V^*)}$ is concentrated in degree n and spanned by the density $\pi^{-n/4}\exp\left(-|Y|^2/2\right)dY$. This fits with the first identity in (1.51).

1.7. Euler characteristic and the Pythagorean theorem

We saw in Subsection 1.1 that the cohomology of the complex $(\mathcal{A}^\cdot(V^*), d)$ is concentrated in degree 0 and is one-dimensional. In particular its Euler characteristic χ is equal to 1.

For $t > 0$, the operator $\exp\left(-tN^{S^\cdot(V^*)}\right)$ is self-adjoint and trace class, and moreover,

$$\operatorname{Tr}\left[\exp\left(-tN^{S^\cdot(V^*)}\right)\right] = \left(1 - e^{-t}\right)^{-n}. \tag{1.60}$$

The supertrace of $\exp\left(-tN^{\Lambda^\cdot(V^*)}\right)$ is given by

$$\operatorname{Tr_s}\left[\exp\left(-tN^{\Lambda^\cdot(V^*)}\right)\right] = \left(1 - e^{-t}\right)^{n}. \tag{1.61}$$

By (1.60), (1.61), we conclude that the operator $\exp\left(-tN^{\mathcal{A}^\cdot(V^*)}\right)$ is trace class, and that

$$1 = \operatorname{Tr_s}\left[\exp\left(-tN^{\mathcal{A}^\cdot(V^*)}\right)\right]. \tag{1.62}$$

Note that because of (1.10), equation (1.62) is natural with respect to direct sums. More precisely, the operator $\exp\left(-tN^{\mathcal{A}^\cdot(V^*)}\right)$ behaves multiplicatively with respect to direct sums, which is also a reflection of the fact that the exponential maps sums into products.

We have the splitting of $(\mathcal{A}^\cdot(V^*), d)$ in (1.14). By (1.15), for $k > 0$, the complex $(\mathcal{A}^\cdot_k(V^*), d)$ is exact. We get

$$\begin{aligned}\operatorname{Tr_s}^{\mathcal{A}^\cdot_k(V^*)}\left[\exp\left(-tN^{\mathcal{A}^\cdot(V^*)}\right)\right] &= 1 \text{ for } k = 0,\\ &= 0 \text{ for } k > 0.\end{aligned} \tag{1.63}$$

Of course (1.63) explains (1.62). The above suggests that equation (1.62) can be rewritten in the form

$$1 = \left[1 + \left(e^{-t} - e^{-t}\right) + \left(e^{-2t} - e^{-2t}\right) + \cdots\right]^n. \tag{1.64}$$

Set

$$D^V = d + i_Y. \tag{1.65}$$

By (1.12), (1.13), we get

$$N^{\mathcal{A}^\cdot(V^*)} = L_Y = D^{V,2}. \tag{1.66}$$

Then (1.62) can be rewritten in the form

$$\chi = \mathrm{Tr_s}\left[\exp\left(-tD^{V,2}\right)\right]. \tag{1.67}$$

Using the Bargmann isomorphism and (1.54), we can rewrite (1.67) in the form

$$\chi = \mathrm{Tr_s}\left[\exp\left(-t\left(\mathcal{H}^V + N^{\Lambda^\cdot(V^*)}\right)\right)\right]. \tag{1.68}$$

Let us now make $t \to 0$ in equation (1.67). By using classical local index theory on the right-hand side of (1.67), that is by working on smooth differential forms on V, we find that in this case the index formula takes the form

$$\chi = \mathrm{Tr_s}\left[\exp\left(-tD^{V,2}\right)\right] = \int_V \exp\left(-\left|Y\right|^2/2\right)\frac{dY}{(2\pi)^{n/2}}. \tag{1.69}$$

It is remarkable that the last two terms in (1.69) have a similar Gaussian character.

1.8. Symmetric algebras, the geodesic flow and Brownian motion

Let X be a smooth manifold of dimension n. The exterior algebra $\Lambda^\cdot(T^*X)$ and the de Rham complex $\left(\Omega^\cdot(X), d^X\right)$ have been universally used in geometry. One cannot say the same for the symmetric algebra $S^\cdot(T^*X)$. Symmetric algebras are important in K-theory and representation theory, but their specific geometric role is difficult to point out.

If TX is equipped with a Riemannian metric, classical Hodge theory on the de Rham complex $\left(\Omega^\cdot(X), d^X\right)$ is inescapable. However, the geometric role of the induced scalar product on $S^\cdot(T^*X)$, of its completion $\overline{S^\cdot(T^*X)}$, and of the Bargmann isomorphism with the Hilbert bundle $L_2(TX)$ are not so much used.

Still, one can say that in geometric analysis, the harmonic oscillator appears recurrently in a number of related questions:

- In problems involving localization. For instance, in Witten's proof of Morse inequalities [W82], the local model of the deformed Witten complex near a critical point is precisely the one described in the right-hand side of (1.54). In complex geometry, the Koszul complex and its Hodge theory [B90, BL91] produce another version of a harmonic oscillator.
- The computation by Getzler [Ge86] of the local index theorem also involves a harmonic oscillator which is responsible for the appearance of the class $\widehat{A}(TX)$ in the Atiyah-Singer index theorem for Dirac operators.
- The evaluation of a number of exotic genera appears in explicit computations involving the harmonic oscillator [B90, B94, BG00, BG01, BG04], still in questions connected with localization.

The fact that the harmonic oscillator appears repeatedly in various unrelated questions has prompted us to try putting it at centre stage in our construction of the hypoelliptic Laplacian, which will be reviewed in Sections 3, 4, and 5.

Let us briefly mention two easy constructions involving $S^{\cdot}(T^*X)$ when TX is equipped with a Riemannian metric. First we proceed by analogy with the case of the de Rham complex. Let d^{X*} be the formal adjoint of d^X with respect to the obvious L_2 scalar product on $\Omega^{\cdot}(X)$. Let $\nabla^{\Lambda^{\cdot}(T^*X)}$ be the connection on $\Lambda^{\cdot}(T^*X)$ which is induced by the Levi-Civita connection ∇^{TX} on TX. If $e_1, \dots, e_n$ is an orthonormal basis of TX, and if $e^1, \dots, e^n$ is the corresponding dual basis, then

$$d^X - d^{X*} = \sum_{i=1}^{n} \left(e^i + i_{e_i}\right) \nabla_{e_i}^{\Lambda^{\cdot}(T^*X)}. \tag{1.70}$$

We will now obtain the $S^{\cdot}(T^*X)$ analogue of (1.70). Let $\nabla^{S^{\cdot}(T^*X)}$ be the connection on $S^{\cdot}(T^*X)$ which is induced by ∇^{TX}. If $e \in TX$, let ∇_e^V denotes differentiation along the fibre TX. Using the results of Subsection 1.1, we find that the analogue of $d^X - d^{X*}$ is the operator E acting on $C^\infty(X, S^{\cdot}(T^*X))$ which is given by

$$E = \sum_{i=1}^{n} \left(\langle e^i, Y\rangle + \nabla_{e_i}^V\right) \nabla_{e_i}^{S^{\cdot}(T^*X)}. \tag{1.71}$$

Let $\pi : \mathcal{X} \to X$ be the total space of TX, with fibre $\widehat{TX}$, another copy of the tangent bundle TX. The notation $\widehat{TX}$ is used instead of TX, because we want to emphasize that $\widehat{TX}$ is distinct from the physical tangent bundle TX. Then $C^\infty\left(X, S^{\cdot}\left(\widehat{T^*X}\right)\right)$ is the vector space of smooth functions on $\mathcal{X}$ which are polynomial along the fibre. Via the Bargmann isomorphism, a proper completion of this vector space can be identified with the Schwartz space $\mathcal{S}(\mathcal{X})$ of smooth rapidly decreasing functions on $\mathcal{X}$ together with their derivatives of arbitrary order.

Let $\widehat{Y}$ be the tautological section of $p^*\widehat{TX}$ on $\mathcal{X}$, and let Y be the corresponding section of p^*TX. Let B be the fibrewise Bargmann transform. Set

$$F = BEB^{-1}. \tag{1.72}$$

Then F acts on the Schwartz space $\mathcal{S}(\mathcal{X})$. In the sequel, if $e \in TX$, we identify e with the corresponding horizontal vector field on $\mathcal{X}$. In particular ∇_e^X denotes horizontal differentiation in the direction e. From (1.43), (1.71), and (1.72), we get

$$F = \sqrt{2} \sum_{i=1}^{n} \langle e^i, Y\rangle \nabla_{e_i}^X. \tag{1.73}$$

By following the previous conventions, ∇_Y^X, which differentiates horizontally, is the geodesic vector field on $\mathcal{X}$. We can rewrite (1.73) in the form

$$F = \sqrt{2}\nabla_Y^X, \tag{1.74}$$

i.e., $F/\sqrt{2}$ is the geodesic vector field ∇_Y^X on $\mathcal{X}$. The geodesic vector field appears then to be a kind of bosonic Dirac operator, similar to the fermionic operator

$d^X - d^{X*}$. This should not be too much of a surprise. After all, the number of canonical objects one can produce by simple algebra is limited.

Observe that the principal symbol $\sigma\left(\nabla_Y^X\right)$ of ∇_Y^X is given by

$$\sigma\left(\nabla_Y^X\right) = i\left\langle \xi, Y\right\rangle, \tag{1.75}$$

with ξ the canonical section of T^*X. Equation (1.75) has an obvious Fourier transform quality. Moreover, when identifying TX and T^*X by the metric, both Y and ξ represent two distinct forms of differentiation on X. One can then view Y and $i\xi$ as lying in the real and imaginary parts of the complexified tangent bundle $T_{\mathbf{C}}X$.

Another construction is related to the embedding of X as the zero section of $\mathcal{X}$. Note that $\left(\mathcal{A}^{\cdot}\left(\widehat{T^*X}\right), d\right)$ is the algebraic de Rham complex along the fibre, and $\left(\mathcal{B}\left(\widehat{TX}\right), Z\wedge\right)$ is the dual complex.

The cohomology of $\left(\mathcal{A}^{\cdot}\left(\widehat{T^*X}\right), d\right)$ is concentrated in degree 0 and represented by $1 \in S^0\left(T^*X\right)$. Via the Bargmann isomorphism, the function 1 is replaced by the fibrewise Gaussian function $\pi^{-n/4}\exp\left(-\left|Y\right|^2/2\right)$, and the compactly supported cohomology of the fibre is represented by the Gaussian measure $dm^{\widehat{TX}}$ along the fibre $\widehat{TX}$.

The family of Gaussian measures $dm^{\widehat{TX}}$ along the fibres $\widehat{TX}$ has been related to the embedding of X as the zero section of $\mathcal{X}$. Now we give a dynamical interpretation of this family of Gaussian measures.

A vector field U on X is a smooth section of TX. When identifying TX and $\widehat{TX}$, we can identify U with the family of Dirac masses $\delta_{U(x)}$ in the fibres $\widehat{TX}$. From the above, we see that the family dm^V represents another kind of family of measures on $\widehat{TX}$. What is its dynamical or vector field counterpart? It has to be Brownian motion on the manifold X, since the speed of Brownian motion, even if it is infinite, can be thought of as being given by independent Gaussians. On the other hand, the natural differential operator associated with Brownian motion is the Laplace-Beltrami operator Δ^X.

It follows from the above that the Laplace-Beltrami operator has been made to be related to the embedding of X into $\mathcal{X}$. As we shall see in Section 3, this idea reappears in the construction of the hypoelliptic Laplacian.

2. Gaussian index theory

The purpose of this section is explain the relation of index theory to the Gaussian formalism. In particular, we extend to general Dirac operators the results of Section 1 on the algebraic de Rham complex.

This section is organized as follows. In Subsection 2.1, we recall elementary facts on the index of Dirac operators, among which the McKean-Singer formula [McS67] for the index.

In Subsection 2.2, using the superconnection formalism of [Q85], we give a Gaussian expression for the Atiyah-Singer index class, which ultimately 'explains' the multiplicativity of the index formula.

In Subsection 2.3, we consider the case of the Lefschetz formulas.

Finally, in Subsection 2.4, we briefly describe other natural multiplicative constructions, which include the Brownian measure.

2.1. The index of the Dirac operator

Let X be a compact even-dimensional Riemannian manifold, which we assume to be oriented and spin. Let ∇^{TX} be the Levi-Civita connection on TX, and let R^{TX} be its curvature. Let $\left(E, g^{E}, \nabla^{E}\right)$ be a complex Hermitian vector bundle on X equipped with a unitary connection, and let R^{E} be the curvature of the connection.

Let $S^{TX} = S^{TX}_{+} \oplus S^{TX}_{-}$ be the Hermitian $\mathbf{Z}_2$-graded vector bundle of $\left(TX, g^{TX}\right)$ spinors on X, and let $\nabla^{S^{TX}} = \nabla^{S^{TX}_{+}} \oplus \nabla^{S^{TX}_{-}}$ be the Levi-Civita connection on this vector bundle.

Let D^{X} be the Dirac operator acting on $C^{\infty}\left(X, S^{TX} \otimes E\right)$. Then D^{X} is an odd elliptic first-order operator, which can be written in matrix form as

$$D^{X} = \begin{bmatrix} 0 & D^{X}_{-} \\ D^{X}_{+} & 0 \end{bmatrix}. \tag{2.1}$$

Moreover, D^{X}_{+} is a Fredholm operator, whose index is denoted $\operatorname{Ind} D^{X}_{+}$.

The McKean-Singer formula [McS67] asserts that if $\operatorname{Tr_s}$ still denotes the supertrace, for any $t > 0$,

$$\operatorname{Ind} D^{X}_{+} = \operatorname{Tr_s}\left[\exp\left(-tD^{X,2}\right)\right]. \tag{2.2}$$

The proof of (2.2) is well known. The eigenspaces of $D^{X,2}$ define an increasing filtration on $C^{\infty}\left(X, S^{TX} \otimes E\right)$, and D^{X} preserves the filtration. The eigenspace splitting of $C^{\infty}\left(X, S^{TX} \otimes E\right)$ is strictly similar to (1.14), and the proof of (2.2) can be obtained by proceeding as in (1.63). The analogy is not perfect in the sense that the set of eigenvalues of $D^{X,2}$ does not have an additive structure.

Let X', E' be another pair similar to X, E. Let p, p' be the projections $X \times X' \to X, X \times X' \to X'$. Then $(X \times X', p^{*}E \otimes p'^{*}E')$ is also such a pair. Moreover,

$$\begin{aligned} C^{\infty}&\left(X \times X', S^{T(X \times X')} \otimes p^{*}E \otimes p'^{*}E'\right) \\ &= C^{\infty}\left(X, S^{TX} \otimes E\right) \widehat{\otimes} C^{\infty}\left(X', S^{TX'} \otimes E'\right). \end{aligned} \tag{2.3}$$

Also,

$$D^{X \times X'} = D^{X} \widehat{\otimes} 1 + 1 \widehat{\otimes} D^{X'}. \tag{2.4}$$

In the sequel, we will write (2.4) in the simpler form

$$D^{X \times X'} = D^{X} + D^{X'}. \tag{2.5}$$

By (2.4), (2.5), we get

$$\operatorname{Ind} D_+^{X\times X'} = \operatorname{Ind} D_+^X \operatorname{Ind} D_+^{X'}. \tag{2.6}$$

Also the anticommutator of D^X and $D^{X\prime}$ vanishes. If $\left[D^X, D^{X\prime}\right]$ is the supercommutator of D^X and $D^{X\prime}$, we have the equivalent identity

$$\left[D^X, D^{X\prime}\right] = 0. \tag{2.7}$$

By (2.7), we get the Pythagorean theorem for Dirac operators,

$$D^{X\times X',2} = D^{X,2} + D^{X',2}. \tag{2.8}$$

By (2.8), we obtain

$$\exp\left(-tD^{X\times X',2}\right) = \exp\left(-tD^{X,2}\right) \widehat{\otimes} \exp\left(-tD^{X',2}\right). \tag{2.9}$$

By (2.9), we get

$$\operatorname{Tr_s}\left[\exp\left(-tD^{X\times X',2}\right)\right] = \operatorname{Tr_s}\left[\exp\left(-tD^{X,2}\right)\right] \operatorname{Tr_s}\left[\exp\left(-tD^{X',2}\right)\right]. \tag{2.10}$$

Of course, (2.2), (2.6), and (2.10) are compatible.

In equation (2.2), one can replace $\exp(-t\cdot)$ by any function $F \in \mathcal{S}(\mathbf{R})$ such that $F(0) = 1$, so that instead of (2.2), we have the identity

$$\operatorname{Ind}\left(D_+^X\right) = \operatorname{Tr_s}\left[F\left(D^{X,2}\right)\right]. \tag{2.11}$$

However, in general, the operator $F\left(D^{X,2}\right)$ does not behave multiplicatively.

The real heroes of equation (2.2) are the integer $\operatorname{Ind} D_+^X$ and the operator $\exp\left(-tD^{X,2}\right)$, which, by (2.6), (2.9), exhibit multiplicativity in their category.

Let $\mathcal{D}(X)$ be the algebra of differential operators acting on $C^\infty(X, E)$, and let $\mathcal{P}(X)$ be the algebra of pseudodifferential operators acting on the same vector space. Note that

$$\mathcal{D}(X \times X') = \mathcal{D}(X) \widehat{\otimes} \mathcal{D}(X'). \tag{2.12}$$

However the algebra $\mathcal{P}(X)$ does not behave properly with respect to products.

The index theorem of Atiyah-Singer [AS68] is formulated for elliptic pseudodifferential operators. However if $P \in \mathcal{P}(X), P' \in \mathcal{P}(X')$, in general $P\widehat{\otimes}1 + 1\widehat{\otimes}P'$ is not an element of $\mathcal{P}(X \times X')$. This means that in the abstract framework of index theory, the considerations we made before for Dirac operators are not valid. It is because Dirac operators are differential operators which are compatible to products of manifolds, and are such that their square verifies the Pythagorean formula in (2.8) that the McKean-Singer index formula, restricted to such operators, has this universal form. Needless to say, the operators we met in Section 1 are also differential operators, and this fact plays a critical role in their naturality with respect to direct sums.

2.2. The index formula as a Gaussian formula

The index formula of Atiyah-Singer [AS68] asserts that

$$\operatorname{Ind} D_+^X = \int_X \widehat{A}(TX)\operatorname{ch}(E). \tag{2.13}$$

The heat equation proof of (2.13) consists in making $t \to 0$ in the McKean-Singer formula (2.2), and to use the 'fantastic cancellation' mechanism anticipated by McKean-Singer [McS67] to show that as $t \to 0$, the local supertrace of the heat kernel associated with $\exp\left(-tD^{X,2}\right)$, instead of being singular as it should be, has a limit as $t \to 0$, which can ultimately be identified with a local Chern-Weil polynomial.

Observe that the right-hand side of (2.13) is compatible to products. In particular, the two following multiplicativity properties play a crucial role:

$$\widehat{A}(F \oplus F') = \widehat{A}(F)\widehat{A}(F'), \qquad \operatorname{ch}(E \otimes E') = \operatorname{ch}(E)\operatorname{ch}(E'). \tag{2.14}$$

Let us think more about (2.14). First note that the Chern character form $\operatorname{ch}\left(E, \nabla^E\right)$ associated with the connection ∇^E is given by

$$\operatorname{ch}\left(E, \nabla^E\right) = \varphi \operatorname{Tr}\left[\exp\left(-R^E\right)\right]. \tag{2.15}$$

In (2.15), φ denotes the standard $2i\pi$ normalization of characteristic forms in Chern-Weil theory. The curvature R^E can be written in the form

$$R^E = \nabla^{E,2}, \tag{2.16}$$

so that (2.15) has the equivalent Gaussian form

$$\operatorname{ch}\left(E, \nabla^E\right) = \varphi \operatorname{Tr}\left[\exp\left(-\nabla^{E,2}\right)\right]. \tag{2.17}$$

Let $\left(E', g^{E'}, \nabla^{E'}\right)$ be another Hermitian vector bundle with connection. Let $\nabla^{E\otimes E'}$ be the connection on $E \otimes E'$ which is induced by $\nabla^E, \nabla^{E'}$, i.e.,

$$\nabla^{E\otimes E'} = \nabla^E \otimes 1 + 1 \otimes \nabla^{E'}. \tag{2.18}$$

Then we have the Pythagorean formula

$$\nabla^{E\otimes E',2} = \nabla^{E,2} \otimes 1 + 1 \otimes \nabla^{E',2}, \tag{2.19}$$

from which we get

$$\operatorname{ch}\left(E \otimes E', \nabla^{E\otimes E'}\right) = \operatorname{ch}\left(E, \nabla^E\right)\operatorname{ch}\left(E', \nabla^{E'}\right), \tag{2.20}$$

which in turn implies the second equation in (2.14).

Let us now concentrate on the first equation in (2.14). This first equation can be refined to Chern-Weil forms, i.e., if $\left(F, g^F, \nabla^F\right), \left(F', g^{F\prime}, \nabla^{F'}\right)$ are real Euclidean vector bundles with metric connections, then

$$\widehat{A}\left(F \oplus F', \nabla^{F\oplus F'}\right) = \widehat{A}\left(F, \nabla^F\right)\widehat{A}\left(F', \nabla^{F'}\right). \tag{2.21}$$

We will now give a Gaussian explanation for (2.21). Let $\left(F, g^F, \nabla^F\right)$ be a real Euclidean oriented spin vector bundle of even dimension n, which is equipped with a metric connection. The curvature of ∇^F is denoted R^F. Let $\mathcal{F}$ be the total space of F. Then $\pi : \mathcal{F} \to X$ defines a fibration with fibre F. Also the fibres are equipped with the metric g^F, and moreover, the connection ∇^F defines a horizontal subbundle $T^H\mathcal{F} \subset T\mathcal{F}$. Let Y be the tautological section of $\pi^* F$ on $\mathcal{F}$.

Let $S^F = S^F_+ \oplus S^F_-$ be the vector bundle of (F, g^F) spinors on X. The above data define unambiguously a Levi-Civita superconnection A^{S^F} [B86] on the vector bundle $C^\infty(F, \pi^* S^F)$ over X. Let D^F be the obvious family of Dirac operators along the fibres F. As explained in [B90, Section 3(b)], with the proper normalization, A^{S^F} takes the form

$$A^{S^F} = \nabla^{C^\infty(F,\pi^* S^F)} + \frac{D^F}{\sqrt{2}} - \frac{c(R^F Y)}{2\sqrt{2}}. \tag{2.22}$$

In (2.22), $\nabla^{C^\infty(F,\pi^* S^F)}$ is the connection on $C^\infty(F, \pi^* S^F)$ which is induced by ∇^F, D^F is the family of Dirac operators along the fibres, and c denotes Clifford multiplication. The discrepancy by a factor $1/\sqrt{2}$ with respect to [B90, eq. (3.12] comes from the fact that in this reference, the family of Dirac operators along the fibres is just $D^F/\sqrt{2}$.

A remarkable fact established in [B90, Theorem 3.6] is that if $e_1, \dots, e_n$ is an orthonormal basis of F,

$$A^{S^F,2} = -\frac{1}{2}\sum_{i=1}^{n}\left(\nabla_{e_i} + \frac{1}{2}\langle R^F Y, e_i\rangle\right)^2. \tag{2.23}$$

Let us comment on equation (2.23). First if $F = TX$, the reader will have recognized the Getzler operator [Ge86], which appears as a renormalized limit of $tD^{X,2}/2$ as $t \to 0$, after rescaling of the local coordinates and of the Clifford variables. The fact that the Getzler operator is a square is a miracle. The fact that it is the curvature of a Levi-Civita superconnection [B86], whose purpose was to prove a local version of the families index theorem of Atiyah-Singer [AS71], is even more surprising. It indicates that when proving the index theorem for a single manifold, the formalism of the families index theorem is already there, even if it is, for the moment, difficult to interpret this coincidence. This is another manifestation of the fact that algebra severely constrains the structure of universal objects. In our opinion, this coincidence is the most important fact in the index theory of Dirac operators[1]. For an interpretation in terms of localization formulas in equivariant cohomology, we refer to [B11c, Remark 1.4 and Section 4.2].

Let us now describe the consequences of (2.23). First, the right-hand side of (2.23) does not contain Clifford multiplication operators. Except for the exterior algebra $\Lambda^\cdot(T^*X)$ which is present in R^F, $A^{S^F,2}$ is a scalar operator. Besides its component of degree 0 in $\Lambda^\cdot(T^*X)$ is just $-\frac{1}{2}\Delta^F$, which indicates that the heat operator $\exp\left(-A^{S^F,2}\right)$ is not trace class. This is not surprising. Indeed the fibres F are noncompact, there is no natural families index associated with the projection

[1]This is, admittedly, a minority view.

$\pi : \mathcal{F} \to X$. Let $P^{S^F}(Y, Y')$ be the smooth kernel along the fibre associated with the operator $\exp\left(-A^{S^F,2}\right)$ with respect to the volume form $dY'/(2\pi)^{n/2}$. As explained before, $P^{S^F}(Y, Y')$ is a differential form on X, and does not contain Clifford variables. Therefore its classical supertrace over S^F vanishes. Moreover, one verifies easily that $P^{S^F}(Y, Y)$ does not depend on Y.

From the above it is natural to view $P^{S^F}(Y, Y) = P^{S^F}(0, 0)$ as a generalized von Neumann supertrace of $\exp\left(-A^{S^F,2}\right)$. We will write this in the form

$$\mathrm{Tr_s}\left[\exp\left(-A^{S^F,2}\right)\right] = P^{S^F}(0,0). \tag{2.24}$$

Now the basic computation of the local index theorem by Getzler [Ge86] says that

$$\widehat{A}\left(F, \nabla^F\right) = \varphi \mathrm{Tr_s}\left[\exp\left(-A^{S^F,2}\right)\right]. \tag{2.25}$$

Equation (2.25) indicates that $\widehat{A}\left(F, \nabla^F\right)$ has a Gaussian expression which in turn implies (2.21).

If in the above construction, we twist S^F by $\left(E, g^E, \nabla^E\right)$, we obtain a superconnection $A^{S^F \otimes E}$ on $C^\infty\left(F, \pi^*\left(S^F \otimes E\right)\right)$ such that

$$A^{S^F \otimes E,2} = A^{F,2} \otimes 1 + 1 \otimes \nabla^{E,2}. \tag{2.26}$$

If $P^{S^F \otimes E}(Y, Y')$ is the kernel associated with $\exp\left(-A^{S^F \otimes E,2}\right)$, then

$$P^{S^F \otimes E}(Y, Y') = P^{S^F}(Y, Y') \otimes \exp\left(-R^E\right). \tag{2.27}$$

Now we define $\mathrm{Tr_s}\left[\exp\left(-A^{S^F \otimes E,2}\right)\right]$ by the formula

$$\mathrm{Tr_s}\left[\exp\left(-A^{S^F \otimes E,2}\right)\right] = \mathrm{Tr}^E\left[P^{S^F \otimes E}(0,0)\right]. \tag{2.28}$$

Then

$$\widehat{A}\left(F, \nabla^F\right) \mathrm{ch}\left(E, \nabla^E\right) = \varphi \mathrm{Tr_s}\left[\exp\left(-A^{S^F \otimes E,2}\right)\right]. \tag{2.29}$$

For $b > 0$, set

$$K_b s(Y) = s(bY). \tag{2.30}$$

Observe that for $t > 0$,

$$K_{\sqrt{t}} t A^{S^F \otimes E,2} K_{1/\sqrt{t}} = -\frac{1}{2}\sum_{i=1}^{n}\left(\nabla_{e_i} + \frac{1}{2}\left\langle tR^{TX}Y, e_i\right\rangle\right)^2 + tR^E. \tag{2.31}$$

For $s \in \mathbf{R}$, let ψ_s the endomorphism of $\Lambda^{\cdot}(T^*X)$ such that if $\alpha \in \Lambda^p(T^*X)$, then

$$\psi_s \alpha = s^p \alpha. \tag{2.32}$$

We can rewrite (2.31) in the form

$$K_{\sqrt{t}} t A^{S^F \otimes E,2} K_{1/\sqrt{t}} = \psi_{\sqrt{t}} A^{S^F \otimes E,2} \psi_{1/\sqrt{t}}. \tag{2.33}$$

We will denote by $P_t^{S^F\otimes E}(Y,Y')$ the smooth kernel associated with

$$\exp\left(-tA^{S^F\otimes E,2}\right)$$

with respect to the volume $dY'/(2\pi)^{n/2}$. By (2.33), we deduce that

$$P_t^{S^F\otimes E}(0,0)=\psi_{\sqrt{t}}\frac{1}{t^{n/2}}P^{S^F\otimes E}(0,0)\,\psi_{1/\sqrt{t}}. \tag{2.34}$$

We define $\mathrm{Tr_s}\left[\exp\left(-tA^{S^F\otimes E,2}\right)\right]$ by the formula

$$\mathrm{Tr_s}\left[\exp\left(-tA^{S^F\otimes E,2}\right)\right]=\mathrm{Tr}^E\left[P_t^{S^F\otimes E}(0,0)\right]. \tag{2.35}$$

By (2.34), (2.35), we get

$$\mathrm{Tr_s}\left[\exp\left(-tA^{S^F\otimes E,2}\right)\right]=\frac{1}{t^{n/2}}\psi_{\sqrt{t}}\mathrm{Tr_s}\left[\exp\left(-A^{S^F\otimes E,2}\right)\right]. \tag{2.36}$$

If (F,g^F,∇^F) is taken to be (TX,g^{TX},∇^{TX}), then the index formula (2.13) takes the form

$$\mathrm{Ind}\,D_+^X=\int_X\varphi\mathrm{Tr_s}\left[\exp\left(-A^{S^{TX}\otimes E,2}\right)\right]. \tag{2.37}$$

More generally, by (2.36), (2.37), for any $t>0$, we get

$$\mathrm{Ind}\,D_+^X=\int_X\varphi\mathrm{Tr_s}\left[\exp\left(-tA^{S^{TX}\otimes E,2}\right)\right]. \tag{2.38}$$

The fact that the right-hand side of (2.38) does not depend on $t>0$ should not come as a surprise, since it just reflects the fact that in the McKean-Singer formula, the function $\exp(-s)$ can be replaced by $\exp(-ts)$ for any $t>0$.

Let us now put together the McKean-Singer formula in (2.2) and (2.38). For any $t>0$, we get

$$\mathrm{Ind}\,D_+^X=\mathrm{Tr_s}\left[\exp\left(-tD^{X,2}/2\right)\right]=\int_X\varphi\mathrm{Tr_s}\left[\exp\left(-tA^{S^{TX}\otimes E,2}\right)\right]. \tag{2.39}$$

Comparing with (1.69), we find that equation (2.39) for $\mathrm{Ind}\,D_+^X$ has exactly the same doubly Gaussian character.

Also, we pointed out before that the hero of the McKean-Singer formula in (2.2) is the operator $\exp\left(-tD^{X,2}/2\right)$, and not the equation itself. Equation (2.39) suggests that $\exp\left(-tA^{S^F\otimes E,2}\right)$ is another such hero. Actually, the equality of the two last expressions in (2.39) suggests that the operators $D^{X,2}/2$ and $A^{S^F\otimes E,2}$ are just the two faces of a Janus like object, which we do not immediately identify as identical because of our imperfect understanding of their similarity.

The Getzler operator $A^{S^{TX}\otimes E,2}$ is obtained via a local rescaled deformation of $tD^{X,2}/2$ as $t\to 0$. The theory of the hypoelliptic Dirac operator is an attempt to reverse the above limit, i.e., to obtain the Dirac operator $D^X/\sqrt{2}$ as a limit of a deformation of a version of $A^{S^{TX}\otimes E}$.

Let $f(t) : \mathbf{R}_+ \to \mathbf{R}$ be a smooth function with compact support not including 0, and such that

$$\int_0^{+\infty} f(t)\,dt = 1. \tag{2.40}$$

Set

$$F(x) = \int_0^{+\infty} \exp(-tx) f(t)\,dt. \tag{2.41}$$

Then $F(x)$ is an analytic function of x such that $F(0) = 1$.

By integrating (2.39), we get

$$\operatorname{Ind} D_+^X = \operatorname{Tr_s}\left[F\left(\frac{D^{X,2}}{2}\right)\right] = \int_X \varphi \operatorname{Tr_s}\left[F\left(A^{S^F \otimes E,2}\right)\right]. \tag{2.42}$$

We have now artificially destroyed the multiplicative nature of the index formula, by replacing the exponential by some other analytic function. Still equation (2.42) indicates how similar $D^{X,2}/2$ and $A^{S^F \otimes E,2}$ are.

Finally, let us point out that there is a Fourier transform quality to the index formula of Atiyah-Singer in (2.13), if only because it is a formula expressing the global quantity $\operatorname{Ind} D_+^X$ in local terms. This facts fits with the arguments developed in subsection 1.5 on the local versus global aspects of the Bargmann isomorphism, as well as with the considerations of Subsection 1.7. For other connections of Fourier transform with index theory, we refer to [B11c].

2.3. The Lefschetz fixed point formulas

We make the same assumptions as in Subsection 2.1. Assume that G is a compact Lie group acting on X and preserving all the data, including the metrics. Then G acts isometrically on $C^\infty(X, S^{TX} \otimes E)$, and this action commutes with D^X. In particular G acts on $\ker D^X$. If $g \in G$, let $\operatorname{Tr_s}^{\ker D^X}[g]$ denote the supertrace of the action of g. This character verifies exactly the same functorial properties as $\operatorname{Ind} D_+^X$. In this case, the McKean-Singer formula asserts that for any $t > 0$,

$$\operatorname{Tr_s}^{\ker D^X}[g] = \operatorname{Tr_s}\left[g \exp\left(-tD^{X,2}\right)\right]. \tag{2.43}$$

Needless to say, for $g = 1$, equations (2.2) and (2.43) coincide.

The Lefschetz formulas of Atiyah-Bott [ABo67, ABo68] for $\operatorname{Tr_s}^{\ker D^X}[g]$ are local formulas for $\operatorname{Tr_s}^{\ker D^X}[g]$, which are integrals of characteristic classes on the fixed point X_g of g in X. They can be obtained by making $t \to 0$ in (2.43).

Let $\mathfrak{g}$ be the Lie algebra of G. If $A \in \mathfrak{g}$, let L_A^X denote the corresponding Lie derivative operator acting on $C^\infty(X, S^{TX} \otimes E)$. Then L_A^X commutes with $D^{X,2}$. Moreover, if $g = e^A$, then

$$\operatorname{Tr_s}^{\ker D^X}[g] = \operatorname{Tr_s}\left[\exp\left(-L_A^X - tD^{X,2}\right)\right]. \tag{2.44}$$

The considerations of Subsection 2.2 remain valid for the Atiyah-Bott formulas.

The above formalism applies to the case where $D^X = d^X + d^{X*}$ or $D^X = \sqrt{2}\left(\overline{\partial}^X + \overline{\partial}^{X*}\right)$. Then $\ker D^X$ can be identified with the cohomology of the corresponding complex, and $\mathrm{Tr_s}^{\ker D^X}[g]$ is just the supertrace of the action of g on the cohomology, which we will also call a Lefschetz trace.

2.4. Other multiplicative constructions

If X is a Riemannian manifold, the Laplace-Beltrami operator Δ^X is natural with respect to products, i.e.,

$$\Delta^{X\times X'} = \Delta^X \otimes 1 + 1 \otimes \Delta^{X'}. \tag{2.45}$$

If $\sigma\left(-\Delta^X\right)$ is the principal symbol of $-\Delta^X$, then

$$\sigma\left(-\Delta^X\right) = |\xi|^2. \tag{2.46}$$

In (2.46), $|\xi|^2$ is just the square of the norm of $\xi \in T^*X$. Taking the principal symbol of (2.45) gives the Pythagorean theorem.

By (2.45), we find that for $t > 0$,

$$\exp\left(t\Delta^{X\times X'}/2\right) = \exp\left(t\Delta^X/2\right) \otimes \exp\left(t\Delta^{X'}/2\right). \tag{2.47}$$

Let LX be the smooth loop space of X, that is LX is the set of smooth maps $x : S^1 \to X$. Let $E^X(x)$ be the energy of $x \in LX$, which is given by

$$E^X(x) = \frac{1}{2}\int_{S^1} |\dot{x}|^2 \, ds. \tag{2.48}$$

Note that

$$L(X \times X') = LX \times LX'. \tag{2.49}$$

Then

$$E^{X\times X'} = p^* E^X + p'^* E^{X'}. \tag{2.50}$$

The associated Brownian motion on X defines a canonical family of measures P_t^X on the continuous loop space L^0X, which is multiplicative with respect to products, i.e.,

$$P_t^{X\times X'} = P_t^X \otimes P_t^{X'}. \tag{2.51}$$

At least formally, one can write P_t^X in the form

$$P_t^X = \exp\left(-\frac{E^X}{t}\right)\mathcal{D}x. \tag{2.52}$$

In (2.52), $\mathcal{D}x$ is a non-existing Lebesgue measure on LX, and E^X takes the value $+\infty$ on the support of P_t^X. Still, the measures P_t^X have the same universal Gaussian flavour as the other formulas we presented before. As was already explained in Subsection 1.8, Brownian motion can be thought of as the integral curve of the family of Gaussian measures m^{TX}. This point will be clarified in equation (5.34).

All these natural constructions are related. Let just give an example. For $t > 0$, let $p_t^X(x,x')$ be the smooth kernel associated with the operator $\exp\left(t\Delta^X/2\right)$. Then

$$\operatorname{Tr}\left[\exp\left(t\Delta^X/2\right)\right] = \int_X p_t^X(x,x)\,dx. \tag{2.53}$$

Now using the Itô calculus, we have the rigorous equality

$$\operatorname{Tr}\left[\exp\left(t\Delta^X/2\right)\right] = \int_X p_t^X(x,x)\,dx = \int_{L^0X} dP_t^X, \tag{2.54}$$

which, using (2.52), can be complemented by

$$\operatorname{Tr}\left[\exp\left(t\Delta^X/2\right)\right] = \int_X p_t^X(x,x)\,dx = \int_{L^0X} dP_t^X = \int_{LX} \exp\left(-\frac{E^X}{t}\right)\mathcal{D}x. \tag{2.55}$$

The objects which appear in (2.55), whether they exist or not, are multiplicative in their category.

As explained in [B11c, Section 3.5], for $t > 0$, $\operatorname{Tr_s}\left[\exp\left(-tD^{X,2}/2\right)\right]$ can be expressed as the integral of a signed measure on L^0X which has a density with respect to P_t^X. Atiyah [A85] has shown how to relate these formulas to the localization formulas of Duistermaat-Heckman [DH82, DH83]. For a systematic exposition of equivariant cohomology on loops spaces, we refer to [B11c], where other natural objects over LX are also constructed. Whether they do exist rigorously or not, they are still naturally multiplicative in their own category, the multiplicativity having the same Gaussian character as before.

In particular, using (2.23), (2.25), the form $\widehat{A}\left(F,\nabla^F\right)$ can be expressed as a Gaussian integral involving the flat Brownian motion, whose explicit evaluation is known as Lévy's stochastic area formula [Le51]. The unavoidability of this identity in connection with the index formula in (2.13), (2.37) and equivariant cohomology is explained in [B11c].

3. The hypoelliptic Laplacian in de Rham theory

The purpose of this section is to briefly describe the construction of the hypoelliptic Laplacian in de Rham theory, which was introduced in [B05].

This section is organized as follows. In Subsection 3.1, we describe the action of the hypoelliptic Laplacian on smooth functions.

In Subsection 3.2, we briefly explain how to extend this action to smooth forms of arbitrary degree.

Finally, in Subsection 3.3, we describe the functorial and functional integral aspects of the hypoelliptic Laplacian in de Rham theory.

3.1. The hypoelliptic Laplacian in degree 0

Let $\left(X, g^{TX}\right)$ be a compact Riemannian manifold of dimension n, and let Δ^X be the Laplace-Beltrami operator. Let $\square^X = \left[d^X, d^{X*}\right]$ be the Hodge Laplacian. Then $-\Delta^X$ is just the restriction of $\square^X$ to differential forms of degree 0, i.e.,

$$-\Delta^X = d^{X*}d^X|_{\Omega^0(X)}. \tag{3.1}$$

Let $\pi : \mathcal{X} \to X$ be the total space of TX. As in subsection 1.8, we denote by $\widehat{TX}$ the fibre of π, and we use the corresponding notation.

In de Rham theory [B05], the hypoelliptic Laplacian L_b^X is a second-order operator acting on smooth forms over $\mathcal{X}$, which is associated with an exotic Hodge theory on $\mathcal{X}$. As $b \to 0$, L_b^X is supposed to converge in the proper sense to $\square^X/2$, and as $b \to +\infty$, one expects L_b^X to converge to minus the Lie derivative operator associated with the geodesic flow.

Let Δ^V be the Laplacian along the fibre $\widehat{TX}$. Up to lower degree terms, L_b^X can be written in the form

$$L_b^X = \frac{1}{2b^2}\left(-\Delta^V + |Y|^2 - n\right) - \frac{1}{b}\nabla_Y^X + \cdots \tag{3.2}$$

In the right-hand side of (3.2), the first operator is a scaled version of the harmonic oscillator along the fibre, and the second operator is the generator of the geodesic flow. The analytic properties of the hypoelliptic Laplacian have been studied in detail in Bismut-Lebeau [BL08]. We refer to the surveys [B08b, B08c] for more details on the hypoelliptic Laplacian in de Rham theory.

For the moment we just consider the scalar version M_b^X of the operator L_b^X in (3.2), i.e.,

$$M_b^X = \frac{1}{2b^2}\left(-\Delta^V + |Y|^2 - n\right) - \frac{1}{b}\nabla_Y^X. \tag{3.3}$$

Using the notation in (1.44), we can write (3.3) in the form

$$M_b^X = \frac{\mathcal{H}^{TX}}{b^2} - \frac{1}{b}\nabla_Y^X. \tag{3.4}$$

Observe that in (3.2)–(3.4), the operator ∇_Y^X appears. As explained in Subsection 1.8, this is a kind of bosonic Hodge de Rham operator, which implements a version of Fourier transform. We will elaborate more on this point in Subsection 5.1.

Here we will limit ourselves to write M_b^X in a form similar to (3.1). Let ∇^{TX} be the Levi-Civita connection on TX. Then $T\mathcal{X}$ splits as

$$T\mathcal{X} = \pi^*\left(TX \oplus \widehat{TX}\right). \tag{3.5}$$

In (3.5), π^*TX is the horizontal subbundle of $T\mathcal{X}$, and $\pi^*\widehat{TX}$ is the tangent bundle to the fibre of π.

Consider the Witten twist $d_b^{\mathcal{X}}$ of the de Rham operator $d^{\mathcal{X}}$ by the function $\exp\left(-|Y|^2/2b^2\right)$, i.e.,

$$d_b^{\mathcal{X}} = d^{\mathcal{X}} + \frac{\widehat{Y}^*}{b^2}\wedge. \tag{3.6}$$

In (3.6), $\widehat{Y}^*$ is just the vertical form along the fibre associated with the tautological section $\widehat{Y}$.

The de Rham operator acting on $\Omega^0(\mathcal{X})$ can be written in the form

$$d^{\mathcal{X}}|_{\Omega^0(\mathcal{X})} = d^H + d^V. \tag{3.7}$$

In (3.7), d^H, d^V correspond to horizontal and vertical differentiation. By (3.7), we get

$$d_b^{\mathcal{X}}|_{\Omega^0(\mathcal{X})} = d^H + d^V + \frac{\widehat{Y}^*}{b^2}\wedge. \tag{3.8}$$

We equip $T\mathcal{X} = \pi^*\left(TX \oplus \widehat{TX}\right)$ with the bilinear symmetric form associated with the matrix $\mathfrak{f}$ given by

$$\mathfrak{f} = \begin{bmatrix} 1 & 1 \\ 1 & 0 \end{bmatrix}. \tag{3.9}$$

Then $\mathfrak{f}$ is a bilinear form of signature (n,n). It induces on $\mathcal{X}$ its obvious volume form $dv_{\mathcal{X}}$. Moreover, it also defines a nondegenerate symmetric bilinear pairing $(\,)$ on the $\Lambda^p(T^*\mathcal{X}), 1 \le p \le 2n$. We equip $\Omega^{\cdot,c}(\mathcal{X})$ with the nondegenerate symmetric bilinear form η given by

$$\eta(s,s') = \int_{\mathcal{X}} \left(s(x,Y), s'(x,-Y)\right) dv_{\mathcal{X}}. \tag{3.10}$$

Let $d_b^{\mathcal{X}*}$ be the formal adjoint of $d_b^{\mathcal{X}}$ with respect to η. Let $e_1,\dots,e_n$ be an orthonormal basis of TX, which we identify with the corresponding basis π^*TX in the right-hand side of (3.5). Let $\widehat{e}_1,\dots,\widehat{e}_n$ be the corresponding basis of $\pi^*\widehat{TX}$. The associated dual bases are denoted with upper indices. Let $\nabla^{\Lambda^\cdot(T^*X)\widehat{\otimes}\Lambda^\cdot\left(\widehat{T^*X}\right)}$ be the connection on $\Lambda^\cdot(T^*X)\widehat{\otimes}\Lambda^\cdot\left(\widehat{T^*X}\right)$ which is induced by ∇^{TX}. Finally, if $\widehat{e} \in \widehat{TX}$, let $\nabla^V_{\widehat{e}}$ denote the corresponding derivative operator along the fibre $\widehat{TX}$. By (3.8), one verifies easily that

$$d_b^{\mathcal{X}*}|_{\Omega^1(X)} = -i_{\widehat{e}_i}\nabla_{e_i}^{\Lambda^\cdot(T^*X)\widehat{\otimes}\Lambda^\cdot\left(\widehat{T^*X}\right)} + i_{e_i-\widehat{e}_i}\nabla^V_{\widehat{e}_i} - i_{Y-\widehat{Y}}/b^2. \tag{3.11}$$

By (3.7), (3.11), we get

$$\frac{1}{2}d_b^{\mathcal{X}*}d_b^{\mathcal{X}}|_{\Omega^0(\mathcal{X})} = \frac{1}{2}\left(-\Delta^V + \frac{1}{b^4}|Y|^2 - \frac{n}{b^2}\right) - \frac{1}{b^2}\nabla_Y^X. \tag{3.12}$$

Let K_b be the map $s(x,Y) \to s(x,bY)$. By (3.12), we get

$$M_b^X = K_b\frac{1}{2}d_b^{\mathcal{X}*}d_b^{\mathcal{X}}K_b^{-1}. \tag{3.13}$$

3.2. The hypoelliptic de Rham Laplacian in arbitrary degree

Equation (3.13) makes clear that $-2M_b^X$ is an exotic Laplace-Beltrami operator on $\mathcal{X}$. But it also suggests how to construct L_b^X. Indeed, (3.13) should extend to

$$L_b^X = K_b \frac{1}{2}\left[d_b^{\mathcal{X}}, d_b^{\mathcal{X}*}\right] K_b^{-1}. \tag{3.14}$$

This is precisely one of the definitions of L_b^X which is taken in [B05, Section 2.7], [BL08, Section 2.1], and [B08c, Section 3.4]. A formula for L_b^X can be written explicitly, which coincides with (3.3) in degree 0. In higher degree, the right-hand side contains a third term where the curvature R^{TX} of ∇^{TX} appears explicitly. This complicates the formula for $d_b^{\mathcal{X}*}$ and for L_b^X. For a detailed discussion, we refer to [B05].

3.3. Functorial aspects of the hypoelliptic Laplacian

First note that M_b^X verifies the same naturality properties as the Laplacian Δ^X in (2.45), i.e.,

$$M_b^{X\times X'} = M_b^X \otimes 1 + 1 \otimes M_b^{X'}. \tag{3.15}$$

Of course the two components of M_b^X in (3.3) verify similar functoriality properties. This is in particular the case for the vector field ∇_Y^X which defines the geodesic flow.

For the functional integral aspects of the hypoelliptic Laplacian, and its connection with Chern-Gauss-Bonnet, we refer to [B06, B08b]. Let us just give the analogue of equation (2.55). By the results of [BL08, Chapter 3], for $t > 0$, the operator $\exp\left(-tM_b^X\right)$ is trace class, and there is an associated smooth kernel $p_{b,t}^X(z, z')$. Then

$$\operatorname{Tr}\left[\exp\left(-tM_b^X\right)\right] = \int_{\mathcal{X}} p_{b,t}^X(z, z)\, dv_{\mathcal{X}}. \tag{3.16}$$

For $b \geq 0, t > 0, x \in LX$, set

$$H_{b,t}^X(x) = \frac{1}{2t}\int_{S^1} |\dot{x}|^2 ds + \frac{b^4}{2t^3}\int_{S^1} |\ddot{x}|^2 ds = \frac{1}{2t}\int_{S^1}\left|\dot{x} + b^2\frac{\ddot{x}}{t}\right|^2 ds. \tag{3.17}$$

Note that

$$H_{0,t}^X(x) = \frac{E^X(x)}{t}. \tag{3.18}$$

For $b > 0, t > 0$, let $P_{b,t}^X$ be the measure on LX which is given formally by

$$P_{b,t}^X = \exp\left(-H_{b,t}^X(x)\right)\mathcal{D}x. \tag{3.19}$$

The measure $P_{b,t}^X$ exists, and is carried by L^1X, the set of C^1 maps $s \in S^1 \to X$. The analogue of (2.55) just says that

$$\begin{aligned}\operatorname{Tr}\left[\exp\left(-tM_b^X\right)\right] &= \int_{\mathcal{X}} p_{b,t}^X((x,Y),(x,Y))\, dv_{\mathcal{X}} \\ &= \int_{LX} dP_{b,t}^X = \int_{LX}\exp\left(-H_{b,t}^X\right)\mathcal{D}x.\end{aligned} \tag{3.20}$$

In (3.20), only the last expression is formal. In Subsection 5.2, we will come back to the interpretation of (3.20).

By (3.18), as $b \to 0$, the quantities in (3.20) should converge to the ones in (2.55). Equivalently, as $b \to 0$, one should have

$$\operatorname{Tr}\left[\exp\left(-tM_b^X\right)\right] \to \operatorname{Tr}\left[\exp\left(t\Delta^X/2\right)\right]. \tag{3.21}$$

The convergence result in (3.21) is proved in [BL08, Section 3.4]. This is an example of the mysterious predictive power of the functional integral, even if the mathematics behind the rigorous proof is much deeper than what the formal arguments suggest. For a probabilistic proof of this result, we also refer to [B11d, Section 12.8].

4. The hypoelliptic Dirac operator for complex manifolds

If X is a complex Kähler manifold, the purpose of this section is to explain the construction of the hypoelliptic deformation of the Dirac operator $D^X = \sqrt{2}\left(\overline{\partial}^X + \overline{\partial}^{X*}\right)$ which is given in [B08a, Section 3]. Roughly speaking, the starting point of the construction is a modified version of the Levi-Civita superconnection $A^{S^{TX}\otimes E}$ (whose square is the Getzler operator in (2.23)). This superconnection is modified for reasons which are explained in some detail. By combining this modified superconnection with the Koszul complex along the fibre, we obtain a family of hypoelliptic operators $\mathcal{A}_b$, which deforms the Dirac operator $D^X/\sqrt{2}$. Also $\mathcal{A}_b^2$ is a hypoelliptic Laplacian, in the sense that it has the same structure as the operators L_b^X, M_b^X in (3.2), (3.3).

This section is organized as follows. In Subsection 4.1, we describe a natural superconnection associated with a holomorphic Hermitian vector bundle.

In Subsection 4.2, we construct the elliptic Dirac operator D^X.

In Subsection 4.3, for $b > 0$, we obtain the hypoelliptic Dirac operator $\mathcal{A}_b$.

In Subsection 4.4, we give formal arguments showing that $\mathcal{A}_b$ is a deformation of $D^X/\sqrt{2}$.

In Subsection 4.5, we relate the operator $\mathcal{A}_b$ to the Levi-Civita superconnection of TX.

Finally, in Subsection 4.6, we construct a suitable deformation of the Levi-Civita superconnection of a vector bundle. This construction produces a microlocal analogue of the hypoelliptic Dirac operator.

For a survey of other properties of the hypoelliptic Dirac operator, we refer to [B08c, Section 4].

4.1. A superconnection associated with a vector bundle

Let X be a compact complex manifold. Let TX be the holomorphic tangent bundle to X, let $T_{\mathbf{C}}X = TX \oplus \overline{TX}$ be the complexification of the real tangent bundle $T_{\mathbf{R}}X$.

Let $\left(F, g^F\right)$ be a holomorphic Hermitian vector bundle of dimension n on X, let ∇^F be the holomorphic Hermitian connection on F, and let R^F be its curvature.

Let $\pi : \mathcal{F} \to X$ be the total space of F. Then $\mathcal{F}$ is a complex manifold. Let $i : X \to \mathcal{F}$ be the embedding of X as the zero section of F. Using the connection ∇^F, we have the identification of smooth vector bundles,

$$T\mathcal{F} \simeq \pi^* \left(TX \oplus F\right). \tag{4.1}$$

From (4.1), we get the smooth identification

$$\Lambda^{\cdot} \left(\overline{T^*\mathcal{F}}\right) = \pi^* \left(\Lambda^{\cdot} \left(\overline{T^*X}\right) \widehat{\otimes} \Lambda^{\cdot} \left(\overline{F}^*\right)\right). \tag{4.2}$$

Let $\left(\Omega^{(0,\cdot)}\left(\mathcal{F}\right), \overline{\partial}^{\mathcal{F}}\right)$ be the Dolbeault complex of smooth antiholomorphic forms on $\mathcal{F}$. Let $\mathbf{I}$ be the vector bundle on X of the smooth sections of $\pi^*\Lambda^{\cdot}\left(\overline{F}^*\right)$ along the fibre F. It is well known that the operator $\overline{\partial}^{\mathcal{F}}$ splits as

$$\overline{\partial}^{\mathcal{F}} = \nabla^{\mathbf{I}\prime\prime} + \overline{\partial}^V. \tag{4.3}$$

In (4.3), $\overline{\partial}^V$ is the Dolbeault operator along the fibre F, and $\nabla^{\mathbf{I}\prime\prime}$ is the horizontal part of $\overline{\partial}^{\mathcal{F}}$. By (4.3), $\overline{\partial}^{\mathcal{F}}$ can also be viewed as a holomorphic superconnection A'' on $\mathbf{I}$.

The adjoint superconnection A' of A'' is given by

$$A' = \nabla^{\mathbf{I}\prime} + \overline{\partial}^{V*}. \tag{4.4}$$

In (4.4), $\nabla^{\mathbf{I}\prime}$ is the holomorphic part of the horizontal connection $\nabla^{\mathbf{I}}$ on $\mathbf{I}$, and $\overline{\partial}^{V*}$ is the fibrewise L_2-adjoint of $\overline{\partial}^V$ with respect to the metric g^F.

Set

$$A = A'' + A'. \tag{4.5}$$

By (4.3)–(4.5), we get

$$A = \nabla^{\mathbf{I}} + \overline{\partial}^V + \overline{\partial}^{V*}. \tag{4.6}$$

Then A is a standard superconnection on $\mathbf{I}$.

Still A is not the Levi-Civita superconnection on $\mathbf{I}$ in the sense of (2.22). We will briefly describe this last superconnection in more detail.

We denote by y the tautological section of π^*F on $\mathcal{F}$, by Y the tautological section of $\pi^*F_{\mathbf{R}}$, so that $Y = y + \overline{y}$, and $|Y|^2_{g^F} = 2\,|y|^2_{g^F}$. Set

$$\omega^{\mathcal{F}} = i\overline{\partial}^{\mathcal{F}} \partial^{\mathcal{F}} |y|^2_{g^F}. \tag{4.7}$$

Let $\omega^{\mathcal{F},V}$ be the Kähler form along the fibres, i.e., $\omega^{\mathcal{F},V}$ is the restriction of $\omega^{\mathcal{F}}$ to the fibres F. Put

$$\omega^{\mathcal{F},H} = i\left\langle R^F y, \overline{y}\right\rangle_{g^F}. \tag{4.8}$$

Then a simple computation shows that

$$\omega^{\mathcal{F}} = \omega^{\mathcal{F},V} + \omega^{\mathcal{F},H}. \tag{4.9}$$

Equation (4.9) gives the splitting of $\omega^{\mathcal{F}}$ into its vertical and horizontal parts. The same equation indicates that $\pi : \mathcal{F} \to X$ is a Kähler fibration in the sense of [BGS88], i.e., the horizontal vector bundle $T^H\mathcal{F}$ is just the orthogonal bundle to the vertical vector bundle F with respect to $\omega^{\mathcal{F}}$.

Definition 4.1. Set

$$\begin{aligned} B'' &= A'', \qquad B' = e^{i\omega^{\mathcal{F},H}} A' \exp\left(-i\omega^{\mathcal{F},H}\right), \\ C'' &= \exp\left(-i\omega^{\mathcal{F},H}/2\right) A'' \exp\left(i\omega^{\mathcal{F},H}/2\right), \\ C' &= \exp\left(i\omega^{\mathcal{F},H}/2\right) A' \exp\left(-i\omega^{\mathcal{F},H}/2\right), \\ B &= B'' + B', \qquad C = C'' + C'. \end{aligned} \tag{4.10}$$

Then B, C are superconnections on $\mathbf{I}$, and moreover,

$$C = \exp\left(-i\omega^{\mathcal{F},H}/2\right) B \exp\left(i\omega^{\mathcal{F},H}/2\right). \tag{4.11}$$

By (4.8), (4.10), we get

$$\begin{aligned} B &= \nabla^{\mathbf{I}} + \overline{\partial}^V + \overline{\partial}^{V*} + i_{\overline{R^F y}}, \\ C &= \nabla^{\mathbf{I}} + \overline{\partial}^V + \overline{\partial}^{V*} - \frac{1}{2}\left(R^F y^* \wedge - i_{\overline{R^F y}}\right). \end{aligned} \tag{4.12}$$

Let $c(F_{\mathbf{R}})$ be the Clifford algebra of $\left(F_{\mathbf{R}}, g^F\right)$. Then $c(F_{\mathbf{R}})$ acts naturally on $\Lambda^{\cdot}\left(\overline{F}^*\right)$. Set

$$D^F = \sqrt{2}\left(\overline{\partial}^V + \overline{\partial}^{V*}\right). \tag{4.13}$$

Then D^F is a standard Dirac operator along the fibre F.

We can rewrite the second identity in (4.12) in the form

$$C = \nabla^{\mathbf{I}} + \frac{D^F}{\sqrt{2}} - \frac{c\left(R^F Y\right)}{2\sqrt{2}}. \tag{4.14}$$

Comparing with (2.22), we find that C is indeed the Levi-Civita superconnection associated with the fibration $\pi : \mathcal{F} \to X$.

Still, as we shall see, C is not the right object to consider in the construction of the hypoelliptic Dirac operator.

4.2. The elliptic Dirac operator

Let E be a complex holomorphic vector bundle on X. Let $\left(\Omega^{(0,\cdot)}(X,E), \overline{\partial}^X\right)$ be the Dolbeault complex of smooth antiholomorphic forms with coefficients in E.

Let g^{TX}, g^E be Hermitian metrics on TX, E. Let $\overline{\partial}^{X*}$ denote the formal adjoint of $\overline{\partial}^X$ with respect to the obvious Hermitian product on $\Omega^{(0,\cdot)}(X,E)$. Set

$$D^X = \sqrt{2}\left(\overline{\partial}^X + \overline{\partial}^{X*}\right). \tag{4.15}$$

If the metric g^{TX} is Kähler, it is well known that D^X is a standard Dirac operator.

4.3. The hypoelliptic Dirac operator

Here we follow [B08a, Section 3]. Let $\widehat{TX}$ be another copy of TX, and let $\mathcal{X}$ be the total space of $\widehat{TX}$. Let $g^{\widehat{TX}}$ be a Hermitian metric on $\widehat{TX}$, let $\nabla^{\widehat{TX}}$ be the corresponding holomorphic Hermitian connection, and let $R^{\widehat{TX}}$ be its curvature. We now use the notation of Subsection 4.1, with $\left(\widehat{TX}, g^{\widehat{TX}}\right) = \left(F, g^F\right)$.

Note that $\Lambda^{\cdot}(T^*X)$ is the holomorphic part of the exterior algebra of X. Recall that y is the tautological section of $\pi^*\widehat{TX}$. Since $\widehat{TX}$ and TX are isomorphic, the operator i_y acts on $\pi^*\Lambda^{\cdot}(T^*X)$. The Koszul complex $(\mathcal{O}_{\mathcal{X}}\pi^*\Lambda^{\cdot}(T^*X), i_y)$ provides a resolution of the sheaf $i_*\mathcal{O}_X$.

Set

$$\mathcal{E} = \pi^*\left(\Lambda^{\cdot}(T^*X) \otimes E\right), \qquad \mathscr{E} = \Lambda^{\cdot}(T^*_{\mathbf{C}}X)\widehat{\otimes}\Lambda^{\cdot}\left(\overline{\widehat{T^*X}}\right) \otimes E. \tag{4.16}$$

In the sequel, our operators will act on $\Omega^{(0,\cdot)}(\mathcal{X}, \pi^*\mathcal{E}) = C^\infty(\mathcal{X}, \pi^*\mathscr{E})$.

Recall that the superconnection $A'' = \overline{\partial}^{\mathcal{X}}$ was defined in Subsection 4.1. For $b > 0$, set

$$A''_b = A'' + i_y/b^2. \tag{4.17}$$

By (4.3), (4.17), we get

$$A''_b = \nabla^{\mathbf{I}''} + \overline{\partial}^V + i_y/b^2. \tag{4.18}$$

Since y is a holomorphic section of π^*TX, we get

$$A''^2_b = 0. \tag{4.19}$$

We will take the 'adjoint' of A''_b, partly in the sense of superconnections. The operator $i_{\overline{y}}$ acts on $\Lambda^{\cdot}\left(\overline{T^*X}\right)$, and so it acts on $\mathscr{E}$. Set

$$A'_b = A' + i_{\overline{y}}/b^2. \tag{4.20}$$

Then A'_b also acts on $\Omega^{(0,\cdot)}(\mathcal{X}, \pi^*\mathcal{E})$. Moreover,

$$A'^2_b = 0. \tag{4.21}$$

Set

$$A_b = A''_b + A'_b. \tag{4.22}$$

When identifying $Y \in \widehat{T_{\mathbf{R}}X}$ to the corresponding section of $T_{\mathbf{R}}X$, we get

$$A_b = A + i_Y/b^2. \tag{4.23}$$

Note that A_b is no longer a superconnection on $\mathbf{I}$, but a genuine operator acting on $\Omega^{(0,\cdot)}(\mathcal{X}, \pi^*\mathcal{E})$.

One verifies easily that A_b^2 is hypoelliptic, because the anticommutator of $\nabla^{\mathbf{I}}$ with i_Y/b^2 produces the critical term $\nabla^{\mathbf{I}}_Y/b^2$. This term is obviously related to Cartan's formula in (1.12). If the metric induced by $g^{\widehat{TX}}$ on TX is Kähler, this is the generator of the geodesic flow on $\mathcal{X}$.

Still A_b is not the right hypoelliptic Dirac operator, because A_b^2 does not contain a potential which is quadratic in Y. Let g^{TX} be a Hermitian metric on

TX, unrelated to $g^{\widehat{TX}}$, and let ω^X be the Kähler form on X associated with the metric g^{TX}. If J is the complex structure of $T_{\mathbf{R}}X$, if $U, V \in T_{\mathbf{R}}X$, then

$$\omega^X (U, V) = \langle U, JV \rangle_{g^{TX}} . \tag{4.24}$$

We will view ω^X as a section of $\Lambda^{\cdot} (T^*_{\mathbf{C}}X)$.

Put

$$\mathcal{A}''_b = A''_b, \qquad \mathcal{A}'_b = e^{i\omega^X} A'_b e^{-i\omega^X}, \qquad \mathcal{A}_b = \mathcal{A}''_b + \mathcal{A}'_b. \tag{4.25}$$

Assume that the metric g^{TX} is Kähler, i.e., ω^X is closed. Let $\overline{y}^* \in T^*X$ be dual to $\overline{y} \in \overline{TX}$ with respect to the metric g^{TX}. Then

$$\mathcal{A}'_b = A'_b + \overline{y}^* \wedge /b^2. \tag{4.26}$$

By (4.23), (4.25), and (4.26), we get

$$\mathcal{A}_b = A_b + \overline{y}^* \wedge /b^2. \tag{4.27}$$

Let $\Delta^V_{g^{\widehat{TX}}}$ be the Laplacian along the fibre $\widehat{TX}$ with respect to $g^{\widehat{TX}}$. Let ∇^V denote differentiation along the fibre $\widehat{TX}$. One then verifies easily that

$$\mathcal{A}^2_b = \frac{1}{2}\left(-\Delta^V_{g^{\widehat{TX}}} + \frac{1}{b^4} |Y|^2_{g^{TX}} \right) + \frac{1}{b^2}\nabla^{\mathbf{I}}_Y - \nabla^V_{R^{\widehat{TX}}Y} + \cdots , \tag{4.28}$$

where $\ldots$ denotes a differential operator of order 0. The effect of the addition of $\overline{y}^* \wedge /b^2$ in (4.26) is precisely to produce the desired $|Y|^2_{g^{TX}} /2b^4$ in $\mathcal{A}^2_b$.

When g^{TX} is Kähler and $g^{\widehat{TX}} = g^{TX}$, the operator $\mathcal{A}^2_b$ is precisely the hypoelliptic Dirac operator constructed in [B08a, Section 3]. It has the preferred structure of a hypoelliptic Laplacian, i.e., its principal part is a scaled sum of a harmonic oscillator and of the generator of the geodesic flow. When $g^{\widehat{TX}}$ is unrelated to g^{TX} and g^{TX} is Kähler, this operator was constructed in [B08a, Section 10]. When no assumption is made on g^{TX} and $g^{\widehat{TX}}$, this construction is developed in [B11b, Section 7].

4.4. The operator $\mathcal{A}_b$ as a deformation of $D^X/\sqrt{2}$

We still follow [B08a, Section 3]. We assume that g^{TX} is Kähler, and that $g^{\widehat{TX}} = g^{TX}$. Let $w_1, \ldots, w_n$ be an orthonormal basis of TX. Let $\widehat{w}_1, \ldots, \widehat{w}_n$ be the corresponding basis of $\widehat{TX}$. Put

$$\Lambda = \sqrt{-1} i_{\overline{w}_i} i_{w_i}. \tag{4.29}$$

Set

$$\mathcal{B}_b = \exp(i\Lambda)\, \mathcal{A}_b \exp(-i\Lambda). \tag{4.30}$$

For $b > 0$, we define K_b as in (2.30). Put

$$\mathcal{C}_b = K_b \mathcal{B}_b K_b^{-1}. \tag{4.31}$$

Let $\nabla^{\mathscr{E}}$ be the unitary connection on $\mathscr{E}$ which is induced by the connections $\nabla^{TX}, \nabla^{\widehat{TX}}, \nabla^E$. Let K be the operator acting on $C^\infty(\mathcal{X}, \mathscr{E})$,

$$K = \left(\overline{w}^i \wedge + i_{w_i}\right) \nabla^{\mathscr{E}}_{\overline{w}_i} + \left(w^i \wedge - i_{\overline{w}_i}\right) \nabla^{\mathscr{E}}_{w_i}. \tag{4.32}$$

In [B08a, Eq. (3.68)], it is shown that

$$\mathcal{C}_b = K + \frac{1}{b}\left(\overline{\partial}^V + i_y + \overline{\partial}^{V*} + \overline{y}^* \wedge\right). \tag{4.33}$$

Set

$$\widehat{\omega}^{\mathcal{X},V} = -\sqrt{-1} w^i \wedge \widehat{\overline{w}}^i. \tag{4.34}$$

By results obtained in [B90, Proposition 1.5 and Theorem 1.6], the kernel of $\overline{\partial}^V + i_y + \overline{\partial}^{V*} + \overline{y}^* \wedge$ acting on the Schwartz space $\mathcal{S}^{(0,\cdot)}\left(\widehat{TX}, \pi^* \Lambda^\cdot (T^*X)\right)$ is one-dimensional and spanned by $\beta = \exp\left(i\widehat{\omega}^{\mathcal{X},V} - \frac{|Y|^2_{g^{\widehat{TX}}}}{2}\right)$. Let P be the orthogonal projection operator on this kernel. We embed $\Omega^{(0,\cdot)}(X,E)$ into $C^\infty(\mathcal{X}, \pi^*\mathscr{E})$ by the embedding $\alpha \to \pi^*\alpha \wedge \beta$. This embedding is isometric for a suitable normalization of the L_2 Hermitian product.

The following result was established in [B08a, Theorem 3.12].

Theorem 4.2. *The following identity holds:*

$$PKP = \overline{\partial}^X + \overline{\partial}^{X*}. \tag{4.35}$$

Proof. Using (4.32), we get easily

$$PKP = \overline{w}^i \nabla^{\Lambda^\cdot(\overline{T^*X}) \otimes E}_{\overline{w}_i} - i_{\overline{w}_i} \nabla^{\Lambda^\cdot(\overline{T^*X}) \otimes E}_{w_i}, \tag{4.36}$$

which is just (4.35). □

The decomposition (4.33) and the identity (4.35) are the key algebraic facts which ultimately explain that as $b \to 0$, $\mathcal{A}_b$ converges in the proper sense to $D^X/\sqrt{2}$. For more details, we refer to [B08a], and also to [BL08, Chapter 3], where a similar problem is dealt with involving de Rham cohomology.

In [B11a, B11b], the above results have been extended to the case where ω^X is not closed. A motivation for doing this is that certain Riemann-Roch-Grothendieck theorems in Bott-Chern cohomology [D09] cannot be proved in the elliptic world, because of the absence of the relevant local cancellations.

4.5. The operator $\mathcal{A}_b$ and the Levi-Civita superconnection of $\widehat{TX}$

If we make $b = +\infty$ in (4.23), (4.27), we get

$$\mathcal{A}_\infty = A. \tag{4.37}$$

Comparing (4.6) and (4.12) shows that $\mathcal{A}_\infty$ does not coincide with the Levi-Civita superconnection C, or with its conjugate B. Let us briefly explain why it would be wrong to try forcing the use of the Levi-Civita superconnection in this context.

Indeed instead of (4.25), set

$$\begin{aligned} \mathscr{A}''_b &= A''_b, \\ \mathscr{A}'_b &= \exp\left(i\omega^X + i\omega^{\mathcal{X},H}\right) A'_b \exp\left(-i\left(\omega^X + \omega^{\mathcal{X},H}\right)\right), \\ \mathscr{A}_b &= \mathscr{A}''_b + \mathscr{A}'_b. \end{aligned} \tag{4.38}$$

By (4.25), (4.38), we get

$$\mathscr{A}'_b = \exp\left(i\omega^{\mathcal{X},H}\right) \mathcal{A}'_b \exp\left(-i\omega^{\mathcal{X},H}\right). \tag{4.39}$$

By (4.8), (4.20), (4.26), and (4.39), we obtain

$$\mathscr{A}'_b = \mathcal{A}'_b + i_{\widehat{R^{TX}y}} + \left\langle R^{TX}\left(\overline{y},\cdot\right) y, \overline{y}\right\rangle_{g^{TX}} / b^2. \tag{4.40}$$

By (4.18), (4.38), and (4.40), in $\mathscr{A}_b'^2$, there is a quartic term

$$\frac{1}{b^4}\left\langle R^{TX}\left(\overline{y}, y\right) y, \overline{y}\right\rangle_{g^{TX}}. \tag{4.41}$$

This term destroys the fibrewise harmonic oscillator character one expects from a hypoelliptic Laplacian. If (4.41) is negative, the operator $\mathscr{A}_b'^2$ is hopeless anyway. This is why the hypoelliptic Dirac operator is defined to be $\mathcal{A}_b$ and not $\mathscr{A}_b$.

Incidentally, in the fibre direction, quartic nonnegative potentials in the variable Y have also their role in the theory of the hypoelliptic Laplacian. In [B11d], the hypoelliptic Laplacian which is used in the evaluation of orbital integrals contains such a quartic term. This is also the case for the version of the hypoelliptic Laplacian which appears in [B11b], a quadratic potential being there more a nuisance than an asset. Such quartic terms are needed in [B11b] in spite of the fact they violate the Pythagorean principle of Section 2.

4.6. The hypoelliptic deformation of the local index theoretic data

As shown in [B08a, Sections 2 and 3], it is also possible to deform the superconnections attached to a family of complex manifolds whose curvature is elliptic to superconnections whose curvature is fibrewise hypoelliptic. This we will only explain in the context of vector bundles.

Let us consider the Levi-Civita superconnection C in (4.12) which is attached to the Hermitian vector bundle $\left(F, g^F\right)$. Its curvature C^2 is an elliptic operator along the fibre F. Let $e_1, \ldots, e_{2n}$ be an orthonormal basis of $F_{\mathbf{R}}$ with respect to g^F. By (2.23),

$$C^2 = -\frac{1}{2}\sum_{i=1}^{2n}\left(\nabla_{e_i} + \frac{1}{2}\left\langle R^F Y, e_i\right\rangle\right)^2 + \frac{1}{2}\operatorname{Tr}\left[R^F\right]. \tag{4.42}$$

The presence of the term $\frac{1}{2}\operatorname{Tr}\left[R^F\right]$ just reflects the fact that $S^{F_{\mathbf{R}}}$ is replaced by $\Lambda^{\cdot}\left(\overline{F}^*\right)$.

Let $\widehat{F}$ denote the tangent bundle to the fibre F, and let $\widehat{\mathcal{F}}$ be the total space of this tangent bundle. Equivalently $\widehat{\mathcal{F}}$ is the total space $F \oplus \widehat{F}$, where $\widehat{F}$ is another copy of F. Let σ be the projection $\widehat{\mathcal{F}} \to X$, and let $p : \widehat{\mathcal{F}} \to \mathcal{F}$ be the obvious projection with fibre $\widehat{F}$. Let y, z be the tautological sections of $\sigma^* F, \sigma^* \widehat{F}$ on $\widehat{\mathcal{F}}$.

Let $g^{\widehat{F}}$ be a Hermitian metric on $\widehat{F}$. Let $\nabla^{\widehat{F}}$ be the holomorphic Hermitian connection on $\widehat{F}$, and let $R^{\widehat{F}}$ be its curvature. We will view $\sigma^* \widehat{F}$ as a holomorphic vector bundle on $\mathcal{F}$, to which we will apply the constructions of Subsection 4.1.

Let $\mathbf{J}^{\cdot}$ be the vector bundle on $\mathcal{F}$ of the smooth sections of $\Lambda^{\cdot}\left(\widehat{\overline{F}}^{*}\right)$ along the fibre $\widehat{F}$. Let $\mathfrak{A}'', \mathfrak{A}', \mathfrak{A}$ be the analogues of A'', A', A. Let $\overline{\partial}^{\widehat{V}}$ be the $\overline{\partial}$ operator along $\widehat{F}$, and let $\overline{\partial}^{\widehat{V}*}$ be its fibrewise adjoint with respect to $g^{\widehat{F}}$. Then

$$\begin{aligned}
\mathfrak{A}'' &= \nabla^{\mathbf{J}^{\cdot}\prime\prime} + \overline{\partial}^{V} + i_{\overline{R^F y}} + \overline{\partial}^{\widehat{V}}, \\
\mathfrak{A}' &= \nabla^{\mathbf{J}^{\cdot}\prime} + \overline{\partial}^{V} + i_{R^{TX} y} + \overline{\partial}^{\widehat{V}*}, \\
\mathfrak{A} &= \mathfrak{A}'' + \mathfrak{A}'.
\end{aligned} \tag{4.43}$$

The first three terms in the right-hand side of the first two equations in (4.43) are the contributions of $\overline{\partial}^{\mathcal{F}}, \partial^{\mathcal{F}}$. Also $\mathfrak{A}^2$ is fibrewise elliptic along $\widehat{F}$.

Let $\mathbf{K}^{\cdot}$ be the vector bundle on X of smooth sections along the fibre $F \oplus \widehat{F}$ of $\sigma^{*}\left(\Lambda^{\cdot}\left(F^{*}_{\mathbf{C}}\right)\widehat{\otimes}\Lambda^{\cdot}\left(\widehat{\overline{F}}^{*}\right)\right)$. We will now imitate (4.17), (4.20), and (4.25). Let $\overline{z}^{*} \in F^{*}$ be dual to $\overline{z} \in \overline{F}$ with respect to g^{F}. Because of the identification $\widehat{F} \simeq F$, the operators $i_{z}, i_{\overline{z}}, \overline{z}^{*}\wedge$ act on $\Lambda^{\cdot}\left(F^{*}_{\mathbf{C}}\right)$. Recall that the form $\omega^{\mathcal{F}}$ on $\mathcal{F}$ was defined in (4.7). Set

$$\begin{aligned}
\mathfrak{A}''_{b} &= \mathfrak{A}'' + i_{z}/b^{2}, \\
\mathfrak{A}'_{b} &= e^{i\omega^{\mathcal{F}}}\left(\mathfrak{A}' + i_{\overline{z}}/b^{2}\right)e^{-i\omega^{\mathcal{F}}}, \\
\mathfrak{A}_{b} &= \mathfrak{A}''_{b} + \mathfrak{A}'_{b}.
\end{aligned} \tag{4.44}$$

By proceeding as in Subsection 4.4, one can show that the operator $\mathfrak{A}_{b}$ is a deformation of the Levi-Civita superconnection C for $\left(F, g^{F}\right)$.

When $F = TX$, this deformation is a microlocal version of the deformation of $D^{X}/\sqrt{2}$ to $\mathcal{A}_{b}$. It deforms local index theoretic data for $D^{X}/\sqrt{2}$ to what turns out to be local index theoretic data for $\mathcal{A}_{b}$. More precisely as shown in [B08a, Theorem 6.12], when ω^{X} is closed and $g^{\widehat{TX}} = g^{TX}$, the superconnection $\mathfrak{A}_{b}$ associated with $\left(TX, g^{TX}, g^{\widehat{TX}}\right)$ appears when computing the limit as $t \to 0$ of $\mathrm{Tr_s}\left[\exp\left(-t\mathcal{A}^{2}_{b\sqrt{t}}\right)\right]$.

5. Orbital integrals and the hypoelliptic heat kernel

The purpose of this section is to survey some aspects of the application of the hypoelliptic Laplacian to the evaluation of semisimple orbital integrals for reductive groups [B11d]. The orbital integrals are the main ingredient which appears in the geometric side of Selberg's trace formula.

One of the main points made in [B11d] is that the evaluation of the trace of a heat kernel can be made formally similar to the evaluation of a Lefschetz formula. We will prove this in the trivial case where the symmetric space is $\mathbf{R}$. Also we describe in some detail the uniform estimates on the hypoelliptic heat kernel which were established in [B11d]. Finally, we explain the link made in [B11d] between the heat equation for the hypoelliptic Laplacian and the wave equation, as a version of quantization of the Hamiltonian-Lagrangian correspondence in the classical calculus of variations.

This section is organized as follows. In Subsection 5.1, we explain the role of the hypoelliptic Laplacian in the evaluation of the orbital integrals on $\mathbf{R}$.

In Subsection 5.2, we describe some results on the hypoelliptic heat kernel in the case where the base manifold is a symmetric space of noncompact type, and the role of corresponding action functionals.

Finally, in Subsection 5.3, we explain the wave like character of the hypoelliptic Laplacian.

5.1. Orbital integrals: the case of the real line

Here, we adopt temporarily the formalism of Sections 2 and 3 with $X = \mathbf{R}$. The fact that X is noncompact will be irrelevant. Note that $\mathcal{X} = \mathbf{R} \times \mathbf{R}$. We will denote by (x, y) the generic element in $\mathcal{X}$. By (3.3), we get

$$\Delta^{\mathbf{R}} = \frac{\partial^2}{\partial x^2}, \qquad M_b^{\mathbf{R}} = \frac{1}{2b^2}\left(-\frac{\partial^2}{\partial y^2} + y^2 - 1\right) - \frac{1}{b} y \frac{\partial}{\partial x}. \tag{5.1}$$

The heat kernel $p_t^{\mathbf{R}}(x, x')$ associated with $\exp\left(t\Delta^{\mathbf{R}}/2\right)$ is given by

$$p_t^{\mathbf{R}}(x, x') = \frac{1}{\sqrt{2\pi t}} \exp\left(-\frac{(x' - x)^2}{2t}\right). \tag{5.2}$$

The heat kernel $p_{b,t}^{\mathbf{R}}((x, y), (x', y'))$ associated with $\exp\left(-tM_b^{\mathbf{R}}\right)$ depends only on $x' - x, y, y'$. It has been computed explicitly in [B11d, Proposition 10.5.1].

Take $a \in \mathbf{R}$. In the sequel we use the notation

$$\begin{aligned} \operatorname{Tr}^a\left[\exp\left(t\Delta^{\mathbf{R}}/2\right)\right] &= p_t^{\mathbf{R}}(0, a), \\ \operatorname{Tr}^a\left[\exp\left(-tM_b^{\mathbf{R}}\right)\right] &= \int_{\mathbf{R}} p_{b,t}^{\mathbf{R}}((0, y), (a, y))\, dy. \end{aligned} \tag{5.3}$$

The following result was established in [B11d, Eq. (10.6.13)].

Proposition 5.1. *The following identity holds:*

$$\operatorname{Tr}^a\left[\exp\left(-tM_b^{\mathbf{R}}\right)\right] = \left(1 - e^{-t/b^2}\right)^{-1} \operatorname{Tr}^a\left[\exp\left(t\Delta^{\mathbf{R}}/2\right)\right]. \tag{5.4}$$

Proof. We give a formal proof which can be ultimately easily justified. Note that

$$M_b^{\mathbf{R}} = \frac{1}{2b^2}\left(-\frac{\partial^2}{\partial y^2} + \left(y - b\frac{\partial}{\partial x}\right)^2 - 1\right) - \frac{1}{2}\frac{\partial^2}{\partial x^2}. \tag{5.5}$$

By (5.5), we get

$$\exp\left(b\frac{\partial^2}{\partial x \partial y}\right) M_b^{\mathbf{R}} \exp\left(-b\frac{\partial^2}{\partial x \partial y}\right) = \frac{1}{2b^2}\left(-\frac{\partial^2}{\partial y^2} + y^2 - 1\right) - \frac{1}{2}\frac{\partial^2}{\partial x^2}. \tag{5.6}$$

Using the fact that $p_{b,t}^{\mathbf{R}}((x,Y),(x',Y'))$ only depends on $x'-x,y,y'$, we deduce from (5.6) that

$$\operatorname{Tr}^{a}\left[\exp\left(-tM_{b}^{\mathbf{R}}\right)\right]=\operatorname{Tr}\left[\exp\left(-\frac{t}{b^{2}}\mathcal{H}^{\mathbf{R}}\right)\right]\operatorname{Tr}^{a}\left[\exp\left(t\Delta^{\mathbf{R}}/2\right)\right]. \tag{5.7}$$

Since the spectrum of the harmonic oscillator $\mathcal{H}^{\mathbf{R}}$ in (1.44) is just $\mathbf{N}$, by (5.7), we get (5.4). □

Remark 5.2. As explained in [B11d, Subsection 10.5], equation (5.6) should be thought of as a version of Egorov's theorem, the main difference being that the conjugating operator $\exp\left(b\frac{\partial^2}{\partial x\partial y}\right)$ is not well defined, because the operator $\frac{\partial^2}{\partial x\partial y}$ is hyperbolic. As was shown in [B11d, Chapter 10], while Egorov's theorem is related to the quantization of symplectic transformations, here the conjugating operator $\exp\left(b\frac{\partial^2}{\partial x\partial y}\right)$ quantizes formally an imaginary symplectic transformation.

The above has nothing to do with the idea of a Wick rotation. In the Wick rotation, a well-defined Schrödinger operator is changed into a well-defined heat operator, by complexifying the time variable. Here one could consider that if b was changed into ib, the conjugating operator would be well defined, and would quantize a real symplectic transformation. However, the operator $M_{ib}^{\mathbf{R}}$ is no longer hypoelliptic, and its conjugate by the conjugation in (5.6) is now hyperbolic! The power of equation (5.6) is that because it does not make literal sense, and still its consequences are true, it can lead to exotic results, which cannot be anticipated by traditional physical considerations.

The role of the operator $y\frac{\partial}{\partial x}$ in $M_b^{\mathbf{R}}$ is to implement a version of Fourier transform. Indeed, we proceed by analogy, by simply giving the main steps in the computation of the Fourier transform of the Gaussian distribution, while introducing also the parameter $b>0$. We get

$$\begin{aligned}\int_{\mathbf{R}}&\exp\left(iy\xi/b-y^2/2b^2\right)\frac{dy}{b\sqrt{2\pi}}\\&=\int_{\mathbf{R}}\exp\left(-\left(y-ib\xi\right)^2/2b^2-\xi^2/2\right)\frac{dy}{b\sqrt{2\pi}}\\&=\int_{\mathbf{R}}\exp\left(-y^2/2b^2-\xi^2/2\right)\frac{dy}{b\sqrt{2\pi}}=\exp\left(-\xi^2/2\right).\end{aligned} \tag{5.8}$$

The analyticity of the function $\exp\left(-y^2/2\right)$ plays a key role in (5.8). The various identities in (5.8) should be compared with (5.5), (5.6). One should also remember that $\exp\left(-y^2/2\right)$ is the ground state of the harmonic oscillator $\mathcal{H}^{\mathbf{R}}$.

Now we introduce the exterior algebra $\Lambda^{\cdot}(\mathbf{R}^*)$ and its corresponding number operator $N^{\Lambda^{\cdot}(\mathbf{R}^*)}$. Set

$$N_b^{\mathbf{R}}=M_b^{\mathbf{R}}+\frac{N^{\Lambda^{\cdot}(\mathbf{R}^*)}}{b^2}. \tag{5.9}$$

Let $q_{b,t}^{\mathbf{R}}\left((x,y),(x',y')\right)$ be the smooth kernel associated with $\exp\left(-tN_b^{\mathbf{R}}\right)$. Then

$$q_{b,t}^{\mathbf{R}}\left((x,y),(x',y')\right) = p_{b,t}^{\mathbf{R}}\left((x,y),(x',y')\right)\exp\left(-tN^{\Lambda^{\cdot}(\mathbf{R}^*)}/b^2\right). \tag{5.10}$$

We define $\mathrm{Tr_s}^a\left[\exp\left(-tN_b^{\mathbf{R}}\right)\right]$ by a formula similar to (5.3).

Theorem 5.3. *For any $a \in \mathbf{R}, b > 0, t > 0$,*

$$\mathrm{Tr}^a\left[\exp\left(t\Delta^{\mathbf{R}}/2\right)\right] = \mathrm{Tr_s}^a\left[\exp\left(-tN_b^{\mathbf{R}}\right)\right]. \tag{5.11}$$

Proof. This is an obvious consequence of Proposition 5.1. □

Since (5.11) is valid for any $a \in \mathbf{R}$, we will rewrite (5.11) in the form

$$\exp\left(t\Delta^{\mathbf{R}}/2\right) = \mathrm{Tr_s}\left[\exp\left(-tN_b^{\mathbf{R}}\right)\right]. \tag{5.12}$$

Equation (5.12) is an equality of operators acting on smooth functions on $\mathbf{R}$. For a detailed analysis of (5.11), (5.12), we refer to [B11d, Chapter 10].

The first striking fact about equation (5.12) is that the right-hand side does not depend on $b > 0$. The second fact is that when $b \to 0$, the right-hand side converges obviously to the left-hand side. Indeed note that

$$N_b^{\mathbf{R}} = \frac{1}{b^2}\left(\mathcal{H}^{\mathbf{R}} + N^{\Lambda^{\cdot}(\mathbf{R}^*)}\right) - \frac{1}{b}y\frac{\partial}{\partial x}. \tag{5.13}$$

As we saw in Subsection 1.6, the kernel of $\mathcal{H}^{\mathbf{R}} + N^{\Lambda^{\cdot}(\mathbf{R}^*)}$ is one-dimensional and spanned by $\pi^{-1/4}\exp\left(-y^2/2\right)$. Also multiplication by y maps this kernel in its L_2 orthogonal. It follows that when splitting the L_2 space of $\mathbf{R}^2$ into the kernel of $\mathcal{H}^{\mathbf{R}} + N^{\Lambda^{\cdot}(\mathbf{R}^*)}$ and its orthogonal, the operator $N_b^{\mathbf{R}}$ has the preferred matrix structure

$$N_b^{\mathbf{R}} = \begin{bmatrix} 0 & -\frac{y}{b}\frac{\partial}{\partial x} \\ -\frac{y}{b}\frac{\partial}{\partial x} & \frac{\mathcal{H}}{b^2} - \frac{1}{b}y\frac{\partial}{\partial x} \end{bmatrix}. \tag{5.14}$$

Now we deal with this matrix as if it were a finite-dimensional matrix. As explained in [B05, Theorem 3.14], [BL08, Section 17.1], and [B08c, Section 3.7], if we consider (5.14) as a finite-dimensional matrix, one finds that if P is orthogonal projection operator on $\ker\left(\mathcal{H}^{\mathbf{R}} + N^{\Lambda^{\cdot}(\mathbf{R}^*)}\right)$, then

$$\exp\left(-tN_b^{\mathbf{R}}\right) \to P\exp\left(t\Delta^{\mathbf{R}}/2\right)P. \tag{5.15}$$

In this case, equation (5.15) can be easily justified by explicit computations.

We define K_b as in equation (2.30). Note that

$$K_b N_b^{\mathbf{R}} K_b^{-1} = \frac{y^2}{2} - y\frac{\partial}{\partial x} + \frac{1}{b^2}\left(-\frac{1}{2b^2}\frac{\partial^2}{\partial y^2} - 1 + N^{\Lambda^{\cdot}(\mathbf{R}^*)}\right). \tag{5.16}$$

By (5.16), as $b \to +\infty$, the heat kernel for the operator in (5.16) propagates more and more along the geodesic flow generated by the vector field $y\frac{\partial}{\partial x}$. As explained in [B05, Section 3.10], from (5.12), one can obtain this way the explicit formula for $p_t^{\mathbf{R}}(x,x')$ in (5.2). In other words, the right-hand side of (5.12) interpolates between the operator $\exp\left(t\Delta^{\mathbf{R}}/2\right)$ for $b \to 0$, and its local expression as a heat kernel for $b \to +\infty$.

Let us now replace $\mathbf{R}$ by S^1. The same argument as in (5.6) shows that

$$\exp\left(b\frac{\partial^2}{\partial x\partial y}\right) N_b^{S^1} \exp\left(-b\frac{\partial^2}{\partial x\partial y}\right) = \frac{1}{b^2}\left(\mathcal{H}^{\mathbf{R}} + N^{\Lambda^{\cdot}(\mathbf{R}^*)}\right) - \frac{1}{2}\Delta^{S^1}. \qquad (5.17)$$

As explained in [B08c, Section 1.2], from (5.17), one can deduce that the operator $N_b^{S^1}$ is isospectral to the operator in the right-hand side of (5.17). This can be done first by using Fourier transform in the variable x, and also by noting that the eigenfunctions of $\mathcal{H}^{\mathbf{R}}$ are analytic in the variable y. Ultimately the operator $N_b^{S^1}$ can be explicitly diagonalized, and its eigenfunctions form a complete set in the corresponding L_2 space.

The consequence is that

$$\operatorname{Sp} N_b^{S^1} = \left\{2\pi^2 k^2\right\}_{k\in\mathbf{Z}} + \frac{\mathbf{N}}{b^2}. \qquad (5.18)$$

The remarkable feature of (5.18) is that $\left\{2\pi^2 k^2\right\}_{k\in\mathbf{Z}}$ remains fixed in $\operatorname{Sp} N_b^{S^1}$ while $\frac{\mathbf{N}}{b^2}$ moves around, essentially disappearing as $b \to 0$, except for 0, and accumulating near 0 as $b \to +\infty$.

From the above, we find that instead of (5.11), we now have

$$\operatorname{Tr}\left[\exp\left(t\Delta^{S^1}/2\right)\right] = \operatorname{Tr_s}\left[\exp\left(-tN_b^{S^1}\right)\right]. \qquad (5.19)$$

Equation (5.19) is an equality of genuine traces or supertraces. Of course (5.19) is weaker than (5.11). If we make $b \to 0$ in (5.19), the right-hand side becomes the left-hand side. Making $b \to +\infty$ localizes the heat kernel along the closed geodesics in S^1, and ultimately expresses the trace of the heat kernel in terms of a Poisson summation formula.

We will now compare the considerations of Subsection 2.3 with the ones we just made. Indeed we can think of $g = \exp\left(t\Delta^{S^1}/2\right)$ as an element of a group. Let $L_2^{S_1}$ be the L_2 space of S^1. Tautologically,

$$\operatorname{Tr}\left[\exp\left(t\Delta^{S^1}/2\right)\right] = \operatorname{Tr}^{L_2^{S^1}}[g]. \qquad (5.20)$$

Now think of $L_2^{S^1}$ as the cohomology of some complex, so that (5.20) is just the computation of a Lefschetz number. Formula (5.19) begs to be compared with the McKean-Singer formula in (2.44) for a Lefschetz number. The fact that the right-hand side does not depend on $b > 0$ is just the analogue of the right-hand side of (2.44) not depending on $t > 0$. When comparing with (2.44), it is natural to take $t = 1/b^2$, so that making $b \to 0$ is like making $t \to +\infty$ in (2.44), and making $b \to +\infty$ is like making $t \to 0$ in (2.44). Making $b \to 0$ localizes $\operatorname{Tr_s}\left[\exp\left(-tN_b^{S^1}\right)\right]$ on $\ker \mathcal{H}^{\mathbf{R}} \simeq L_2^{S^1}$. We saw that making $b \to +\infty$ localizes the right-hand side of (5.19) near the closed geodesics, which can be thought of as the fixed points of the return map for the geodesic flow in $S^1 \times \mathbf{R}$.

From this point of view, $L_2^{S_1}$ should be identified with $\ker D_+^X$, and $t\Delta^{S^1}/2$ with $-L_A^X$. The spectral formula in (5.18) should be viewed as the analogue of the computation of the spectrum of $\mathrm{Sp}\left(L_A^X + tD^{X,2}\right)$ which also depends on $t > 0$, but is such that the supertrace of the associated heat kernel does not depend on $t > 0$. A minor discrepancy is that in (2.44), L_A^X contributes to the spectrum by imaginary eigenvalues, while the contribution of $-t\Delta^{S^1}/2$ is real. This is because $g = \exp\left(t\Delta^{S^1}/2\right)$ is not unitary but self-adjoint.

Ultimately, we hope to have shown that equation (5.19) has all the features of a McKean-Singer formula for a Lefschetz number, except that we have not produced the associated Dirac operator. Let us just do that.

Let d_x, d_y be the de Rham operators acting on $C^\infty\left(\mathbf{R}, \Lambda^\cdot\left(\mathbf{R}^*\right)\right)$, the first operator acting on the variable $x \in \mathbf{R}$, the second on the variable $y \in \mathbf{R}$, the exterior algebra being the same algebra in both cases. Let d_x^*, d_y^* be the formal adjoints of d_x, d_y. The operators $y\wedge, i_y$ also act on $\Lambda^\cdot\left(\mathbf{R}^*\right)$.

Set

$$\widehat{D}^{\mathbf{R}} = d_x - d_x^*. \tag{5.21}$$

Then

$$\widehat{D}^{\mathbf{R},2} = \Delta^{\mathbf{R}}. \tag{5.22}$$

For $b > 0$, put

$$\mathfrak{D}_b^{\mathbf{R}} = \frac{1}{\sqrt{2}}\left(-\widehat{D}^{\mathbf{R}} + \frac{1}{b}\left(d_y + y\wedge + d_y^* + i_y\right)\right). \tag{5.23}$$

An easy computation shows that

$$N_b^{\mathbf{R}} = \mathfrak{D}_b^{\mathbf{R},2} - \frac{1}{2}\Delta^{\mathbf{R}}. \tag{5.24}$$

By (5.23),

$$\left[\mathfrak{D}_b^{\mathbf{R}}, \Delta^{\mathbf{R}}\right] = 0, \qquad \left[\mathfrak{D}_b^{\mathbf{R}}, N_b^{\mathbf{R}}\right] = 0. \tag{5.25}$$

Then equation (5.12) can be written in the form

$$\exp\left(t\Delta^{\mathbf{R}}/2\right) = \mathrm{Tr_s}\left[\exp\left(\frac{t}{2}\Delta^{\mathbf{R}} - t\mathfrak{D}_b^{\mathbf{R},2}\right)\right]. \tag{5.26}$$

Similarly, if we replace $\mathbf{R}$ by S_1, we can still define the operator $\mathfrak{D}_b^{S_1}$ by a formula like (5.23). Equation (5.19) is just the identity

$$\mathrm{Tr}\left[\exp\left(t\Delta^{S_1}/2\right)\right] = \mathrm{Tr_s}\left[\exp\left(\frac{t}{2}\Delta^{S_1} - t\mathfrak{D}_b^{S_1,2}\right)\right]. \tag{5.27}$$

A direct proof of the fact that the right-hand sides of (5.12), (5.13) do not depend on $b > 0$ can be given, which is based on (5.26), (5.27), and on classical arguments of index theory. In the case of $\mathbf{R}$, we get the chain of equalities,

$$\begin{aligned}
&\frac{\partial}{\partial b}\mathrm{Tr_s}\left[\exp\left(\frac{t}{2}\Delta^{\mathbf{R}}-t\mathfrak{D}_b^{\mathbf{R},2}\right)\right]\\
&=-t\mathrm{Tr_s}\left[\left[\mathfrak{D}_b^{\mathbf{R}},\frac{\partial}{\partial b}\mathfrak{D}_b^{\mathbf{R}}\right]\exp\left(\frac{t}{2}\Delta^{\mathbf{R}}-t\mathfrak{D}_b^{\mathbf{R},2}\right)\right]\\
&=-t\mathrm{Tr_s}\left[\left[\mathfrak{D}_b^{\mathbf{R}},\frac{\partial}{\partial b}\mathfrak{D}_b^{\mathbf{R}}\exp\left(\frac{t}{2}\Delta^{\mathbf{R}}-t\mathfrak{D}_b^{\mathbf{R},2}\right)\right]\right]=0.
\end{aligned}\tag{5.28}$$

In (5.28), we have used the first identity in (5.25), and also the fact that supertraces vanish on supercommutators [Q85].

In [B11d], we have shown that if G is a reductive group, if K is a maximal compact subgroup, if $X=G/K$ is the corresponding symmetric space, there is an analogue of Theorem 5.3, which becomes an equality of semisimple orbital integrals. It is obtained by extending the construction of the operator $\mathfrak{D}_b^{\mathbf{R}}$, with $\widehat{D}^{\mathbf{R}}$ replaced by the Dirac operator of Kostant [K97] which is associated with G. Making $b\to+\infty$ then leads to an explicit local evaluation of these orbital integrals. The analogue of (5.19) produces a version of the Selberg trace formula. We refer to [B11d] for more details.

5.2. The hypoelliptic heat kernel and the action functional

We use here the notation of Subsections 2.4 and 3.3. Also n denotes the dimension of X.

For $t>0, b\geq 0$, if $s\in[0,t]\to x_s\in X$ be a smooth path, set

$$\begin{aligned}
E_t^X(x)&=\frac{1}{2}\int_0^t|\dot{x}|^2\,ds,\\
H_{b,t}^X(x)&=\frac{1}{2}\int_0^t\left(|\dot{x}|^2+b^4|\ddot{x}|^2\right)ds,\\
K_{b,t}^X(x)&=\frac{1}{2}\int_0^t\left|b^2\ddot{x}+\dot{x}\right|^2ds.
\end{aligned}\tag{5.29}$$

Then

$$H_{0,t}^X=K_{0,t}^X(x)=E_t^X,\qquad K_{b,t}(x)=H_{b,t}(x)+\frac{b^2}{2}\left(|\dot{x}_t|^2-|\dot{x}_0|^2\right).\tag{5.30}$$

Note that the map $x_\cdot\to x_{t\cdot}$ maps paths parametrized by $[0,t]$ to paths parametrized by $[0,1]$.

For $t>0$, let L_tX be the set of smooth maps $s\in\mathbf{R}/t\mathbf{Z}\to X$. For $t=1$, L_tX is just the space LX defined in Subsection 2.4. Moreover, there is a canonical identification $L_tX\simeq LX$ via the map $x_\cdot\in L_tX\to x_{t\cdot}\in LX$. Via this map, E_t^X corresponds to E^X/t, and $H_{b,t}^X$ is precisely the functional in (3.17). If $x,x'\in X$, we denote by $L_{t,x,x'}X$ the set of $x_\cdot\in L_tX$ such that $x_0=x, x_t=x'$.

Assume that X is compact. Let d be the distance function on X. Recall that for $t>0$, $p_t^X(x,x')$ is the smooth kernel associated with the operator $\exp\left(t\Delta^X/2\right)$.

For $M > 0$, for $t \in]0, M]$, we have the uniform estimate,

$$p_t^X(x, x') \leq \frac{C_M}{t^{n/2}} \exp\left(-d^2(x, x')/2t\right), \tag{5.31}$$

The theory of Brownian motion shows that there is a canonical positive measure $P^X_{t,x,x'}$ on $L^0_{t,x,x'}X$, which is the obvious set of continuous paths, such that

$$p_t^X(x, x') = \int_{L^0_{t,x,x'}X} dP^X_{t,x,x'}, \tag{5.32}$$

Also $P^X_{t,x,x'}$ has the formal representation

$$P^X_{t,x,x'} = \exp\left(-E_t^X\right) \mathcal{D}x. \tag{5.33}$$

The content of (5.33) is that the Brownian motion $x.$ verifies the stochastic differential equation

$$\dot{x} = \dot{w}, \tag{5.34}$$

where $w.$ is a standard Brownian motion in TX which is transported along $x.$ using the Levi-Civita connection. Equation (2.55) can be viewed as a consequence of (5.32), (5.33). Also equation (5.31) can be proved to be a consequence of (5.32).

Given $b > 0$, $z = (x, Y), z' = (x', Y') \in \mathcal{X}$, let $L_{t,z,z'}X$ be the set of smooth maps $s \in [0, t] \to x_s \in X$ such that $(x_0, \dot{x}_0) = z$, $(x_t, \dot{x}_t) = z'$. Recall that $p^X_{b,t}(z, z')$ is the smooth kernel on $\mathcal{X}$ which is associated with $\exp\left(-tM_b^X\right)$. Let $L^1_{t,z,z'}$ be the obvious C^1 loop space in X. There is a canonical positive measure $P^X_{b,t,z,z'}$ on $L^1_{t,z,z'}X$ such that

$$b^n p^X_{b,t}((x, bY), (x', bY')) = \int_{L^1_{t,z,z'}X} dP^X_{b,t,z,z'}. \tag{5.35}$$

Also we have the formal representation

$$P^X_{b,t,z,z'} = \exp\left(-H^X_{b,t}\right) \mathcal{D}x. \tag{5.36}$$

Set

$$M_b^{X\prime} = \exp\left(|Y|^2/2\right) M_b^X \exp\left(-|Y|^2/2\right). \tag{5.37}$$

Then

$$M_b^{X\prime} = \frac{1}{2b^2}\left(-\Delta^V + 2\nabla^V_{\widehat{Y}}\right) - \frac{1}{b}\nabla^X_Y. \tag{5.38}$$

Let $p^{X\prime}_{b,t}(z, z')$ be the smooth kernel associated with $\exp\left(-tM^{X\prime}_{b,t}\right)$. Then

$$p^{X\prime}_{b,t}(z, z') = \exp\left(|Y|^2/2\right) p^X_{b,t}(z, z') \exp\left(-|Y'|^2/2\right). \tag{5.39}$$

There is a canonical positive measure $P^{X\prime}_{b,t,z,z'}$ on $L^0_{t,z,z'}X$ such that

$$b^n p^{X\prime}_{b,t}((x, bY), (x', bY')) = \int_{L^0_{t,z,z'}X} dP^{X\prime}_{b,t,z,z'}. \tag{5.40}$$

Also the measure $P^{X\prime}_{b,t,z,z'}$ can be formally represented in the form

$$P^{X\prime}_{b,t,z,z'} = \exp\left(-K^{X}_{b,t}\right)\mathcal{D}x. \tag{5.41}$$

By (5.30), (5.36), and (5.41), we obtain

$$P^{X\prime}_{b,t,z,z'} = \exp\left(\frac{b^2}{2}\left(|Y|^2 - |Y'|^2\right)\right) P^{X}_{b,t,z,z'}. \tag{5.42}$$

Equation (5.39) is compatible with (5.35), (5.40), and (5.42).

If $z_\cdot = (x_\cdot, Y_\cdot)$ is the stochastic process associated with the heat semigroup $\exp\left(-tM^{X\prime}_b\right)$, an elementary application of the Itô calculus shows that it verifies the stochastic differential equation

$$\dot{x} = \frac{Y}{b}, \qquad \dot{Y} = -\frac{Y}{b^2} + \frac{\dot{w}}{b}. \tag{5.43}$$

By (5.43), we get

$$b^2\ddot{x} = -\dot{x} + \dot{w}. \tag{5.44}$$

The formal content of (5.40), (5.41) is just (5.44). Also (3.20) is a consequence of (5.35), (5.36).

If we make $b = 0$ in (5.44), we recover (5.34). We already warned the reader against believing that making $b \to 0$ in (5.44) is innocuous.

Given $z, z' \in \mathcal{X}$, let $H^X_{b,t}(z, z')$ be the minimum of $H^X_{b,t}(x)$ over the paths $x_\cdot \in L_{t,z,z'}X$. Using large deviations, $H^X_{b,t}((x, Y/b), (x', Y'/b))$ should play for $p^X_{b,t}(z, z')$ the same role as $d^2/2$ for $p^X_t(x, x')$. This question has been studied by Lebeau [L05] on a compact manifold X for fixed $b > 0$.

An elementary observation is that the function $H^X_{b,t}$ is not symmetric. More precisely,

$$H^X_{b,t}((x', Y'), (x, Y)) = H^X_{b,t}((x, -Y), (x', -Y')). \tag{5.45}$$

Moreover, the function $H^X_{b,t}$ does not vanish on the diagonal. This makes that contrary to the function $d^2/2$, the function $H^X_{b,t}$ is very far from defining a distance over $\mathcal{X}$. It is because it is as far as possible from a distance on $\mathcal{X}$ that it will be of any use, geometrically and analytically. This makes the operator M^X_b fundamentally different from more classical hypoelliptic operators like the hypoelliptic Laplacian on the Heisenberg group. On such a space, the action functional still defines a Carnot-Carathéodory distance.

If X is a locally symmetric space of non compact type, for a given $t > 0$, we have the global estimate,

$$p^X_t(x, x') \leq c\exp\left(-Cd^2(x, x')\right). \tag{5.46}$$

One way of proving (5.31) is to use finite propagation speed for the wave equation on X. Also equations (5.32)–(5.34) are still valid.

When X is a symmetric space, estimates similar to (5.46) have been established in [B11d, Chapter 14] for the hypoelliptic heat kernel $p^X_{b,t}$. Let us briefly explain these estimates.

First of all, assume temporarily that $X = \mathbf{R}^n$. An elementary result established in [B11d, Eq. (10.3.48)] says that as $b \to 0$,

$$\begin{aligned} H_{b,t}^{\mathbf{R}^n}\left(\left(x, Y/b\right), \left(x', Y'/b\right)\right) \to \underline{H}_{0,t}^{X}\left(\left(x, Y\right), \left(x', Y'\right)\right) \\ = \frac{1}{2t}\left|x'-x\right|^2 + \frac{1}{2}\left(\left|Y\right|^2 + \left|Y'\right|^2\right). \end{aligned} \tag{5.47}$$

The proof of (5.47) is completely elementary. In spite of the fact that as $b \to 0$, $H_{b,t}^{\mathbf{R}^n} \to E_t^{\mathbf{R}^n}$, still the corresponding minimum values do not converge, the defect being $\frac{1}{2}\left(\left|Y\right|^2 + \left|Y'\right|^2\right)$. A simple result established in [B11d, Proposition 10.3.3] is that given $0 < \epsilon < M < +\infty$, there exists $C > 0$ such that for $0 \leq b \leq M, \epsilon \leq t \leq M$,

$$H_{b,t}^{\mathbf{R}^n}\left(\left(x, Y/b\right), \left(x', Y'/b\right)\right) \geq C\left(\left|x'-x\right|^2 + \left|Y\right|^2 + \left|Y'\right|^2\right). \tag{5.48}$$

Equations (5.47), (5.48) show how far the function $H_{b,t}^{\mathbf{R}^n}\left(\left(x, Y/b\right), \left(x', Y'/b\right)\right)$ is from a distance on $\mathcal{X}$. With the exception of the defect $\frac{1}{2}\left(\left|Y\right|^2 + \left|Y'\right|^2\right)$, as $b \to 0$, it collapses to the normalized square of the distance on X.

No attempt has been made in [B11d] to establish similar results on the functional $H_{b,t}^X$ when X is instead a Riemannian manifold. However, it has been shown in [B11d, Theorem 13.2.4] that when X is a symmetric space of noncompact type, given $M > 0$, there exist $C > 0, C' > 0$ such that for $b \in]0, M], \epsilon \leq t \leq M$,

$$p_{b,t}^X\left(\left(x, Y\right), \left(x', Y'\right)\right) \leq C \exp\left(-C'\left(d^2\left(x, x'\right) + \left|Y\right|^2 + \left|Y'\right|^2\right)\right), \tag{5.49}$$

Proving (5.49) for a given $b > 0$ is rather easy. The real challenge is to prove uniformity in $b \in]0, M]$.

In the case where X is a compact manifold, it was shown in Bismut-Lebeau [BL08, Section 3.4] that as $b \to 0$,

$$\begin{aligned} p_{b,t}^X\left(\left(x, Y\right), \left(x', Y'\right)\right) \to p_{0,t}^X\left(\left(x, Y\right), \left(x', Y'\right)\right) \\ = p_t^X\left(x, x'\right)\pi^{-n/2}\exp\left(-\frac{1}{2}\left(\left|Y\right|^2 + \left|Y'\right|^2\right)\right). \end{aligned} \tag{5.50}$$

When X is a symmetric space, this result was established in [B11d, Theorem 12.8.1]. Observe that in the last exponential, what we called the defect in the right-hand side of (5.47) appears. This is no accident.

The proof of the above results is difficult. In the case where X is compact, the proof of (5.50) in [BL08] uses pseudodifferential operators. When X is a symmetric space, the arguments in [B11d] are mostly probabilistic. Finite propagation speed arguments cannot be used on $M_{b,t}^X$, because this operator does not have a wave equation. The proof in [B11d] uses instead (5.47), (5.48), the Malliavin calculus [M78], and also the wave-like character of the hypoelliptic Laplacian, which we will review next.

5.3. The hypoelliptic Laplacian and the wave equation

Let us briefly consider the parabolic heat operator

$$P = \frac{\partial}{\partial t} - \Delta^X, \tag{5.51}$$

and also, for $b > 0$, its hyperbolic deformation

$$P_b = b^2 \frac{\partial^2}{\partial t^2} + \frac{\partial}{\partial t} - \Delta^X. \tag{5.52}$$

In the naive sense, P_b deforms P. Also the operator P_b has finite propagation speed $1/b$. As $b \to 0$, the propagation speed tends to $+\infty$, which is compatible with the fact that the heat equation has infinite propagation speed.

Consider the stochastic differential equations in (5.34), (5.44). Let us rewrite (5.44) by a simple change of notation

$$\left(b^2 \frac{D^2}{Dt^2} + \frac{D}{Dt}\right) x - \dot{w} = 0. \tag{5.53}$$

In (5.53), $\frac{D}{Dt}$ denotes covariant differentiation. The similarity between equation (5.52) for the operator P_b and the dynamical equation (5.53) is impossible to miss. What the formal similarity betrays is a hidden wave-like character in the operator $M_b^{X\prime}$.

Indeed observe that contrary to Brownian motion in (5.34), the solution of the second-order differential equation (5.44), (5.53) has finite speed. More precisely $\dot{x}_\cdot$ is a continuous process. This means that on a given time interval, $x_\cdot$ cannot escape too much at infinity. However, this argument is misleading, since while Brownian motion in (5.53) has infinite speed, it does not escape at infinity either. The wave-like character of the hypoelliptic Laplacian is subtler than that.

Let us give an operator theoretic version of (5.53). Let $f : X \to \mathbf{R}$ be a smooth function. Then f lifts to a smooth function $\mathcal{X} \to \mathbf{R}$. One verifies easily that

$$\left(b^2 M_b^{X\prime,2} - M_b^{X\prime}\right) f = \nabla_Y^{TX} \nabla_Y f. \tag{5.54}$$

For $t > 0$, put

$$S_t = \exp\left(-t M_b^{X\prime}\right). \tag{5.55}$$

If $f(x)$ is a smooth bounded function on X, by (5.54), we get

$$\left(b^2 \frac{\partial^2}{\partial t^2} + \frac{\partial}{\partial t}\right) S_t f = S_t \nabla_Y^{TX} \nabla_Y f. \tag{5.56}$$

Equation (5.56) is just the analytic counterpart to (5.53). Set

$$U_t\left((x,Y), x'\right) = \int_{T_{x'}X} S_t\left((x,Y),(x',Y')\right) dY'. \tag{5.57}$$

By (5.56), as a function of x', U_t verifies a nonautonomous wave equation on X.

These considerations are developed in [B11d, Chapter 12]. They play an important role in establishing the estimate (5.49).

Also observe the first-order Hamiltonian differential equation for the geodesic flow on $\mathcal{X}$,

$$\dot{x} = Y, \qquad \dot{Y} = 0, \tag{5.58}$$

projects on X to the second-order differential equation

$$\ddot{x} = 0. \tag{5.59}$$

That the parabolic equation associated with the heat flow for $M_b^{X\prime}$ projects approximately to a wave-like equation on X should be viewed as a sort of quantization of the Hamiltonian-Lagrangian correspondence. The fact that for $b \to 0$, $M_b^{X\prime}$ collapses in the proper way to the classical $-\frac{1}{2}\Delta^X$, which is the genuine quantization of the geodesic flow, makes this intermediate quantization especially relevant.

Index

References

[A85] M.F. Atiyah, *Circular symmetry and stationary-phase approximation*, Astérisque (1985), no. 131, 43–59, Colloquium in honor of Laurent Schwartz, Vol. 1 (Palaiseau, 1983). MR **87h:**58206

[ABo67] M.F. Atiyah and R. Bott, *A Lefschetz fixed point formula for elliptic complexes. I*, Ann. of Math. (2) **86** (1967), 374–407. MR 35 #3701

[ABo68] ———, *A Lefschetz fixed point formula for elliptic complexes. II. Applications*, Ann. of Math. (2) **88** (1968), 451–491. MR 38 #731

[ABoP73] M.F. Atiyah, R. Bott, and V.K. Patodi, *On the heat equation and the index theorem*, Invent. Math. **19** (1973), 279–330. MR 58 #31287

[AS68] M.F. Atiyah and I.M. Singer, *The index of elliptic operators. I*, Ann. of Math. (2) **87** (1968), 484–530. MR 38 #5243

[AS71] ———, *The index of elliptic operators. IV*, Ann. of Math. (2) **93** (1971), 119–138. MR 43 #5554

[B86] J.-M. Bismut, *The Atiyah-Singer index theorem for families of Dirac operators: two heat equation proofs*, Invent. Math. **83** (1986), no. 1, 91–151. MR **87g:**58117

[B90] ———, *Koszul complexes, harmonic oscillators, and the Todd class*, J. Amer. Math. Soc. **3** (1990), no. 1, 159–256, With an appendix by the author and C. Soulé. MR **91b:**58245

[B94] ———, *Equivariant short exact sequences of vector bundles and their analytic torsion forms*, Compositio Math. **93** (1994), no. 3, 291–354. MR **96g:**58201

[B05] ———, *The hypoelliptic Laplacian on the cotangent bundle*, J. Amer. Math. Soc. **18** (2005), no. 2, 379–476 (electronic). MR MR2137981

[B06] ———, *The hypoelliptic Laplacian and Chern-Gauss-Bonnet*, Differential geometry and physics, Nankai Tracts Math., vol. 10, World Sci. Publ., Hackensack, NJ, 2006, pp. 38–52. MR **MR2322387 (2008e:**58045)

[B08a] ———, *The hypoelliptic Dirac operator*, Geometry and dynamics of groups and spaces, Progr. Math., vol. 265, Birkhäuser, Basel, 2008, pp. 113–246. MR MR2402405

[B08b] ———, *Loop spaces and the hypoelliptic Laplacian*, Comm. Pure Appl. Math. **61** (2008), no. 4, 559–593. MR MR2383933

[B08c] ———, *A survey of the hypoelliptic Laplacian*, Astérisque (2008), no. 322, 39–69, Géométrie différentielle, physique mathématique, mathématiques et société. II. MR MR2521653

[B11a] ———, *Laplacien hypoelliptique et cohomologie de Bott-Chern*, C. R. Acad. Sci. Paris Sér. I Math. **349** (2011), 75–80.

[B11b] ———, *Hypoelliptic Laplacian and Bott-Chern cohomology*, preprint (Orsay) (2011).

[B11c] ———, *Duistermaat-Heckman formulas and index theory*, Geometric Aspects of Analysis and Mechanics (Johan A.C. Kolk and Erik P. van den Ban, eds.), Progr. Math., vol. 292, Birkhäuser/Springer, New York, 2011, pp. 1–55. MR 2809466

[B11d] ———, *Hypoelliptic Laplacian and orbital integrals*, Annals of Mathematics Studies, vol. 177, Princeton University Press, Princeton, NJ, 2011. MR 2828080

[BGS88] J.-M. Bismut, H. Gillet, and C. Soulé, *Analytic torsion and holomorphic determinant bundles. II. Direct images and Bott-Chern forms*, Comm. Math. Phys. **115** (1988), no. 1, 79–126. MR **89g:**58192b

[BG00] J.-M. Bismut and S. Goette, *Holomorphic equivariant analytic torsions*, Geom. Funct. Anal. **10** (2000), no. 6, 1289–1422. MR 1 810 746

[BG01] ———, *Families torsion and Morse functions*, Astérisque (2001), no. 275, x+293. MR **2002h:**58059

[BG04] ———, *Equivariant de Rham torsions*, Ann. of Math. (2) **159** (2004), no. 1, 53–216. MR **2051391** **(2005f:**58059)

[BL91] J.-M. Bismut and G. Lebeau, *Complex immersions and Quillen metrics*, Inst. Hautes Études Sci. Publ. Math. (1991), no. 74, ii+298 pp. (1992). MR **94a:**58205

[BL08] ———, *The hypoelliptic Laplacian and Ray-Singer metrics*, Annals of Mathematics Studies, vol. 167, Princeton University Press, Princeton, NJ, 2008. MR MR2441523

[BZ92] J.-M. Bismut and W. Zhang, *An extension of a theorem by Cheeger and Müller*, Astérisque (1992), no. 205, 235, With an appendix by François Laudenbach. MR **93j:**58138

[D09] J.-P. Demailly, *Complex analytic and differential geometry*, OpenContent Book, http://www-fourier.ujf-grenoble.fr/ demailly/manuscripts/agbook.pdf, 2009.

[DH82] J.J. Duistermaat and G.J. Heckman, *On the variation in the cohomology of the symplectic form of the reduced phase space*, Invent. Math. **69** (1982), no. 2, 259–268. MR **84h:**58051a

[DH83] ______, *Addendum to: "On the variation in the cohomology of the symplectic form of the reduced phase space"*, Invent. Math. **72** (1983), no. 1, 153–158. MR **84h:**58051b

[Ge86] E. Getzler, *A short proof of the local Atiyah-Singer index theorem*, Topology **25** (1986), no. 1, 111–117. MR **87h:**58207

[Gi74] P.B. Gilkey, *The index theorem and the heat equation*, Publish or Perish Inc., Boston, Mass., 1974, Notes by Jon Sacks, Mathematics Lecture Series, No. 4. MR MR0458504 (56 #16704)

[K97] B. Kostant. Clifford algebra analogue of the Hopf-Koszul-Samelson theorem, the ρ-decomposition $C(\mathfrak{g}) = \operatorname{End} V_\rho \otimes C(P)$, and the $\mathfrak{g}$-module structure of $\bigwedge \mathfrak{g}$. *Adv. Math.*, 125(2):275–350, 1997.

[L05] G. Lebeau, *Geometric Fokker-Planck equations*, Port. Math. (N.S.) **62** (2005), no. 4, 469–530. MR MR2191631

[Le51] P. Lévy, *Wiener's random function, and other Laplacian random functions*, Proceedings of the Second Berkeley Symposium on Mathematical Statistics and Probability, 1950 (Berkeley and Los Angeles), University of California Press, 1951, pp. 171–187. MR MR0044774 (13,476b)

[M78] P. Malliavin, *Stochastic calculus of variation and hypoelliptic operators*, Proceedings of the International Symposium on Stochastic Differential Equations (Res. Inst. Math. Sci., Kyoto Univ., Kyoto, 1976) (New York), Wiley, 1978, pp. 195–263. MR **81f:**60083

[McS67] H.P. McKean, Jr. and I.M. Singer, *Curvature and the eigenvalues of the Laplacian*, J. Differential Geometry **1** (1967), no. 1, 43–69. MR 36 #828

[Q85] D. Quillen, *Superconnections and the Chern character*, Topology **24** (1985), no. 1, 89–95. MR **86m:**58010

[W82] E. Witten, *Supersymmetry and Morse theory*, J. Differential Geom. **17** (1982), no. 4, 661–692 (1983). MR **84b:**58111

Jean-Michel Bismut
Département de Mathématique
Université Paris-Sud
Bâtiment 425
F-91405 Orsay, France
e-mail: `Jean-Michel.Bismut@math.u-psud.fr`

Adiabatic Limit, Heat Kernel and Analytic Torsion

Xianzhe Dai and Richard B. Melrose

Dedicated to Jeff Cheeger for his 65th birthday

Abstract. We study the uniform behavior of the heat kernel under the adiabatic limit using microlocal analysis and apply it to derive a formula for the analytic torsion.

Mathematics Subject Classification (2000). 58Jxx, 35K08, 57Q10.

Keywords. Adiabatic limit, heat kernel, singularity, microlocal analysis, analytic torsion.

Introduction

The adiabatic limit refers to the geometric degeneration in which the metric is being blown up along certain directions, typically the base directions of a fibration. The study of the adiabatic limit of geometric invariants is initiated by E. Witten [40], who relates the adiabatic limit of the η-invariant to the holonomy of determinant line bundle, the so-called "global anomaly". In this case the manifold is fibered over a circle and the metric is being blown up along the circle direction. Witten's result was given full mathematical treatment in [8], [9] and [13], see also [16]. In [4], J.-M. Bismut and J. Cheeger studied the adiabatic limit of the eta invariant for a general fibration of closed manifolds. Assuming the invertibility of the Dirac family along the fibers, they showed that the adiabatic limit of the η-invariant of a Dirac operator on the total space is expressible in terms of a canonically constructed differential form, $\tilde{\eta}$, on the base. The Bismut-Cheeger $\tilde{\eta}$ form is a higher-dimensional analogue of the η-invariant and it is exactly the boundary correction term in the families index theorem for manifolds with boundary,

Research supported in part by the National Science Foundation

A version of this manuscript has been in circulation for many years. Some parts of the material, e.g., Section 4 on the bundle rescaling, has appeared in print, see [30].

[5], [6]. The families index theorem for manifolds with boundary has since been established in full generality by Melrose-Piazza in [31], [32]. The general adiabatic limit formula for eta invariant is proved in [14].

Around the same time, Mazzeo and the second author took on the analytic aspect of the adiabatic limit [27] and studied the uniform structure of the Green's operator of the Laplacian in the adiabatic limit. Their analysis enables the first author to prove the general adiabatic limit formula in [14]. The adiabatic limit formula is used in [7] to prove a generalization of the Hirzebruch conjecture on the signature defect (cf. [1],[35]). Other applications of adiabatic limit technique can be found in [41], [17] and [37].

The main purpose of this paper is to study the uniform behavior of the heat kernel in the adiabatic limit. The adiabatic limit introduces degeneracy along the base directions and gives rise to a new singularity for the heat kernel which interacts in a complicated way with the usual diagonal singularity. We resolve this difficulty by lifting the heat kernel to a larger space obtained by blowing up certain submanifolds of the usual carrier space of the heat kernel (times the adiabatic direction). The new space is a manifold with corners and the uniform structure of the adiabatic heat kernel can be expressed by stating that it gives rise to a polyhomogeneous conormal distribution on the new space.

More precisely, if $\phi : M \longrightarrow Y$ is a fibration with typical fibre F, the adiabatic metric is a one-parameter family of metrics $x^{-2}g_x$, with $g_x = \phi^*h + x^2g$, on M, where h is a metric on Y and g a symmetric 2-tensor on M which restricts to Riemannian metrics on the fibers. Note that g_x collapses the fibration to the base space in the limit $x \to 0$. Our main object of study is the regularity of the heat kernel $\exp(-t\Delta_x)$ of the Laplacian for the metric g_x. We prefer to write the heat kernel as $\exp(-\frac{t}{x^2}\,{}^a\!\Delta)$ where ${}^a\!\Delta$ is the Laplacian of the adiabatic metric g_x/x^2. The techniques of [27] are extended to construct the 'adiabatic heat calculus,' of which this heat kernel is a fairly typical element. In particular ${}^a\!\Delta$ is considered as an operator on the rescaled bundle ${}^a\!\Lambda^*$ which is the bundle of exterior powers of the adiabatic cotangent bundle ${}^aT^*M_a$, $M_a = M \times [0,1]$. This rescaling and parabolic blow-up methods are used to define the 'adiabatic heat space,' M_A^2, from $[0,\infty) \times M^2 \times [0,1]$. The adiabatic heat calculus, $\Psi_A^{*,*,*}(M; {}^a\!\Lambda^*)$, is defined in terms of the Schwartz kernels of its elements which are smooth sections over M_A^2 of the kernel bundle, a weighted version of the lift of the homomorphism bundle tensored with a density bundle. The operator $\exp(-t\Delta_x)$ is constructed in this calculus directly using the three symbol homomorphisms. Each of these maps is defined by evaluation of the Schwartz kernel at one of the boundary hypersurfaces of M_A^2.

The first of the symbol maps is just a parametrized version of the corresponding map for the ordinary heat calculus and is used in exactly the same way; in this case it takes values in the fibrewise operators on the bundle ${}^aTM_a$ over M_a. The second map is more global and takes values in fibrewise operators on the space $[0,\infty) \times \phi^*(TY)$ as a bundle over Y. In fact the image consists of elements of the heat calculus for the fibres $T_yY \times F_y = {}^YTM_{(y,z)}$ for each $y \in Y, z \in F$. The ordinary heat calculus can be used to invert these operators. The third map is the 'obvious' boundary map obtained by setting $x = 0$. In practice it is necessary to consider a 'reduced' normal operator at this face.

Our first result is

Theorem 0.1. *The heat kernel is an element*

$$\exp(-x^{-2}t\,{}^a\!\Delta) \in \Psi^{-2,-2,0}_{A,E}(M;{}^a\!\Lambda^k) \tag{0.1}$$

with normal operators

$$N_{h,-2} = (4\pi)^{-\frac{n}{2}} \exp\left(-\frac{1}{4}|v|_a^2\right) \tag{0.2}$$

$$N_{A,-2} = \exp(-T\Delta_A),\ T = x^{-2}t, \tag{0.3}$$

$$N_{a,0} = \exp(-t\Delta_Y) \tag{0.4}$$

where Δ_A is the fibrewise Laplacian on the bundle ${}^YTM$ and Δ_Y is the reduced Laplacian on Y, that is, the Laplacian on Y twisted by the flat bundle of the fiber cohomologies.

Here the subscript E indicates a refinement of the adiabatic heat calculus which will be discussed in §3. Note that the right-hand side of (0.2) is the Euclidean heat kernel on the tangent space evaluated at time $t = 1$ and $(v, 0)$. In (0.3) the fibers of ${}^YTM$ are ${}^YTM_{(y,z)} = T_yY \times F_y$, hence the fiberwise Laplacian Δ_A consists of the Euclidean Laplacian on T_yY together with the Laplacian of F_y. The heat kernel for the Eucildean Laplacian here should be evaluated at $(v, 0)$. As we will see later, Theorem 0.1 contains all the usual uniform regularity properties of the adiabatic heat kernel.

However, in applications to geometric problems such as the study of the eta invariant and the analytic torsion, one needs to incorporate the Getzler's rescaling [22], which has important implications for supertrace cancellations. We show how to build it into the calculus using bundle filtrations, resulting in the rescaled adiabatic heat calculus $\Psi^*_{A,G}(M;{}^a\!\Lambda^*)$. This enables us to refine Theorem 0.1 to obtain our main result.

Theorem 0.2. *The heat kernel is an element of the rescaled adiabatic heat calculus $\Psi^*_{A,G}(M;{}^a\!\Lambda^*)$,*

$$\exp\left(-\frac{t}{x^2}\,{}^a\!\Delta\right) \in \Psi^{-2,-2,0}_{A,G}(M;{}^a\!\Lambda^*). \tag{0.5}$$

Moreover, the normal operator at the temporal front face is given by

$$N_{h,G,-2} = e^{-[\mathcal{H}-\frac{1}{8}C(R)]} \tag{0.6}$$

where $\mathcal{H}$ is the generalized harmonic oscillator on the tangent space T_pM defined in (5.39) *and $C(R)$ is the quantization of the curvature tensor R defined in* (5.38). *The normal operator at the adiabatic front face is*

$$N_{A,G,-2} = \exp(-\mathcal{H}_Y)\exp\left(\frac{1}{8}C(R_Y)\right)\exp(-\mathcal{A}_T^2). \tag{0.7}$$

Here $\mathcal{H}_Y$ is the generalized harmonic oscillator on the fibres of ${}^xTY$, and $\mathcal{A}_T$ is the rescaled Bismut superconnection:

$$\mathcal{A}_T^2 = -T\left[\nabla_{e_i} + \frac{1}{2}T^{-\frac{1}{2}}\langle\nabla_{e_i}e_j, f_\alpha\rangle cl(e_i)cl(f_\alpha) + \frac{1}{4}\langle\nabla_{e_i}f_\alpha, f_\beta\rangle cl(f_\alpha)cl(f_\beta)\right]^2 + \frac{1}{4}TK_F, \tag{0.8}$$

where e_i is an orthonormal basis of the fibers and f_α that of the base, and K_F denotes the scalar curvature of the fibers.

We then apply this to the study of the adiabatic limit of the analytic torsion. The analytic torsion, $T_\rho(M,g)$, introduced by Ray and Singer [38], is a geometric invariant associated to each orthogonal representation, ρ, of the fundamental group of a compact manifold M with Riemann metric g (later extended to more general representations such as unimodular ones [34], [11]). It depends smoothly on g and is identically equal to 1 in even dimensions. As conjectured in [38] it has been identified with the Reidemeister torsion and this is the celebrated Cheeger-Müller theorem ([12], [33], see also [34], [11] for generalizations.). Using the uniform behavior of the heat kernel, we show that the torsion (our normalization corresponds to the square of that in [38]) in the adiabatic limit satisfies

$$T_\rho(M,g_x) = x^{-2\alpha}b(x),\ \alpha\in\mathbb{N},\ b\in\mathcal{C}^\infty([0,1]). \tag{0.9}$$

Thus, whilst not necessarily smooth in the adiabatic limit, $x\downarrow 0$, the analytic torsion behaves quite simply.

The characteristic exponent in (0.9) can be expressed in terms of the Leray spectral sequence for the cohomology twisted by ρ as

$$\alpha = -\chi_2(M) + \chi_2(Y,\mathcal{H}^*(F)) + \sum_{r\geq 2}(r-1)[\chi_2(E_r) - \chi_2(E_{r+1})]. \tag{0.10}$$

Here if β_j is the dimension in degree j of the cohomology then

$$\chi_2(E_r) = \sum_j j(-1)^j\beta_j.$$

For $\chi_2(M)$ the β_j are the Betti numbers of M, for $\chi_2(Y,\mathcal{H}^*(F))$ they are the dimensions of the twisted cohomology spaces of Y and for E_r they are the dimensions, $\beta_{j,r}$, of the rth term, (E_r,d_r), of the Leray spectral sequence.

The limiting value of the smooth factor in (0.9) depends on the parity of the dimension of the fibres. If the fibres are even-dimensional then

$$b(0) = \prod_{j=1}^{\dim F}\left[T_{\phi_*(\rho)\otimes\rho_j'}(Y,h)\right]^{(-1)^j j}\prod_{r\geq 2}\tau(E_r,d_r),\ \dim Y \text{ odd}, \tag{0.11}$$

where ρ_j' is the representation of $\pi_1(Y)$ associated to the flat bundle given by the fibre cohomology in dimension j, twisted by $\rho_{\restriction\pi_1(F)}$, and $\tau(E_r,d_r)$ is the torsion of the finite complex. The other case, when the fibres are odd-dimensional, is only a little more complicated. If R_Y is the curvature operator of the metric h on Y then the Gauss-Bonnet theorem states that the Euler characteristic of Y is given by the integral over Y of the Pfaffian density $(2\pi)^{-n}\operatorname{Pf}(R_Y)$, $n = \dim Y$. Consider the weighted integral

$$\chi_\rho(Y,\phi,g,h) = (2\pi)^{-n}\int_Y \operatorname{Pf}(R_Y)\log T_{\rho(y)}(F_y, g_y) \tag{0.12}$$

where g_y is the restriction of g to the fibre $F_y = \phi^{-1}(y)$, $y \in Y$, and $\rho(y)$ is the representation of $\pi_1(F_y)$ induced by ρ. Then (0.10) still holds and

$$b(0) = e^{\chi_\rho(Y,\phi,g,h)}\prod_{r\geq 2}\tau(E_r,d_r),\ \dim Y \text{ even.} \tag{0.13}$$

If $\dim Y$ is even the twisted torsion factor in (0.11) reduces to 1 and if $\dim Y$ is odd the weighted Euler characteristic in (0.12) is zero, so these two formulæ can be combined to give one in which the parity does not appear explicitly.

Theorem 0.3. *The analytic torsion of an adiabatic metric g_x for a fibration satisfies*

$$\begin{aligned}
&\log T_\rho(M,g_x)\\
&= -2\left(-\chi_2(M) + \chi_2(Y,\mathcal{H}^*(F)) + \sum_{r\geq 2}(r-1)[\chi_2(E_r)-\chi_2(E_{r+1})]\right)\log x\\
&\quad + \chi_\rho(Y,\phi,g,h) + \sum_{j=1}^{\dim F}(-1)^j j\log T_{\phi_*(\rho)\otimes\rho'}(Y,h) + \sum_{r\geq 2}\log\tau(E_r,d_r) + xb'(x),\\
&\quad b'\in\mathcal{C}^\infty([0,1]).
\end{aligned} \tag{0.14}$$

Remark 1. For the holomorphic analogue see [3]. It should be pointed out that, in our case, the analytic torsion form of [10] did appear in the formula. However the higher degree terms in the torsion form are cancelled by the Pffafian term. Our proof also extends to the holomorphic case.

Remark 2. Under appropriate acyclicity conditions formula (0.14) reduces to the purely topological formulas for the Reidemeister torsion obtained by D. Fried [20], D. Freed [21], and Lück-Schick-Thielman [26].

The analytic torsion is defined in terms of the torsion zeta function

$$\log T_\rho(M,g) = \zeta_T'(0) \tag{0.15}$$

where $\zeta_T(s)$ is a meromorphic function of $s\in\mathbb{C}$ which is regular at $s=0$. For $\operatorname{Re} s >> 0$

$$\zeta_T(s) = \frac{1}{\Gamma(s)}\sum_{j=1}^{\dim M}(-1)^j j\int_0^\infty t^s \operatorname{Tr}\left(\exp(-t\underline{\Delta}_j)\right)\frac{dt}{t} \tag{0.16}$$

where $\underline{\Delta}_j$ is the Laplacian on j-forms with null space removed (i.e., acting on the orthocomplement of the harmonic forms). The proof of (0.14) is thus reduced to a sufficiently fine understanding of the heat kernel in the adiabatic limit.

The paper is organized as follows. In §1 we recall the construction of the heat kernel in the standard case of a compact manifold. This is done to introduce, in a simple context, the approach via parabolic blow-up, which is used here. The appropriate notion of parabolic blow-up is described in §2. The important finite time properties are then summarized by the statement that the heat kernel is an element of order -2 of the even part of the heat calculus acting on the exterior bundle

$$\exp(-t\Delta) \in \Psi_{h,E}^{-2}(M;\Lambda^*). \tag{0.17}$$

This in turn is a regularity statement for the lift of the Schwartz kernel from the space $[0,\infty) \times M^2$, where it is usually defined, to the heat space, M_h^2, obtained by t-parabolic blow-up of the diagonal at $t=0$. This is discussed in §3. The heat calculus has a 'symbol map', the normal homomorphism, into the homogeneous and translation-invariant part of the heat calculus on (the compactification of) the fibres of the tangent bundle to M. Under this map the heat kernel is carried to the family (over M) of heat kernels for the fibrewise Laplacian on TM :

$$\begin{gathered} N_{h,-2}(\exp(-t\Delta))(m,v) = \exp(-\Delta_m)(v), \\ \Delta_m = \sum_{j,k=1}^{\dim M} g^{jk}(m) D_{v^j} D_{v^k} \text{ on } T_m M. \end{gathered} \tag{0.18}$$

Conversely (0.18) allows an iterative construction of the heat kernel. The appropriate composition properties for the heat calculus, allowing this iterative approach, are also discussed in §3.

For a general Laplacian, P, without null space, the heat kernel is, for $t > 0$, a smoothing operator which decreases exponentially as $t \to \infty$. As a result the zeta function, obtained by Mellin transform of the heat kernel is, following Seeley [39], meromorphic with poles in $\operatorname{Re} s \geq 0$ only at $s = \frac{1}{2}\dim M - k$, $k = 0, 1, \ldots$. These poles come from the short-time asymptotics of the heat kernel. If Δ is the Laplacian on forms then (0.16) can be rewritten

$$\zeta_T(s) = \frac{1}{\Gamma(s)} \int_0^\infty t^s \operatorname{STr}\left(N \exp(-t\underline{\Delta})\right) \frac{dt}{t} \tag{0.19}$$

where N is the number operator, acting as k on Λ^k, and STr is the supertrace functional, i.e.,

$$\operatorname{STr}(A) = \operatorname{Tr}(QA),\ Q = (-1)^k \text{ on } \Lambda^k. \tag{0.20}$$

The algebraic properties of the supertrace imply that there are actually no poles of $\zeta_T(s)$ in $s > \frac{1}{2}$.

To see this, in §5, the rescaling argument of Getzler [22] (see also [2]) is formulated in terms of the heat calculus. This is done by defining a 'rescaled' version of the homomorphism bundle over M_h^2, of which the heat kernel is a smooth section, and then showing that the kernel is again a smooth section of the rescaled bundle. The general process of rescaling a bundle is discussed in §4. Under this rescaling (of length $\dim M + 1$), defined by extension of the Clifford degree of a homomorphism of the exterior algebra, the (pointwise) supertrace functional lies in the maximal graded quotient. The number operator has degree two so it follows that

$$\begin{aligned} \operatorname{str}(N\exp(-t\Delta)) &\in t^{-\frac{1}{2}}\mathcal{C}^\infty([0,\infty)\times M;\Omega M),\\ \operatorname{STr}(N\exp(-t\Delta)) = \int_M \operatorname{str}(N\exp(-t\Delta)) &\in t^{-\frac{1}{2}}\mathcal{C}^\infty([0,\infty)). \end{aligned} \tag{0.21}$$

Moreover the leading term can be deduced from the normal operator for the rescaled calculus:

$$\begin{gathered} t^{\frac{1}{2}}\operatorname{str}(N\exp(-t\Delta_j))_{\restriction t=0} = c(n)\sum_{k=1}^{\dim M}(-1)^k\operatorname{Pf}(R_k)\wedge\omega_k \in \mathcal{C}^\infty(M;\Omega M)\\ c(n) = 2i(-1)^{\frac{1}{2}(n+1)}(16\pi)^{-\frac{1}{2}n} \end{gathered} \tag{0.22}$$

where with respect to any local orthonormal frame, ω_k of T^*M, R_k is the curvature operator with kth row and column deleted,

$$R_k = \sum_{i,j\neq k} R_{ijk\ell}\,\omega_i\omega_j, \tag{0.23}$$

and $\operatorname{Pf}(R_k)$ is its Pfaffian as an antisymmetric matrix. The cancellation formula (0.21) shows that $\zeta_T(s)$ has only one pole in $\operatorname{Re}s \geq 0$, at $s = \frac{1}{2}$, and (0.22) gives its residue. It is thus straightforward to give a formula for $\zeta_T(s)$ which is explicitly regular near $s = 0$; see Corollary 5.2.

This representation is used to obtain (0.9). As already noted, the main step is a clear analysis of the regularity of the heat kernel $\exp(-t\Delta_x)$ of the Laplacian for the metric g_x; this is carried out in §6–§9. In §10 the incorporation of Getzler's rescaling into the ordinary heat calculus is extended to the adiabatic heat calculus to give the cancellation effects for the supertrace. Two related but distinct rescalings are required, one just as in the standard case and the other at the adiabatic front face. For finite times this results in the following description of the regularity of the supertrace. Consider the space $Q = [0,\infty)_t \times [0,1]_x$ on the interior of which $\operatorname{STr}(\exp(-t\Delta_x))$ is defined and $\mathcal{C}^\infty$. Let Q_2 be obtained by t-parabolic blow-up of the corner $\{t = x = 0\}$ with $\beta_1 : Q_2 \longrightarrow Q$ the blow-down map; the subscripts 1 and 2 here refer to the Leray spectral sequence. Then Q_2 has boundary lines tf, arising from $t = 0$, af from the blow-up, ab from $x = 0$ and ef from $x = 1$. For appropriate defining functions, ρ_F, for the boundary lines, F, the lift to Q_2 satisfies

$$\beta_1^*\operatorname{STr}(\exp(-t\Delta_x)) \in \begin{cases} \rho_{\mathrm{tf}}^{-1}\rho_{\mathrm{af}}^{-1}\mathcal{C}^\infty(Q_2), & \dim Y \text{ odd}\\ \rho_{\mathrm{tf}}^{-1}\mathcal{C}^\infty(Q_2), & \dim Y \text{ even}. \end{cases} \tag{0.24}$$

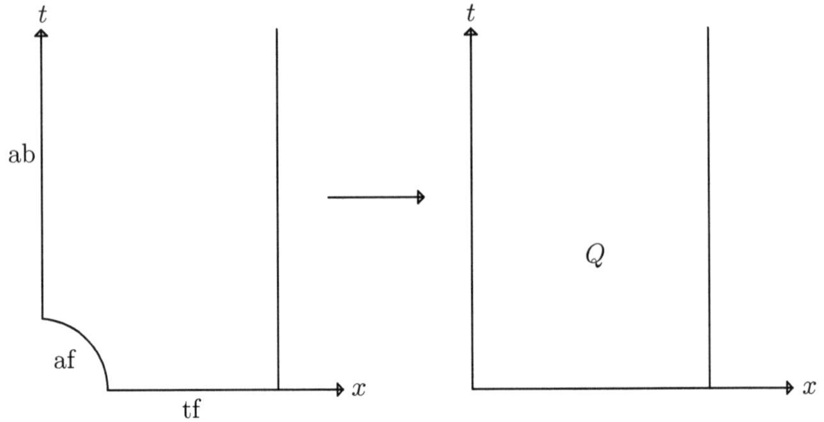

FIGURE 1. $\beta_1 : Q_2 \to Q$

We also need to discuss the behaviour of the heat kernel as $t \to \infty$. To do so we use results from [14]. The kernel decomposes as $t \to \infty$ into a part which is rapidly decreasing, and uniformly smoothing, plus finite rank parts corresponding to the small eigenvalues of the Laplacian. These are in turn associated to the individual terms (E_r, d_r), $r \geq 2$, of the Leray spectral sequence.

The leading terms at the boundary faces in (0.24) can be deduced from the construction of the heat kernel, and ultimately therefore from the solutions of the three model problems arising from the rescaled symbol maps. Together with the behaviour of the small eigenvalues this leads directly to (0.9), (0.10) and (0.14); the final derivation is given in §11.

1. Hadamard's construction

To orient the reader towards our detailed description of the behaviour in the adiabatic limit of the heat kernel, and in particular its trace, we shall first recall the 'classical' case. Thus we shall show how to construct the heat kernel for an elliptic differential operator, P, of second order on a compact $\mathcal{C}^\infty$ manifold under the assumption that P has positive, diagonal, principal symbol. Such a construction is well known and can certainly be carried out by the method developed by Hadamard [23] for the wave equation. The construction below proceeds rather formally, in terms of the heat calculus. The only novelty here is in the definition, and discussion, of the calculus itself in §3 in which a systematic use of the process of parabolic blow-up is made. As opposed to the standard construction of Hadamard this allows us to generalize, to the adiabatic limit and, later, to the case of boundary problems with limited changes.

The basic model for the heat kernel is the Euclidean case, $\Delta_E = D_1^2 + \cdots + D_n^2$ where $D_j = -i\partial/\partial x_j$ on $\mathbb{R}^n$. Then the function

$$\Phi'(t,x) = (4\pi t)^{-\frac{n}{2}} \exp\left(-\frac{|x|^2}{4t}\right), \ t > 0, \ x \in \mathbb{R}^n \tag{1.1}$$

has a unique extension to $\widetilde{\Phi}' \in \mathcal{S}'(\mathbb{R}^{1+n})$ which is locally integrable in t with values in $\mathcal{S}'(\mathbb{R}^n)$ and vanishes in $t < 0$. Acting as a convolution operator this is the unique tempered forward inverse of $\partial_t + \Delta_E$.

The convolution operator defined by $\widetilde{\Phi}'$ can be embedded into a graded algebra of operators by generalizing (1.1). For $p < 0$ the operators of order p in this Euclidean heat calculus, or perhaps more correctly homogeneous and translation-invariant heat calculus, have kernels of the form

$$K(t,x) = \begin{cases} t^{-\frac{n}{2}-1-\frac{p}{2}} \kappa(\frac{x}{t^{\frac{1}{2}}})|dx|, & t > 0 \\ 0, & t \leq 0 \end{cases} \tag{1.2}$$

where $\kappa \in \mathcal{S}(\mathbb{R}^n)$. Again K is locally integrable in t as a function with values in $\mathcal{S}'(\mathbb{R}^{n+1})$ and so fixes a convolution operator on $\mathbb{R}^{1+n}$. We shall denote this space of operators as $\Psi^p_{\text{th}}(\mathbb{R}^n)$. For $p = 0$ there is a similar, but slightly more subtle, definition of the space of operators $\Psi^0_{\text{th}}(\mathbb{R}^n)$. Namely this space is the span of the identity and the convolution operators given by distributions as in (1.2) with $p = 0$ and satisfying in addition the condition

$$\int_{\mathbb{R}^n} \kappa(x)dx = 0. \tag{1.3}$$

In this case the kernel is not locally integrable as a function of t; nevertheless the mean value condition (1.3) means that

$$\langle K, \phi \rangle = \lim_{\delta \downarrow 0} \int_{|x|^2 + t > \delta^2, \ t \geq 0} \phi(t,x)\kappa\left(\frac{x}{t^{\frac{1}{2}}}\right) t^{-\frac{n}{2}-1} dx dt, \ \phi \in \mathcal{S}(\mathbb{R}^{1+n}) \tag{1.4}$$

defines $K \in \mathcal{S}'(\mathbb{R}^{1+n})$. It is also straightforward to check that these spaces of operators are invariant under linear transformations of $\mathbb{R}^n$, so the spaces $\Psi^p_{\text{th}}(V)$ are well defined for any vector space V and $p \leq 0$. Essentially by definition we have 'normal operators' for these spaces,

$$\begin{aligned} N_{h,p} : \Psi^p_{\text{th}}(V) \ni K &\longrightarrow \kappa \in \mathcal{S}(V;\Omega), \ p < 0 \\ N_{h,0} : \Psi^0_{\text{th}}(V) \ni (c\,\mathrm{Id} + K) &\longrightarrow (c,\kappa) \in \mathbb{C} \oplus \overline{\mathcal{S}}(V;\Omega), \end{aligned} \tag{1.5}$$

where $\overline{\mathcal{S}}(V;\Omega)$ is the null space of the integral on $\mathcal{S}(V;\Omega)$. The normal operators here are isomorphisms.

For Schwartz densities on a vector space, V, we can define a two-parameter family of products in any linear coordinates by

$$a \star_{p,q} b(x)|dx| \\ = \int_0^1 \int_V a((1-s)^{-\frac{1}{2}}x-y)b(s^{-\frac{1}{2}}y)dy(1-s)^{-\frac{n}{2}-1+p}s^{-\frac{n}{2}-1+q}ds|dx|,\ p,q<0. \tag{1.6}$$

If a (resp b) has mean value 0 the definition extends to $p=0$ (resp $q=0$); in case both p and q are zero the result has mean value zero. The product is extended to the sum in (1.5) by letting $\mathbb{C}$ act as multiples of the identity. These products give the composition law for the normal operators in the Euclidean heat calculus

$$\begin{aligned} &\Psi^p_{\text{th}}(V)\circ\Psi^q_{\text{th}}(V)\subset\Psi^{p+q}_{\text{th}}(V),\ p,q\le 0 \\ &N_{h,p+q}(A\circ B)=N_{h,p}(A)\star_{p,q}N_{h,q}(B) \end{aligned} \tag{1.7}$$

as follows by a simple computation. Since the calculus and products are invariant under linear transformations both extend to the fibres of any vector bundle and can be further generalized to act on sections of the lift of another vector bundle over the base.

Replacing the Euclidean Laplacian by a differential operator P, as described above, acting on sections of an Hermitian vector bundle over the Riemann manifold M and having principal symbol $|\xi|_x^2\,\mathrm{Id}$ at $(x,\xi)\in T^*M$ we wish to construct an analogue of $\widetilde{\Phi}'$; this will be the kernel of an operator which is still a convolution operator in t but not in the spatial variables. These operators, which will be discussed in §3, are defined directly in terms of their Schwartz' kernels. The kernels are sections of a density bundle over the manifold M_h^2, which is the product $[0,\infty)_t\times M^2$ with the product manifold $\{0\}\times\mathrm{Diag}$ blown up. Here $\mathrm{Diag}\subset M^2$ is the diagonal, so functions on M_h which are smooth up to the new boundary hypersurface (the 'front face') produced by the blow-up are really singular, in the manner appropriate for the heat kernel, at $\{0\}\times\mathrm{Diag}$ when considered as functions on $(0,\infty)\times M^2$.

If U is a vector bundle over M we denote by $\Psi_h^{-k}(M;U)$ this space of heat operators, discussed in detail in §3, of order $-k$, for $k\in\mathbb{N}_0=\{0,1,\dots\}$. These operators act on $\dot{\mathcal{C}}^\infty([0,\infty)\times M;U)$, which is the space of $\mathcal{C}^\infty$ sections of U, lifted to $[0,\infty)\times M$, vanishing with all derivatives at $\{0\}\times M$. There is a well-defined *normal operator*:

$$N_{h,-k}:\ \Psi_h^{-k}(M;U)\twoheadrightarrow\mathcal{S}(TM;\Omega_{\text{fibre}}\otimes\pi_M^*\hom(U)),\ k\in\mathbb{N}. \tag{1.8}$$

Here $\mathcal{S}$ denotes the (fibre) Schwartz space on TM. This normal operator is determined by the leading coefficient of the Scwartz kernel at the front face of M_h^2. For operators of order 0 the normal operator becomes

$$N_{h,0}:\ \Psi_h^0(M;U)\twoheadrightarrow\mathbb{C}\oplus\overline{\mathcal{S}}(TM;\Omega_{\text{fibre}}\otimes\pi_M^*\hom(U)), \tag{1.9}$$

where $\overline{\mathcal{S}}(TM;\Omega_{\text{fibre}})$ denotes the space of Schwartz fibre densities with mean value zero on each fibre. The following result is proved in §3 after developing the necessary tools in §2.

Proposition 1.1. *The maps* (1.8) *and* (1.9) *extend to normal homomorphisms which filter* $\Psi_h^0(M;U)$ *as an asymptotically complete algebra of operators on* $\dot{\mathcal{C}}^\infty([0,\infty)\times M;U)$*, i.e., the null space of* $N_{h,p}$ *is exactly* $\Psi_h^{p-1}(M;U)$ *and*

$$\begin{aligned} A\in\Psi_h^{-k}(M;U),\ B\in\Psi_h^{-j}(M;U)\Longrightarrow A\circ B\in\Psi_h^{-k-j}(M;U),\ j,k\in\mathbb{N},\ \textit{with}\\ N_{h,-k-j}(A\circ B)=N_{h,-k}(A)\star_{k,j}N_{h,-j}(B). \end{aligned}\tag{1.10}$$

Any element of $\mathrm{Id}+\Psi_h^{-1}(M;U)$ *is invertible with inverse in the same space.*

In view of (1.5), the maps in (1.8) and (1.9) can be interpreted as homomorphisms into the homogeneous and translation-invariant heat calculus on the fibres of TM.

This calculus of operators of non-positive order can be extended to non-positive real orders and positive orders as well, but all we need is the composition properties with differential operators. The following result, which follows easily from the definition, is also proved in §3.

Proposition 1.2. *If* $P\in\mathrm{Diff}^k(M;U)$ *and* $j\geq k$ *composition gives*

$$\begin{aligned} \Psi_h^{-j}(M;U)\ni A\longmapsto P\circ A\in\Psi_h^{k-j}(M;U)\\ N_{k-j}(P\circ A)=\sigma_k(P)N_{-j}(A) \end{aligned}\tag{1.11}$$

where the symbol of P *is considered as a homogeneous differential operator with constant coefficients on the fibres of* TM*; similarly if* V_r *is the radial vector field on the fibres of* TM *then*

$$\begin{aligned} &\Psi_h^{-j}(M;U)\ni A\longmapsto D_t\circ A\in\Psi_h^{-j+2}(M;U)\ \textit{if}\ j\geq 2\ \textit{with}\\ &N_{h,-j+2}(D_tA)=\frac{i}{2}(V_r+n-j+2)N_{h,-j}(A),\ j>2\\ &D_tA=a\,\mathrm{Id}+\overline{B},\ \overline{B}\in\overline{\Psi}_h^0(M;U),\ \textit{if}\ A\in\Psi_h^{-2}(M;U)\ \textit{with}\\ &a=-i\int_{\text{fibre}}N_{h,-2}(A),\ N_0(\overline{B})=\frac{i}{2}(V_r+n)N_{h,-2}(A). \end{aligned}\tag{1.12}$$

This calculus allows us to give a direct construction of the heat kernel. Namely we look for $E\in\Psi_h^{-2}(M;U)$ satisfying

$$(\partial_t+P)E=\mathrm{Id}.\tag{1.13}$$

Here $\mathrm{Id}\in\Psi_h^0(M;U)$ has symbol 1. From (1.12) this imposes conditions on the 'normal operator' $N_{h,-2}(E_1)$, *viz*

$$\int_{\text{fibre}}N_{h,-2}(E_1)=1,\ \left[\sigma_2(P)-\frac{1}{2}(V_r+n)\right]N_{h,-2}(E_1)=0.\tag{1.14}$$

This has a unique Schwartz solution, namely that derived from (1.1) in any linear coordinates on TM induced by a local orthonormal basis:

$$N_{h,-2}(E_1) = (4\pi)^{-\frac{n}{2}} \exp\left(-\frac{1}{4}|v|_x^2\right) \mathrm{Id}_U, \text{ at } (x,v) \in TM. \tag{1.15}$$

The surjectivity of the map (1.8) means that we can indeed find E_1 with properties (1.14) which therefore satisfies

$$(\partial_t + P)E_1 = \mathrm{Id} + R,\ R \in \Psi_h^{-1}(M;U). \tag{1.16}$$

From the last part of Proposition 1.1 we conclude that the inverse exists

$$(\mathrm{Id} + R)^{-1} = \mathrm{Id} + S$$

with $S \in \Psi_h^{-1}(M;U)$ and so

$$E = E_1 \circ (\mathrm{Id} + S) = E_1 + E_1 \circ S \in \Psi_h^{-2}(M;U) \tag{1.17}$$

satisfies (1.13).

There is a further refinement of the calculus. Namely let $\Psi^0_{h,E}(M;U) \subset \Psi^0_h(M;U)$ be the subspace which extends (by duality) to define an operator on $\mathcal{C}^\infty([0,\infty)\times M)$. This space is characterized in terms of kernels in §3. It is a filtered subalgebra with normal homomorphism taking values in the even subspace of the Schwartz space under the involution $v \longmapsto -v$ on the fibres of TM. Proposition 1.1 and Proposition 1.2 extend directly to this smaller algebra, so the construction of the heat kernel above actually shows that

$$\exp(-tP) \in \Psi^{-2}_{h,E}(M;U). \tag{1.18}$$

The point of this improvement is that the trace tensor on the kernel restricted to the diagonal gives in general

$$\mathrm{tr} : \Psi_h^{-k}(M;U) \longrightarrow t^{-\frac{1}{2}(\dim M - k + 2)}\mathcal{C}^\infty([0,\infty)_{\frac{1}{2}} \times M) \tag{1.19}$$

where the subscript denotes that $t^{\frac{1}{2}}$ is the $\mathcal{C}^\infty$ variable. For the even subspace however

$$\mathrm{tr} : \Psi^{-k}_{h,E}(M;U) \longrightarrow t^{-\frac{1}{2}(\dim M - k + 2)}\mathcal{C}^\infty([0,\infty) \times M) \tag{1.20}$$

and apart from the singular factor the trace is $\mathcal{C}^\infty$ in the usual sense up to $t = 0$. This gives a rather complete description of the heat kernel for finite times.

We are also interested in the behaviour as $t \to \infty$. In the case of a compact manifold this is easily described. The variable $t/(1+t) \in [0,1]$ can be used to compactify the heat space near $t = \infty$. Thus $1/t$ is a $\mathcal{C}^\infty$ defining function for 'temporal infinity,' ti $= \{1\} \times M^2$ in the new heat space; this leads to a new heat calculus $\Psi^0_H(M;U)$ with smoothness as $t \to \infty$ added. This refined heat calculus has another 'normal operator,' with values in the smoothing operators on M, namely

$$N_{h,\infty} : \Psi^0_H(M;U) \longrightarrow \Psi^{-\infty}(M;U). \tag{1.21}$$

Proposition 1.3. *If $P \in \mathrm{Diff}^2(M;U)$ is a differential operator on sections of U over M which is elliptic, non-negative self-adjoint and with diagonal principal symbol the heat kernel*

$$\exp(-tP) \in \Psi^{-2}_{H,E}(M;U) \tag{1.22}$$

has normal operator (1.15) *at* tf *and at* ti *has normal operator the orthogonal projection onto the zero eigenspace of P.*

It is this (well-known) result which we wish to generalize in various ways, in particular to the adiabatic limit.

2. Parabolic blow-up

Since it is used extensively in the discussion of the heat kernel we give a brief description of the notion of the parabolic blow-up of a submanifold of a manifold with corners. More specifically we define the notation

$$[X;Y,S], \quad \beta\colon [X;Y,S] \longrightarrow X, \quad \beta = \beta[X;Y,S], \tag{2.1}$$

where $Y \subset X$ is a submanifold and $S \subset N^*Y$ is a subbundle, satisfying certain extra conditions. The result, $[X;Y,S]$, is "X blown up along Y with parabolic directions S." A more extensive discussion of this operation will be given in [18], see also [19], [30].

The basic model case we consider is $X = \mathbb{R}^{n,k} = [0,\infty)^k \times \mathbb{R}^{n-k}$ with coordinates x_i, $i = 1,\dots,k$, y_j, $j = 1,\dots,n-k$,

$$Y = Y_{l,p} = \{x_1 = \cdots x_l = 0,\ y_1 = \cdots y_p = 0\} \tag{2.2}$$

for some $1 \le l \le k$, $1 \le p \le n-k$ and

$$S = S_r = \mathrm{sp}\{dx_1,\dots,dx_r\} \subset \mathbb{R}^n \tag{2.3}$$

for some $r \le l$. Then the blown-up manifold is, by definition, the product

$$[\mathbb{R}^{n,k}; Y_{l,p}, S_r] = \mathbb{S}^{l+p-1,l,r}_{\frac12} \times [0,\infty) \times \mathbb{R}^{n-p-l,k-l} \tag{2.4}$$

where

$$\mathbb{S}^{l+p-1,l,r}_{\frac12} = \Big\{(x',x'',y') \in \mathbb{R}^{l+p,l};\ \sum_{1\le i\le r} x_i^2 + \sum_{r<i\le l} x_i^4 + \sum_{1\le j\le p} y_j^4 = 1\Big\}. \tag{2.5}$$

The 'blow-down map' is by definition

$$\begin{gathered}\beta\colon [\mathbb{R}^{n,k}; Y_{l,p}, S_r] \longrightarrow \mathbb{R}^{n,k},\ \beta((x',x'',y'),r,(x''',y'')) = (r^2x', rx'', x''', ry', y''),\\ \text{where}\ \ x' = (x_1,\dots,x_r),\ x'' = (x_{r+1},\dots,x_l),\ x''' = (x_{l+1},\dots,x_k),\\ y' = (y_1,\dots,y_p)\ \text{and}\ y'' = (y_{p+1},\dots,y_{n-k}).\end{gathered} \tag{2.6}$$

Let us note some properties of the triple of blown-up space, blow-down map and original space which are easily checked by direct computation. Both $\mathbb{R}^{n,k}$ and $[\mathbb{R}^{n,k}; Y_{l,p}, S_r]$ are manifolds with corners, $[\mathbb{R}^{n,k}; Y_{l,p}, S_r]$ has one more boundary hypersurface than $\mathbb{R}^{n,k}$. Clearly β is smooth and surjective, it is a diffeomorphism from $\mathbb{S}^{l+p-1,l,r} \times (0,\infty) \times \mathbb{R}^{n-p-l,k-l}$ onto $\mathbb{R}^{n,k} \setminus Y_{l,p}$ and a fibration from the 'front face'

$$\beta \colon \mathrm{ff}[\mathbb{R}^{n,k}; Y_{l,p}, S_r] = \mathbb{S}_{\frac{1}{2}}^{l+p-1,l,r} \times \{0\} \times \mathbb{R}^{n-p-l,k-l} \longrightarrow \mathbb{R}^{n-p-l,k-l}. \tag{2.7}$$

Any smooth function on $\mathbb{R}^{n,k} \setminus Y_{l,p}$ which is homogeneous of non-negative integral degree under the $\mathbb{R}^+$ action

$$(x', x'', x''', y', y'') \longmapsto (s^2 x', s x'', x''', s y', y''), \ s \in (0,\infty) \tag{2.8}$$

lifts to be $\mathcal{C}^\infty$ on $[\mathbb{R}^{n,k}; Y_{l,p}, S_r]$. Moreover these functions generate the $\mathcal{C}^\infty$ structure, i.e., give local coordinates near each point on the blown-up manifold. In fact the front face, $\mathrm{ff}[\mathbb{R}^{n,k}; Y_{l,p}, S_r]$ can be identified with the quotient of $\mathbb{R}^{n,k} \setminus Y_{l,p}$ under this $\mathbb{R}^+$ action. As a set the blown-up manifold can then be written

$$[\mathbb{R}^{n,k}; Y_{l,p}, S_r] = \mathrm{ff}[\mathbb{R}^{n,k}; Y_{l,p}, S_r] \sqcup \left(\mathbb{R}^{n,k} \setminus Y_{l,p}\right). \tag{2.9}$$

Any smooth vector field on $\mathbb{R}^{n,k}$, which is homogeneous of non-negative integral degree under (2.8) lifts to be smooth on $[\mathbb{R}^{n,k}; Y_{l,p}, S_r]$; if the vector field is tangent to all boundary hypersurfaces of $\mathbb{R}^{n,k}$ then the lift is tangent to all boundary hypersurfaces (including the new front face) of $[\mathbb{R}^{n,k}; Y_{l,p}, S_r]$. The lifts of these homogeneous vector fields tangent to all boundary hypersurfaces of $\mathbb{R}^{n,k} \setminus Y_{l,p}$ lift to span, over $\mathcal{C}^\infty([\mathbb{R}^{n,k}; Y_{l,p}, S_r])$, all smooth vector fields tangent to the boundary hypersurfaces of $[\mathbb{R}^{n,k}; Y_{l,p}, S_r]$. It follows from these results, or direct computation, that any local diffeomorphism on $\mathbb{R}^{n,k}$, $F \colon O \longrightarrow O' = F(O)$, which preserves both $Y_{l,p}$ and S_r in the sense that

$$F(O \cap Y_{l,p}) = O' \cap Y_{l,p}, \ F^*(S_r) = S_r, \tag{2.10}$$

lifts to a diffeomorphism on the blown-up space, i.e., there is a uniquely defined smooth diffeomorphism $\tilde{F} \colon \beta^{-1}(O) \longrightarrow \beta^{-1}(O')$ giving a commutative diagram

$$\begin{array}{ccccccc}
[\mathbb{R}^{n,k}; Y_{l,p}, S_r] & \hookleftarrow & \beta^{-1}(O) & \xrightarrow[\tilde{F}]{} & \beta^{-1}(O') & \hookrightarrow & [\mathbb{R}^{n,k}; Y_{l,p}, S_r] \\
\downarrow \beta & & \downarrow \beta & & \downarrow \beta & & \downarrow \beta \\
\mathbb{R}^{n,k} & \hookleftarrow & O & \xrightarrow[F]{} & O' & \hookrightarrow & \mathbb{R}^{n,k}
\end{array} \tag{2.11}$$

This invariance allows the blow-up to be defined more generally. Suppose that X is a manifold with corners (in particular each of the boundary hypersurfaces should be embedded.) Let $Y \subset X$ be a closed embedded submanifold which is of product type (a p-submanifold), in the sense that near each point of Y there is a local diffeomorphism of X to a neighborhood of $0 \in \mathbb{R}^{n,k}$ which reduces Y locally to some $Y_{l,p}$. In particular this means that the conormal bundle of Y is reduced to a product. Let $S \subset N^*Y$ be a subbundle which can be simultaneously reduced to S_r in (2.3) by such a diffeomorphism. The blown-up manifold is then given as a set by the extension of (2.9)

$$[X; Y, S] = \mathrm{ff}[X; Y, S] \sqcup (X \setminus Y). \tag{2.12}$$

Here the front face can be defined as the set of equivalence classes of curves with initial point on Y which are S-tangent to it. That is, consider the set of all curves

$$\begin{gathered}\chi\,[0,\epsilon),\ \epsilon>0,\ \mathcal{C}^\infty \text{ with}\\ \chi(0)\in Y,\ \frac{d(\chi^*f)}{ds}(0)=0 \text{ if } f\in\mathcal{C}^\infty(X),\ df(y)\in S_y\ \forall\ y\in Y.\end{gathered}\tag{2.13}$$

The first equivalence relation imposed on this set is that

$$\begin{gathered}\chi_1\sim\chi_2 \text{ if } \chi_1(0)=\chi_2(0),\ \frac{d(\chi_1^*f-\chi_2^*f)}{ds}(0)=0 \text{ and}\\ \frac{d^2(\chi_1^*f-\chi_2^*f)}{ds^2}(0)=0 \text{ if } f\in\mathcal{C}^\infty(X) \text{ has } df(y)\in S_y \text{ for } y\in Y.\end{gathered}\tag{2.14}$$

For each $y\in Y$ the curve with $\chi(s)\equiv y$ gives a base, or zero section. The second equivalence relation is on the curves which are non-zero in this sense, in which $\chi(ts)\sim\chi(s)$ for any $t>0$. The resulting space $\mathrm{ff}[X;Y,S]$ is a fibre bundle over Y with fibre diffeomorphic to $\mathbb{S}_{\frac12}^{p-1,r}$. In particular it reduces to $\mathbb{S}_{\frac12}^{l+p-1,l,r}$ in the model case discussed above. The invariance properties just described show that local identification with $[\mathbb{R}^{n,k};Y_{l,p},S_r]$ leads to a $\mathcal{C}^\infty$ structure on $[X;Y,S]$. The blow-down map is the obvious map from $[X;Y,S]$ to X, it has similar properties to those described above in the model case.

If $Y'\subset X$ is a closed submanifold the lift $\beta^*(Y')\subset[X;Y,S]$ is defined if $Y'\subset Y$ (respectively $Y'=\mathrm{cl}(Y'\setminus Y)$) to be $\beta^{-1}(Y')$ (resp. $\mathrm{cl}(\beta^{-1}(Y'\setminus Y))$. The lift of a subbundle $S'\subset N^*Y'$, denoted $\beta^*(S')$, is defined in these two cases as, respectively, $\beta^*(S')$ and the closure in $T^*[X;Y,S]$ of $\beta^*(S'\restriction(Y'\setminus Y)$. If this lifted manifold and the lift of S' satisfy the decomposition conditions introduced above then the iterated blow-up is defined. In this case we use the notation

$$[X;Y,S;Y',S']=[[X;Y,S];\beta^*(Y'),\beta^*(S')].\tag{2.15}$$

3. Heat calculus

To define the heat calculus we shall extrapolate from the properties of the model convolution operator $\widetilde{\Phi}'$ considered in §1. The kernel of this operator, Φ' from (1.1), is 'simple' in a sense that is related to homogeneity under the transformation

$$\mu_s:\ (t,x)\longmapsto(s^2t,sx),\ \ s\in\mathbb{R}^+.\tag{3.1}$$

Thus, set $Z=[0,\infty)\times M^2$ and consider its t-parabolic blow-up along the submanifold

$$B=\{(0,x,x)\in Z;x\in M\}.\tag{3.2}$$

Following the notation for parabolic blow-up in §2 above, this can be written

$$M_h^2=[Z;B,S]\xrightarrow{\beta_h}Z \text{ where } S=\mathrm{sp}(dt)\subset N^*B.\tag{3.3}$$

In case $M = \mathbb{R}^n$ the space M^2_h is easily identified. Let $\sim_\mu$ be the equivalence relation on $Z\setminus B$ generated by (3.1) in the sense that $p = (t, x, x') \sim_\mu p' = (r, y, y')$ if and only if $\mu_s(t, x) = (r, y)$ and $\mu_s(t, x') = (r, y')$ for some $s > 0$. Then

$$M^2_h = [(Z \setminus B)/ \sim_\mu] \sqcup [Z \setminus B]. \tag{3.4}$$

For $M = \mathbb{R}^n$ this space has a natural $\mathcal{C}^\infty$ structure, as a manifold with corners, which restricts to that on $Z \setminus B$ and which is generated by those $\mathcal{C}^\infty$ functions on $Z \setminus B$ which are homogeneous of non-negative integral degrees under μ_s (meaning this space of functions includes local coordinate systems). This $\mathcal{C}^\infty$ structure is independent of the coordinates in $\mathbb{R}^n$ and so is defined in the general case of a manifold M. In terms of the definition (2.12)

$$M^2_h = {}^+SN\{Z; B, S\} \sqcup [Z \setminus B] \tag{3.5}$$

where ${}^+SN\{Z; B, S\}$ is the inward-pointing part of the S-parabolic normal bundle to B in Z. The first term in (3.5) forms the front face, denoted tf, the other boundary face will be denoted tb . Defining functions for these faces will be written ρ_{tf} and ρ_{tb}. Notice that

$$\beta^*_h t = \rho^2_{\mathrm{tf}} \rho_{\mathrm{tb}} \tag{3.6}$$

for an appropriate choice of these defining functions.

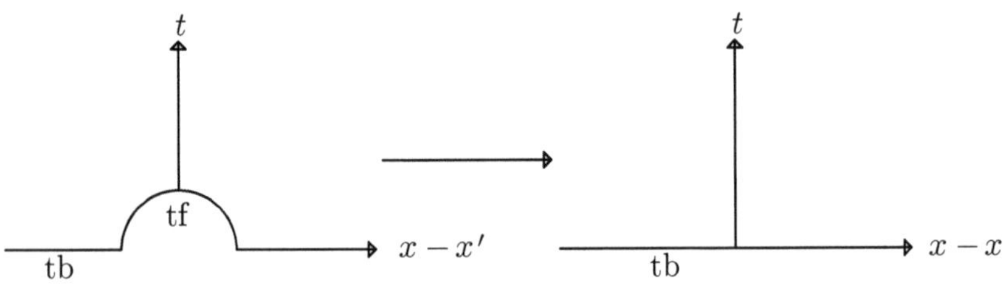

FIGURE 2. $\beta_h : Z_h = M^2_h \to Z$

Since we wish to consider $\widetilde{\Phi}'$ as a convolution operator we need to consider density factors; for the usual reasons of simplicity we work with half-densities as the basic coefficient bundle. In case $M = \mathbb{R}^n$ the half-density

$$\Phi'' = \Phi'(t, x - x')|dtdxdx'|^{\frac{1}{2}}, \ x, x' \in \mathbb{R}^n, \tag{3.7}$$

lifts to M^2_h to a smooth half-density away from tf which extends to be of the form

$$\Phi = \beta^*_h \Phi'' \in \rho_{\mathrm{tf}}^{-\frac{n}{2}+\frac{1}{2}} \rho^\infty_{\mathrm{tb}} \mathcal{C}^\infty(M^2_h; \Omega^{\frac{1}{2}}). \tag{3.8}$$

The notation here means that, for every $k \in \mathbb{N}$, $\rho_{\mathrm{tf}}^{\frac{n}{2}-\frac{1}{2}} \rho_{\mathrm{tb}}^{-k} \Phi \in \mathcal{C}^\infty(M^2_h; \Omega^{\frac{1}{2}})$. In particular Φ vanishes to infinite order at tb . We shall hide the singular factor of ρ_{tf} in (3.8), since it is of geometric origin, by defining a new bundle, the kernel density bundle KD, by the prescription

$$\mathcal{C}^\infty(M^2_h; \mathrm{KD}) = \rho_{\mathrm{tf}}^{-\frac{n}{2}-\frac{3}{2}} \mathcal{C}^\infty(M^2_h; \Omega^{\frac{1}{2}}). \tag{3.9}$$

The weighting here is chosen so that the identity is (for the moment formally) of order 0. Then we can write (3.8) in the form $\Phi \in \rho_{\mathrm{tf}}^2 \rho_{\mathrm{tb}}^\infty \mathcal{C}^\infty(M_h^2; \mathrm{KD})$. It is important to note that (3.9) does indeed define a new vector bundle. Since we shall use such constructions significantly below we give a general result of this type in §4. In particular Proposition 4.1 applied with the trivial filtration of $\Omega^{\frac{1}{2}}$, shows that (3.9) defines the vector bundle KD. In this same sense we can write

$$\mathrm{KD} = \rho_{\mathrm{tb}}^{\frac{n}{2}+1} \beta_h^* \left[t^{-\frac{n}{2}-1} \Omega^{\frac{1}{2}}(Z) \right]. \tag{3.10}$$

The singularity of KD at tb is not very important, since all the kernels vanish to infinite order there.

This discussion is all for the case $M = \mathbb{R}^n$. However the definition (3.9) extends to the general case and we simply set

$$\Psi_h^{-k}(M; \Omega^{\frac{1}{2}}) = \rho_{\mathrm{tf}}^k \rho_{\mathrm{tb}}^\infty \mathcal{C}^\infty(M_h^2; \mathrm{KD}) \text{ for } k \in \mathbb{N}. \tag{3.11}$$

These are to be the elements of the 'heat calculus' of negative integral order. To define the elements of order zero observe that we can define a leading part of any element $A \in \rho_{\mathrm{tf}}^k \rho_{\mathrm{tb}}^\infty \mathcal{C}^\infty(M_h^2; \mathrm{KD})$ by setting

$$N_{\mathrm{tf}}(A) = t^{-k/2} A_{\restriction \mathrm{tf}} \in \dot{\mathcal{C}}^\infty(\mathrm{tf}; \mathrm{KD}) \tag{3.12}$$

where the dot indicates that the resulting section of KD vanishes to infinite order at the boundary of the compact manifold with boundary $\mathrm{tf}(M_h^2)$. From the definition, (3.5), the front face fibres over $B \cong M$. The fibres are half-spheres (or balls) of dimension $n = \dim M$:

$$\begin{array}{ccc} \mathbb{S}_+^n & \longrightarrow & \mathrm{tf}(M_h^2) \\ & & \big\downarrow \pi_{\mathrm{tf}} \\ & & B \cong M. \end{array} \tag{3.13}$$

In fact the interiors of the fibres of (3.13) have natural linear structures, coming from the definition of ${}^+SN\{Z; B, S\}$. Namely tf is a compactification of the normal bundle to the diagonal, $\mathrm{Diag} \subset M^2$, which in turn is naturally isomorphic to TM so $TM \hookrightarrow \mathrm{tf}(M_h^2)$ is the interior. Using t, as in (3.10), to remove the singular powers from the kernel density bundle and noting that the lift of the density bundle on B is naturally isomorphic to the density bundle on TM this allows us to identify $\dot{\mathcal{C}}^\infty(\mathrm{tf}(M_h^2); \mathrm{KD}) \longleftrightarrow \mathcal{S}(TM; \Omega_{\mathrm{fibre}})$. Thus for each k the normal map (3.12) can be regarded as a map

$$N_{h,-k} : \Psi_h^{-k}(M; \Omega^{\frac{1}{2}}) \twoheadrightarrow \mathcal{S}(TM; \Omega_{\mathrm{fibre}}), \ k \in \mathbb{N}. \tag{3.14}$$

The map also extends to the space defined by the right side of (3.11) for $k = 0$. However, using the fact that the fibre integral is well defined on the right image of (3.14), we actually set

$$\begin{aligned}\Psi^0_h(M;\Omega^{\frac12}) &= \mathcal{C}^\infty(M)\,\mathrm{Id}\oplus\overline{\Psi}^0_h(M;\Omega^{\frac12}),\\ \overline{\Psi}^0_h(M;\Omega^{\frac12}) &= \Big\{A\in\rho_{\mathrm{tb}}^\infty\mathcal{C}^\infty(M^2_h;\mathrm{KD});\ \int\limits_{\text{fibre}} N_{h,0}(A)=0\Big\}.\end{aligned}\tag{3.15}$$

This of course is just the analogue of the usual 'mean value zero' condition for singular integrals.

To see how the operators $\Psi^k_h(M;\Omega^{\frac12})$ act, let $M_h=[0,\infty)\times M$ and consider the bilinear map

$$\begin{aligned}&\dot{\mathcal{C}}^\infty_c(M_h;\Omega^{\frac12})\times\dot{\mathcal{C}}^\infty(M_h;\Omega^{\frac12})\ni(\phi,\psi)\\ &\qquad\longmapsto \phi\hat{*}_t\psi=\int\limits_0^\infty\phi(t+t',x)\psi(t',x')dt'\in\mathcal{C}^\infty_c(Z;\Omega^{\frac12}).\end{aligned}\tag{3.16}$$

Lifting to M^2_h we can define

$$\langle A\psi,\phi\rangle=\int\limits_{M^2_h}A\cdot\beta^*_h(\phi\hat{*}_t\psi),\ A\in\Psi^{-k}_h(M;\Omega^{\frac12}),\ k\in\mathbb{N}\tag{3.17}$$

since the integrand is in the product

$$\rho^k_{\mathrm{tf}}\rho^\infty_{\mathrm{tb}}\mathcal{C}^\infty(M^2_h;\mathrm{KD})\cdot\beta^*_h\mathcal{C}^\infty_c(Z;\Omega^{\frac12})\subset\rho^{k-1}_{\mathrm{tf}}\mathcal{C}^\infty_c(M^2_h;\Omega)\subset L^1(M^2_h;\Omega)\text{ if }k\geq1.\tag{3.18}$$

For operators of order 0 the limiting form of the same definition applies to the second term in (3.15) since

$$A\in\overline{\Psi}^0_h(M;\Omega^{\frac12})\Longrightarrow\langle A\psi,\phi\rangle=\lim_{\epsilon\downarrow0}\int\limits_{\{p\in M^2_h;\rho_{\mathrm{tf}}\geq\epsilon\}}A\cdot\beta^*_h(\phi\hat{*}_t\psi)\tag{3.19}$$

exists, independent of the choice of ρ_{tf} (which can be replaced by $t^{\frac12}$). Of course $\mathrm{Id}\in\Psi^0_h(M;\Omega^{\frac12})$ acts as the identity. Thus we find that

$$\begin{aligned}&A\in\Psi^0_h(M;\Omega^{\frac12})\text{ defines an operator}\\ &A:\dot{\mathcal{C}}^\infty([0,\infty)\times M;\Omega^{\frac12})\longrightarrow\mathcal{C}^{-\infty}([0,\infty)\times M;\Omega^{\frac12}).\end{aligned}\tag{3.20}$$

The image space here is just the dual space to $\dot{\mathcal{C}}^\infty([0,\infty)\times M;\Omega^{\frac12})$ and contains it as a dense subspace in the weak topology. In fact the range of A in (3.20) is always contained in $\dot{\mathcal{C}}^\infty([0,\infty)\times M;\Omega^{\frac12})$, as is shown in Lemma 3.1 below, so these operators can be composed. Before stating these results we note how to extend the discussion to general vector bundle coefficients.

Suppose that U and W are vector bundles over M. The (diagonal) homomorphism bundle from U to V over M is denoted $\hom(U,V)$, the (full) homomorphism bundle over M^2 is denoted $\operatorname{Hom}(U,V)$:

$$\hom(U,W) \cong \bigsqcup_{x\in M} W_x \otimes U'_x,\ \operatorname{Hom}(U,W) \cong \bigsqcup_{x,x'\in M} W_x \otimes U'_{x'}. \tag{3.21}$$

To 'reduce' general operators to operators on half-densities consider the bundle

$$\operatorname{Hom}_\Omega(U,W) = \bigsqcup_{x,x'\in M} (W_x \otimes \Omega_x^{-\frac{1}{2}}) \otimes (U'_{x'} \otimes \Omega_{x'}^{\frac{1}{2}}) \equiv \operatorname{Hom}(U\otimes\Omega^{-\frac{1}{2}}, W\otimes\Omega^{-\frac{1}{2}}) \tag{3.22}$$

with the half-density bundles those on M. We define the general kernels by taking the tensor product, over $\mathcal{C}^\infty(M_h^2)$ of the space of $\mathcal{C}^\infty$ section of the lift of this bundle and the kernels already discussed:

$$\begin{aligned}\Psi_h^{-k}(M;U,W) &= \Psi_h^{-k}(M;\Omega^{\frac{1}{2}}) \otimes_{\mathcal{C}^\infty(M_h^2)} \mathcal{C}^\infty(M_h^2; \beta_h^* \operatorname{Hom}_\Omega(U,W)) \\ &= \rho_{\mathrm{tf}}^{-k}\rho_{\mathrm{tb}}^\infty \mathcal{C}^\infty(M_h^2; \operatorname{KD}(U,W))\end{aligned} \tag{3.23}$$

with the modified kernel density bundle

$$\operatorname{KD}(U,W) = \operatorname{KD} \otimes \beta_h^* \operatorname{Hom}_\Omega(U,W). \tag{3.24}$$

If $U = W$, which is often the case, we denote the space as $\Psi_h^{-k}(M;U)$.

The boundary hypersurface tf lies above the diagonal so the additional density factors in (3.23) cancel there. The normal operator therefore extends to a surjective linear map

$$N_{h,-k} : \Psi_h^{-k}(M;U,W) \twoheadrightarrow \mathcal{S}(TM; \Omega_{\mathsf{fibre}} \otimes \pi^* \hom(U,W)),\ k \in \mathbb{N}. \tag{3.25}$$

We shall make a further small, but significant, refinement of this construction. The Taylor series at tf of $\mathcal{C}^\infty$ functions on M_h^2 are generated by the homogeneous functions under (3.1) in any local coordinates. Now we can choose these local coordinates to be t, $x_j - y_j$ and $x_j + y_j$ where x and y are the same local coordinates in the two factors of M. Under the involution, J, on M^2, which interchanges the factors, $x_j - y_j$ is odd and $x_j + y_j$ is invariant. So consider the subspace $\mathcal{C}_E^\infty(M_h^2) \subset \mathcal{C}^\infty(M_h^2)$ fixed by the condition that its elements have Taylor series at tf with terms of even homogeneity invariant under J and terms of odd homogeneity odd under J. If $\rho = (t + |x-y|^2)^{\frac{1}{2}}$ then the Taylor series at tf of a general $\mathcal{C}^\infty$ function on M_h^2 is of the form

$$\sum_{k=0}^\infty \rho^k F_k\left(\frac{t}{\rho^2}, \frac{x-y}{\rho}, x+y\right) \tag{3.26}$$

where the F_k are $\mathcal{C}^\infty$ functions on $\mathbb{R}^{2n+1}$ away from 0. It is therefore clear that the space $\mathcal{C}_E^\infty(M_h^2)$ is well defined independent of the choice of coordinates. Similarly the space $\mathcal{C}_O^\infty(M_h^2) \subset \mathcal{C}^\infty(M_h^2)$ is fixed by requiring the F_k to be odd or even in the second variables for k even or odd respectively. Then $\mathcal{C}_E^\infty(M_h^2) + \mathcal{C}_O^\infty(M_h^2) = \mathcal{C}^\infty(M_h^2)$ and the intersection $\mathcal{C}_E^\infty(M_h^2) \cap \mathcal{C}_O^\infty(M_h^2) = \rho_{\mathrm{tf}}^\infty \mathcal{C}^\infty(M_h^2)$ consists of the functions with trivial Taylor series at tf.

Notice that $\mathcal{C}^\infty(Z)$ lifts under β_h into $\mathcal{C}^\infty_E(M^2_h)$. This means that we can define the spaces $\mathcal{C}^\infty_E(M^2_h;\beta_h^*U)$ and $\mathcal{C}^\infty_O(M^2_h;\beta_h^*U)$ for any vector bundle over M^2. Since we can certainly choose defining functions $\rho_{\mathrm{tf}} \in \mathcal{C}^\infty_O(M^2_h)$, $\rho_{\mathrm{tb}} \in \mathcal{C}^\infty_E(M^2_h)$ and also $t^{\frac12} \in \mathcal{C}^\infty_O(M^2_h)$ this means we can define the odd and even parts of the heat calculus using (3.10) and (3.11). We define

$$\Psi^{-k}_{h,E}(M;U,W) = \begin{cases} \rho^k_{\mathrm{tf}}\rho^\infty_{\mathrm{tb}}\mathcal{C}^\infty_E(M^2_h;\mathrm{KD}(U,W)), & k \text{ even} \\ \rho^k_{\mathrm{tf}}\rho^\infty_{\mathrm{tb}}\mathcal{C}^\infty_O(M^2_h;\mathrm{KD}(U,W)), & k \text{ odd}. \end{cases} \tag{3.27}$$

For $k = 0$ we define

$$\begin{aligned} \Psi^0_{h,E}(M;U,V) &= \mathcal{C}^\infty(M)\,\mathrm{Id} \oplus \overline{\Psi}^0_{h,E}(M;U,V) \text{ with} \\ \overline{\Psi}^0_{h,E}(M;U,V) &= \overline{\Psi}^0_h(M;U,V) \cap \mathcal{C}^\infty_E(M^2_h;\mathrm{KD}(U,W)). \end{aligned} \tag{3.28}$$

Let $[0,\infty)_{\frac12}$ be the half line with $t^{\frac12}$ as smooth variable.

Lemma 3.1. *Each element $A \in \Psi^{-k}_h(M;U,W)$, for a compact manifold M and any $k \geq 0$, defines a continuous linear map*

$$A : \dot{\mathcal{C}}^\infty([0,\infty)\times M;U) \longrightarrow \dot{\mathcal{C}}^\infty([0,\infty)\times M;V) \tag{3.29}$$

and the same pairing, (3.17), *leads to a continuous linear map*

$$A : \mathcal{C}^\infty([0,\infty)\times M;U) \longrightarrow t^{\frac{k}{2}}\mathcal{C}^\infty([0,\infty)_{\frac12}\times M;V). \tag{3.30}$$

For an element $A \in \Psi^{-k}_{h,E}(M;U,V)$ the operator (3.30) *has range in*

$$t^{[\frac{k}{2}]}\mathcal{C}^\infty([0,\infty)\times M;V).$$

Here $S = [s]$ is the largest integer satisfying $S \leq s$.

Proof. We give a rather 'geometric' proof of this regularity result, in the spirit of [28]. That is we introduce the singular coordinates needed to analyze the integral in the action of the operators by defining certain blown-up spaces. In the process of showing (3.30) we shall in essence work with the t-variable coefficient heat calculus.

Thus consider the product

$$\begin{aligned} Z_2 = [0,\infty)^2 \times M^2 &= [0,\infty)_{t-t'} \times M \times [0,\infty)_{t'} \times M \\ &= \{(t,t',q); q \in M^2, t,t' \in \mathbb{R}, t \geq t' \geq 0\}, \end{aligned} \tag{3.31}$$

with two 'time' variables. There are the three obvious projections,

$$\begin{aligned} &\pi_L(t,t',m,m') = (t,m),\ \pi_R(t,t',m,m') = (t',m') \text{ and} \\ &\pi_K(t,t',m,m') = (t-t',m,m'). \end{aligned} \tag{3.32}$$

These combine to give a diagram:

$$\begin{array}{ccc} & & [0,\infty)\times M \\ & \nearrow^{\pi_R} & \\ Z \xleftarrow{\ \pi_K\ } Z_2 & & \\ & \searrow_{\pi_L} & \\ & & [0,\infty)\times M. \end{array} \tag{3.33}$$

The space $[0,\infty)^2 \times M^2$ is not symmetric in t, t' and this is reflected in the fact that the right projection is a fibration whereas the map π_L is not; in fact it is not even a b-map. To compensate for this asymmetry we need only blow up the submanifold (the corner) $t = t' = 0$; set

$$M_2^2 = [Z_2; \{t = t' = 0\}], \ \beta_2 : M_2^2 \longrightarrow Z_2. \tag{3.34}$$

After this blow-up none of the three lifted projections is a fibration but all three are now b-fibrations.

Consider the lift under π_K of the submanifold B, in (3.2), blown up in (3.3). Under the blow-up of $t = t' = 0$ this further lifts to two submanifolds:

$$\begin{aligned} B_1 &= \operatorname{cl}\big\{\big(\pi_K^{-1}(B)\big) \setminus \{t = t' = 0\}\big\} \text{ in } M_2^2 \text{ and} \\ B_2 &= \beta_2^{-1}(\{(0,0,q); q \in \operatorname{Diag} \subset M^2\}. \end{aligned} \tag{3.35}$$

These submanifolds are each contained in one boundary hypersurface; $B_1 \subset \mathrm{df}$ and $B_2 \subset \mathrm{ff}$ where df is the lift of $t = t'$. Then we further blow up the space, along these submanifolds and parabolically in the direction of the conormal to the respective boundary hypersurfaces:

$$M_{2h}^2 = [M_2^2; B_2, N^* \mathrm{ff}; B_1, N^* \mathrm{df}]. \tag{3.36}$$

The order of blow-up here is important.

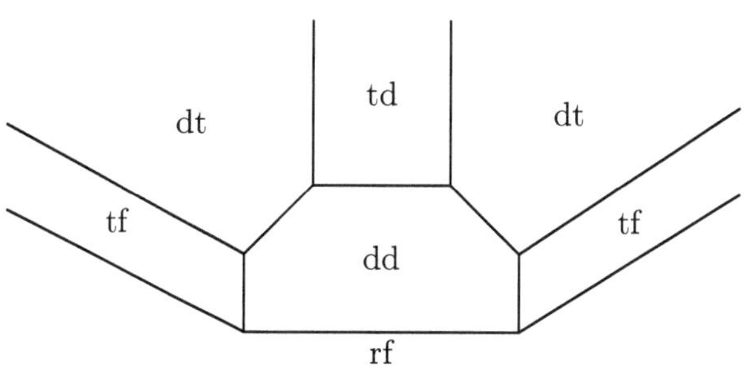

FIGURE 3. Faces of M_{2h}^2

Now the diagram of maps (3.33) lifts to a triple of b-fibrations:

$$\begin{array}{ccccc} & & & & [0,\infty)\times M \\ & & & \nearrow^{\pi_{2,R}} & \\ M_h^2 & \xleftarrow{\pi_{2,K}} & M_{2h}^2 & & \\ & & & \searrow_{\pi_{2,L}} & \\ & & & & [0,\infty)\times M. \end{array} \tag{3.37}$$

Certainly the left and right lifted projections exist, since they are just composites of the blow-down map from M_{2h}^2 with the maps in (3.33). That they are b-fibrations follows from the fact that the lifts to M_2^2 are b-fibrations and that each of the two subsequent blow-ups in (3.37) is of a submanifold to which the corresponding map is b-transversal. Similarly the map to M_h^2 arising from the lift of π_K is a b-fibration because of the lifing theorem for b-fibrations in [28], i.e., because the b-fibration from M_2^2 to $[0,\infty)\times M^2$ coming from the lift of π_K is transversal to B and the components of the lift of B under it, just B_1 and B_2, are blown up with the appropriate, lifted, parabolic directions.

Thus (3.37) is a diagram of b-fibrations. The space M_{2h}^2 has five boundary hypersurfaces, two of them arising from the lifts of $t'=0$ and of $t=t'$ and the remaining three produced by the blow-ups of $t=t'=0$, of B_2 and of B_1. We shall denote them rf, dt, tf, dd and td respectively. The lifts of defining functions are then easily computed and for appropriate choices of defining functions

$$\begin{aligned} \pi_{2,K}^*\rho_{\mathrm{tf}} &= \rho_{\mathrm{dd}}\rho_{\mathrm{td}}, & \pi_{2,K}^*\rho_{\mathrm{tb}} &= \rho_{\mathrm{dt}}\rho_{\mathrm{tf}} \\ \pi_{2,R}^*t' &= \rho_{\mathrm{rf}}\rho_{\mathrm{tf}}\rho_{\mathrm{dd}}^2 \quad\text{and} & \pi_{2,L}^*t &= \rho_{\mathrm{tf}}\rho_{\mathrm{dd}}^2. \end{aligned} \tag{3.38}$$

Similarly the product of the lift of smooth positive half-densities from each of the images $[0,\infty)\times M$, in (3.37) with the lift of a smooth section of the kernel density bundle on M_2^2 and of $|dt|^{\frac12}$ is easily computed. It follows that

$$\pi_{2,K}^*\,\mathrm{KD}\cdot\pi_{2,L}^*(\Omega^{\frac12}\otimes|dt|^{\frac12})\cdot\pi_{2,R}^*\Omega^{\frac12} = \rho_{\mathrm{tf}}\rho_{\mathrm{dd}}\rho_{\mathrm{td}}^{-1}\Omega \tag{3.39}$$

Now if $\phi\in\dot{\mathcal{C}}^\infty([0,\infty)\times M;\Omega^{\frac12})$ and $\psi\in\mathcal{C}^\infty([0,\infty)\times M;\Omega^{\frac12})$ the action of $A\in\Psi_h^k(M;\Omega^{\frac12})$, where for the moment we assume that $k>0$, on ϕ can be written

$$A\phi\cdot\psi = (\pi_{2,L})_*\left((\pi_{2,K})^*A\cdot(\pi_{2,L})^*(\psi)\cdot(\pi_{2,R})^*\phi\right). \tag{3.40}$$

Applying (3.39) and (3.38) it can be seen that

$$A\phi\cdot\psi\subset(\pi_{2,L})_*\left(\rho_{\mathrm{tf}}^\infty\rho_{\mathrm{dt}}^\infty\rho_{\mathrm{dd}}^\infty\rho_{\mathrm{td}}^{k-1}\Omega)\right)\subset\dot{\mathcal{C}}^\infty([0,\infty)\times M;\Omega) \tag{3.41}$$

which gives (3.29). The inclusion in (3.41) follows from the push-forward theorem for conormal distributions under b-fibrations, from [29]. If instead $\phi\in\mathcal{C}^\infty([0,\infty)\times M;\Omega^{\frac12})$, still with $k>0$, then a similar computation and the same theorem shows that

$$A\phi\cdot\psi\subset(\pi_{2,L})_*\left(\rho_{\mathrm{tf}}^\infty\rho_{\mathrm{dd}}^{k+2}\rho_{\mathrm{td}}^{k-1}\Omega)\right)\subset t^{\frac{k}{2}}\mathcal{C}^\infty([0,\infty)_{\frac12}\times M;\Omega) \tag{3.42}$$

which proves (3.30).

The case $k = 0$ is similar except that the integral implicit in (3.41) or (3.42) is not absolutely convergent at $t = t'$, i.e., $\mathrm{dd}(M^2_{2h})$. However the mean value condition in (3.15) makes the integral conditionally convergent and the same results, (3.29) and (3.30), follow.

The improved regularity in the case of an operator in the even part of the calculus follows from the fact that non-integral powers of t in the Taylor series expansion in $t^{\frac{1}{2}}$ would arise from the odd part of the integrand and hence vanish. □

We now turn to the proofs of the composition results in §1.

Proof of Proposition 1.1 . To prove this composition result we proceed very much as above in the proof of Lemma 3.1. Thus we first construct a 'triple' space to which the two kernels can be simultaneously lifted. Set

$$Z_3 = \{(t,t') \in \mathbb{R}^2; t' \geq 0, t \geq t'\} \times M^3 \tag{3.43}$$

and consider the three maps:

$$\begin{aligned} \pi_o &: Z_3 \longrightarrow Z, \ o = f, c, s \\ \pi_f(t,t',x,x',x'') &= (t',x',x'') \\ \pi_s(t,t',x,x',x'') &= (t-t',x,x') \\ \pi_c(t,t',x,x',x'') &= (t,x,x''). \end{aligned} \tag{3.44}$$

the first two of which are projections. The diagram:

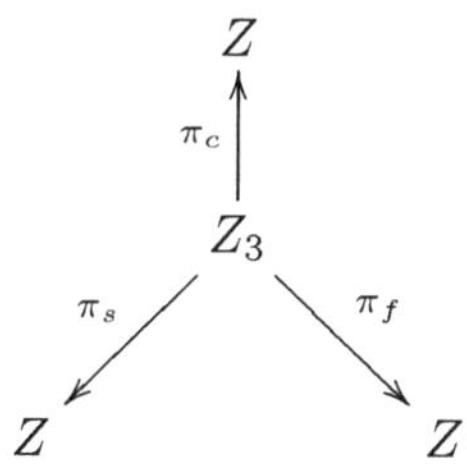

(3.45)

is a symbolic representation of the composition of operators $A, B \in \Psi^{-\infty}_H(M; \Omega^{\frac{1}{2}})$ in the sense that if $C = A \circ B$ then

$$C = (\pi_c)_* \left[(\pi_s)^* A \cdot (\pi_f)^* B\right]. \tag{3.46}$$

We define blown-up versions of Z_3 by defining the three partial diagonals:

$$B_o = \pi_o^{-1}(B), \ o = f, c, s \tag{3.47}$$

and the triple surface, which is the intersection of any pair in (3.47):

$$B_3 = \{(0,0,x,x,x) \in Z_3\}. \tag{3.48}$$

Similarly set

$$S_o = \pi_o^*(S) \subset N^*(B_o) \Longrightarrow S_f = \mathrm{sp}(dt'), \ S_s = \mathrm{sp}(dt - dt'), \ S_c = \mathrm{sp}(dt) \tag{3.49}$$

and

$$S_3 = \mathrm{sp}(dt, dt') \text{ over } B_3. \tag{3.50}$$

Consider first the manifold with corners defined by iterated parabolic blow-up (this is discussed in [18]Appendix B):

$$Z_{3,1} = [Z_3; B_3, S_3; B_f, S_f; B_s, S_s] \xrightarrow{\beta_{3,1}} Z_3. \tag{3.51}$$

The order of blow-up amongst the last two submanifolds is immaterial since they lift to be disjoint in $[Z_3; B_3, S_3]$. In fact, since we can also interchange the blow-up of B_3 and either B_s or B_f, we have natural $\mathcal{C}^\infty$ maps

$$\begin{gathered} Z_{3,1} \equiv [Z_3; B_f, S_f; B_3, S_3; B_s, S_s] \xrightarrow{\pi_{2,f}} Z_h (= M_h^2) \\ \pi_{2,f} : [Z_3; B_f, S_f; B_3, S_3; B_s, S_s] \longrightarrow [Z_3; B_f, S_f] \equiv Z_h \times [0,\infty) \times M \longrightarrow Z_h \\ Z_{3,1} \equiv [Z_3; B_s, S_s; B_3, S_3; B_f, S_f] \xrightarrow{\pi_{2,s}} Z_h \\ \pi_{2,s} : [Z_3; B_s, S_s; B_3, S_3; B_f, S_f] \longrightarrow [Z_3; B_s, S_s] \equiv Z_h \times [0,\infty) \times M \longrightarrow Z_h. \end{gathered} \tag{3.52}$$

These maps give a commutative diagram with the bottom part of (3.45):

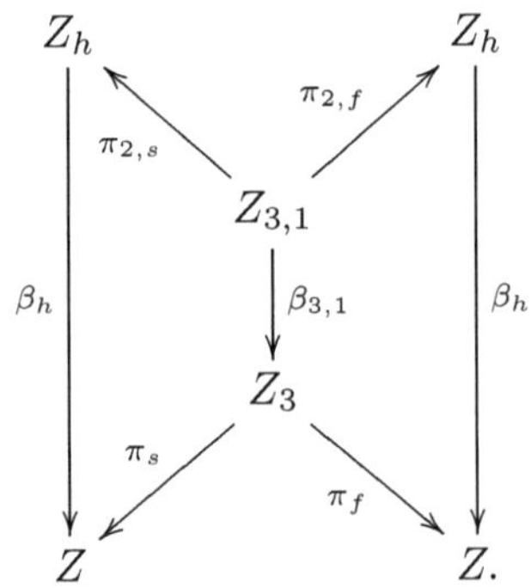

(3.53)

This allows us to lift the product of the kernels in (3.46) to $Z_{3,1}$ by lifting the individual kernels under $\pi_{2,f}$ and $\pi_{2,s}$:

$$\beta_{3,1}^* [(\pi_s)^* A \cdot (\pi_f)^* B] = (\pi_{2,s})^* A \cdot (\pi_{2,f})^* B. \tag{3.54}$$

Using (3.10) we can write the kernel as

$$B = bt^{-\frac{n}{2}-1+k/2}\nu,\ \nu \in \mathcal{C}^\infty(Z; \Omega^{\frac{1}{2}}),\ b \in \mathcal{C}^\infty(Z_h),\ b \equiv 0 \text{ at tb}. \tag{3.55}$$

The manifold $Z_{3,1}$ has five boundary hypersurfaces, the two 'trivial' faces tr and tl arising from the lifts of $t' = 0$ and $t = t'$ respectively and the three faces created by blow-up; namely tt arising from the blow-up of B_3, sf arising from the blow-up of $B_{2,f}$ and ss arising from the blow-up of $B_{2,s}$. Clearly

$$\begin{aligned} \mathcal{C}^\infty(Z_{3,1}) \ni (\pi_{2,f})^* b \equiv 0 \text{ at tr} \\ \mathcal{C}^\infty(Z_{3,1}) \ni (\pi_{2,s})^* a \equiv 0 \text{ at tl} \end{aligned} \tag{3.56}$$

Thus the product vanishes to infinite order at two of the boundary hypersurfaces, i.e., has non-trivial Taylor series only at sf, ss and tt. If we take into account the fact that $\pi_{2,f}t'$ and $\pi_{2,s}(t-t')$ vanish to second order at tt we conclude that the product in (3.54) is of the form

$$\begin{gathered}\beta_{3,1}^*\left[(\pi_s)^*A\cdot(\pi_f)^*B\right]=\rho_{\mathrm{tt}}^{-2n-4+j+k}\rho_{\mathrm{sf}}^{-n-1+k}\rho_{\mathrm{ss}}^{-n-1+j}c(\pi_{2,f})^*\nu(\pi_{2,s})^*\nu,\\ c\in\mathcal{C}^\infty_{\mathrm{tt}}(Z_{3,1}),\ \text{i.e.,}\ \mathcal{C}^\infty(Z_{3,1})\ni c\equiv 0\ \text{at}\ \mathrm{tr}\cup\mathrm{tl}.\end{gathered}\tag{3.57}$$

In particular the product of the kernels vanishes to infinite order at the corner, B', produced by the intersection of tl and tr in $Z_{3,1}$. Consider the manifold, $Z_{3,2}$ defined by blowing this up, parabolically with respect to both normal directions:

$$Z_{3,2}=[Z_{3,1};B',N^*B'],\ B'=\mathrm{tr}\cap\mathrm{tl}.\tag{3.58}$$

This adds another boundary hypersurface, td, but makes no essential difference to the kernel so that (3.57) becomes, with the same notation used for the other boundary hypersurfaces and their lifts,

$$\begin{gathered}\beta_{3,2}^*\left[(\pi_s)^*A\cdot(\pi_f)^*B\right]=\rho_{\mathrm{tt}}^{-2n-4+j+k}\rho_{\mathrm{sf}}^{-n-1+k}\rho_{\mathrm{ss}}^{-n-1+j}c(\pi_{2,f})^*\nu(\pi_{2,s})^*\nu,\\ c\in\mathcal{C}^\infty_{\mathrm{tt}}(Z_{3,1}),\ \text{i.e.,}\ \mathcal{C}^\infty(Z_{3,2})\ni c\equiv 0\ \text{at}\ \mathrm{tr}\cup\mathrm{tl}\cup\mathrm{td}.\end{gathered}\tag{3.59}$$

Having arrived at $Z_{3,2}$ with a 'simple' kernel we need to map back to Z_h. The manifold $Z_{3,2}$ can be constructed in another way, using the commutability of appropriate blow-ups. Thus, the final blow-up in (3.58) does not meet ss or sf so can be performed after that of B_3 in (3.51). Furthermore, B_3 is then a submanifold of the corner, $Y=\{t=t'=0\}$ being blown up, with the same parabolic directions. The order can therefore be interchanged and so

$$Z_{3,2}=[Z_3;B',S_3;B_3,S_3;B_f,S_f;B_s,S_s]\overset{\beta_{3,2}}{\longrightarrow}Z_3\tag{3.60}$$

This means that the third map in (3.44) lifts into a b-fibration from $Z_{3,2}$ to Z_h as we proceed to show.

Indeed, consider the blown-up space

$$\beta':\ Z_3'=[Z_3;B']\longrightarrow Z_3,\ B'=\{t=t'=0\}.\tag{3.61}$$

The composite map is then a fibration

$$\pi_c'=\pi_c\cdot\beta':\ Z_3'\longrightarrow Z.\tag{3.62}$$

Parabolically blowing up the lift, which we can denote $B_{2,c}$, to Z_3', of the submanifold B, in Z gives a further fibration

$$\pi_c'':Z_{3,1}'=[Z_3;B';B_{2,c},S_{2,c}]\longrightarrow Z_h\tag{3.63}$$

where the lift of $S_{2,c}$ is just the intersection of the conormal bundle of $(\pi_c^1)^{-1}(B)$ with the conormal bundle to the front face of Z_3'. Consider next the blow-up in $Z_{3,1}'$ of the lift of B_3, which is a submanifold of $B_{2,c}$:

$$\beta_{3,2}':Z_{3,2}'=[Z_3;B';B_{2,c},S_{2,c};B_3,S_3]\longrightarrow Z_{3,1}'.\tag{3.64}$$

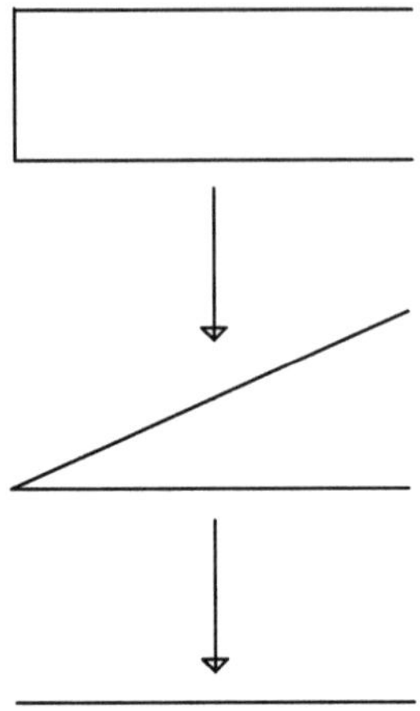

FIGURE 4. $\pi'_c = \pi_c \circ \beta' : Z'_3 \to Z$

Now we can also blow up the other two partial diagonals, lifted to $Z'_{3,2}$, and again use commutation for non-intersecting submanifolds to write

$$\begin{aligned} Z''_{3,2} &= [Z'_{3,2}; B_{2,s}, S_{2,s}; B_{2,f}, S_{2,f}] \\ &\equiv [Z_3; B'; B_3, S_3; B_{2,s}, S_{2,s}; B_{2,f}, S_{2,f}; B_{2,c}, S_{2,c}]\,. \end{aligned} \tag{3.65}$$

This means that there is a blow-down map (for the lift of $B_{2,c}$)

$$Z''_{3,2} \longrightarrow Z_{3,2}, \ Z''_{3,2} \equiv [Z_{3,2}; B_{2,c}, S_{2,c}]. \tag{3.66}$$

Since the density in (3.59) vanishes to infinite order at the submanifold, $B_{2,c} \subset B'$, blown up in (3.66) we also conclude that

$$\begin{aligned} [(\pi_s)^*A \cdot (\pi_f)^*B] \ \text{lifts to } & \rho_{\mathrm{tt}}^{-2n-4+j+k} \rho_{\mathrm{sf}}^{-n-1+k} \rho_{\mathrm{ss}}^{-n-1+j} c(\pi_{2,f})^*\nu(\pi_{2,s})^*\nu, \\ & \mathcal{C}^\infty(Z''_{3,2}) \ni c \equiv 0 \ \text{at} \ \mathrm{tr} \cup \mathrm{tl} \cup \mathrm{td} \cup \mathrm{sc}, \end{aligned} \tag{3.67}$$

where sc is the hypersurface produced by the blow-up of $B_{2,c}$.

The last step is to consider the push-forward of this density under the map from $Z''_{3,2}$ to Z_h given by (3.63). We wish to consider the image, a half-density on Z_h, as a multiple of the lift of a smooth half-density on Z, as in (3.10), so simply multiply by the lift to Z_3 of ν under $\pi_{2,c}$. Lifting to $Z'_{3,1}$ this gives

$$\gamma(\pi_{2,c})^*\nu = (\pi'_c)^*(t^{-n-2+\frac{j+k}{2}})c''\mu \tag{3.68}$$

where μ is the lift to $Z'_{3,1}$ of the product of ν from Z under the maps π_o. Thus it is a non-vanishing density on Z_3 and lifted to $Z'_{3,1}$ it is an element of

$$\rho_{\mathrm{tt}}^{n+1}\mathcal{C}^\infty(Z'_{3,1};\Omega). \tag{3.69}$$

Inserting this into (3.68), using the rapid vanishing at all faces except tt shows that

$$\gamma(\pi_{2,c})^*\nu \in (\pi'_c)^*\left(t^{-(n+3)/2+(k+j)/2}\right)\dot{\mathcal{C}}^\infty_{\mathrm{tt}}(Z'_{3,1};\Omega). \tag{3.70}$$

The map π'_c in (3.63) is not a fibration but it is a b-fibration and from the push-forward results in [29] it follows that

$$(\pi'_c)_* : \dot{\mathcal{C}}^\infty_{\mathrm{tt}}(Z'_{3,1};\Omega) \longrightarrow \dot{\mathcal{C}}^\infty_{\mathrm{tf}}(Z_h;\Omega). \tag{3.71}$$

This proves the composition formula, since it shows that composite kernel is an element of $\Psi_h^{k+j}(M;\Omega^{\frac{1}{2}})$.

To see the last statement note that the composition formula shows that the Neumann series can be summed modulo a rapidly vanishing term. This reduces the consideration to $\mathrm{Id}+\Psi_h^{-\infty}(M;U)$, for which Duhamel's principle finishes the proof. □

Proof of Proposition 1.2. It suffices to prove (1.11) for vector fields. Namely, we need to show that if V is a vector field acting through the connection on U then $\Psi_h^{-j}(M;U) \ni A \to V\circ A \in \Psi_h^{-j+1}(M;U)$, and $N_{-j+1}(V\circ A) = \sigma_1(V)N_{-j}(A)$.

We first assume that U is the trivial bundle $\mathbb{C}$. The projective coordinates

$$t,\ X = \frac{x-x'}{t^{\frac{1}{2}}},\ x, \tag{3.72}$$

give a valid coordinate system near the front face, except at the corner, where $X=\infty$. Any $A\in\Psi_h^{-j}(M;U)$ can be written as

$$A = t^{\frac{j}{2}-\frac{n+2}{2}+\frac{n}{4}}\tilde{A}(t,X,x)|dtdXdx|^{\frac{1}{2}}, \tag{3.73}$$

where $\tilde{A}$ is smooth and vanishes rapidly as $X\to\infty$.

Now if $\phi=\phi_0|dtdx|^{\frac{1}{2}}\in\dot{\mathcal{C}}^\infty_c(M_h;\Omega^{\frac{1}{2}})$ and $\psi=\psi_0|dtdx|^{\frac{1}{2}}\in\dot{\mathcal{C}}^\infty(M_h;\Omega^{\frac{1}{2}})$, we have

$$\langle A\psi,\phi\rangle = \int_{M_h^2} t^{\frac{j}{2}-1}\tilde{A}\beta_h^*(\phi_0\hat{*}_t\psi_0)|dtdXdx|, \tag{3.74}$$

and $N_{-j}(A)=\tilde{A}(0,X,x)|dXdx|^{\frac{1}{2}}$.

Also, if we let $V=a(x)\partial_x$ be a smooth vector field and V' denote its transpose:

$$\langle V\psi,\phi\rangle = -\langle\psi,V'\phi\rangle, \tag{3.75}$$

then $V'=\partial_x a(x)$, and $\beta_h^*(t^{\frac{1}{2}}V') = t^{\frac{1}{2}}\partial_x a(x) + a(x)\partial_X$.

Now

$$\begin{aligned}\langle (V\circ A)\psi,\phi\rangle &= -\langle A\psi, V'\phi\rangle\\ &= -\int_{M_h^2} t^{\frac{j}{2}-1}\tilde{A}\beta_h^*((V'\phi_0)\hat{*}_t\psi_0)|dtdXdx|\\ &= -\int_{M_h^2} t^{\frac{j}{2}-1}\tilde{A}\beta_h^*(V')[\beta_h^*(\phi_0\hat{*}_t\psi_0)]|dtdXdx|\\ &= -\lim_{\epsilon\downarrow 0}\int_{t\geq\epsilon} t^{\frac{j-1}{2}-1}\tilde{A}\beta_h^*(t^{\frac{1}{2}}V')[\beta_h^*(\phi_0\hat{*}_t\psi_0)]|dtdXdx|,\end{aligned} \tag{3.76}$$

since $j\geq 1$. Integration by part gives

$$\langle (V\circ A)\psi,\phi\rangle = \lim_{\epsilon\downarrow 0}\int_{t\geq\epsilon} (\beta_h^*(t^{\frac{1}{2}}V'))'[t^{\frac{j-1}{2}-1}\tilde{A}]\beta_h^*(\phi_0\hat{*}_t\psi_0)|dtdXdx|, \tag{3.77}$$

where there is no boundary contribution because there is no integration by parts in the t direction. Since $(\beta_h^*(t^{\frac{1}{2}}V'))' = t^{\frac{1}{2}}a(x)\partial_x + a(x)\partial_X$, it follows this integral reduces to

$$\begin{aligned}&\lim_{\epsilon\downarrow 0}\int_{t\geq\epsilon} (t^{\frac{j}{2}-1}a(x)\partial_x\tilde{A} + t^{\frac{j-1}{2}-1}a(x)\partial_X\tilde{A})\beta_h^*(\phi_0\hat{*}_t\psi_0)|dtdXdx|\\ &= \int_{M_h^2} (t^{\frac{j}{2}-1}a(x)\partial_x\tilde{A} + t^{\frac{j-1}{2}-1}a(x)\partial_X\tilde{A})\beta_h^*(\phi_0\hat{*}_t\psi_0)|dtdXdx|,\end{aligned} \tag{3.78}$$

where the integral converges since

$$\int a(x)\partial_X\tilde{A}\beta_h^*(\phi_0\hat{*}_t\psi_0)|_{t=0}|dX| = 0. \tag{3.79}$$

Therefore

$$V\circ A = t^{\frac{j-1}{2}-\frac{n+2}{2}+\frac{n}{4}}(a(x)\partial_X\tilde{A} + t^{\frac{1}{2}}a(x)\partial_x\tilde{A})|dtdXdx|^{\frac{1}{2}} \tag{3.80}$$

is an element of $\Psi_h^{-j+1}(M;\Omega^{\frac{1}{2}})$. Moreover

$$N_{-j+1}(V\circ A) = a(x)\partial_X\tilde{A}(0,X,x)|dXdx|^{\frac{1}{2}} = \sigma_1(V)N_{-j}(A). \tag{3.81}$$

A similar computation works for D_t, except that

$$\beta_h^*(tD_t) = tD_t + \frac{i}{2}X\partial_X. \tag{3.82}$$

Therefore the integration by part will produce a boundary term, and

$$(\beta_h^*(tD_t))' = D_t t + \frac{i}{2}\partial_X X = D_t t + \frac{i}{2}(n + X\partial_X). \tag{3.83}$$

In fact, by carrying out the above computation for D_t, one find, for $k \geq 2$,

$$\begin{aligned}\langle (D_t \circ A)\psi, \phi\rangle = & \lim_{\epsilon\downarrow 0} \int_{t\geq\epsilon} (\beta_h^*(tD_t))'[t^{\frac{j-2}{2}-1}\tilde{A}]\beta_h^*(\phi_0 \hat{*}_t \psi_0)|dtdXdx| \\ & - \lim_{\epsilon\downarrow 0} \int_{t=\epsilon} it[t^{\frac{j-2}{2}-1}\tilde{A}]\beta_h^*(\phi_0 \hat{*}_t \psi_0)|dXdx|.\end{aligned} \tag{3.84}$$

The second term vanishes if $j > 2$, and becomes

$$-i\left\langle \left(\int N_{-2}(\tilde{A})|dX| \right) \psi, \phi \right\rangle, \tag{3.85}$$

if $j = 2$. This proves (1.12).

Now if U is not trivial, by linearity, we can assume that ϕ, ψ and A are supported in a small neighborhood where U is trivialized by an orthonormal basis $\{s_i\}$. Write

$$\phi = \phi_i s_i, \ \psi = \psi_i s_i, \ A = A_{ij} s_i^* \otimes s_j. \tag{3.86}$$

Then

$$\begin{aligned}\langle (V \circ A)\psi, \phi\rangle &= - \langle A\psi, \nabla_V \phi\rangle \\ &= - \langle A_{ij}\psi_i, V\phi_j + \Gamma_{kj}(V)\phi_k\rangle,\end{aligned} \tag{3.87}$$

where $\Gamma_{kj}(V) = \langle \nabla_V s_i, s_j\rangle$. This reduces to the scalar case and one sees that the connection produces only a lower-order term. □

4. Bundle filtrations

Systematic use is made here of the 'geometrization' of a bundle filtration. Recall that for a vector bundle, E, over a compact manifold (possibly with corners) a *filtration* is a finite non-decreasing sequence of subbundles:

$$E^0 \subset E^1 \subset E^2 \subset \cdots \subset E^N = E \tag{4.1}$$

where the length of the filtration is N, so $N = 0$ corresponds to the trivial filtration. In particular we allow the same subbundle to be repeated. Another filtration $F^0 \subset F^1 \subset F^{N'} = E$ of E is said to be a *refinement* of (4.1) if there is a strictly increasing map $I : \{0, 1, \ldots, N\} \longrightarrow \{0, 1, \ldots, N'\}$ such that for each j $E^j = F^{I(j)}$. Near any point $p \in M$ we can always find a basis, $e_1, \ldots, e_n$, $n = \dim_{\text{fibre}} E$, of E which is *compatible* with the filtration in the sense that for each $k = 0, \ldots N$ there is a subset $I(k) \subset \{1, \ldots, n\}$ such that the e_i for $i \in I(k)$ span E^k. Of course we can even arrange that $I(k) = \{1, \ldots, \dim_{\text{fibre}} E^k\}$ but it is more convenient not to demand this. Any collection of filtrations of a given bundle is said to be compatible if there is one filtration which is a refinement of each of them (it need not be one of the given filtrations).

We are typically interested in a vector bundle E, over a manifold with corners M, which is such that for a particular boundary hypersurface, $H \subset M$, the bundle

$E_H = E_{\restriction H}$ has a filtration E^k. We then wish to define a 'rescaled' version of E, i.e., a new vector bundle $\widetilde{E}$ with the properties:

$$\widetilde{E} \cong E \text{ over } X \setminus H$$
$$\widetilde{E}_{\restriction H} \cong \bigoplus_{k=0}^{N} [E^k/E^{k-1}] \otimes [N^*H]^k, \ E^{-1} = \{0\}. \tag{4.2}$$

The second condition means that $\widetilde{E}$ is (naturally) isomorphic to the graded bundle associated to the filtration of E.

In fact the filtration alone does not fix the bundle $\widetilde{E}$ in a differential sense, except in the (important) case of filtrations of length one. To construct $\widetilde{E}$ the filtration should be extended to a *jet-filtration.* By a k-jet of subbundles of E at H we mean an equivalence class of subbundles in neighborhoods of H where the equivalence relation is $F \sim G$ if there is some neighborhood, P, of H in X such that

$$\mathcal{I}(E,F) \overset{\text{def}}{=} \mathcal{C}^\infty(P;F) + \rho_H^k \mathcal{C}^\infty(P;E) = \mathcal{C}^\infty(P;G) + \rho_H^k \mathcal{C}^\infty(P;E). \tag{4.3}$$

Here $\rho_H \in \mathcal{C}^\infty(X)$ is a defining function for H. If F is a k-jet of subbundle then the space $\mathcal{I}(E,F)$ determines F. If F and G are respectively a k-jet and a p-jet of subbundle of E at H then we write $F \subset G$ to mean that $k \geq p$ and F and G have representatives subbundles, F' and G', in some neighborhood of H with $F' \subset G'$. This relation can also be written $\mathcal{I}(E,F) \subset \mathcal{I}(E,G)$. By a jet-filtration of E at H we mean a sequence E^j of $N-j$-jets of subbundle satisfying (4.1) in this sense of inclusion.

Suppose that E^j is such a jet-filtration of the bundle E at H. Consider the space of sections of E :

$$\mathcal{D} = \sum_{p=0}^{N} \rho_H^p \mathcal{I}(E;E^p) \subset \mathcal{C}^\infty(X;E). \tag{4.4}$$

Away from H this consists, locally, of all sections of E. Thus if $\mathcal{I}_p \subset \mathcal{C}^\infty(X)$ is the ideal of functions vanishing at $p \in X$ then the vector spaces

$$\widetilde{E}_p = \mathcal{D}/\mathcal{I}_p \cdot \mathcal{D} \tag{4.5}$$

are canonically isomorphic to the fibres of E for $p \notin H$. Since $\mathcal{D} \subset \mathcal{C}^\infty(X;E)$ for any p there is a natural map

$$\widetilde{E}_p \longrightarrow E_p. \tag{4.6}$$

Proposition 4.1. *If E is a $\mathcal{C}^\infty$ vector bundle over a manifold with corners, X, with a jet-filtration at a boundary hypersurface H then*

$$\widetilde{E} = \bigsqcup_{p \in X} \widetilde{E}_p, \tag{4.7}$$

defined using (4.5), *has a unique structure as a* $\mathcal{C}^\infty$ *vector bundle over* X *such that the map* $\iota : \widetilde{E} \longrightarrow E$ *defined by* (4.6) *is a* $\mathcal{C}^\infty$ *bundle map,*

$$\iota^* \mathcal{D} = \mathcal{C}^\infty(X; \widetilde{E}) \tag{4.8}$$

and (4.2) *holds.*

Proof. Suppose $F \subset G$ are respectively a k-jet and a $k-1$-jet of subbundle of E at H. Then given any representative of F as a subbundle of E near H we can find a representative of G which contains it. Thus, starting at the bottom of the filtration we can find for each j a representative F^j of the $N-j$-jet of subbundle E^j such that $F^j \subset F^{j-1}$ as subbundles near H. The definition of $\mathcal{D}$ in (4.8) then becomes

$$\mathcal{D} = \left\{ u \in \mathcal{C}^\infty(X; E);\ \text{near } H,\ u = \sum_{j=0}^{N} \rho_H^j u_j,\ u_j \in \mathcal{C}^\infty(P; F^j) \right\} \tag{4.9}$$

where P is some neighborhood of H. Locally near any $p \in H$ we can choose a basis $e_1, \dots, e_N$ of E such that $e_1, \dots, e_{R(j)}$ is a basis of F^j, where $R(j)$ is the rank of F^j. Then, from (4.9), $\rho_H^j e_p$, where j is the smallest index such that $p \leq R(j)$, is a basis for $\widetilde{E}$. This gives $\widetilde{E}$ its structure as a $\mathcal{C}^\infty$ vector bundle; it is clearly independent of choices and (4.8) holds by construction. □

In the main application above we need to carry out two such rescalings at two intersecting boundary hypersurfaces. Let H_1 and H_2 be the two boundary hypersurfaces of X equipped with the jet-filtrations E_1^j, and E_2^p. Naturally some compatibility conditions are required between the two. The rescaling at H_1 will be carried out first, so the compatibility conditions is just that the rescaling must induce a jet-filtration at H_2 of the rescaled bundle.

To see what this amounts to suppose first that E itself has a filtration, G^j over X. If this filtration is to induce a filtration on the rescaled bundle $\widetilde{E}$ with respect to some jet-filtration at a boundary hypersurface, H_1, F^p, it is necessary and sufficient that

$$G^j \cap F^p,\ p = 1, \dots, N \text{ be a jet-filtration of } G^j \text{ at } H_1. \tag{4.10}$$

In case the G^j only constitute a jet-filtration of E at a boundary hypersurface, H_2, we demand that (4.10) hold in the sense that the G^j have representative subbundles of E near H_2 which filter E and on which the F^p induce jet-filtrations at H_1 near H_2. If these conditions hold then we can define the doubly-rescaled bundle $\widetilde{\widetilde{E}}$ by first defining $\widetilde{E}$ with respect to the rescaling at H_1 and then rescaling $\widetilde{E}$ with respect to the jet-filtration on it at H_2 induced by the rescaling of the jet-filtration of E.

In practice the jet-filtrations are defined from local filtrations of the bundle obtained by normal translation of a filtration from the boundary hypersurface H. Thus suppose that E has a connection and that V is a real vector field which is transversal to H. Then any filtration E^j of E on H can be extended to a filtration near H by taking F^j to be the subbundle of E which is spanned (over $\mathcal{C}^\infty(X)$) by the sections satisfying

$$\nabla_V e = 0 \text{ near } H,\ e_{\restriction H} \in \mathcal{C}^\infty(H, E^j). \tag{4.11}$$

The connection will be a natural one, but the choice of normal vector field is less natural. It is also of interest to know the extent to which the rescaled bundle inherits a connection. The obvious condition is that the connection should preserve the filtration:

$$\nabla_W e_j \in \mathcal{C}^\infty(H; E^j), \quad \forall\, W \in \mathcal{C}^\infty(H; TH),\ e_j \in \mathcal{C}^\infty(H; E^j). \tag{4.12}$$

Proposition 4.2. *Suppose E is a vector bundle with connection over the $\mathcal{C}^\infty$ manifold with corners X and that on a boundary hypersurface H the connection preserves a filtration E^j in the sense of* (4.12)*, then if the covariant derivatives of the curvature of the connection satisfy*

$$\begin{gathered}(\nabla_{U_1} \cdots \nabla_{U_k} R)(W_1, W_2) \colon \mathcal{C}^\infty(H; E^j) \longrightarrow \mathcal{C}^\infty(H; E^{j+k-p+2}) \\ \forall\, k \le N - j - 2 + p, \text{ where } p = 0, 1 \text{ and if } p = 1,\ W_2 \text{ is tangent to } H\end{gathered} \tag{4.13}$$

the jet-filtration defined by (4.12) *is independent of the choice of normal vector field and the rescaled bundle has a b-connection, i.e.,*

$$\nabla_W : \mathcal{C}^\infty(X; \widetilde{E}) \longrightarrow \mathcal{C}^\infty(X; \widetilde{E}) \text{ provided } W \text{ is tangent to } H. \tag{4.14}$$

Proof. If the jet-filtration is defined by (4.12) then $\mathcal{D} \subset \mathcal{C}^\infty(X; E)$ is characterized by the Taylor series of the action of the chosen normal vector field:

$$(\nabla_V)^j u_{\restriction H} \in \mathcal{C}^\infty(H; E^j) \text{ for } j = 0, \ldots, N-1. \tag{4.15}$$

Suppose $W \in \mathcal{C}^\infty(X, TX)$ is tangent to H. Then the Taylor series of $\nabla_W u$, for $u \in \mathcal{D}$, in the sense of (4.15) can be written

$$\nabla_V^j(\nabla_W u) = \nabla_W(\nabla_V^j u) + \sum_{p<j} R_p(V, W_2^p) \nabla_V^p u \tag{4.16}$$

where R_p is a covariant derivative of order $s \le j - p - 1$ of the curvature operator and if $s = j - p - 1$ then $W_2^p = W$ is tangent to H. Thus from (4.15) and (4.13) it follows that $\nabla_W u$ also satisfies (4.15), i.e., (4.14) holds. Changing V by a non-vanishing multiple clearly does not change the jet filtration. If any vector field tangent to H is added to V it follows, using (4.13), that the content of (4.15) is unchanged. Thus the rescaling is independent of the normal vector field used to define it. □

5. Analytic torsion

Let M be a compact Riemann manifold, of odd dimension, with metric tensor g. If $\rho : \pi_1(M) \longrightarrow U(k)$ is a unitary representation of the fundamental group let

$$L_\rho = \tilde{M} \otimes_\rho \mathbb{C}^k, \tag{5.1}$$

where $\tilde{M}$ is the universal cover of M, be the associated locally flat Hermitian bundle over M. Exterior differentiation extends to differential forms twisted by ρ

$$d : \mathcal{C}^\infty(M; \Lambda^* M \otimes L_\rho) \longrightarrow \mathcal{C}^\infty(M; \Lambda^* M \otimes L_\rho). \tag{5.2}$$

Using the Hermitian inner product on L, metric inner product on $\Lambda^* M$ and volume form on M the adjoint, δ, and hence the twisted Laplacian can be defined

$$\Delta = d\delta + \delta d,\ \Delta : \mathcal{C}^\infty(M; \Lambda^* \otimes L_\rho) \longrightarrow \mathcal{C}^\infty(M; \Lambda^* M \otimes L_\rho). \tag{5.3}$$

Let $Q \in \mathcal{C}^\infty(M; \hom(\Lambda^* M \otimes L_\rho))$ be the parity involution defined by $Q = (-1)^p$ on $\Lambda^p M \otimes L_\rho$ and let $\operatorname{str} A = \operatorname{tr} QA$, for $A \in \mathcal{C}^\infty(M; \hom(\Lambda^* M \otimes L_\rho))$ be the associated supertrace tensor. For a smoothing operator, defined by its Schwartz kernel

$$\begin{aligned} &B \in \Psi^{-\infty}(M; \Lambda^* M \otimes L_\rho) \\ &\iff B \in \mathcal{C}^\infty(M^2; \operatorname{Hom}(\Lambda^* M \otimes L_\rho) \otimes \pi_R^* \Omega M) \end{aligned} \tag{5.4}$$

the 'big' supertrace is defined by

$$\operatorname{STr}(B) = \int_M \operatorname{str}(B_{\restriction \mathrm{Diag}}). \tag{5.5}$$

By Lidsky's theorem the supertrace of a smoothing operator is given in terms of the operator trace by $\operatorname{Tr}(QB)$.

Consider the number operator $N = p$ on $\Lambda^p M \otimes L_\rho$. The supersymmetric zeta function is defined by

$$\zeta_T(s) = \frac{1}{\Gamma(s)} \int_0^\infty t^s \operatorname{STr}(N e^{-t\underline{\Delta}}) \frac{dt}{t},\ \operatorname{Re} s \gg 0 \tag{5.6}$$

where $\underline{\Delta}$ is the Laplacian restricted to the orthocomplement of its null space. That is, if Π_N is orthogonal projection off the null space of Δ then

$$\zeta_T(s) = \frac{1}{\Gamma(s)} \int_0^\infty t^s \operatorname{STr}(N e^{-t\Delta} \Pi_N) \frac{dt}{t}. \tag{5.7}$$

This zeta function extends to a meromorphic function on the entire complex plane with $s = 0$ a regular value as indeed follows from (1.20). We define the analytic torsion of Ray and Singer by

$$\log T_\rho(M) = \zeta_T'(0). \tag{5.8}$$

As defined here $T_\rho(M)$ is the square of the torsion defined in [38]. Formally it is a ratio of powers of determinants for the Laplacians $\underline{\Delta}_j$, restricted to $\Lambda^j M \otimes L_\rho$ and with null space removed:

$$T_\rho(M) \sim \prod_{j=1}^n \left[\det \underline{\Delta}_j\right]^{(-1)^j j},\ n = \dim M. \tag{5.9}$$

To analyze $\zeta_T(s)$ near $s = 0$ the right side of (5.6) needs to be continued analytically. The integral decays exponentially as $t \to \infty$ so only the behaviour near 0 needs to be considered. In fact there is only one obstruction to convergence for s near 0:

Theorem 5.1. *If M is an odd-dimensional Riemann manifold, as above, the pointwise supertrace of the weighted heat kernel has a uniform asymptotic expansion as $t \downarrow 0$*

$$\operatorname{str}(Ne^{t\Delta}) \sim a_{-\frac{1}{2}} t^{-\frac{1}{2}} + \sum_{j\geq 1, odd} t^{\frac{j}{2}} a_{\frac{j}{2}} \tag{5.10}$$

with coefficients $a_k \in \mathcal{C}^\infty(M;\Omega M)$ and leading term

$$a_{-\frac{1}{2}} = c(n) \sum_{k=1}^{n} (-1)^k \operatorname{Pf}(R_k) \wedge \omega_k, \; c(n) = 2i(-1)^{\frac{1}{2}(n+1)}(16\pi)^{-\frac{1}{2}n}. \tag{5.11}$$

Here ω_k is an orthonormal frame for T^*M, R_k is obtained by deleting the kth row and column from the curvature matrix R in this frame and $\operatorname{Pf}(R_k)$ is its Pfaffian.

Corollary 5.2. *For any $\delta > 0$*

$$\log T_\rho(M) = \int_0^\delta \left[\operatorname{STr}(Ne^{-t\Delta}) - a_{-\frac{1}{2}}(M,g)t^{-\frac{1}{2}}\right]\frac{dt}{t} \tag{5.12}$$

$$+ \int_\delta^\infty \operatorname{STr}(Ne^{-t\Delta})\frac{dt}{t} - 2\delta^{-\frac{1}{2}} a_{-\frac{1}{2}}(M,g) - (c + \log\delta)\chi_2(M,\rho)$$

where $a_{-\frac{1}{2}}(M,g) = \int_M a_{-\frac{1}{2}}$ is given by (5.11), χ_2 is the twisted, weighted Euler characteristic

$$\chi_2(M,\rho) = \sum_{k=0}^{N} (-1)^k k b_k, \; b_k = \dim H^k(M;\rho) \tag{5.13}$$

and c is Euler's constant.

Proof of Corollary. Writing (5.6) in the form

$$\zeta_T(s) = \frac{1}{\Gamma(s)} \int_\delta^\infty t^s \operatorname{STr}\left[Ne^{-t\Delta}\right]\frac{dt}{t} + \frac{1}{\Gamma(s)} \int_0^\delta t^s \left[\operatorname{STr}(Ne^{-t\Delta}) - a_{-\frac{1}{2}}t^{-\frac{1}{2}}\right]\frac{dt}{t}$$

$$+ \frac{1}{\Gamma(s)} \frac{\delta^{s-\frac{1}{2}}}{(s-\frac{1}{2})} a_{-\frac{1}{2}}(M,g) - \frac{1}{\Gamma(s+1)}\delta^s \chi_2(M;\rho) \tag{5.14}$$

gives an explicitly regular formula near $s = 0$ from which (5.12) follows by differentiation and evaluation at $s = 0$. □

To prove Theorem 5.1 we shall adapt Getzler's scaling argument to the odd-dimensional case, leading to the cancellation inherit in (5.10). We do so by making a global rescaling of the homomorphism bundle of $\Lambda^* M \otimes L_\rho$ near the front face of the heat space defined above. Since this is localized near the diagonal, L_ρ does not appear in the discussion. To get (5.10) we then show that the heat kernel lifts to the rescaled bundle.

Getzler's rescaling is defined by a decomposition of the homomorphism bundle in terms of Clifford multiplication. Let V be any Euclidean vector space. Let $C\ell(V)$ be the associated Clifford algebra, the tensor algebra of V with the one relation

$$e \cdot f + f \cdot e = -2\langle e, f\rangle \operatorname{Id} \ \forall\ e, f \in V. \tag{5.15}$$

This algebra acts by Clifford multiplication on the exterior algebra, $\Lambda^* V$:

$$c_l(e) : \Lambda^* V \longrightarrow \Lambda^* V,\ c_l(e) = \operatorname{ext}(e) - \operatorname{int}(e),\ e \in V, \tag{5.16}$$

where $\operatorname{ext}(e)$ is exterior (wedge) product with e and $\operatorname{int}(e)$ is contraction with the dual vector $v \in V^*$ to $e \in V$. This is *left* Clifford multiplication, we also consider right Clifford multiplication

$$c_r(e) = (\operatorname{ext}(e) + \operatorname{int}(e)) \cdot Q \tag{5.17}$$

where Q is the parity operator for the natural grading of $\Lambda^* V$; the left and right actions commute.

In case W is an even-dimensional Euclidean vector space the complexified Clifford algebra $\mathbb{C}\ell(W) = C\ell(W) \otimes_{\mathbb{R}} \mathbb{C}$ is isomorphic to $\mathrm{Gl}(2^k; \mathbb{C})$, $\dim W = 2k$. If V is odd-dimensional we shall exploit this by extending the left Clifford action on $\mathbb{C}\Lambda^* V = \Lambda^* V \otimes_{\mathbb{R}} \mathbb{C}$ to an action of $\mathbb{C}\ell(V \oplus \mathbb{R})$. Let $e_1, \dots, e_n$ be an orthonormal basis for V and, setting $e_0 = 1 \in \mathbb{R}$, consider the operator on $\mathbb{C}\Lambda^*(V \oplus \mathbb{R})$

$$\tilde{\tau} = i^{\frac{n+1}{2}} c_r(e_0) \cdot c_r(e_1) \dots c_r(e_n). \tag{5.18}$$

Lemma 5.3. *If V is an odd-dimensional Euclidean vector space and $\tilde{\tau}$ is defined by* (5.18) *the map*

$$E : \mathbb{C}\Lambda^* V \ni \omega \longmapsto \frac{1}{2}(\omega + \tilde{\tau}\omega) \in \mathbb{C}\Lambda^*(V \oplus \mathbb{R}), \tag{5.19}$$

where $\Lambda^* V \hookrightarrow \Lambda^*(V \oplus \mathbb{R})$ is the natural embedding, embeds $\mathbb{C}\Lambda^* V$ as a subspace invariant under the left Clifford action of $\mathbb{C}\ell(V \oplus \mathbb{R})$ such that

$$E \cdot c_l(e) = c_l(e) \cdot E,\ \forall\ e \in V,\ \textit{and}\ E \cdot Q = Q \cdot E. \tag{5.20}$$

Proof. Clearly $\tilde{\tau}$ is an involution. Moreover E is injective and has range precisely the 1-eigenspace of $\tilde{\tau}$. The range of E is invariant under left Clifford multiplication by $\mathbb{C}\ell(V \oplus \mathbb{R})$ and E intertwines the action of $\mathbb{C}\ell(V)$ on $\mathbb{C}\Lambda^* V$ and as a subspace of $\mathbb{C}\ell(V \oplus \mathbb{R})$. Since $\tilde{\tau}\omega$ is a form of the same parity as ω, E also intertwines the super symmetries, Q, on $\Lambda^* V$ and $\Lambda^*(V \oplus \mathbb{R})$. □

The Clifford action gives a decomposition of the endomorphism space:

Lemma 5.4. *For any odd-dimensional Euclidean vector space*

$$\hom(\mathbb{C}\Lambda^* V) = \mathbb{C}\ell(V \oplus \mathbb{R}) \otimes \hom'(\mathbb{C}\Lambda^* V) \tag{5.21}$$

where the second factor is the subspace commuting with the action of $\mathbb{C}\ell(V \oplus \mathbb{R})$, it is generated by the right Clifford action of $\mathbb{C}\ell(V)$.

Proof. For the even-dimensional case

$$\hom(\mathbb{C}\Lambda^* W) = \mathbb{C}\ell(W) \otimes \mathbb{C}\ell(W) \tag{5.22}$$

with the two factors acting by left and right Clifford multiplication. For $W = V \oplus \mathbb{R}$ we deduce (5.21) with the right factor being the subspace which preserves the 1-eigenspace of $\tilde{\tau}$. This is generated by the elements $c_l(e_0) \cdot c_l(e_j)$ $i, j = 1, \dots, n$, and this is the right Clifford action by $\mathbb{C}\ell(V)$.

Notice that the involution (5.18) depends only on the choice of orientation of V. Switching orientation replaces (5.19) by the embedding of $\mathbb{C}\Lambda^* V$ as the -1-eigenspace of $\tilde{\tau}$. However $c_r(e_0)$ interchanges the ± 1-eigenspaces of $\tilde{\tau}$ and intertwines the left Clifford actions on them, so the decomposition (5.21) is completely natural. □

Using (5.21) the filtration of the Clifford algebra, by minimal degree in the generators, induces a filtration of the endomorphism space

$$\hom^{[k]}(\mathbb{C}\Lambda^* V) = \mathbb{C}\ell^{[k]}(V \oplus \mathbb{R}) \otimes \hom'(\mathbb{C}\Lambda^* V), \ k = 0, \dots, n+1. \tag{5.23}$$

To find the decomposition of operators on $\mathbb{C}\Lambda^* V$ in this sense we only need find their action, on the image of E in (5.19), in terms of left and right Clifford multiplication on $\Lambda^*(V \oplus \mathbb{R})$.

For any orthonormal basis e_i, $i = 1, \dots, n$,

$$E \cdot \operatorname{ext}(e_i) = [\operatorname{ext}(e_i) \operatorname{int}(e_0) \operatorname{ext}(e_0) - \operatorname{int}(e_i) \operatorname{ext}(e_0) \operatorname{int}(e_0)] \cdot E. \tag{5.24}$$

On $\Lambda^*(V \oplus \mathbb{R})$ we have

$$\operatorname{ext}(e_i) = \frac{1}{2}\left[c_l(e_i) + c_r(e_i) Q\right], \quad \operatorname{int}(e_i) = \frac{1}{2}\left[-c_l(e_i) + c_r(e_i) Q\right] \tag{5.25}$$

Inserting these in (5.24) gives the decompositions

$$\begin{aligned} E \cdot \operatorname{ext}(e_i) &= \left[\frac{1}{2} c_l(e_i) \otimes \operatorname{Id} - \frac{1}{2} c_l(e_0) \otimes (c_r(e_i) c_r(e_0))\right] \cdot E \\ E \cdot \operatorname{int}(e_i) &= \left[-\frac{1}{2} c_l(e_i) \otimes \operatorname{Id} - \frac{1}{2} c_l(e_0) \otimes (c_r(e_i) c_r(e_0))\right] \cdot E. \end{aligned} \tag{5.26}$$

Thus both exterior and interior multiplication are operators of order 1,

$$\operatorname{ext}(v), \ \operatorname{int}(v) \in \hom^{[1]}(\mathbb{C}\Lambda^* V), \ \forall \ v \in V. \tag{5.27}$$

Similarly we decompose the number operator by writing it on the image of E,

$$N = \sum_{k=1}^{n} [\operatorname{ext}(e_i) \operatorname{int}(e_i) \operatorname{int}(e_0) \operatorname{ext}(e_0) + \operatorname{int}(e_i) \operatorname{ext}(e_i) \operatorname{ext}(e_0) \operatorname{int}(e_0)]. \tag{5.28}$$

Again using (5.25) this becomes

$$N = \frac{1}{2} \sum_{k=1}^{n} (\operatorname{Id} - c_l(e_k) c_l(e_0) c_r(e_k) c_r(e_0)). \tag{5.29}$$

Thus $N \in \hom^{[2]}(\mathbb{C}\Lambda^* V)$.

The parity involution, Q, can be written

$$Q = c_l(e_0)c_l(e_1)\dots c_l(e_n)c_r(e_0)c_r(e_1)\dots c_r(e_n). \tag{5.30}$$

That this involution has maximal order is, together with the following fundamental observation of Patodi, the main reason for introducing the filtration.

Lemma 5.5. *The supertrace functional annihilates* $\hom^{[n]}(\mathbb{C}\Lambda^*V)$ *in* (5.23) *and*

$$\operatorname{str}(Q) = 2^n. \tag{5.31}$$

Proof. Of course (5.31) is immediate. Taking an orthonormal basis for V and considering the basis elements of $\mathbb{C}\ell(V \oplus \mathbb{R})$, the odd elements anticommute with Q and hence have zero trace after composition with Q. For an element $\mu = \prod_{1\le r\le k} c_l(e_{j_r}) \otimes A$, $0 \le j_1 < j_2 < \dots < j_k$, with $k \le n$ there exists e_q, $q \ne j_\ell$ for $1 \le \ell \le k$. Then μ commutes with Q and e_q, which interchanges the $+1$ and -1-eigenspaces of Q, so $\operatorname{tr}(Q\mu) = 0$. □

If M is an odd-dimensional Riemann manifold the naturality of (5.23) means that it extends to give a filtration of the endomorphism bundle

$$\hom^{[k]}(\Lambda^*M \otimes L_\rho) = \mathbb{C}\ell^{[k]}(T^*M \oplus \mathbb{R}) \otimes \hom'(\Lambda^*M \otimes L_\rho) \tag{5.32}$$

where $\hom'$ is the subbundle of elements commuting with the Clifford action. The 'full' homomorphism bundle over M^2

$$\operatorname{Hom}(\Lambda^*M \otimes L_\rho) = \bigsqcup_{(x,x')\in M^2} \hom\left((\Lambda^*M \otimes L_\rho)_{x'}, (\Lambda^*M \otimes L_\rho)_x\right) \tag{5.33}$$

has the property that its restriction to the diagonal is canonically isomorphic to $\hom(\Lambda^*M \otimes L_\rho)$ over M.

Thus over the diagonal $\operatorname{Hom}(\Lambda^*M \otimes L_\rho)$ has the filtration (5.32). The extension of the filtration off the diagonal is discussed in §4. In order to apply Proposition 4.2 we need to show that the curvature functional has the appropriate order with respect to the filtration. That is, if R is the curvature operator on $\operatorname{Hom}(\Lambda^*M \otimes L_\rho)$ and $V, W, U_1, \dots, U_p$ are $\mathcal{C}^\infty$ vector fields on M^2 near the diagonal we need to show that

$$\nabla_{U_1} \dots \nabla_{U_p} R(V,W)|_{\mathrm{Diag}} \text{ has order } p + 2 - \ell \tag{5.34}$$

where $\ell \le p + 2$ is the number of vector fields which are tangent to the diagonal.

Since the action of the curvature operator, and its covariant derivatives, is always given by a sum of products of interior and exterior multiplication it follows from (5.26) that its order can never be greater than 2. Thus (5.34) certainly holds when $p+2-\ell \ge 2$, i.e., $p \ge \ell$. It therefore suffices to consider the case when either all the vector fields are tangent to the diagonal, or all but one are so tangent. In the first case the curvature operator and its covariant derivatives are of order zero, since the Levi-Civita connection on M preserves the filtration (5.32). In the second case the fact that the diagonal is geodesically flat means that the operator (5.34) vanishes.

Thus Proposition 4.2 applies and the rescaled bundle, GHom $(\Lambda^*M \otimes L_\rho)$, and rescaled heat calculus, $\Psi^*_{h,G}(M; \Lambda^*M \otimes L_\rho)$, are therefore defined. We wish to show that the heat kernel

$$\exp(-t\Delta) \in \Psi^{-2}_{h,G}(M; \Lambda^*M \otimes L_\rho), \tag{5.35}$$

for the twisted Laplacian.

Following the discussion in §4 it suffices to show that Δ acts on the rescaled bundle and to compute the normal operator in the rescaled calculus. This follows from the Lichnerowicz/Weitzenböck formula.

Proposition 5.6. *For the twisted Laplacian*

$$\Psi^{-k}_{h,G}(M; \Lambda^*M \otimes L_\rho) \ni A \longmapsto \Delta \cdot A \in \Psi^{-k+2}_{h,G}(M; \Lambda^*M \otimes L_\rho) \tag{5.36}$$

and

$$N_{h,G,-k+2}(\Delta A) = \left[\mathcal{H} - \frac{1}{8}C(R)\right] \cdot N_{h,G,-k}(A) \tag{5.37}$$

where

$$C(R) = \sum_{i,j,s,t} R_{ijst} c_l(e_i) c_l(e_j) c_r(e_s) c_r(e_t) \tag{5.38}$$

with respect to any orthonormal frame of T^*M, and $\mathcal{H}$ is the generalized harmonic oscillator:

$$\mathcal{H} = -\sum_i \left(\sigma_1(e_i) + \frac{1}{8}R(e_i, V_r)\right)^2. \tag{5.39}$$

Here V_r denotes the radial vector field on TM.

Proof. The Weitzenböck formula for the action of the Laplacian on Λ^*M is

$$\Delta = \Delta^c - \sum_{i,j,k,\ell} R_{ijk\ell} \operatorname{ext}(e_i) \operatorname{int}(e_j) \operatorname{ext}(e_k) \operatorname{int}(e_\ell). \tag{5.40}$$

Here Δ^c is the connection Laplacian; with respect to any local orthonormal frame of TM it is

$$\Delta^c = -\sum_i \nabla^2_{v_i}. \tag{5.41}$$

Inserting (5.26) into the tensorial term in (5.40) and using the symmetries of the Riemann curvature tensor we find

$$\begin{aligned} &\sum_{i,j,k,\ell} R_{ijk\ell} \operatorname{ext}(e_i) \operatorname{int}(e_j) \operatorname{ext}(e_k) \operatorname{int}(e_\ell) \qquad (5.42)\\ &= \frac{1}{16} \sum_{i,j,k\ell} R_{ijk\ell} [c_l(e_i)c_l(e_j) + c_r(e_i)c_r(e_j)][c_l(e_k)c_l(e_\ell) + c_r(e_k)c_r(e_\ell)] \end{aligned}$$

is equal to $\frac{1}{8}C(R) - \frac{1}{4}S$ where the first term is given by (5.38) and S is the scalar curvature. As a scalar S is of order 0 with respect to the filtration so does not contribute to the normal operator.

As for the connection Laplacian term, we first show that

$$\Psi^{-k}_{h,G}(M; \Lambda^*M \otimes L_\rho) \ni A \longmapsto \nabla_V \cdot A \in \Psi^{-k+1}_{h,G}(M; \Lambda^*M \otimes L_\rho) \tag{5.43}$$

and

$$N_{h,G,-k+1}(\nabla_V \cdot A) = \left(\sigma_1(V) + \frac{1}{8}R(V, V_r)\right) N_{-k}(A). \tag{5.44}$$

The proof is similar to the proof of Proposition 1.2 in §3. In fact, from the computation there we obtain the following formula for the action of A. If we write

$$A = t^{\frac{k}{2}-\frac{n+2}{2}+\frac{n}{4}}\tilde{A}|dtdXdx|^{\frac{1}{2}},\ \Psi = \Psi_0|dtdx|^{\frac{1}{2}} \text{ and } A\Psi = (A\Psi)_0|dtdx|^{\frac{1}{2}} \tag{5.45}$$

then

$$(A\Psi)_0(t,x) = \int (t')^{\frac{k}{2}-1}\tilde{A}(t', X, x)\Psi_0(t-t', x-(t')^{\frac{1}{2}}X)|dt'dx|. \tag{5.46}$$

Since the rescaling does not involve L_ρ, the same argument as in the proof of Proposition 1.2 shows that it produces only a lower-order term. So we need only deal with Λ^*M.

For this we trivialize Λ^*M near the diagonal by parallel translating from each x $(= (x,x) \in \Delta(M) \subset M \otimes M)$ along the radial direction. This gives an identification

$$\mathrm{Hom}(\Lambda^*M) \equiv \mathrm{hom}(\Lambda^*M) \tag{5.47}$$

near the front face. Now if $\{e_i\}$ is an orthonormal frame at x, parallel translated to a neighborhood around x, and

$$\alpha = (\alpha_1, \dots, \alpha_n) \tag{5.48}$$

is a multi-index, then we can write

$$\tilde{A} = \tilde{A}_{\alpha\beta}t^{\frac{|\alpha|}{2}}c_l(e_\alpha)c_r(e_\beta), \tag{5.49}$$

where $\tilde{A}_{\alpha\beta}(t, X, x)$ is a smooth function vanishing rapidly as $X \to \infty$.

Similarly if we use $\{s_i\}$ to denote the corresponding orthonormal basis for Λ^*M, we can write $\Psi_0 = \Psi_i s_i$. With this notation, we have

$$(A\Psi)_0(t,x) = \int (t')^{\frac{k}{2}-1+\frac{|\alpha|}{2}}\tilde{A}_{\alpha\beta}(t', X, x)\Psi_i(t-t', x-(t')^{\frac{1}{2}}X)c_l(e_\alpha)c_r(e_\beta)s_i dt'dX. \tag{5.50}$$

Therefore

$$\begin{aligned}((\nabla_V \cdot A)\Psi)_0(t,x) = &\int (t')^{\frac{k}{2}-1+\frac{|\alpha|}{2}}(V\tilde{A}_{\alpha\beta}\Psi_i + \tilde{A}_{\alpha\beta}V\Psi_i)c_l(e_\alpha)c_r(e_\beta)s_i \\ &+ \int (t')^{\frac{k}{2}-1+\frac{|\alpha|}{2}}\tilde{A}_{\alpha\beta}\Psi_i\nabla_V(c_l(e_\alpha)c_r(e_\beta)s_i)dt'dX.\end{aligned} \tag{5.51}$$

The first term can be handled exactly as in the proof of Proposition 1.2, which produces, in the normal operator, the term $\sigma_1(V)N_{-k}(A)$.

We now look at the second term. We have

$$\nabla_V(c_l(e_\alpha)c_r(e_\beta)s_i) = c_l(\nabla_V e_\alpha)c_r(e_\beta)s_i + c_l(e_\alpha)c_r(\nabla_V e_\beta)s_i + c_l(e_\alpha)c_r(e_\beta)\nabla_V s_i. \tag{5.52}$$

Clearly the first two terms will only produce lower-order terms so we can happily ignore them. Now

$$\begin{aligned}\nabla_V s_i &= -\langle \nabla_V e_k, e_l\rangle \operatorname{ext}(e_k)\operatorname{int}(e_l)s_i \\ &= \frac{1}{4}\Gamma_{kl}(V)[c_l(e_k)c_l(e_l) + c_r(e_k)c_r(e_l)]s_i.\end{aligned} \tag{5.53}$$

Once again, only the first term matters so we need to consider

$$\int (t')^{\frac{k}{2}-1+\frac{|\alpha|}{2}} \tilde{A}_{\alpha\beta}(t',X,x)\Psi_i(t-t',x-(t')^{\frac{1}{2}}X)\frac{1}{4}\Gamma_{kl}(V)(x-(t')^{\frac{1}{2}}X) \\ \times c_l(e_\alpha)c_r(e_\beta)c_l(e_k)c_l(e_l)s_i dt'dX. \tag{5.54}$$

This appears to be an operator of order $-k+2$ but is really of order $-k+1$ since

$$\Gamma_{kl}(V)(x-(t')^{\frac{1}{2}}X) = \frac{1}{2}R(V,V_r)(t')^{\frac{1}{2}}X + O(|(t')^{\frac{1}{2}}X|^2). \tag{5.55}$$

Moreover its contribution to the normal operator is $\frac{1}{8}R(V,V_r)N_{-k}(A)$. □

Recall that the normal operator in the rescaled heat calculus is a section of the rescaled homomorphism bundle over the front face of the heat space. This is just the associated graded bundle to the filtration (5.31), i.e.,

$$\operatorname{GHom}(\mathrm{tf}; \Lambda^* M \otimes L_\rho) = \mathbb{C}\Lambda^*(M \oplus \mathbb{R}) \otimes \hom'(\Lambda^* M \otimes L_\rho) \tag{5.56}$$

lifted from M to $\mathrm{tf}(M_h^2) \cong TM$, over its interior. Thus the left Clifford multiplication in (5.38) acts as exterior multiplication and $C(R)$ is therefore nilpotent.

Moreover ∂_t acts in the same way as before. From the discussion in §4 we conclude not only that (5.35) holds but that its normal operator is given by

$$N_{h,G,-2} = \exp(-\mathcal{H}) \cdot \exp\left(\frac{1}{8}C(R)\right) \tag{5.57}$$

since the two terms commute.

Now we can finally turn to the

Proof of Theorem 5.1. Recalling the formula, (5.29), for the number operator we see that $tNe^{-t\Delta} \in \Psi_{h,G}^{-2}(M; \Lambda^* M \otimes L_\rho)$ has normal operator

$$-\frac{1}{2}\sum_{k=1}^{n} c_l(e_k)c_l(e_0)c_r(e_k)c_r(e_0)\exp\left(\frac{1}{8}C(R)\right) \times \exp(-\mathcal{H}). \tag{5.58}$$

As noted in Lemma 5.5 only the maximal-order term contributes to the supertrace. Thus we conclude directly that

$$\operatorname{str}(tNe^{-t\Delta}) \sim t^{\frac{n+1}{2}-\frac{n}{2}}\left(a_{-\frac{1}{2}} + \sum_{j\geq 1 \text{ odd}} t^{\frac{j}{2}+\frac{1}{2}} a_{\frac{j}{2}}\right) \tag{5.59}$$

where the factor of $t^{\frac{n+1}{2}}$ comes from the rescaling of the bundle and $t^{-\frac{n}{2}}$ from the normalization in the heat calculus. We also use Proposition 1.3 to deduce that there are only odd powers of $t^{\frac{1}{2}}$ in the expansion. Dividing by t gives (5.10).

Furthermore the leading part in (5.58), in terms of the filtration, is a multiple of Q. Using (5.31) we find

$$a_{-\frac{1}{2}} = c(n)\sum_{k=1}^{n}(-1)^k \operatorname{Pf}(R_k)e_1\wedge\cdots\wedge e_n. \tag{5.60}$$

This is a well-defined density, the Pfaffian being (by definition) the term of degree $n-1$ in $\exp(R_k)$, $R_k = \sum_{p,q\neq k} R_{ijpq}e^p\wedge e^q$ as an operator on span $\{e_j, j\neq k\}$. This completes the proof of Theorem 5.1. □

6. Adiabatic scaling

As in the introduction, consider a fibration of compact manifolds

$$\begin{array}{ccc} F & \text{——} & M \\ & & \big\downarrow \phi \\ & & Y. \end{array} \tag{6.1}$$

On M consider the 1-parameter family of Riemannian metrics $g_x = \phi^*h + x^2 g$ and the conformal metric

$$^a g = g + \frac{1}{x^2}\phi^* h \tag{6.2}$$

where g is a metric on M (or at least a non-negative smooth 2-cotensor inducing a metric on each fibre of ϕ) and h is a metric on Y.

As in [27] we first rescale the vector bundle to make $^a g$ a fibre metric. Thus on the manifold $M_a = M\times[0,1]_x$ consider first the lift of the tangent bundle from M, the sections of which are simply vector fields on M depending on x as a parameter; we shall denote this bundle $^M TM_a$. At the boundary hypersurface $\mathrm{ab} = \{x=0\}\subset M_a$, which we identify with M, consider the filtration given by the subspace of fibre vector fields

$$^\phi TM = \bigsqcup_{p\in M} T_p\phi^{-1}(\phi(p)) \subset TM \equiv {}^M T_{\mathrm{ab}} M_a \tag{6.3}$$

and let $\pi_a : {}^a TM_a \longrightarrow M_a$ be the vector bundle over M_a defined by Proposition 4.1 from $^M TM_a$ and this filtration. Thus in local coordinates in M_a, (x, y, z) where (y, z), are coordinates in M with the y_i coordinates in Y, $\partial_{z_k}, x\partial_{y_i}$ is a local basis for $^a TM_a$. We shall denote the space of $\mathcal{C}^\infty$ sections of $^a TM_a$ by

$$\mathcal{V}_a(M_a) = \mathcal{C}^\infty(M_a; {}^a T) = \left\{u\in\mathcal{C}^\infty(M_a; {}^M TM_a); u_{\restriction \mathrm{ab}}\in\mathcal{C}^\infty(M; {}^\phi TM)\right\}. \tag{6.4}$$

An a-differential operator on a vector bundle F over M_a is one which can be written in a (any) local basis of F as a matrix of operators each entry of which is a sum of up to k-fold products of elements of $\mathcal{V}_a(M_a)$. The order is then k; the space of these is denoted $\operatorname{Diff}^k_{\mathrm{a}}(M_a; F)$ and more generally $\operatorname{Diff}^k_{\mathrm{a}}(M_a; E, F)$ is the space of such operators from sections of E to sections of F. The principal symbol of such an operator is a homogeneous polynomial of degree k on ${}^aT^*M_a$, the dual of ${}^aTM_a$, with values in the lift of the homomorphism bundle of F

$$\sigma^a_k : \mathrm{Diff}^k_{\mathrm{a}}(M_a; F) \twoheadrightarrow P^k({}^aT^*M_a; \pi^*_a \hom(F)). \tag{6.5}$$

An a-differential operator is elliptic if $\sigma^a_k(P)$ is invertible on ${}^aT^*M_a\setminus 0$. The basic example of an a-differential operator is the Laplacian, discussed in [27]. Let ${}^a\Lambda^k M_a$, $k = 1,\dots,\dim M$ be the exterior powers of ${}^aT^*M_a$. These bundles can also be identified with bundles constructed using Proposition 4.1. In $x > 0$ the Laplacian of the metric (6.2) acts on these bundles and in fact

Lemma 6.1 ([27]). *The Laplacian* ${}^a\Delta$ *of the metric* (6.2) *is an elliptic element of the ring of a-differential operators,* $\mathrm{Diff}^2_{\mathrm{a}}(M_a; {}^a\Lambda^k M_a)$.

Proof. One can write exterior differentiation on the x-fibres of M_a as

$$df = \sum_{i=1}^{\dim Y} x\partial_{y_i} f \frac{dy_i}{x} + \sum_{j=1}^{\dim F} \partial_{z_j} f dz_j. \tag{6.6}$$

This shows that $d \in \mathrm{Diff}^1_{\mathrm{a}}(M_a; {}^a\Lambda^0 M_a, {}^a\Lambda^1 M_a)$ and by Leibniz' formula d extends to an element of $\mathrm{Diff}^1_{\mathrm{a}}(M_a; {}^a\Lambda^k M_a, {}^a\Lambda^{k+1} M_a)$ for each k. Directly from the definition of $\mathcal{V}_a(M_a)$ it follows that on taking adjoints with respect to ${}^ag$, for any Hermitian bundles E and F, $A \longmapsto A^*$ is an isomorphism of $\mathrm{Diff}^m_{\mathrm{a}}(M_a; E, F)$ onto $\mathrm{Diff}^m_{\mathrm{a}}(M_a; F, E)$. Thus, for any k,

$$\delta \in \mathrm{Diff}^1_{\mathrm{a}}(M_a; {}^a\Lambda^{k+1} M_a, {}^a\Lambda^k M_a).$$

Since composition gives

$$\mathrm{Diff}^m_{\mathrm{a}}(M_a; E, F) \cdot \mathrm{Diff}^{m'}_{\mathrm{a}}(M_a; G, E) \subset \mathrm{Diff}^{m+m'}_{\mathrm{a}}(M_a; G, F) \tag{6.7}$$

we conclude that $\Delta = d\delta + \delta d \in \mathrm{Diff}^2_{\mathrm{a}}(M_a; {}^a\Lambda^k M_a)$. Ellipticity is a consequence of the usual computation of symbols, that of d being $i\xi\wedge$, $\xi \in {}^aT^*M_a$, so the symbol of δ is $-i\,\mathrm{int}(\xi)$ and hence ${}^a\sigma_2(\Delta) = |\xi|^2$ on ${}^aT^*M_a$. □

Another way to prove Lemma 6.1 is to observe that the Levi-Civita connection on the x-fibres of M_a, for $x > 0$, extends by continuity to a connection on ${}^aT^*_{\mathrm{ab}}M_a$. That it extends to an a-connection, i.e., defines covariant differentiation by elements of $\mathcal{V}_a$ over ab is immediate; the fact that covariant differentiation is a differential operator

$$\nabla : \mathcal{C}^\infty(\mathrm{ab}; {}^aT^*M_a) \longrightarrow \mathcal{C}^\infty(\mathrm{ab}; {}^aT^*M_a \otimes T^*\,\mathrm{ab}) \tag{6.8}$$

in the usual sense follows from the product nature of the metric (6.2). Since the Hodge $*$ operator is well defined on the a-form bundles the fact that both d and δ are a-differential operators follows, even from the weaker result that the Levi-Civita connection is an a-connection.

Next we recall, and slightly refine, the results of [27] which follow from this description of ${}^a\Delta$ and the use of a-pseudodifferential calculus introduced there. We can suppress the factor L_ρ since it makes no difference, except notational, to the discussion. The fibre cotangent bundle over M, with fibre over $p \in M$ equal to T^*F_y, $y = \phi(p)$, is a natural subbundle of ${}^aT^*_{\mathrm{ab}}M_a$, the restriction of ${}^aT^*M_a$ to $\mathrm{ab}(M_a) = \{x = 0\} \equiv M$. Since ${}^a g$ defines a non-degenerate metric on ${}^aT^*M_a$ the orthocomplement of the fibre cotangent bundle is a bundle which is naturally identified with $\frac{1}{x}\phi^*T^*Y$. This gives the decomposition of the a-form bundle as

$$ {}^a\Lambda^*_{\mathrm{ab}}M_a = {}^F\Lambda^* \otimes \phi^* \left(\frac{1}{x}\Lambda^*Y\right). \tag{6.9} $$

This decomposition is preserved by the Levi-Civita connection (6.8). For any smooth section $u \in \mathcal{C}^\infty(M_a; {}^a\Lambda^*)$

$$ ({}^a\Delta u)\restriction_{\mathrm{ab}} = {}^F\Delta(u\restriction_{\mathrm{ab}}) \tag{6.10} $$

is given by the fibre Laplacian, acting as ${}^F\Delta \otimes 1$ in terms of (6.9). Thus

$$ E_1 = \{v \in \mathcal{C}^\infty(\mathrm{ab}(M_a); {}^a\Lambda^*); \exists\ u \in \mathcal{C}^\infty(M_a; {}^a\Lambda^*), u\restriction_{\mathrm{ab}} = v,\ {}^a\Delta u \in x\mathcal{C}^\infty(M_a; {}^a\Lambda^*)\} \tag{6.11} $$

is the space of fibre-harmonic forms. It is important that E_1 can be realized as a $\mathcal{C}^\infty$ vector bundle over Y :

$$ E_1 = H^*_{\mathrm{Ho}}(F) \otimes \mathcal{C}^\infty(Y; \Lambda^*) \tag{6.12} $$

where the fibre of $H^*_{\mathrm{Ho}}(F)$ at $y \in Y$ is $H^*_{\mathrm{Ho}}(F_y)$, the Hodge cohomology of F_y with respect to the metric g_y.

Using formal Hodge theory it can be seen that the space (6.11) can also be obtained as the case $k = 1$ of

$$ E_k = \big\{v \in \mathcal{C}^\infty(\mathrm{ab}(M_a); {}^a\Lambda^*); \exists\ u \in \mathcal{C}^\infty(M_a; {}^a\Lambda^*), \\ u\restriction_{\mathrm{ab}} = v,\ {}^a\Delta u \in x^{2k}\mathcal{C}^\infty(M_a; {}^a\Lambda^*)\big\}, \tag{6.13} $$

i.e., the error term in (6.11) can always be improved to $O(x^2)$. These spaces give a Hodge-theoretic form of the Leray spectral sequence for the cohomology of M :

Proposition 6.2 ([27]). *For k sufficiently large E_k is isomorphic to $H^*(M)$, the deRham cohomology of the total space M of the fibration.*

In fact (see [27]) for each $k \geq 0$ one obtains the same space in (6.13) by weakening the condition to ${}^a\Delta u \in x^{2k-1}\mathcal{C}^\infty(M_a; {}^a\Lambda^*)$. For each k, let Π_k be the orthogonal projection with respect to ${}^a g$ from $E_0 = L^2(\mathrm{ab}; {}^a\Lambda^*)$ to the closure of the subspace E_k in L^2. The Hodge-theoretic arguments in [27] show that $\Pi_k\,\mathcal{C}^\infty(\mathrm{ab}; {}^a\Lambda^*) \longrightarrow E_k$ for each k. Moreover if $v \in E_k$ then choosing u as in (6.13) it follows that $du, \delta u \in x^k\mathcal{C}^\infty(M_a; {}^a\Lambda^*)$ and the operators

$$ d_k v = \Pi_k(x^{-k}du\restriction_{\mathrm{ab}}),\ \delta_k v = \Pi_k(x^{-k}\delta u\restriction_{\mathrm{ab}}) \tag{6.14} $$

are well defined, independent of the choice of u, are adjoints of each other with respect to the L^2 inner product on E_k and are such that

$$ d_k^2 = 0,\ \delta_k^2 = 0,\ \text{and} \\ E_{k+1} = \{v \in E_k; d_k v = \delta_k v = 0\} = \{v \in E_k; \Delta_k v = 0, \Delta_k = d_k\delta_k + \delta_k d_k\}. \tag{6.15} $$

For $k = 1$ the operator d_1 is just the differential on Y, in the sense of (5.2), for the representation of $\pi_1(Y)$ on the fibre cohomology. The differential complexes (E_k, d_k) are precisely the Leray spectral sequence for the cohomology of the fibration. For $k \geq 2$ the spaces E_k are finite-dimensional. If $E_k^j = E_k \cap \mathcal{C}^\infty(\mathrm{ab}; {}^a\Lambda^j)$ is the part of E_k in degree j then the torsion for the complex

$$E_k^0 \xrightarrow{d_k} E_k^1 \cdots \xrightarrow{d_k} E_k^{\dim M} \tag{6.16}$$

is by definition

$$\tau(E_k, d_k) = \prod_{j=0}^{\dim M} (\det \Delta_k')^{-1^j j} \tag{6.17}$$

where Δ_k' is the restriction of Δ_k to $E_k \ominus E_{k+1}$.

From [14] it follows that the E_k have another representation in terms of the Laplacian Δ_x. Namely, for $\epsilon > 0$ small enough and each $k \geq 2$,

$$(\widetilde{E}_k)_x = \mathrm{sp}\{u \in \mathcal{C}^\infty(M; \Lambda^*); \Delta_x u = \lambda_x u,\ \lambda_x \in \mathbb{R}, 0 \leq \lambda_x < \epsilon^{-2} x^k\},\ 0 < x < \epsilon \tag{6.18}$$

is a vector space of dimension independent of x. Thought of as subspaces of $\mathcal{C}^\infty(M_a; {}^a\Lambda^*)$ over $(0,\epsilon) \times M$ these form subbundles which are smooth down to $x = 0$, with the limiting space exactly E_k. That is each element of E_k can be extended to a smooth a-form over $[0,\epsilon) \times M$ which is a sum of eigenvectors of Δ_x with eigenvalues $O(x^{2k})$ as $x \downarrow 0$. All other eigenvalues of Δ_x are bounded away from 0. Moreover

$$\lim_{x \downarrow 0} x^{-2k}\, {}^a\Delta_{\restriction \widetilde{E}_k} = \Delta_k,\ k \geq 2. \tag{6.19}$$

This alternative representation of E_k as the span of the boundary values at $x = 0$ of the eigenforms of Δ_x corresponding to x^{2k}-small eigenvalues arises in the long-time asymptotics of the heat kernel in §11.

7. Heat kernel for the adiabatic metric

We wish to consider the heat kernel of $x^{-2}P$ where P is a self-adjoint elliptic a-differential operator, acting on some bundle F, with diagonal principal part with symbol dual to (6.2), i.e., given by the fibre metric on ${}^aT^*M_a$. Thus we seek a distribution

$$\begin{gathered} E \in \mathcal{C}^{-\infty}(\mathbb{R}_t \times M^2 \times [0,1]_x; \mathrm{Hom}(F) \otimes \pi_R^*\Omega) \text{ satisfying} \\ \left(\partial_t + \frac{1}{x^2} P\right) E = \delta(t) \otimes \mathrm{Id}_F,\ E = 0 \text{ in } t < 0. \end{gathered} \tag{7.1}$$

As is usual in such analysis we treat the case of the half-density bundle first, to get the bundles right, and then comment on the changes needed for the general case. In §10 the further modifications corresponding to Getzler's rescaling are considered.

To construct, and analyze, E we first guess the space on which it should be reasonably simple. Set $Z = [0,\infty) \times M^2 \times [0,1]$ and consider the submanifolds

$$\begin{aligned} B_h &= \{(0,m,m,x) \in Z; m \in M\} \\ B_a &= \{(0,m,m',0); m,m' \in M,\ \phi(m) = \phi(m')\}\,. \end{aligned} \tag{7.2}$$

In both cases consider $S = \mathrm{sp}(dt)$ as a subbundle of the conormal bundle. Then, in terms of parabolic blow-up as described in §2 and [18], we put

$$M_A^2 = [Z; B_a, S; B_h, S]\,,\ \beta_A : M_A^2 \longrightarrow Z. \tag{7.3}$$

Thus M_A^2 is a manifold with corners, having five boundary hypersurfaces:

$$\begin{aligned} &\mathrm{eb}(M_A^2) = \beta_A^{-1}\{x=1\} && \text{the `extension' or trivial boundary} \\ &\mathrm{tb}(M_A^2) = \mathrm{cl}\,\beta_A^{-1}(\{t=0\}\setminus B_h) && \text{the temporal boundary} \\ &\mathrm{ab}(M_A^2) = \mathrm{cl}\,\beta_A^{-1}(\{x=0\}\setminus (B_a\cap B_h)) && \text{the adiabatic boundary} \\ &\mathrm{tf}(M_A^2) = \beta_A^*(B_h) && \text{the temporal front face} \\ &\mathrm{af}(M_A^2) = \beta_A^*(B_a) && \text{the adiabatic front face} \end{aligned} \tag{7.4}$$

at each of which there will be a model operator. Of course eb can be freely ignored. Moreover all the kernels we shall consider vanish to infinite order at $\mathrm{tb}(M_A^2)$ so we shall build this into the calculus. As usual we let ρ_F denote a defining function for the boundary hypersurface F for $F = \mathrm{tb}$, ab, tf or af.

On M_A^2 consider the kernel density bundle

$$\mathrm{KD}_A = \rho_{\mathrm{af}}^{-\frac{n}{2}-1}\rho_{\mathrm{tf}}^{-\frac{N}{2}-\frac{3}{2}}\Omega^{\frac12},\ n = \dim Y,\ N = \dim M \tag{7.5}$$

and the spaces of kernels

$$\Psi_A^{-j,-k,-p}(M;\Omega^{\frac12}) = \rho_{\mathrm{tf}}^{j}\rho_{\mathrm{af}}^{k}\rho_{\mathrm{ab}}^{p}\rho_{\mathrm{tb}}^{\infty}\mathcal{C}^\infty(M_A^2;\mathrm{KD}_A),\ j,k,p\in\mathbb{N}. \tag{7.6}$$

As in the ordinary heat calculus there are invariantly defined subspaces of the space of $\mathcal{C}^\infty$ functions on M_A^2 corresponding to involutions around the submanifolds which are blown up to define it. If y, z are coordinates in M, near p with the y_j coordinates in Y and y', z' are coordinates near p', with $\phi(p) = \phi(p')$ and $y = y'$ as coordinates in Y, consider the coordinates t, x, y, z, y', z' in Z near $(0,0,p,p') \in B_a$. Then, with $\rho_{\mathrm{af}} = (t + x^2 + |y-y'|^2)^{\frac12}$, we can consider the space of $\mathcal{C}^\infty$ functions on $[Z; B_a, S]$ with Taylor series at af, the front face defined in the blow-up, of the form

$$\sum_k \rho_{\mathrm{af}}^k F_k\left(\frac{t}{\rho_{\mathrm{af}}^2}, \frac{x}{\rho_{\mathrm{af}}}, \frac{y-y'}{\rho_{\mathrm{af}}}, y+y', z, z'\right) \tag{7.7}$$

with F_k even or odd in the second two sets of variables as k is even or odd. The lift of any $\mathcal{C}^\infty$ function on Z satisfies this condition. The further blow-up of B_h is just a parametrized form of the definition of the ordinary heat space, and so even functions at tf can be defined as before. Again the $\mathcal{C}^\infty$ functions on $[Z; B_a, S]$ all lift to be even. Moreover the evenness conditions at the two front faces are independent so four subspaces $\mathcal{C}^\infty_{E,E}(M_A^2)$, $\mathcal{C}^\infty_{E,O}(M_A^2)$, $\mathcal{C}^\infty_{O,E}(M_A^2)$ and

$\mathcal{C}^\infty_{O,O}(M^2_A) \subset \mathcal{C}^\infty(M^2_A)$ are all well defined, where the first subscript refers to tf and the second to af. Choosing, as we can

$$\rho_{\mathrm{af}} \in \mathcal{C}^\infty_{E,O}(M^2_A) \text{ and } \rho_{\mathrm{tf}} \in \mathcal{C}^\infty_{O,E}(M^2_A) \tag{7.8}$$

gives

$$\begin{gathered}\mathcal{C}^\infty_{O,E}(M^2_A) = \rho_{\mathrm{tf}}\mathcal{C}^\infty_{E,E}(M^2_A),\ \mathcal{C}^\infty_{E,O}(M^2_A) = \rho_{\mathrm{af}}\mathcal{C}^\infty_{E,E}(M^2_A),\\ \mathcal{C}^\infty_{O,O}(M^2_A) = \rho_{\mathrm{tf}}\rho_{\mathrm{af}}\mathcal{C}^\infty_{E,E}(M^2_A).\end{gathered} \tag{7.9}$$

As already noted, $\mathcal{C}^\infty(Z)$ lifts into $\mathcal{C}^\infty_{E,E}(M^2_A)$ so we can define the corresponding space of sections $\mathcal{C}^\infty_{E,E}(M^2_A;\beta_A^*U)$ for any $\mathcal{C}^\infty$ vector bundle over Z. Then we refine (7.6) to

$$\Psi^{-j,-k,-p}_{A,E}(M;\Omega^{\frac12}) = \rho^j_{\mathrm{tf}}\rho^k_{\mathrm{af}}\rho^p_{\mathrm{ab}}\rho^\infty_{\mathrm{tb}}\mathcal{C}^\infty_{E,E}(M^2_A;\mathrm{KD}_A),\ j,k,p\in\mathbb{N} \tag{7.10}$$

subject to (7.8), and similarly for action on general vector bundles over M. In §9 it is shown that these kernels define operators (by convolution in t) on $\dot{\mathcal{C}}^\infty(X;\Omega^{\frac12})$, with $X = [0,\infty)\times M\times[0,1]$. Composition results for these, and related, operators are presented (although to get a general composition formula we allow logarithmic terms at ab). This allows the solution to (7.1) to be constructed in the same spirit as in §1.

To describe the results of this construction, which is actually carried out in §9, consider the normal operators associated to an element of $\Psi^{-j,-k,-p}_{A}(M;\Omega^{\frac12})$ at the boundary hypersurfaces tf, af and ab; by *fiat* the normal operator at tb is trivial. These give maps into simpler calculi.

The normal operator at tf is just a parametrized version of the normal operator in the heat calculus discussed in §1 and §3. To see this, first consider the structure of tf. This is the boundary face produced, in (7.3), by the blow-up of the lift of B_h, which we denote for the moment by B'_h. The submanifold B'_h lies in the lift, to $[Z;B_a,S]$, of $t=0$ and the parabolic direction for the blow-up is just the conormal bundle to this boundary hypersurface. Thus tf can be canonically identified as a fibre-by-fibre compactification of the normal bundle to B'_h in the boundary hypersurface. Within $t=0$, B_h is the diagonal and the blow-up of B_a is the blow-up of the fibre diagonal over $x=0$, just as in the definition of M^2_A. From this it follows that the normal bundle to B'_h is canonically identified with ${}^aTM_a$ so tf is the fibre-by-fibre compactification of the vector bundle ${}^aTM_a$, using as 'trivial' time variable $T = x^{-2}t$. Note that $T^{\frac12}$ is a defining function for $\mathrm{tf}(M^2_A)$ in a neighborhood of tf except at tb; apart from a square-root singularity at tb, it blows up as ρ_{ab}^{-1} at $\mathrm{ab}(M^2_A)$, but this hypersurface is disjoint from $\mathrm{tf}(M^2_A)$. The boundary hypersurface $\mathrm{tf}(M^2_A)$ has boundary hypersurfaces which we can denote eb, af and tb from their intersections with the boundary faces of M^2_A. If $A\in\Psi^{-j,-k,-p}_{A}(M;\Omega^{\frac12})$, multiplication of the kernel by $T^{\frac12(N-j)+1}$, followed by evaluation at tf gives

$$N_{h,-j}:\Psi^{-j,-k,-p}_{A}(M;\Omega^{\frac12}) \twoheadrightarrow \rho^k_{\mathrm{af}}\rho^\infty_{\mathrm{tb}}\mathcal{C}^\infty({}^aTM_a;\Omega_{\mathrm{fibre}}); \tag{7.11}$$

obviously the null space of this map is $\Psi^{-j-1,-k,-p}_{A}(M;\Omega^{\frac12})$.

At af the normal operator maps into the fibre heat calculus, which is described in §8. The boundary hypersurface af is just the lift from $[Z; B_a, S]$ of the boundary hypersurface, af$'$, produced by this first blow-up. Consider the ϕ-fibred product of M with itself, this is the manifold M^2_ϕ which is fibred over Y with fibres $F_y \times F_y$. Clearly $B_a \equiv M^2_\phi$. The interior of af$'$ is canonically isomorphic to ${}^YT(M^2_\phi) \times (0, \infty)_T$ where the first factor is the lift of TY under the projection and in the second factor the global variable is $T = t/x^2$. The boundary hypersurface af$'$ of $[Z; B_a, S]$ is a fibre-by-fibre compactification of ${}^YT(M^2_\phi) \times [0, \infty)$ over M^2_ϕ. The fibre $T_yY \times (0, \infty)$ is compactified to a non-round quarter sphere which can be identified smoothly with

$$\begin{aligned} \mathrm{HM}(T_yY) &= ([0, \infty) \times [0, \infty) \times T_yY \setminus \{0\}) \,/ \sim, \\ (T, x, v) \sim (T', x', v') &\Longrightarrow (T', x', v') = (s^2T, sx, sv) \text{ for some } s > 0 \end{aligned} \tag{7.12}$$

As discussed in §8 this quarter-sphere is closely associated to the Euclidean heat space. Thus af$'$ is quarter-sphere bundle over M^2_ϕ :

$$\begin{array}{ccc} F \times F & \longrightarrow & \mathrm{af}' \\ & & \Big\downarrow \phi\times\phi \\ \mathrm{HM}_q & \longrightarrow & \mathrm{HM}(TY) \\ & & \Big\downarrow \pi_Y \\ & & Y. \end{array} \tag{7.13}$$

Now, the effect on af$'$ of the additional blow-up of B_h, to define M^2_A, reduces to the parabolic blow-up of the surface B'_h which is the intersection of af$'$ and the lift of B_h. Explicitly, in terms of the projective coordinates $T, Y = (y-y')/x, y, Z = z - z', z, x$, this is the part of $\{Y = 0, Z = 0\}$ lying above the diagonal part of the fibration of M^2_ϕ over Y. This then describes the boundary hypersurface af of M^2_A :

$$\mathrm{af} = [\mathrm{af}'; B'_h, S]. \tag{7.14}$$

It has three boundary hypersurfaces, coming from intersections with the other boundary hypersurface of M^2_A and denoted accordingly tf, tb and ab.

Thus if we consider the heat calculus, $\Psi^*_{h,\mathrm{fibre}}({}^YTM; \Omega^{\frac12})$ on the fibres of $[0, \infty) \times {}^YTM$ as a bundle over Y, with the action being invariant under translations we get a map

$$N_{A,-k} : \Psi_A^{-j,-k,-p}(M; \Omega^{\frac12}) \twoheadrightarrow \Psi^{-j,-p+k}_{H,\mathrm{fibre}}({}^YTM; \Omega^{\frac12}). \tag{7.15}$$

Here, for simplicity, we have denoted ${}^YTM = {}^YT(M^2_\phi)$.

At the end of §8 the heat calculus on the base of a fibration, with values in the smoothing operators on the fibres, is discussed. The normal operator at ab takes values in this calculus.

Proposition 7.1. *If ${}^a\Delta$ is the Laplacian of an adiabatic metric,* (6.2)*, as in Lemma* 6.1 *then the heat kernel, the unique solution to* (7.1) *in* $x > 0$*, is an element*

$$\exp(-x^{-2}t\,{}^a\Delta) \in \Psi_{A,E}^{-2,-2,0}(M;{}^a\Lambda^k) \tag{7.16}$$

with normal operators

$$N_{h,-2} = (4\pi)^{-\frac{n}{2}} \exp\left(-\frac{1}{4}|v|_a^2\right) \tag{7.17}$$

$$N_{A,-2} = \exp(-T\Delta_A),\ T = x^{-2}t, \tag{7.18}$$

$$N_{a,0} = \exp(-t\Delta_Y) \tag{7.19}$$

where Δ_A is the fibrewise Laplacian on the bundle ${}^YTM$ and Δ_Y is the reduced Laplacian on Y.

The proof of this main regularity result for the adiabatic heat kernel is given §9, after some preparation in the next section.

Consider what this result shows about the restriction of the heat kernel to the spatial diagonal. The spatial diagonal is embedded by

$$[0,\infty) \times M \times [0,1] = \widetilde{\mathrm{Diag}} \hookrightarrow Z = [0,\infty) \times M^2 \times [0,1],\ (t,m,x) \longmapsto (t,m,m,x). \tag{7.20}$$

Let tb, ab and eb be the three boundary hypersurfaces of $\widetilde{\mathrm{Diag}}$, equal to the intersections of $\widetilde{\mathrm{Diag}}$ in the image of (7.20) with the corresponding boundary hypersurfaces of Z. The first blow-up in (7.3), of B_a, results in the blow-up of $\widetilde{\mathrm{Diag}}$ by

$$\widetilde{\mathrm{Diag}} \cap B_a = \{0\} \times M \times \{0\} \tag{7.21}$$

parabolically in the t-direction. The second blow-up is the parabolic blow-up of the boundary surface, $t = 0$, so if we set

$$\widetilde{\mathrm{Diag}}_A = [\widetilde{\mathrm{Diag}}; \{0\} \times M \times \{0\}, S; \mathrm{tb}, S];\ \tilde{\beta}_A : \widetilde{\mathrm{Diag}}_A \longrightarrow \widetilde{\mathrm{Diag}} \tag{7.22}$$

and $Z_A = [Z, B_a; S]$ the first blow-up in producing M_A^2, then embedding (7.20) lifts to give a commutative diagram:

$$\begin{array}{ccc} \widetilde{\mathrm{Diag}}_A & \xrightarrow{\iota_A} & Z_A \\ \Big\downarrow \tilde{\beta}_A & & \beta_A \Big\downarrow \\ \widetilde{\mathrm{Diag}} & \xrightarrow[\iota]{} & Z. \end{array} \tag{7.23}$$

This results in $\widetilde{\mathrm{Diag}}_A$ having four bounding hypersurfaces, tf, af, ab and eb where af results from the first blow-up and tf is only different from tb in that the manifold has the square root $\mathcal{C}^\infty$ structure there. Again the bounding hypersurfaces are equal to the intersections of the image of $\widetilde{\mathrm{Diag}}_A$ under ι_A with the corresponding boundary hypersurfaces of Z_A.

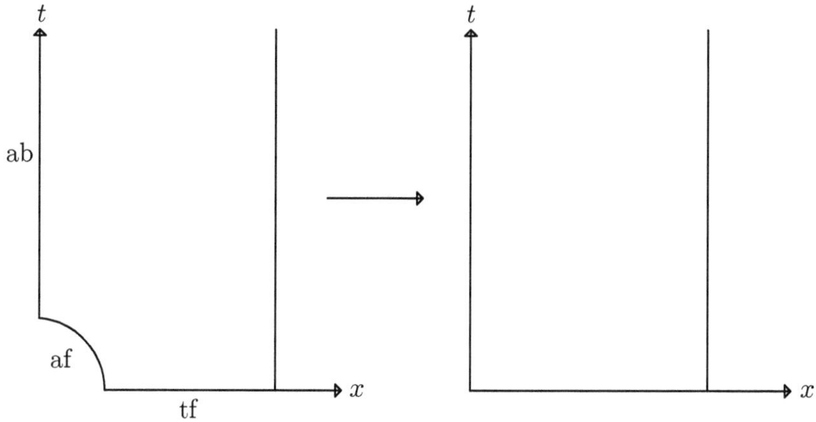

FIGURE 5. $\tilde{\beta}_A : \widetilde{\mathrm{Diag}}_A \to \widetilde{\mathrm{Diag}}$

Directly from Proposition 7.1 we conclude that the restriction to the spatial diagonal is such that, for each k,

$$\widetilde{\beta}_A^* \left[\exp(-x^{-2}tP)_{\restriction \widetilde{\mathrm{Diag}}} \otimes |dt|^{\frac{1}{2}}\right] \in \rho_{\mathrm{tf}}^{-N} \rho_{\mathrm{af}}^{-n} \mathcal{C}^\infty(\widetilde{\mathrm{Diag}}_A; {}^a\Lambda^k \otimes \Omega). \tag{7.24}$$

To see this note that the half-density bundle on Z_A at $\widetilde{\mathrm{Diag}}_A$ decomposes into the normal half-density bundle to $\widetilde{\mathrm{Diag}}_A$ tensored with the half-density bundle on $\widetilde{\mathrm{Diag}}_A$ itself. This gives the natural identification

$$\Omega^{\frac{1}{2}}(Z_A)_{\restriction \widetilde{\mathrm{Diag}}_A} \otimes |dt|^{\frac{1}{2}} \equiv \rho_{\mathrm{tf}}^{-\frac{N}{2}-\frac{1}{2}} \rho_{\mathrm{af}}^{-\frac{n}{2}-1} \Omega(\widetilde{\mathrm{Diag}}_A) \tag{7.25}$$

which leads to (7.24). This is certainly a uniform expansion for the restriction to the diagonal and is optimal for the Laplacian in general.

Notice that $\widetilde{\mathrm{Diag}}_A = Q_2 \times M$ where Q_2 is defined at the end of the introduction. Integration of (7.24) will therefore give

$$\mathrm{Tr}\left(\exp\left(-\frac{t}{x^2}P\right)\right) \in t^{-\frac{N}{2}} \rho_{\mathrm{af}}^{-n} \mathcal{C}^\infty(Q_2). \tag{7.26}$$

This has to be considerably improved to get (0.24).

8. Euclidean and fibre heat calculus

In §1 the Euclidean heat calculus is briefly described. A slightly different description of the global regularity of these kernels is useful below.

Consider again the function Φ' in (1.1). The regularity of this kernel can be described in terms of a blown-up version of the space introduced in (7.12). Let the two boundary hypersurfaces of $\mathrm{HM}(\mathbb{R}^n)$ be denoted tb and ti, where the first arises from $t = 0$ and the second from '$t = \infty$.' Set $Y = (0, 1, 0) \subset \mathrm{HM}(\mathbb{R}^n)$ and $S_Y = \mathrm{sp}(dt)$. The compact manifold with corners

$$\mathrm{HHM}\,\mathbb{R}^n = [\mathrm{HM}(\mathbb{R}^n); Y, S_Y] \tag{8.1}$$

is the natural carrier of the Euclidean heat kernel, as a convolution operator. Thus, denoting by tf the new boundary hypersurface produced by the blow-up in (8.1), the heat kernel lifts under blow-up to an element

$$\Phi' \in \rho_{\mathrm{tf}}^{-n} \rho_{\mathrm{tb}}^{\infty} \rho_{\mathrm{ti}}^{\frac{n}{2}} \mathcal{C}^{\infty}(\mathrm{HHM}(\mathbb{R}^n)). \tag{8.2}$$

More generally the convolution kernels in $\Psi^p_{\mathrm{th}}(\mathbb{R}^n)$ can be identified as the subspaces

$$\Psi^p_{\mathrm{th}}(\mathbb{R}^n) \subset \rho_{\mathrm{tf}}^{-n-2-p} \rho_{\mathrm{tb}}^{\infty} \rho_{\mathrm{ti}}^{\frac{n}{2}+1+\frac{p}{2}} \mathcal{C}^{\infty}(\mathrm{HHM}(\mathbb{R}^n)),\ p < 0, \tag{8.3}$$

consisting of the elements which are homogeneous of degree $-(n + 2 + p)$ under the global $\mathbb{R}^+$ action.

Notice that this construction is independent of the basis of $\mathbb{R}^n$ so is defined for any vector space. Indeed if V is a $\mathcal{C}^\infty$ Euclidean vector bundle of rank n over some compact manifold Y then the construction can be carried out fibre-by-fibre to give a compact manifold $\mathrm{HHM}(V)$ which fibres over Y with fibre diffeomorphic to $\mathrm{HHM}(\mathbb{R}^n)$. This manifold is the natural carrier for the collective heat kernels of the flat Laplacians on the fibres, in the sense that, with the same notation for boundary faces

$$\exp(-t\Delta_{\mathrm{fibre}}) \in \rho_{\mathrm{tf}}^{-n} \rho_{\mathrm{tb}}^{\infty} \rho_{\mathrm{ti}}^{\frac{n}{2}} \mathcal{C}^{\infty}(\mathrm{HHM}(V)). \tag{8.4}$$

This function can also be considered as the kernel for the heat semigroup acting on half-densities, with the fibre-metric half-density used to trivialize the bundle of half-densities.

For the product $\mathbb{R}^n \times F$ of Euclidean space and a compact manifold without boundary the natural heat space is obtained by combining this construction with that of §3. Thus consider the product $\mathrm{HM}(\mathbb{R}^n) \times F^2$. The appropriate heat space is

$$\mathrm{HHM}(\mathbb{R}^n \times F) = [\mathrm{HM}(\mathbb{R}^n) \times F^2; B_h, S_h] \tag{8.5}$$

where $B_h = \mathrm{tb}(\mathrm{HM}(\mathbb{R}^n)) \times \mathrm{Diag}$ and S_h is the conormal to the boundary hypersurface of $\mathrm{HM}(\mathbb{R}^n)$. For the product metric, coming from the Euclidean metric on $\mathbb{R}^n$ and the metric on F, the heat kernel is the product $\exp(-t\Delta) = \exp(-t\Delta_F) \cdot \exp(-t\Delta_E)$.

Lemma 8.1. *The heat kernel on $\mathbb{R}^n \times F$, as a convolution kernel in the first variables, lifts to an element*

$$\exp(-t\Delta) \in \rho_{\mathrm{tf}}^{-\frac{N}{2}+1} \rho_{\mathrm{tb}}^{\infty} \rho_{\mathrm{ti}}^{\frac{n}{2}-2} \mathcal{C}^{\infty}(\mathrm{HHM}(\mathbb{R}^n \times F); \Omega^{\frac{1}{2}}),\ N = n + \dim F. \tag{8.6}$$

Proof. Since the heat kernel is the product of the heat kernels, the regularity of the lifted kernel away from $t = 0$ is immediate from the separate discussions in the Euclidean and compact cases. □

To describe the normal operator at the ab face of the adiabatic heat calculus, we discuss here the heat calculus on the base with values in the smoothing operators on the fibers. Thus let M be the total space of a fibration, with the base Y and the fibers F, and set $Z = [0, \infty) \times M^2$. Now consider its t-parabolic blow-up along the submanifold B_a (defined in (7.2)) instead of the usual diagonal:

$$M^2_{h,\phi} = [Z; B_a, S]\,. \tag{8.7}$$

Note that B_a is the fibered diagonal of the fibration, M^2_ϕ.

As usual we denote by tf and tb its temporal front face and temporal boundary face respectively. Recall that n is the dimension of the base manifold. The kernel density bundle KD is now defined by the prescription

$$\mathcal{C}^\infty(M^2_{h,\phi}; \mathrm{KD}) = \rho_{\mathrm{tf}}^{-\frac{n}{2}-\frac{3}{2}} \mathcal{C}^\infty(M^2_{h,\phi}; \Omega^{\frac{1}{2}}). \tag{8.8}$$

Finally the heat calculus on the base with values in the smoothing operators on the fibers is now defined by

$$\Psi_h^{-k}(M, \phi; \Omega^{\frac{1}{2}}) = \rho_{\mathrm{tf}}^k \rho_{\mathrm{tb}}^\infty \mathcal{C}^\infty(M^2_{h,\phi}; \mathrm{KD}) \text{ for } k \in \mathbb{N}. \tag{8.9}$$

By definition the normal operator at ab of the adiabatic heat calculus, which is the multiplication of the kernel by X^{-p} ($X = x/t^{\frac{1}{2}}$) followed by the evaluation at ab, takes values in this calculus:

$$N_{a,p} : \Psi_A^{-j,-k,-p}(M; \Omega^{\frac{1}{2}}) \twoheadrightarrow \Psi_h^{-k}(M, \phi; \Omega^{\frac{1}{2}}). \tag{8.10}$$

Lemma 8.2. *The heat kernel of the reduced Laplacian lifts to an element*

$$\exp(-t\Delta_Y) \in \Psi_h^{-2}(M, \phi; \Omega^{\frac{1}{2}}). \tag{8.11}$$

Proof. The heat kernel of the reduced Laplacian is an element of the heat calculus of the base manifold Y. By using a partition of unity we can decompose it into the sum of two parts; the first is supported away from the front face and the second near the front face. Both can be lifted, fiberwise constantly, to an element of the base heat calculus, as described above. The first part is an effectively a smoothing operator on Y, that is, its Schwartz kernel is a smooth function on $Y \times Y$. Clearly this lifts to a smooth function on $M \times M$. Similarly, since the second piece is supported near the front face we can effectively think of it as a smooth function on af$\times [0, 1]$ (say) multiplied by a singular density factor. Thus, once again it lifts to a function of the same type near the front face of the base heat space (cf. the analysis of the adiabatic front face of the adiabatic heat calculus). □

9. Adiabatic heat calculus

In this section we generalize the results of §3 for heat calculus to the adiabatic heat calculus. As discussed in §4, the adiabatic heat calculus is defined so that the statement, in Proposition 7.1, that the heat kernel for the adiabatic Laplacian lies in the calculus gives a rather precise description of the degeneracy at $x = 0$.

Formulæ (3.17) and (3.19) still define the action of $\Psi_A^{-j,-k,-p}(M,\Omega^{\frac12})$ on $\dot{\mathcal{C}}^\infty(X,\Omega^{\frac12})$, where $X \equiv [0,\infty)\times M\times[0,1]$. We first note the result of composing these operators with differential operators. For simplicity of notation here we write the action of a vector field V through Lie derivation of half-densities simply as V.

Proposition 9.1. *Let $A\in\Psi_A^{-j,-k,-p}(M,\Omega^{\frac12})$. If V is any smooth vector field, then*

$$(t^{\frac12}V)\circ A\in\Psi_A^{-j,-k,-p}(M,\Omega^{\frac12}) \tag{9.1}$$

and with $\sigma_1(xV)$ the symbol of xV as an adiabatic vector field,

$$\begin{aligned} N_{h,-j}(t^{\frac12}V\circ A)&=\sigma_1(xV)N_{h,-j}(A),\\ N_{A,-k}(t^{\frac12}V\circ A)&=\sigma_1(xV)N_{A,-k}(A),\\ N_{a,-p}(t^{\frac12}V\circ A)&=(t^{\frac12}V)N_{a,-p}(A).\end{aligned} \tag{9.2}$$

If W is a vertical vector field and $T=t/x^2$, then

$$(T^{\frac12}W)\circ A\in\Psi_A^{-j,-k,-p}(M,\Omega^{\frac12}) \tag{9.3}$$

$$\begin{aligned} N_{h,-j}[(T^{\frac12}W)\circ A]&=\sigma_1(W)N_{h,-j}(A),\\ N_{A,-k}[(T^{\frac12}W)\circ A]&=(W)N_{A,-k}(A),\\ N_{a,-p}[(T^{\frac12}W)\circ A]&=(T^{\frac12}W)N_{a,-p}(A).\end{aligned} \tag{9.4}$$

Finally, $(t\partial_t\circ A)\in\Psi_A^{-j,-k,-p}(M,\Omega^{\frac12})$ and

$$\begin{aligned} N_{h,-j}(t\partial_t\circ A)&=\left[\frac{j}{2}-1-\frac12(N+R_M)\right]N_{h,-j}(A),\\ N_{A,-k}(t\partial_t\circ A)&=T\partial_T N_{A,-k}(A),\\ N_{a,-p}(t\partial_t\circ A)&=t\partial_t N_{a,-p}(A).\end{aligned} \tag{9.5}$$

Proof. As a vector field on the left factor of M, $t^{\frac12}V$ lifts from Z to M_H^2 to a vector field of the form $\rho_{\mathrm{tb}}\tilde V$ where $\tilde V\in\mathcal{V}_{\mathrm b}(M_H^2)$. From this (9.1) follows. Similarly if W is a vertical vector field, i.e., is tangent to the fibres, then the same lift is of the form $\rho_{\mathrm{tb}}\rho_{\mathrm{af}}\tilde W$ with $\tilde W\in\mathcal{V}_{\mathrm b}(M_H^2)$, from which (9.3) follows. Similarly $t\partial_t$ lifts into $\mathcal{V}_{\mathrm b}(M_H^2)$.

To compute the normal operator at tf, we can use the projective coordinates

$$s=t^{\frac12}/x=T^{\frac12},\quad y,\quad \bar Y=\frac{y-y'}{t^{\frac12}},\quad z,\quad \bar Z=\frac{z-z'}{t^{\frac12}/x},\quad x \tag{9.6}$$

Then $A=T^{\frac{j}{2}-\frac{N+4}{4}}x^{k-\frac{n}{2}-1}a(T,y,\bar Y,z,\bar Z,x)|dTdyd\bar Ydzd\bar Zdx|^{\frac12}$, where a is a smooth function of $(T^{\frac12},y,\bar Y,z,\bar Z,x)$ and vanishes rapidly as $\bar Y\to\infty$ or $\bar Z\to\infty$.

Since $\beta_A^*(t^{\frac12}V)=t^{\frac12}V+\sigma_1(xV)$, a computation completely similar to that of the proof of Proposition 1.2 shows that

$$\begin{aligned} &(t^{\frac12}V)\circ A\in\Psi_A^{-j,-k,-p}(M,\Omega^{\frac12})\text{ and}\\ &N_{h,-j}(t^{\frac12}V\circ A)=\sigma_1(xV)N_{h,-j}(A).\end{aligned} \tag{9.7}$$

Similarly we have for a vertical vector field W

$$\begin{aligned}&(T^{\frac{1}{2}}W)\circ A\in\Psi_A^{-j,-k,-p}(M,\Omega^{\frac{1}{2}})\text{ and}\\&N_{h,-j}(T^{\frac{1}{2}}W\circ A)=\sigma_1(W)N_{h,-j}(A).\end{aligned}\tag{9.8}$$

and

$$\begin{aligned}&t\partial_t\circ A\in\Psi_A^{-j,-k,-p}(M,\Omega^{\frac{1}{2}})\\&N_{h,-j}(t\partial_t\circ A)=\left[\frac{j}{2}-1-\frac{1}{2}(N+R_M)\right]N_{h,-j}(A),\end{aligned}\tag{9.9}$$

where R_M denote the radial vector field on TM.

To compute the normal operator at af, we use the coordinates

$$T=t/x^2,\quad y,\quad Y=\frac{y-y'}{x},\quad z,\quad z',\quad x.\tag{9.10}$$

The same computation shows that

$$\begin{aligned}N_{A,-k}(t^{\frac{1}{2}}V\circ A)&=\sigma_1(xV)N_{A,-k}(A),\\N_{A,-k}[(T^{\frac{1}{2}}W)\circ A]&=(W)N_{A,-k}(A),\\N_{A,-k}(t\partial_t\circ A)&=T\partial_TN_{A,-k}(A).\end{aligned}\tag{9.11}$$

This finishes the proof of Proposition 9.1. □

To prove composition results for adiabatic heat operators we shall use an 'adiabatic triple space'. In

$$W=\mathbb{R}^2\times M^3\times[0,1]\tag{9.12}$$

consider the three adiabatic fibre diagonals with the associated parabolic directions:

$$\begin{aligned}B_F&=\{(t,0,m,m',m'',0);\phi(m')=\phi(m'')\},\quad &S_F&=\mathrm{sp}(dt')\\B_S&=\{(t,t,m,m',m'',0);\phi(m)=\phi(m')\},\quad &S_S&=\mathrm{sp}(dt-dt')\\B_C&=\{(0,t',m,m',m'',0);\phi(m)=\phi(m'')\},\quad &S_C&=\mathrm{sp}(dt)\end{aligned}\tag{9.13}$$

and the triple fibre diagonal

$$B_T=\{(0,0,m,m',m'',0);\phi(m)=\phi(m')=\phi(m'')\},\ S_T=\mathrm{sp}(dt,dt').\tag{9.14}$$

Then set

$$W_A=[W;B_T,S_T;B_F,S_F;B_S,S_S;B_C,S_C],\ \beta_A^3:W_A\longrightarrow W.\tag{9.15}$$

For this triple product we have the 'usual' results (recall that $Z=[0,\infty)\times M^2\times[0,1]$ and $Z_A=[Z,B_a;S]$):

Proposition 9.2. *The three projections*

$$\begin{aligned}\pi_F^2(t,t',m,m',m'',x)&=(t',m',m'',x)\\\pi_S^2(t,t',m,m',m'',x)&=(t-t',m,m',x)\\\pi_C^2(t,t',m,m',m'',x)&=(t,m,m'',x)\end{aligned}\tag{9.16}$$

lift to b-fibrations

$$\pi^2_{O,A} : W_A \longrightarrow Z_A,\ O = F, S, C \tag{9.17}$$

giving a commutative diagram

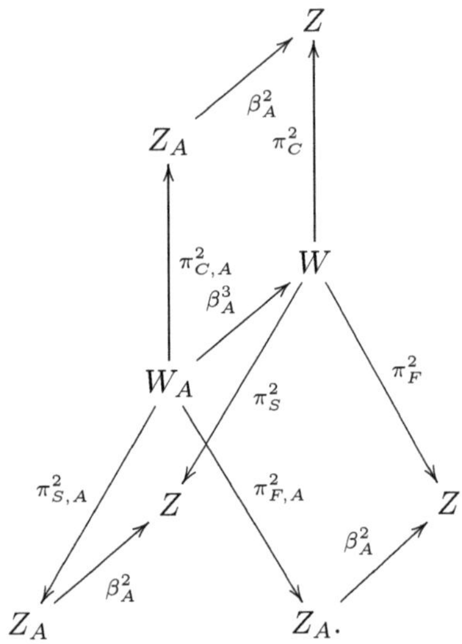

(9.18)

Proof. We need first to show that the 'stretched projections' $\pi^2_{O,A}$, for $O = F, S, C$ exist and are $\mathcal{C}^\infty$. Then we need to check that they are b-fibrations according to the definition given in [28]. There is sufficient symmetry (using $t \longleftrightarrow t'$ and sign reversal) that it is enough to consider one case, say $O = C$. The existence of the stretched projection follows from results on the commutation of blow-up. In this case [18, Appendix C] can be used to rewrite the definition, (9.15), in the form:

$$W_A = [W; B_C, S_C; B_T, S_T; B_F, S_F; B_S, S_S]\,. \tag{9.19}$$

The intermediate space

$$[W; B_C, S_C] \equiv Z_A \times (\mathbb{R} \times M) \tag{9.20}$$

so the iterated blow-up (9.19) gives $\pi^2_{C,A}$ as the product of a blow-down map and the projection from (9.20)

$$[W; B_C, S_C; B_T, S_T; B_F, S_F; B_S, S_S] \longrightarrow [W; B_C, S_C] \longrightarrow Z_A. \tag{9.21}$$

Since both the blow-down map and the projection are b-maps so is $\pi^2_{C,A}$; clearly it is surjective. Thus it remains only to show that $\pi^2_{C,A}$ is a b-fibration.

Let $f : X \longrightarrow Y$ be a b-map between manifolds with corners, i.e., a $\mathcal{C}^\infty$ map such that if $\rho'_i \in \mathcal{C}^\infty(Y)$, $i = 1, \ldots, N'$ are defining functions for the boundary hypersurfaces of Y and $\rho_j \in \mathcal{C}^\infty(X)$, $j = 1, \ldots, N$ are defining functions for the boundary hypersurfaces of X then

$$f^*\rho'_i = a_i \prod_{j=1} \rho_j^{k(i,j)},\ 0 < a_i \in \mathcal{C}^\infty(X). \tag{9.22}$$

The non-negative integers $k(i,j)$ are the boundary exponents of f. The condition that f be a b-fibration can be expressed as two conditions on the map, that it be a 'tangential submersion' and 'b-normal'. For any point $p \in X$ in a manifold with boundary let $\mathrm{BH}_X(p)$ be the smallest boundary face containing p. The first condition is the requirement that

$$f_* (T_p \mathrm{BH}_X(p)) = T_{f(p)} \mathrm{BH}_Y(f(p)) \quad \forall \, p \in X. \tag{9.23}$$

The condition of b-normality is the requirement on the boundary indices:

$$\text{For each } j,\ k(i,j) \neq 0 \text{ for at most one } i. \tag{9.24}$$

To check (9.23) for $\pi^2_{C,A}$ we note that under the iterated blow-down map in (9.21) the image of $T_p\,\mathrm{BH}(p)$ is always the tangent space at the image point to the smallest submanifold formed by the intersection of the boundary faces of the image and the submanifolds blown up. It follows easily that $\pi^2_{C,A}$ is a tangential submersion. To check that it is b-normal we simply compute the boundary indices. In Table 1 the boundary exponents of all three of the stretched projections are recorded. This completes the proof of the proposition.

		at	asF	asC	asS	ab
$\pi^2_{F,A}$	af	1	1	0	0	0
	ab	0	0	1	1	1
$\pi^2_{S,A}$	af	1	0	0	1	0
	ab	0	1	1	0	1
$\pi^2_{C,A}$	af	1	0	1	0	0
	ab	0	1	0	1	1
	ν	0	0	n+2	0	0

Table 1 : Boundary exponents □

Also in Table 1 there is a 'density row', labelled 'ν' which is important in the description of the composition results. These exponents are fixed by the natural identification of density bundles:

$$(\pi^2_{F,A})^* \,\mathrm{KD} \otimes (\pi^2_{S,A})^* \,\mathrm{KD} \otimes (\pi^2_{C,A})^* (\mathrm{KD}') \otimes |dt|^{\frac{1}{2}} \cong \prod_{F=\mathrm{at},\mathrm{asF},\mathrm{asC},\mathrm{asS},\mathrm{ab}} \rho_F^{\nu_F} \cdot \Omega \text{ on } W_A. \tag{9.25}$$

Here KD' is the half-density bundle with the opposite weighting to KD so that $\mathrm{KD}' \otimes \mathrm{KD} \cong \Omega$. A straightforward computation gives the results as stated, i.e.,

$$\nu_{\mathrm{at}} = \nu_{\mathrm{asF}} = \nu_{\mathrm{asS}} = \nu_{\mathrm{ab}} = 0, \ \nu_{\mathrm{asC}} = n+2. \tag{9.26}$$

The table can be used to give an 'upper bound' for the singularities of the composite of two operators using a general push-forward theorem from [28] (see also [18]) which applies because of Proposition 9.2. Thus if $\mathcal{E} = (E_{\rm af}, E_{\rm ab})$ is an index family for M_A^2, assumed trivial at tf and tb then let

$$\Psi_A^{-\infty,\mathcal{E}}(M,\Omega^{\frac12}) = \mathcal{A}_{\rm phg}^{(\infty,\infty,E_{\rm af},E_{\rm ab})}(M_A^2;{\rm KD}) \tag{9.27}$$

be the space of polyhomogeneous conormal distributions on M_A^2 which vanish rapidly at tf and tb and have expansions at af and ab with exponents from $E_{\rm af}$ and $E_{\rm ab}$ respectively.

Proposition 9.3. *Composition, being convolution in t, gives, for any index families $\mathcal{E}$ and $\mathcal{F}$*

$$\Psi_A^{-\infty,\mathcal{E}}(M,\Omega^{\frac12}) \circ \Psi_A^{-\infty,\mathcal{F}}(M,\Omega^{\frac12}) \subset \Psi_A^{-\infty,\mathcal{G}}(M,\Omega^{\frac12}) \tag{9.28}$$

where

$$\begin{aligned} G_{\rm af} &= [E_{\rm af}+F_{\rm af}]\overline{\cup}[E_{\rm ab}+F_{\rm ab}+n+2] \\ G_{\rm ab} &= [E_{\rm af}+F_{\rm ab}]\overline{\cup}[E_{\rm ab}+F_{\rm af}]\overline{\cup}[E_{\rm ab}+F_{\rm ab}] \end{aligned} \tag{9.29}$$

Proof. Let $A \in \Psi_A^{-\infty,\mathcal{E}}(M;\Omega^{\frac12})$, $B \in \Psi_A^{-\infty,\mathcal{F}}(M,\Omega^{\frac12})$. The composition $C = A\circ B$ can be written in terms of their Schwartz kernels via the b-fibrations $\pi_{O,A}^2$:

$$\kappa_C KD' = (\pi_{C,A}^2)_*[(\pi_{S,A}^2)^*\kappa_A\cdot(\pi_{F,A}^2)^*\kappa_B\cdot(\pi_{C,A}^2)^*(KD')\cdot(\tilde\pi_{C,A}^2)^*(|dt'|^{\frac12}|dx|^{-1/2})], \tag{9.30}$$

where $\tilde\pi_{C,A}^2 = \pi_C^2\circ\beta_A^3 : W_A \to Z$.

By the push-forward theorem [28] (see also [18])

$$C \in \Psi_A^{-\infty,\mathcal{G}}(M,\Omega^{\frac12}) \tag{9.31}$$

for some index set $\mathcal{G}$. The index set $\mathcal{G} = (G_{af}, G_{ab})$ can be computed by the Mellin transform

$$\begin{aligned} \langle\kappa_C\cdot KD', \rho_{af}^{z_1}\rho_{ab}^{z_2}\rangle &= \langle(\pi_{F,A}^2)^*(\tilde\kappa_A\cdot(\pi_{S,A}^2)^*(\tilde\kappa_B) \\ &\times \prod_{F=at,asF,asC,asS,ab} \rho_F^{\nu_F}\cdot\Omega(W_A),(\pi_{C,A}^2)^*(\rho_{af}^{z_1}\rho_{ab}^{z_2})\rangle, \end{aligned} \tag{9.32}$$

where $\kappa_A = \tilde\kappa_A\cdot KD$, $\kappa_B = \tilde\kappa_B\cdot KD$. From Table 1 and (9.26) we obtain (9.29). □

We now have the tools to prove our first main result on the uniform behavior of the heat kernel in the adiabatic limit, stated in the introductory section.

Proof of Theorem 0.1. Clearly (7.17), (7.18), (7.19) are compatible therefore there is $G_1 \in \Psi_A^{-2,-2,0}(M,{}^a\Lambda^k)$ whose normal operators are given by (7.17)–(7.19). Now by the composition formulas,

$$\left(t\partial_t + \frac{t}{x^2}\Delta_x\right)G_1 = t\,{\rm Id} - tR_1, \tag{9.33}$$

where $tR_1 \in \Psi_A^{-3,-3,-1}(M,{}^a\Lambda^k)$ or $R_1 \in \Psi_A^{-1,-1,-1}(M,{}^a\Lambda^k)$. Thus G_1 is already a parametrix. We now modify G_1. Using the heat calculus we can find a $G_0 \in$

$\Psi_A^{-2,-2,-\infty}(M,{}^a\Lambda^k)$ such that

$$\left(t\partial_t + \frac{t}{x^2}\Delta_x\right) G_0 = t\operatorname{Id} - tR_0, \tag{9.34}$$

where $R_0 \in \Psi_A^{-\infty,0,-\infty}(M,{}^a\Lambda^k)$.

It follows that there is a correction term $G_0' \in \Psi_A^{-3,-1,-1}(M,{}^a\Lambda^k)$ such that the modification of the parametrix $G_2 = G_1 - G_0'$ still has the normal operator (7.17)–(7.19) and is a parametrix in the strong sense that

$$\left(t\partial_t + \frac{t}{x^2}\Delta_x\right) G_2 = t\operatorname{Id} - tR_2, \tag{9.35}$$

where $R_2 \in \Psi_A^{-\infty,-1,-1}(M,{}^a\Lambda^k)$.

By Proposition 9.3,

$$(R_2)^k \in \Psi_A^{-\infty,-k,-\alpha(k)}(M,{}^a\Lambda^k), \tag{9.36}$$

where $\alpha(k) = \{(-k,k-1)\}$. Thus the Neumann series $\sum_{k=0}^{\infty}(R_2)^k$ can be summed modulo a term vanishing rapidly at both af and ab, i.e., there exists $S' \in \Psi_A^{-\infty,-1,A}$ with $A = \cup\alpha(k)$ such that

$$(\operatorname{Id} - S')(\operatorname{Id} - R_2) = \operatorname{Id} - R_3, \quad R_3 \in \Psi_A^{-\infty,-\infty,-\infty}. \tag{9.37}$$

In other words, R_3 is a Volterra operator vanishing rapidly at all the boundaries. Thus $\operatorname{Id} - R_3$ can be inverted with an operator of the same type. It follows that $\operatorname{Id} - R_2$ has a two-side inverse $\operatorname{Id} - S$, $S \in \Psi_A^{-\infty,-1,A}(M,{}^a\Lambda^k)$. This in turn means we have

$$\exp\left(-\frac{t}{x^2}\Delta_x\right) = G_1(\operatorname{Id} - S) = G_1 - G_1 \circ S \tag{9.38}$$

and by Proposition B.2, $G_1 \circ S \in \Psi_A^{-\infty,-1,A}(M,{}^a\Lambda^k)$. That is, $\mathcal{C}^\infty$ except for logarithmic terms at ab.

To show that $\exp(-\frac{t}{x^2}\Delta_x) \in \Psi_A^{-2,-2,0}(M,{}^a\Lambda^k)$ we show that the logarithmic terms are actually zero. To this end we consider the behavior of the leading log term at the boundary of ab. Near ab, we can use the coordinates

$$t, y, Y = \frac{y-y'}{t^{\frac12}}, z, z', X = \frac{x}{t^{\frac12}}. \tag{9.39}$$

The boundary of ab is $\{t=0, X=0\}$. Let $t^{\frac12(k-\frac{n}{2}-1)}X^p(\log X)^l u(y,Y,z,z')$ be the leading log term of $\exp(-\frac{t}{x^2}\Delta_x)$. By its explicit construction we have $1 \le p$ and $k \ge p+2$. Since $\exp(-\frac{t}{x^2}\Delta_x)$ satisfies the heat equation, we find that the Taylor coefficients of u at $t=0$ (we still denote by u) satisfies

$$\left(\Delta_Y - \frac12 Y\partial_Y + \frac12(k-1) - \frac{n}{4} - \frac{p}{2}\right) u = 0, \tag{9.40}$$

where $u = u(y, Y, z, z')$ is smooth in all variables and vanishes rapidly as $Y \to \infty$. Multiplying the equation by u and integrate by part

$$\int_R |\nabla u|^2 - \frac{1}{2}\int (Y\partial_Y u)u + d\int u^2 = 0. \tag{9.41}$$

But $\int (Y\partial_Y u)u = \int u\partial_Y(Yu) = -\int Yu\cdot\partial_Y u - n\int u^2$. Therefore

$$\int_R |\nabla u|^2 + c\int u^2 = 0 \tag{9.42}$$

with $c = \frac{1}{2}(k-1) - \frac{p}{2} \geq \frac{1}{2}$. Hence $u \equiv 0$.

This shows that the leading log term vanishes rapidly at the boundary of ab and therefore can be blown down to a smooth solution of the heat equation for the base manifold with zero initial data. It must be zero identically. □

10. The rescaled adiabatic calculus and supertrace

With the hard work done for constructing the adiabatic heat calculus, we are now ready to show how to modify it to incorporate the Getzler's rescaling. We will first indicate the modification necessary in constructing the rescaled adiabatic heat calculus. Then a proof is given for Theorem 0.2. Finally we turn to the two lemmas in preparation of the application to the analytic torsion.

To construct the rescaled adiabatic heat calculus, the only thing different from the discussions in the previous sections is that we rescale the homomorphism bundle at both af and tf. Following the discussion in §4 we do so by giving filtrations for the homomorphism bundle over both the submanifolds, B_a and B_h, which are blown up in the construction of M^2_A. These filtrations need to satisfy the appropriate compatibility condition at $B_a \cap B_h$.

As noted in §6 over $\mathrm{ab}(M_a) = \{x = 0\}$ the exterior algebra decomposes into

$${}^a\Lambda^*_p M = {}^x\Lambda^*_y Y \otimes \Lambda^*_p F_y,\ p \in \mathrm{ab}(M_a),\ y = \phi(p). \tag{10.1}$$

Here ${}^x\Lambda^*_y Y = \Lambda^*({}^xT^*_y Y)$ so ${}^x\Lambda^k_y Y = x^{-k}\Lambda^k_y Y$. Since B_a lies above the fibre diagonal (10.1) leads to a decomposition of the homomorphism bundle

$$\mathrm{Hom}_q({}^a\Lambda^* M) = \mathrm{hom}({}^x\Lambda^*_y Y) \otimes \mathrm{Hom}_q(\Lambda^* F_y) \text{ at } q \in F_y \times F_y \subset B_a. \tag{10.2}$$

We now separate into two cases as the parity of the base dimension makes a difference here. If Y is odd-dimensional, the discussion in §2, leading to (5.32), applies to Clifford multiplication by ${}^xT^*Y$ and gives the filtration

$$\mathrm{hom}^{[k]}({}^x\Lambda^*_y Y) = \mathbb{C}\ell^{[k]}({}^xT^*_y Y \oplus \mathbb{R}) \otimes \mathrm{hom}'({}^x\Lambda^*_y Y),\ y \in Y. \tag{10.3}$$

This lifts to give the desired filtration over B_a :

$$\mathrm{Hom}^{[k]}_{B_a}({}^a\Lambda^* M) = \mathbb{C}\ell^{[k]}({}^xT^* Y \oplus \mathbb{R}) \otimes \mathrm{Hom}'({}^a\Lambda^* M). \tag{10.4}$$

which has length $\dim Y + 1$.

Since B_h, defined in (7.2), lies over the diagonal we can use the natural extension of the filtration (5.32). Namely left Clifford multiplication by the rescaled bundle ${}^aT^*M$ extends to give

$$\hom({}^a\Lambda^*M) = \mathbb{C}\ell^{[k]}({}^aT^*M \oplus \mathbb{R}) \otimes \hom'({}^a\Lambda^*M) \text{ over } B_h \tag{10.5}$$

where this filtration has length $\dim M + 1$. This second filtration is consistent with the first over the intersection $B_a \cap B_h$ in the sense that (10.5) induces a filtration on each of the subspaces (10.4).

For the discussion in §4 to apply we need to show that the extension of the connection in (6.8) preserves the filtrations and that the curvature operator has the order property (4.13). These conditions follow as in §5, from the fact that exterior and interior multiplication have order one. Thus, the rescaled adiabatic heat calculus $\Psi^*_{A,G}(M; {}^a\Lambda^*)$ is defined in this case.

If Y is even-dimensional, we use a different filtration of the homomorphism bundle over B_a. Suppose first that W is an even-dimensional Euclidean vector space. The complexified homomorphism bundle has the decomposition (5.22) in terms of left and right Clifford multiplication, defined by (5.16) and (5.17). Consider a different 'right' Clifford multiplication defined by

$$\tilde{c}_r(e) = (\operatorname{ext}(e) + \operatorname{int}(e)) \cdot \tau, \ \tau = \tau_l = i^{n(2n-1)} c_l(e_1) \cdots c_l(e_{2n}) \tag{10.6}$$

in terms of an orthonormal basis $e_1, \ldots, e_{2n}$ of W. The involution τ_l is, up to a power of i, the Hodge $*$ operator and so is independent of the choice of basis; we shall use it as the parity operator defining a (new) superbundle structure on the exterior algebra. Since $\tilde{c}_r(e)$ again commutes with the left Clifford multiplication this action actually gives the same decomposition (5.22) but we write it

$$\hom(\mathbb{C}\Lambda^*W) = \mathbb{C}\ell(W) \otimes \widetilde{\mathbb{C}\ell}(W) \tag{10.7}$$

to emphasize that the action is through (10.6). We consider the filtration corresponding to this action:

$$\hom^{[k]}(\mathbb{C}\Lambda^*W) = \mathbb{C}\ell^{[k]}(W) \otimes \widetilde{\mathbb{C}\ell}(W). \tag{10.8}$$

The true parity operator on forms can be written

$$\begin{aligned} Q &= c_l(e_1) \cdots c_l(e_{2n}) c_r(e_1) \cdots c_r(e_{2n}) = \tau_l \tilde{\tau}_r \\ \tilde{\tau}_r &= i^{n(2n-1)} \tilde{c}_r(e_1) \cdots \tilde{c}_r(e_{2n}) \end{aligned} \tag{10.9}$$

which shows it to be a homomorphism of maximal order.

The filtration (10.8) is independent of the choice of orientation, so extends to the homomorphism bundle of any even-dimensional Riemann manifold. For the fibration with even-dimensional base we consider in place of (10.3)

$$\hom^{[k]}({}^x\Lambda^*_yY) = \mathbb{C}\ell^{[k]}({}^xT^*Y) \otimes \hom'({}^x\Lambda^*_yY), \ y \in Y. \tag{10.10}$$

and then, in place of (10.4)

$$\operatorname{Hom}^{[k]}_{B_a}({}^a\Lambda^*M) = \mathbb{C}\ell^{[k]}({}^xT^*Y) \otimes \operatorname{Hom}'({}^a\Lambda^*M). \tag{10.11}$$

Hence, the rescaled adiabatic heat calculus $\Psi^*_{A,G}(M; {}^a\Lambda^*)$ is also defined in this case.

To prove Theorem 0.2, it remains to analyze the behaviour of the Laplacian and to compute its normal operators. From the Weitzenböck formula (5.40) it again follows that the Laplacian acts on the rescaled bundle and hence

$$\exp\left(-\frac{t}{x^2}\,{}^a\Delta\right) \in \Psi^{-2,-2,0}_{A,G}(M; {}^a\Lambda^*). \tag{10.12}$$

The rescaled normal operator at tf is still given by (5.57). Moreover the rescaled normal operator of the Laplacian at af is just given by (5.57), but with $n = \dim Y$ and with an additional term coming from the fibre, as

$$N_{A,G,-k}\left(\left(\frac{t}{x^2}\,{}^a\Delta\right)A\right) = \left[\mathcal{A}_T^2 + \mathcal{H}_Y + \frac{1}{2}(V_r + n + k - 2) - \frac{1}{8}C(R_Y)\right] \cdot N_{A,G,-k}(A). \tag{10.13}$$

Here $\mathcal{H}_Y$ is the generalized harmonic oscillator on the fibres of ${}^xTY$, and $\mathcal{A}_T$ is the rescaled Bismut superconnection:

$$\mathcal{A}_T^2 = -T\left[\nabla_{e_i} + \frac{1}{2}T^{-\frac{1}{2}}\langle\nabla_{e_i}e_j, f_\alpha\rangle c_l(e_i)c_l(f_\alpha) + \frac{1}{4}\langle\nabla_{e_i}f_\alpha, f_\beta\rangle c_l(f_\alpha)c_l(f_\beta)\right]^2 + \frac{1}{4}TK_F, \tag{10.14}$$

where e_i is an orthonormal basis of the fibers and f_α that of the base, and K_F denotes the scalar curvature of the fibers. We remark in passing that in the above formula the Clifford action of the base variables is acting really by exterior multiplication since at the front face they act on the graded space of the filtration. From this it follows that at the adiabatic front face the rescaled normal operator is just

$$\exp\left(-\mathcal{H}_Y\right)\exp\left(\frac{1}{8}C(R_Y)\right)\exp\left(-\mathcal{A}_T^2\right). \tag{10.15}$$

We have now finished the proof of Theorem 0.2. In what follows we shall show, by use of the rescaled adiabatic heat calculus, that not only does the analogue of (5.10) hold uniformly in x but there is additional cancellation at the adiabatic front face when (7.24) is used to compute the supertrace. The parity of the dimension of fibre and base makes a considerable difference to the argument so we treat the two cases separately.

Lemma 10.1. *If the fibres of* (6.1) *are even-dimensional then*

$$h = \operatorname{str}\left[N\exp\left(-\frac{t}{x^2}\,{}^a\Delta\right)\right] \in \rho_{\mathrm{tf}}^{-1}\rho_{\mathrm{af}}^{-1}\mathcal{C}^\infty_{E,E}(\widetilde{\mathrm{Diag}_A}; \Omega M) \tag{10.16}$$

and if $a_{-\frac{1}{2}} \in \mathcal{C}^\infty(M \times [0,1]; \Omega M)$ *is given by* (5.11) *then*

$$h - a_{-\frac{1}{2}}t^{-\frac{1}{2}} \in \rho_{\mathrm{tf}}\rho_{\mathrm{af}}\mathcal{C}^\infty_{E,E}(\widetilde{\mathrm{Diag}_A}; \Omega M). \tag{10.17}$$

Proof. Observe that the number operator and involution decompose over af as

$$N = N_Y \otimes 1 + 1 \otimes N_F,\ Q = Q_Y \otimes Q_F. \tag{10.18}$$

It follows as before that N has order 2 in terms of the rescaling at af and hence that (10.16) follows. Moreover the leading term at af is

$$t^{-\frac{1}{2}} \operatorname{tr}(Q_Y \operatorname{Pf}(R_k)) \operatorname{tr}(Q_F \exp(-T^F \Delta)). \tag{10.19}$$

Since the fibres are even-dimensional

$$\int_{F_y} \operatorname{tr}(Q_F \exp(-T^F \Delta)) = \chi(F) \tag{10.20}$$

is independent of both T and y. Thus (10.19) is independent of T and it must therefore be just $t^{-\frac{1}{2}} a_{-\frac{1}{2}}$. This proves (10.17) and the lemma. □

Turning to the case where the base is even-dimensional we have a similar result except that the supertrace is less singular at af :

Lemma 10.2. *If the fibres of* (6.1) *are odd-dimensional then*

$$h = \operatorname{str}\left[N \exp\left(-\frac{t}{x^2}\, {}^a\Delta\right)\right] \in \rho_{\mathrm{tf}}^{-1} \mathcal{C}^\infty_{E,E}(\widetilde{\operatorname{Diag}_A}; \Omega M) \tag{10.21}$$

and

$$h_{\restriction \mathrm{af}} = \operatorname{Pf}(R_Y) \operatorname{tr}_s(N_F \exp(-T\Delta_F)). \tag{10.22}$$

Proof. We proceed as for Lemma 10.1. As in the odd-dimensional case, we still have $\exp(-x^{-2} t\, {}^a\Delta) \in \Psi^{-2,-2,0}_{A,G}(M; {}^a\Lambda^*)$ with the rescaled normal operator

$$\exp(-\mathcal{H}_Y) \exp\left(\frac{1}{8} C(R_Y)\right) \exp(-\mathcal{A}_T^2), \tag{10.23}$$

since the rescaled normal operator of the Laplacian is

$$N_{h,G,-k}(x^{-2} t\, {}^a\Delta) = \mathcal{A}_T^2 + \mathcal{H}_Y - \frac{1}{8} C(R_Y). \tag{10.24}$$

On the other hand, and this is the difference between even and odd-dimensional cases, the number operator in this case has order 1 since

$$\begin{aligned} N_Y =& \sum_i \operatorname{ext}(e_i) \operatorname{int}(e_i) \\ =& \frac{1}{4} \sum_i (c_l(e_i) + c_r(e_i) Q)(-c_l(e_i) + c_r(e_i) Q) \\ =& \frac{1}{2} \sum_i (1 + c_l(e_i) c_r(e_i) Q) \\ =& \frac{1}{2} \sum_i (1 + c_l(e_i) \tilde{c}_r(e_i) \tilde{\tau}_r). \end{aligned} \tag{10.25}$$

It follows that

$$\operatorname{str}\left[N_Y \exp\left(-\frac{t}{x^2}\,{}^a\!\Delta\right)\right] \in \rho_{\mathrm{tf}}^{-1}\rho_{\mathrm{af}}^{-1}\mathcal{C}^\infty_{E,E}(\widetilde{\mathrm{Diag}}_A;\Omega M) \tag{10.26}$$

with the leading term at af equal to

$$\rho_{\mathrm{af}}^{-1}\operatorname{tr}\left(\sum_i c_l(e_i)\tilde{c}_r(e_i)\tilde{\tau}_r \exp\left(\frac{1}{8}C(R_Y)\right)\exp(-T\Delta_F)\right) \equiv 0 \tag{10.27}$$

since the whole expression involves an odd number of factor of $c_l(e_i)$. This proves (10.21). Moreover, $\operatorname{tr}_s(\sum_i c_l(e_i)\tilde{c}_r(e_i)\tilde{\tau}_r \exp(-x^{-2}t\,{}^a\!\Delta))$ contributes no constant term at af as it follows from (10.26), leaving us with the simpler

$$\operatorname{tr}_s\left[\left(\frac{n}{2}\operatorname{Id}+1\otimes N_F\right)\exp(-x^{-2}t\,{}^a\!\Delta)\right] \tag{10.28}$$

to evaluate. The leading term for this is

$$\begin{aligned}
&\operatorname{tr}_s\left[\left(\frac{n}{2}\operatorname{Id}+1\otimes N_F\right)\exp\left(\frac{1}{8}C(R_Y)\right)\exp(-T\Delta_F)\right]\\
&= \frac{n}{2}\operatorname{Pf}(R_Y)\chi(F) + \operatorname{Pf}(R_Y)\operatorname{tr}_s(N_F\exp(-T\Delta_F))\\
&= \operatorname{Pf}(R_Y)\operatorname{tr}_s(N_F\exp(-T\Delta_F)),
\end{aligned} \tag{10.29}$$

giving (10.22). □

11. Adiabatic limit of analytic torsion

We show the existence of the expansions (0.14) separately in the two cases, starting with the assumption that $\dim Y$ is odd.

Consider the application of (5.12) to Δ_x as $x \downarrow 0$. Of the four terms let us start with the second. Since $t \geq \delta$ on the integrand the supertrace is smooth down to $x = 0$, locally uniformly in t. We are therefore mainly concerned with the long-time behaviour. Let Π_1 be orthogonal projection onto the null space of ${}^F\!\Delta$, Π_2 the orthogonal projection onto the null space of Δ_Y and in general for $k \geq 3$ let Π_k be the orthogonal projection onto E_k, i.e., the null space of Δ_{k-1}. Thus for small x

$${}^a\underline{\Delta}\Pi_1 = x^2\underline{\Delta}_Y + \prod_{k\geq 2} x^{2k}\underline{\Delta}_k \tag{11.1}$$

and hence

$$\begin{aligned}
&\int_\delta^\infty \operatorname{STr}(Ne^{-t\,{}^a\underline{\Delta}})\frac{dt}{t}\\
&= \int_\delta^\infty \operatorname{STr}(Ne^{-t\underline{\Delta}_Y})\frac{dt}{t} + \sum_{k\geq 2}\int_\delta^\infty \operatorname{STr}(Ne^{-tx^{2(k-1)}\underline{\Delta}_k})\frac{dt}{t} + O(x).
\end{aligned} \tag{11.2}$$

The terms in the sum each have an expansion

$$\int_{x^{2(k-1)}\delta}^{\infty} \mathrm{STr}(Ne^{-T\Delta_k})\frac{dT}{T} \tag{11.3}$$
$$= -\log(\delta x^{2(k-1)})[\chi_2(E_k, d_k) - \chi_2(E_{k+1}, d_{k+1})] + \log\tau(E_k, d_k) + O(x).$$

Next consider the first term in (5.12). Dividing it at some arbitrary point

$$\int_0^{\delta} \left[\mathrm{STr}(Ne^{-t\Delta_x}) - a_{-\frac{1}{2}}(M, g_x)t^{-\frac{1}{2}}\right]\frac{dt}{t}$$
$$= \int_0^{\lambda x^2} \left[\mathrm{STr}(Ne^{-t\Delta_x}) - a_{-\frac{1}{2}}(M, g_x)t^{-\frac{1}{2}}\right]\frac{dt}{t} \tag{11.4}$$
$$+ \int_{\lambda x^2}^{\delta} \left[\mathrm{STr}(Ne^{-t\Delta_x}) - a_{-\frac{1}{2}}(M, g_x)t^{-\frac{1}{2}}\right]\frac{dt}{t}$$

allows either the coordinates $t/x^2, x$ or $t^{\frac{1}{2}}, xt^{-\frac{1}{2}}$ to be used in the two pieces. It then follows directly from Lemma 10.1 that the first term on the right is of the form

$$\int_0^{\lambda x^2} \left[g(\frac{t^{\frac{1}{2}}}{x}, x)\right]\frac{dt}{t} = \int_0^{\lambda} \left[g(T^{\frac{1}{2}}, x)\right]\frac{dT}{T} \tag{11.5}$$

where g is $\mathcal{C}^\infty$ and vanishes if either the first or the second argument vanishes. The integral is therefore $\mathcal{C}^\infty$ in x and vanishes at $x = 0$. Similarly the second is term in (11.4) can be written

$$\int_{\lambda x^2}^{\delta} t^{\frac{1}{2}} g'(t, \frac{x}{t^{\frac{1}{2}}})\frac{dt}{t} \tag{11.6}$$

where g' is $\mathcal{C}^\infty$ and vanishes where the first argument vanishes. This is again $\mathcal{C}^\infty$ in x and converges to

$$\int_0^{\delta} \sum_{j=1}^{\dim F} \left[\mathrm{STr}_Y(N_Y e^{-t\Delta_{Y,j}}) - a_{-\frac{1}{2}}(Y)t^{-\frac{1}{2}}\right]\frac{dt}{t} \tag{11.7}$$

where $a_{-\frac{1}{2}}(Y)$ is necessarily the coefficient which makes the integral converge. Here $\Delta_{Y,j}$ is the Laplacian acting on the ρ-twisted fibre cohomology in dimension j. There is an extra term involving N_F which however vanishes by Poincaré duality.

Since the terms involving $\delta^{-\frac{1}{2}}$ and $\log\delta$ in (5.12) are just those needed to ensure the independence of δ, the first term on the right in (11.2) combines with (11.7) and the remaining two terms to give, in the limit as $x \downarrow 0$ the logarithm of the first factor in (0.11). In brief we have proved (0.10) and (0.11) in case Y is odd-dimensional.

The case of an even-dimensional base is quite similar. The analysis of the second term in (5.12) is exactly the same, so consider the first term. The kernel certainly behaves differently. Taking the decomposition (11.4) we get in place of (11.5) an integral

$$\int_0^{\lambda x^2} g\left(\frac{t^{\frac{1}{2}}}{x}, x\right)\frac{dt}{t} = \int_0^{\lambda} g(T^{\frac{1}{2}}, x)\frac{dT}{T} \tag{11.8}$$

where now g is $\mathcal{C}^\infty$ but vanishes only when the first argument vanishes.

Acknowledgement

The authors would like to thank Jeff Cheeger, Charlie Epstein, Ezra Getzler, Rafe Mazzeo, Paolo Piazza and Is Singer for helpful discussion during the preparation of this paper. The authors are especially grateful to Pierre Albin, Paolo Piazza and Eric Leichtnam for reading through the manuscript and for their many valuable suggestions. We thank the referee for careful reading and thoughtful suggestions.

References

[1] M. Atiyah, H. Donnelly, I. Singer, *Eta invariants, signature defect of cusps and values of L-functions*, Ann. of Math., 118(1983), pp. 131–177.

[2] N. Berline, E. Getzler, M. Vergne, Heat kernels and Dirac operators, Grundlehren der Mathematischen Wissenschaften, Springer-Verlag, 1992.

[3] A. Berthomieu, J.-M. Bismut, *Formes de torsion analytique et metriques de Quillen*, C. R. Acad. Sci., Paris, 315(1992), pp. 1071–1077.

[4] J.-M. Bismut, J. Cheeger, *η-invariants and their adiabatic limits*, J. Amer. Math. Soc., 2(1989), pp. 33–70.

[5] J.-M. Bismut, J. Cheeger, *Families index for manifolds with boundary, superconnections and cones. I.*, J. Funct. Anal., 89(1990), pp. 313–363.

[6] J.-M. Bismut, J. Cheeger, *Families index for manifolds with boundary, superconnections and cones. II.*, J. Funct. Anal., 90(1990), pp. 306–354.

[7] J.-M. Bismut, J. Cheeger, *Transgressed Euler classes of $SL(2n, \mathbf{Z})$ vector bundles, adiabatic limits of eta invariants and special values of L-functions*, Ann. Sci. cole Norm. Sup., 25(1992), pp. 335–391.

[8] J.-M. Bismut, D. Freed, *The analysis of elliptic families. I*, Comm. Math. Phys., 106(1986), pp. 159–176.

[9] J.-M. Bismut, D. Freed, *The analysis of elliptic families. II*, Comm. Math. Phys., 107(1986), pp. 103–163.

[10] J.-M. Bismut, J. Lott, *Fibres plats, images directes et forms de torsion analytique*, C. R. Acad. Sci., Paris, 316(1993), pp. 477–482.

[11] J.-M. Bismut, W. Zhang, *An extension of a theorem by Cheeger and Müller*, Astérisque, 205(1992).

[12] J. Cheeger, *Analytic torsion and the heat equation*, Ann. Math., 109(1979), pp. 259–322.

[13] J.Cheeger, *η-invariants, the adiabatic approximation and conical singularities*, J. Diff. Geom., 26(1987), pp. 175–221.

[14] X. Dai, *Adiabatics limits, the non-multiplicativity of signature and Leray spectral sequence*, J.A.M.S., 4(1991), pp. 265–321.

[15] X. Dai, *Geometric invariants and their adiabatic limits*, Proc. Symp. in Pure Math., 54(1993), pp. 145–156.

[16] X. Dai, D. Freed, *η-invariants and determinant lines*, J. Math. Phys., 35(1994), pp. 5155–5194.

[17] X. Dai, W. Zhang, *Circle bundles and the Kreck-Stolz invariant*, Trans. Amer. Math. Soc., 347(1995), pp. 3587–3593.

[18] C. Epstein, R.B. Melrose, *Shrinking tubes and the $\overline{\partial}$-Neumann problem,*

[19] C.L. Epstein, R.B. Melrose and G. Mendoza, *Resolvent of the Laplacian on strictly pseudoconvex domains*, Acta Math.

[20] D. Fried, *Lefschetz formulas for flows*, Contem. Math., 58(1987), pp. 19–69.

[21] D.S. Freed, *Reidemeister torsion, spectral sequences, and Brieskorn spheres*, J. Reine Angew. Math., 429(1992), pp. 75–89.

[22] E. Getzler, *A short proof of the Atiyah-Singer index theorem*, Topology, 25(1986), pp. 111–117.

[23] J. Hadamard, *The problem of diffusion of waves*, Ann. of Math., 43(1942). pp. 510–522.

[24] A.W. Hassell, *Analytic surgery and analytic torsion*, Ph.D. Thesis, MIT, 1994.

[25] W. Lück, *L^2-torsion and 3-manifolds*, Low-dimensional Topology, Conf. Proc. 1992.

[26] W. Lück, T. Schick, T. Thielman, *Torsion and fibrations*, J. Reine Angew. Math. 498(1998), pp. 1–33.

[27] R.R. Mazzeo and R.B. Melrose, *The adiabatic limit, Hodge cohomology and Leray's spectral sequence for a fibration*, J. Diff. Geom., 31(1990), pp. 185–213.

[28] R. Melrose, Differential analysis on manifolds with corner, in preparation, 1996.

[29] R.B. Melrose, *Calculus of conormal distributions on manifolds with corners*, Int. Math. Res. Notices, 3(1992), pp. 51–61.

[30] R.B. Melrose, The Atiyah-Patodi-Singer index theorem, A.K. Peters, Ltd. 1993.

[31] R.B. Melrose, P. Piazza, *Families of Dirac operators, boundaries and the b-calculus*, J. Diff. Geom., 46(1997), pp. 99–180.

[32] R.B. Melrose, P. Piazza, *An index theorem for families of Dirac operators on odd-dimensional manifolds with boundary*, J. Diff. Geom., 46(1997), pp. 287–334.

[33] W. Müller, *Analytic torsion and R-torsion of Riemannian manifolds*, Adv. in Math., 28(1978), pp. 233–305.

[34] W. Müller, *Analytic torsion and R-torsion for unimodular representations*, J. A. M. S., 6(1993), pp. 721–753.

[35] W. Müller, *Signature defects of cusps of Hilbert modular varieties and values of L-series at $s = 1$*, J. Diff. Geom., 20(1994), pp. 55–119.

[36] M. Nagase, *Twistor spaces and the general adiabatic expansions*, J. Funct. Anal. 251 (2007), no. 2, 680–737.

[37] L. Nicolaescu, *Adiabatic limits of the Seiberg-Witten equations on Seifert manifolds*, Comm. Anal. Geom., 6(1998), pp. 331–392.

[38] D.B. Ray and I.M. Singer, *R-torsion and the Laplacian on Riemannian manifolds*, Adv. in Math., 7(1971), pp. 145–210.

[39] R.T. Seeley, *Topics in pseudo-differential operators*, CIME, Edizioni Cremonese, Roma. 1969, pp. 169–305.

[40] E. Witten, *Global gravitational anomalies*, Comm. Math. Phys., 100(1985), pp. 197–229.

[41] W. Zhang, *Circle bundles, adiabatic limits of η-invariants and Rokhlin congruences*, Ann. Inst. Fourier (Grenoble), 44(1994), pp. 249–270.

Xianzhe Dai
Math. Dept., UCSB
Santa Barbara, CA 93106, USA
e-mail: `dai@math.ucsb.edu`

Richard B. Melrose
Math. Dept., MIT
Cambridge, MA 02139,USA

Transversal Index and L^2-index for Manifolds with Boundary

Xiaonan Ma and Weiping Zhang

Dedicated to Jeff Cheeger for his 65th birthday

Abstract. For manifolds with boundary, we present a self-contained proof of Braverman's result which gives an alternative interpretation of the transversal index through certain kind of L^2-indices.

Mathematics Subject Classification (2000). Primary 58J20; Secondary 53D50.
Keywords. Index theory, Dirac operator, geometric quantization.

1. Introduction

In her ICM 2006 plenary lecture [15], Michèle Vergne formulated a conjecture on "quantization commutes with reduction" for non-compact symplectic manifolds. This conjecture extends the original Guillemin-Sternberg geometric quantization conjecture on compact symplectic manifolds to the non-compact setting. Vergne's conjecture [15] is stated in terms of the indices of transversally elliptic symbols on possibly non-compact manifolds in the sense of Atiyah [1] and Paradan [9], and these indices coincide with the indices of Spin^c Dirac operators for compact manifolds. For a survey on the Guillemin-Sternberg conjecture, see [14].

In [7], [8], we established an extended version of Vergne's conjecture, in the sense that we did not make any extra assumptions besides the properness of the associated moment map. One important step of our approach is to establish an alternative interpretation of the transversal index appearing in the Vergne conjecture by using the Atiyah-Patodi-Singer type index. In this step, we used Braverman's interpretation of the transversal index for manifolds with boundary through certain kind of L^2-indices [2, §5]. The purpose of this note is to give a self-contained proof of Braverman's result.

Let M be an even-dimensional compact oriented Spin^c-manifold with non-empty boundary ∂M. Let $n = \dim M$. Let E be a complex vector bundle over M.

The work of X.M. was partially supported by Institut Universitaire de France. The work of W.Z. was partially supported by MOEC and NNSFC.

Let G be a compact connected Lie group. Let $\mathfrak{g}$ be the Lie algebra of G and $\mathfrak{g}^*$ its dual, and let G act on $\mathfrak{g}$ by Ad_G-action.

Let $\Lambda^*_+ \subset \mathfrak{g}^*$ be the set of dominant weights. For $\gamma \in \Lambda^*_+$, we denote by V^G_γ the irreducible G-representation with highest weight γ. Then V^G_γ, $\gamma \in \Lambda^*_+$, form a $\mathbb{Z}$-basis of the representation ring $R(G)$.

We assume that G acts on the left on M and that this action lifts on E and on the spin^c structure of the tangent bundle $\pi : TM \to M$.

Let g^{TM} be a G-invariant Riemannian metric on TM, and we identify TM and T^*M via g^{TM}. For any $K \in \mathfrak{g}$, let K^M be the vector field generated by K on M. Following [1, p. 7] (cf. [9, §3]), set

$$T_G M = \big\{(x,v) \in TM : x \in M, v \in T_xM \text{ such that } \langle v, K^M(x)\rangle = 0 \text{ for all } K \in \mathfrak{g}\big\}. \tag{1.1}$$

Let $\Psi : M \to \mathfrak{g}$ be a G-equivariant map. Let Ψ^M denote the vector field on M such that

$$\Psi^M(x) := (\Psi(x))^M(x) \quad \text{for any } x \in M, \tag{1.2}$$

where $(\Psi(x))^M$ is the vector field over M generated by $\Psi(x) \in \mathfrak{g}$.

We make the fundamental assumption that Ψ^M is nowhere zero on ∂M.

Let $S(TM) = S_+(TM) \oplus S_-(TM)$ be the bundle of spinors associated to the spin^c-structure on TM and g^{TM} (cf. [5, Appendix D]). For $V \in TM$, let $c(V)$ be the Clifford action of V on $S(TM)$ which exchanges the $\mathbb{Z}_2$-grading $S_\pm(TM)$ of $S(TM)$. Let $\sigma^M_{E,\Psi} \in \mathrm{Hom}(\pi^*(S_+(TM)\otimes E), \pi^*(S_-(TM)\otimes E))$ denote the symbol defined by

$$\sigma^M_{E,\Psi}(x,v) = \pi^*\left(\sqrt{-1}c(v+\Psi^M)\otimes \mathrm{Id}_E\right)\big|_{(x,v)} \quad \text{for } x \in M,\ v \in T_xM. \tag{1.3}$$

Since Ψ^M is nowhere zero on ∂M, the subset $\{(x,v) \in T_GM : \sigma^M_{E,\Psi}(x,v)$ is non-invertible$\}$ of T_GM is contained in a compact subset of $T_G\widehat{M}$ (where $\widehat{M} = M \setminus \partial M$ is the interior of M). Thus, $\sigma^M_{E,\Psi}$ defines a G-transversally elliptic symbol on $T_G\widehat{M}$ in the sense of Atiyah [1, §1, §3] and Paradan [9, §3], [10, §3], which in turn determines a transversal index in the formal representation ring $R[G]$ of G,

$$\mathrm{Ind}\left(\sigma^M_{E,\Psi}\right) = \bigoplus\nolimits_{\gamma\in\Lambda^*_+} \mathrm{Ind}_\gamma\left(\sigma^M_{E,\Psi}\right)\cdot V^G_\gamma \in R[G], \tag{1.4}$$

obtained by embedding M into a compact G-manifold of the same dimension as that of M (cf. [1, §1]). For any $\gamma \in \Lambda^*_+$, $\mathrm{Ind}_\gamma(\sigma^M_{E,\Psi}) \in \mathbb{Z}$ is the multiplicity of V^G_γ in the transversal index $\mathrm{Ind}(\sigma^M_{E,\Psi})$ which depends on the homotopy class of Ψ such that Ψ^M is nowhere zero on ∂M, and which does not depend on g^{TM}. Moreover, the character of $\mathrm{Ind}(\sigma^M_{E,\Psi})$ is a distribution on G. Note that the set of $\gamma \in \Lambda^*_+$ such that $\mathrm{Ind}_\gamma(\sigma^M_{E,\Psi}) \neq 0$ could be infinite.

To compute $\mathrm{Ind}(\sigma^M_{E,\Psi})$, we deform $\Psi : M \to \mathfrak{g}$ inside M (leaving $\Psi|_{\partial M}$ unchanged) to a G-equivariant map $\Psi' : M \to \mathfrak{g}$ with product structure near ∂M, then by the homotopy invariance of the transversal index (cf. [1, Theorems

2.6, 3.7], [9, §3]), $\operatorname{Ind}(\sigma^M_{E,\Psi}) = \operatorname{Ind}(\sigma^M_{E,\Psi'})$. Thus we can and we will assume that $\Psi : M \to \mathfrak{g}$ has a product structure near ∂M.

Let $g^{TM}, h^{S(TM)}, h^E$ be metrics on $TM, S(TM), E$ and let $\nabla^{S(TM)}$ be the canonically induced Clifford connection on $(S(TM), h^{S(TM)})$ and ∇^E be a Hermitian connection on (E, h^E). We assume that the metrics and connections involved are G-invariant, and have a product structure near the boundary ∂M, and the G-action on objects such as $E, S(TM)$ near ∂M is the product of the G-action on their restrictions to ∂M and the identity in the normal direction to ∂M.

We attach now an infinite cylinder $\partial M \times (-\infty, 0]$ to M along the boundary ∂M and extend trivially all objects on M to $\widetilde{M} = M \cup (\partial M \times (-\infty, 0])$. We decorate the extended objects on $\widetilde{M}$ by a "$\sim$".

Let f be a G-invariant smooth real function on $\widetilde{M}$ such that there exists a smooth function $\varrho : (-\infty, 0] \to \mathbb{R}$ (for example, $\varrho(u) = e^{-2u}$) verifying that for $(y, x_n) \in \partial M \times (-\infty, 0]$,

$$f(y, x_n) = \varrho(x_n), \tag{1.5}$$

$$\lim_{x_n \to -\infty} \varrho(x_n) = +\infty \quad \text{and} \quad \lim_{x_n \to -\infty} \frac{\varrho^2}{|\varrho'| + \varrho}(x_n) = +\infty. \tag{1.6}$$

Then f is an admissible function on $\widetilde{M}$ for the triple $(S(T\widetilde{M}) \otimes \widetilde{E}, \nabla^{S(T\widetilde{M}) \otimes \widetilde{E}}, \widetilde{\Psi})$ in the sense of [2, Definition 2.6] (cf. Remark 2.2).

Let $D^{\widetilde{E}}_f$ be the operator acting on $\mathscr{C}^\infty_0(\widetilde{M}, S(T\widetilde{M}) \otimes \widetilde{E})$ to be defined by (2.6). Let $D^{\widetilde{E}}_{\pm,f}$ be the restrictions of $D^{\widetilde{E}}_f$ to the spaces associated to $S_\pm(T\widetilde{M}) \otimes \widetilde{E}$.

The purpose of this note is to give a self-contained proof of the following result of Braverman [2, Theorems 2.9, 5.5].

Theorem 1.1.

a) *For any $\gamma \in \Lambda^*_+$, the multiplicity of V^G_γ in $\operatorname{Ker}(D^{\widetilde{E}}_f)$ is finite.*

b) *The following identity holds:*

$$\operatorname{Ker}(D^{\widetilde{E}}_{+,f}) - \operatorname{Ker}(D^{\widetilde{E}}_{-,f}) = \operatorname{Ind}(\sigma^M_{E,\Psi}) \in R[G]. \tag{1.7}$$

Equivalently, let $\operatorname{Ind}_\gamma(D^{\widetilde{E}}_{+,f})$ be the multiplicity of V^G_γ in $\operatorname{Ker}(D^{\widetilde{E}}_{+,f}) - \operatorname{Ker}(D^{\widetilde{E}}_{-,f})$, then

$$\operatorname{Ind}_\gamma(D^{\widetilde{E}}_{+,f}) = \operatorname{Ind}_\gamma(\sigma^M_{E,\Psi}). \tag{1.8}$$

In particular, $\operatorname{Ind}_\gamma(D^{\widetilde{E}}_{+,f})$ does not depend on $g^{TM}, h^{S(TM)}, h^E, \nabla^{S(TM)}, \nabla^E$, f and it depends only on the homotopy class of Ψ such that Ψ^M is nowhere zero on ∂M.

Certainly all argument here works for any Clifford module without the Spin^c assumption on M.

This paper is organized as follows: In Section 2, we establish Theorem 1.1a). In Section 3, we obtain (1.7).

2. L^2-index

This section is organized as follows: In Section 2.1, we recall the definition of Spin^c Dirac operators. In Section 2.2, we explain the self-adjoint extension of $D_f^{\widetilde{E}}$. In Section 2.3, we prove Theorem 1.1a).

We will use the notation and assumption in Introduction.

2.1. Spin^c Dirac operator

We recall first our set-up. The manifold M is a compact G-manifold with boundary ∂M, and $\widetilde{M}$ is an oriented G-Spin^c manifold such that $M \subset \widetilde{M}$.

Fix $\varepsilon_1 > 0$. We assume that there exists a neighborhood $\partial M \times (-\infty, \varepsilon_1]$ of ∂M in $\widetilde{M}$, where we identify $\partial M \times \{0\}$ to ∂M, such that $\widetilde{M} = M \cup (\partial M \times (-\infty, 0])$.

Let $S(T\widetilde{M}) = S_+(T\widetilde{M}) \oplus S_-(T\widetilde{M})$ be the bundle of spinors associated to the spin^c-structure on $T\widetilde{M}$ and a G-invariant Riemannian metric $g^{T\widetilde{M}}$.

Let $\widetilde{E}$ be a G-complex vector bundle over $\widetilde{M}$. Let $h^{\widetilde{E}}$ be a G-invariant Hermitian metric on $\widetilde{E}$, $\nabla^{\widetilde{E}}$ a G-invariant Hermitian connection on $(\widetilde{E}, h^{\widetilde{E}})$. Let $h^{S(T\widetilde{M})}$ be the G-invariant Hermitian metric on $S(T\widetilde{M})$ induced by $g^{T\widetilde{M}}$ and a G-invariant metric on the line bundle defining the spin^c structure (cf. [5, Appendix D]). Let $h^{S(T\widetilde{M})\otimes \widetilde{E}}$ be the metric on $S(T\widetilde{M}) \otimes \widetilde{E}$ induced by the metrics on $S(T\widetilde{M})$ and on $\widetilde{E}$.

Let $\nabla^{S(T\widetilde{M})}$ be the Clifford connection on $S(T\widetilde{M})$ induced by the Levi-Civita connection $\nabla^{T\widetilde{M}}$ of $g^{T\widetilde{M}}$ and a G-invariant Hermitian connection on the line bundle defining the spin^c structure. Let $\nabla^{S(T\widetilde{M})\otimes \widetilde{E}}$ be the Hermitian connection on $S(T\widetilde{M}) \otimes \widetilde{E}$ obtained by the tensor product of the connections $\nabla^{S(T\widetilde{M})}$ and $\nabla^{\widetilde{E}}$. Let $\widetilde{\Psi} : \widetilde{M} \to \mathfrak{g}$ be a G-equivariant map.

Let $g^{T\partial M}$ be the Riemannian metric on ∂M induced by $g^{T\widetilde{M}}$. We denote the restriction of the objects on $\widetilde{M}$ to M by canceling the superscript "$\sim$".

We assume that for $(y, x_n) \in \partial M \times (-\infty, \varepsilon_1]$, we have

$$\begin{aligned}
&\widetilde{\Psi}(y, x_n) = \Psi(y, 0) \in \mathfrak{g}, \quad g^{T\widetilde{M}}_{(y,x_n)} = g_y^{T\partial M} + (dx_n)^2, \\
&(S(T\widetilde{M}), h^{S(T\widetilde{M})}, \nabla^{S(T\widetilde{M})})|_{\partial M\times(-\infty,\varepsilon_1]} = \pi_1^*((S(TM), h^{S(TM)}, \nabla^{S(TM)})|_{\partial M}), \\
&(\widetilde{E}, h^{\widetilde{E}}, \nabla^{\widetilde{E}})|_{\partial M\times(-\infty,\varepsilon_1]} = \pi_1^*((E, h^{E}, \nabla^{E})|_{\partial M}),
\end{aligned} \tag{2.1}$$

with $\pi_1 : \partial M \times (-\infty, \varepsilon_1] \to \partial M$ the natural projection. Moreover, the G-action on objects such as $\widetilde{M}$, $\widetilde{E}$, $S(T\widetilde{M})$ on $\partial M \times (-\infty, \varepsilon_1]$ is the product of the G-action on their restriction to ∂M and the identity in the direction $(-\infty, \varepsilon_1]$.

Let $\mathscr{C}^\infty(\widetilde{M}, S(T\widetilde{M}) \otimes \widetilde{E})$ be the space of smooth sections of $S(T\widetilde{M}) \otimes \widetilde{E}$ on $\widetilde{M}$, and let $\mathscr{C}_0^\infty(\widetilde{M}, S(T\widetilde{M}) \otimes \widetilde{E})$ be the subspace of smooth sections with compact support. Let $dv_{\widetilde{M}}$ be the Riemannian volume form on $(\widetilde{M}, g^{T\widetilde{M}})$. For

$s \in \mathscr{C}_0^\infty(\widetilde{M}, S(T\widetilde{M}) \otimes \widetilde{E})$, the L^2-norm $\|s\|_{\widetilde{M},0}$ is defined by

$$\|s\|_{\widetilde{M},0}^2 = \int_{\widetilde{M}} |s|^2(x) dv_{\widetilde{M}}(x). \tag{2.2}$$

Let $\langle \cdot, \cdot \rangle_{\widetilde{M}}$ be the Hermitian product on $\mathscr{C}_0^\infty(\widetilde{M}, S(T\widetilde{M}) \otimes \widetilde{E})$ corresponding to $\| \cdot \|_{\widetilde{M},0}^2$. Let $L^2(\widetilde{M}, S(T\widetilde{M}) \otimes \widetilde{E})$ be the L^2-completion of $(\mathscr{C}_0^\infty(\widetilde{M}, S(T\widetilde{M}) \otimes \widetilde{E}), \| \cdot \|_{\widetilde{M},0})$.

Let $\{e_i\}$ be an orthonormal frame of $T\widetilde{M}$. The Spin^c-Dirac operator $D^{\widetilde{E}}$ on $\mathscr{C}_0^\infty(\widetilde{M}, S(T\widetilde{M}) \otimes \widetilde{E})$ is defined by (cf. [5, Appendix D])

$$D^{\widetilde{E}} = \sum_{i=1}^{n} c(e_i) \nabla_{e_i}^{S(T\widetilde{M}) \otimes \widetilde{E}}. \tag{2.3}$$

Then $D^{\widetilde{E}}$ is G-equivariant and formally self-adjoint.

Let e_n be the inward unit normal vector field perpendicular to ∂M. Let $e_1, \ldots, e_{n-1}$ be an oriented orthonormal frame of $T\partial M$ so that $e_1, \ldots, e_{n-1}$, e_n is an oriented orthonormal frame of $TM|_{\partial M}$. Set

$$D_{\partial M}^{E} = -\sum_{j=1}^{n-1} c(e_n) c(e_j) \nabla_{e_j}^{(S(TM) \otimes E)|_{\partial M}}. \tag{2.4}$$

Then $D_{\partial M}^{E}$ is the Dirac operator on $((S(TM) \otimes E)|_{\partial M}, \nabla^{(S(TM) \otimes E)|_{\partial M}})$. By (2.1), (2.3) and (2.4), we have on $\partial M \times (-\infty, \varepsilon_1]$,

$$D^{\widetilde{E}} = c(e_n) D_{\partial M}^{E} + c(e_n) \frac{\partial}{\partial x_n}. \tag{2.5}$$

2.2. Self-adjoint extension of $D_f^{\widetilde{E}}$

Let f be a G-invariant smooth real function on $\widetilde{M}$. In this subsection we need not assume that f verifies (1.5) and (1.6). Let $D_f^{\widetilde{E}}$ be the operator acting on $\mathscr{C}_0^\infty(\widetilde{M}, S(T\widetilde{M}) \otimes \widetilde{E})$ defined by (cf. (1.2), (2.3))

$$D_f^{\widetilde{E}} = D^{\widetilde{E}} + \sqrt{-1} f\, c(\widetilde{\Psi}^{M}). \tag{2.6}$$

By definition, the graph of the minimal extension $(D_f^{\widetilde{E}})_{\min}$ of $D_f^{\widetilde{E}}$ is the closure of the graph of $D_f^{\widetilde{E}}$, i.e.,

$$\begin{aligned} \operatorname{Dom}\left((D_f^{\widetilde{E}})_{\min}\right) = \Big\{ s \in L^2\left(\widetilde{M}, S(T\widetilde{M}) \otimes \widetilde{E}\right) &: \text{ there exists a sequence} \\ s_k \in \mathscr{C}_0^\infty\left(\widetilde{M}, S(T\widetilde{M}) \otimes \widetilde{E}\right) &\text{ such that} \\ \lim_{k \to +\infty} s_k = s, \text{ and } \lim_{k \to +\infty} D_f^{\widetilde{E}} s_k &\text{ exists in } L^2\left(\widetilde{M}, S(T\widetilde{M}) \otimes \widetilde{E}\right) \Big\}, \end{aligned} \tag{2.7}$$

and for $s \in \operatorname{Dom}((D_f^{\widetilde{E}})_{\min})$, $(D_f^{\widetilde{E}})_{\min} s$ is defined by $\lim_{k \to +\infty} D_f^{\widetilde{E}} s_k$ in (2.7).

Since the Riemannian manifold $(\widetilde{M}, g^{T\widetilde{M}})$ is complete, by [4] (cf. also [6, §3.1, §3.3]), the minimal and the maximal extensions of $D_f^{\widetilde{E}}$ coincide, and form a self-adjoint operator. We still denote by $D_f^{\widetilde{E}}$, $(D_f^{\widetilde{E}})^2$ the corresponding maximal extensions, whose domains are, by definition,

$$\begin{aligned} \mathrm{Dom}\left(D_f^{\widetilde{E}}\right) &= \left\{s :\ s,\ D_f^{\widetilde{E}} s \in L^2\left(\widetilde{M}, S\left(T\widetilde{M}\right) \otimes \widetilde{E}\right)\right\}, \\ \mathrm{Dom}\left((D_f^{\widetilde{E}})^2\right) &= \left\{s :\ s,\ \left(D_f^{\widetilde{E}}\right)^2 s \in L^2\left(\widetilde{M}, S\left(T\widetilde{M}\right) \otimes \widetilde{E}\right)\right\}. \end{aligned} \tag{2.8}$$

Note that $\mathrm{Dom}(D_f^{\widetilde{E}})$ is a Hilbert space endowed with the graph-norm $(\|D_f^{\widetilde{E}} s\|^2_{\widetilde{M},0} + \|s\|^2_{\widetilde{M},0})^{1/2}$, for $s \in \mathrm{Dom}(D_f^{\widetilde{E}})$.

Let $\varphi : \mathbb{R} \to [0,1]$ be a smooth even function such that $\varphi(u) = 1$ for $|u| < 1/2$, and $\varphi(u) = 0$ for $|u| > 1$. For $k \geqslant 1$, let $\varphi_k : \widetilde{M} \to \mathbb{R}$ be defined by

$$\varphi_k = 1 \ \text{ on } M, \quad \varphi_k(y, x_n) = \varphi\left(x_n/k\right) \ \text{for } (y, x_n) \in \partial M \times (-\infty, 0]. \tag{2.9}$$

Then each φ_k is G-invariant, smooth, and has a compact support on $\widetilde{M}$.

If $s \in \mathrm{Dom}((D_f^{\widetilde{E}})^2)$, then by the basic elliptic estimate, $\varphi_k s$ lies in the local Sobolev space of second order on $\widetilde{M}$. From (2.5), (2.9), there exists $C > 0$ such that for any $k \geqslant 1$, we have

$$\begin{aligned} \mathrm{Re}\left\langle \varphi_k^2 (D_f^{\widetilde{E}})^2 s, s\right\rangle_{\widetilde{M}} &= \mathrm{Re}\left\langle D_f^{\widetilde{E}}\left(\varphi_k^2 D_f^{\widetilde{E}} s\right) - 2\varphi_k c\left((d\varphi_k)^*\right) D_f^{\widetilde{E}} s, s\right\rangle_{\widetilde{M}} \\ &= \left\|\varphi_k D_f^{\widetilde{E}} s\right\|^2_{\widetilde{M},0} + 2\,\mathrm{Re}\left\langle \varphi_k D_f^{\widetilde{E}} s, c\left((d\varphi_k)^*\right) s\right\rangle_{\widetilde{M}} \\ &\geqslant \frac{1}{2}\left\|\varphi_k D_f^{\widetilde{E}} s\right\|^2_{\widetilde{M},0} - \frac{C}{2k}\|s\|^2_{\widetilde{M},0}, \end{aligned} \tag{2.10}$$

where $(d\varphi_k)^* \in T\widetilde{M}$ denotes the metric dual of $d\varphi_k$ with respect to $g^{T\widetilde{M}}$.

By taking $k \to \infty$ in (2.10), we get $D_f^{\widetilde{E}} s \in L^2(\widetilde{M}, S(T\widetilde{M}) \otimes \widetilde{E})$. Thus $s \in \mathrm{Dom}((D_f^{\widetilde{E}})^2)$ implies $s \in \mathrm{Dom}(D_f^{\widetilde{E}})$. From this, we obtain as in [6, §3.3] that the maximal extension of $(D_f^{\widetilde{E}})^2$ is self-adjoint, and coincides also with the minimal extension of the original operator $(D_f^{\widetilde{E}})^2$. For $s \in \mathrm{Dom}((D_f^{\widetilde{E}})^2)$, we get from (2.10), by letting $k \to \infty$,

$$\left\|\left(1 + (D_f^{\widetilde{E}})^2\right)^{\frac{1}{2}} s\right\|^2_{\widetilde{M},0} = \left\langle\left(1 + (D_f^{\widetilde{E}})^2\right) s, s\right\rangle_{\widetilde{M}} = \left\|D_f^{\widetilde{E}} s\right\|^2_{\widetilde{M},0} + \|s\|^2_{\widetilde{M},0}. \tag{2.11}$$

Finally, from the von Neumann Lemma (cf. [6, Lemma C.1.3]) about self-adjoint operators, $1 + (D_f^{\widetilde{E}})^2 : \mathrm{Dom}((D_f^{\widetilde{E}})^2) \to L^2(\widetilde{M}, S(T\widetilde{M}) \otimes \widetilde{E})$ is bijective and has bounded inverse.

Relation (2.11) shows also that $\mathrm{Dom}((1 + (D_f^{\widetilde{E}})^2)^{\frac{1}{2}}) = \mathrm{Dom}(D_f^{\widetilde{E}})$. From the above discussion, we see that $(1 + (D_f^{\widetilde{E}})^2)^{-1} : L^2(\widetilde{M}, S(T\widetilde{M}) \otimes \widetilde{E}) \to \mathrm{Dom}((D_f^{\widetilde{E}})^2)$

and $(1+(D_f^{\widetilde{E}})^2)^{-\frac{1}{2}} : L^2(\widetilde{M}, S(T\widetilde{M})\otimes \widetilde{E}) \to \mathrm{Dom}(D_f^{\widetilde{E}})$ are bounded bijective linear operators. Moreover,

$$D_f^{\widetilde{E}}\left(1+(D_f^{\widetilde{E}})^2\right)^{-1/2} = \left(1+(D_f^{\widetilde{E}})^2\right)^{-1/2} D_f^{\widetilde{E}} \quad \text{on } \mathrm{Dom}(D_f^{\widetilde{E}}). \tag{2.12}$$

2.3. L^2-index of $D_f^{\widetilde{E}}$

For $\gamma \in \Lambda_+^*$, recall that V_γ^G is the irreducible representation of G with highest weight γ. For V, W two G-vector spaces, let $\mathrm{Hom}_G(V,W)$ denote the linear space of G-equivariant homomorphisms. By the Peter-Weyl theorem, we have the Hilbert space direct sum decomposition

$$L^2\left(\widetilde{M}, S(T\widetilde{M})\otimes \widetilde{E}\right) = \bigoplus_{\gamma\in\Lambda_+^*} L^2\left(\widetilde{M}, S(T\widetilde{M})\otimes \widetilde{E}\right)^\gamma, \tag{2.13}$$

where each $L^2\left(\widetilde{M}, S(T\widetilde{M})\otimes \widetilde{E}\right)^\gamma$ is a multiple of V_γ^G.

Let $D_f^{\widetilde{E}}(\gamma)$ be the restriction of $D_f^{\widetilde{E}}$ to the γ-component. Let $D_{\pm,f}^{\widetilde{E}}(\gamma)$ be the restrictions of $D_f^{\widetilde{E}}(\gamma)$ to the spaces associated to $S_\pm(T\widetilde{M})\otimes\widetilde{E}$. Then by (2.8), we have

$$\mathrm{Dom}\left(D_f^{\widetilde{E}}(\gamma)\right) = \mathrm{Dom}\left(D_f^{\widetilde{E}}\right) \cap L^2\left(\widetilde{M}, S\left(T\widetilde{M}\right)\otimes\widetilde{E}\right)^\gamma. \tag{2.14}$$

Clearly, $\mathrm{Ker}\left(D_f^{\widetilde{E}}\right)$ is closed in $L^2\left(\widetilde{M}, S(T\widetilde{M})\otimes\widetilde{E}\right)$, thus it is a Hilbert space with norm $\| \quad \|_{\widetilde{M},0}$.

The following result is a reformulation of [2, Theorem 2.9].

Theorem 2.1. *Assume that* (2.1) *holds and that $\widetilde{\Psi}^{\widetilde{M}}$ is nowhere zero on ∂M, and that f is a G-invariant real function on $\widetilde{M}$ verifying* (1.5) *and* (1.6). *Then for any $\gamma\in\Lambda_+^*$,*

$$D_f^{\widetilde{E}}(\gamma) : \mathrm{Dom}\left(D_f^{\widetilde{E}}(\gamma)\right) \to L^2\left(\widetilde{M}, S\left(T\widetilde{M}\right)\otimes\widetilde{E}\right)^\gamma$$

is a Fredholm operator, and

$$\mathrm{Ker}\left(D_f^{\widetilde{E}}(\gamma)\right) = \mathrm{Hom}_G\left(V_\gamma^G, \mathrm{Ker}(D_f^{\widetilde{E}})\right)\otimes V_\gamma^G, \tag{2.15a}$$

$$\mathrm{Ker}(D_f^{\widetilde{E}}) = \bigoplus_{\gamma\in\Lambda_+^*} \mathrm{Ker}(D_f^{\widetilde{E}}(\gamma)). \tag{2.15b}$$

Proof. Let $\mathfrak{g}$ be equipped with an Ad_G-invariant metric. Let $V_1,\ldots,V_{\dim G}$ be an orthonormal basis of $\mathfrak{g}$. Then one has

$$\widetilde{\Psi}(x) = \sum_{i=1}^{\dim G} \widetilde{\Psi}_i(x)V_i \quad \text{for } x\in\widetilde{M}, \tag{2.16}$$

where $\widetilde{\Psi}_i : \widetilde{M} \to \mathbb{R}$, $1 \leqslant i \leqslant \dim G$, are bounded smooth functions (together with their derivatives). From (2.16), one gets

$$\widetilde{\Psi}^{\widetilde{M}}(x) = \sum_{i=1}^{\dim G} \widetilde{\Psi}_i(x) V_i^{\widetilde{M}}(x). \tag{2.17}$$

From (2.3) and (2.6), we have

$$\begin{aligned}\left(D_f^{\widetilde{E}}\right)^2 = \left(D^{\widetilde{E}}\right)^2 + \sqrt{-1}\sum_{i=1}^{n} c(e_i)\, c\left(\nabla_{e_i}^{T\widetilde{M}}\left(f\widetilde{\Psi}^{\widetilde{M}}\right)\right)&\\ - 2\sqrt{-1}\nabla_{f\widetilde{\Psi}^{\widetilde{M}}}^{S(T\widetilde{M})\otimes\widetilde{E}} + f^2\left|\widetilde{\Psi}^{\widetilde{M}}\right|^2.&\end{aligned} \tag{2.18}$$

For any $K \in \mathfrak{g}$, let L_K denote the Lie derivative of K acting on $\mathscr{C}^\infty(\widetilde{M}, S(T\widetilde{M}) \otimes\widetilde{E})$, and we set

$$\mu^{S\otimes\widetilde{E}}(K) := \nabla_{K^{\widetilde{M}}}^{S(T\widetilde{M})\otimes\widetilde{E}} - L_K \in \mathscr{C}^\infty\left(\widetilde{M}, \operatorname{End}\left(S(T\widetilde{M}) \otimes \widetilde{E}\right)\right). \tag{2.19}$$

By (2.1), for $K \in \mathfrak{g}$ fixed, $\mu^{S\otimes\widetilde{E}}(K)$ is bounded on $\widetilde{M}$. Set

$$B_f = \sqrt{-1}\sum_{i=1}^{n} c(e_i)\, c\left(\nabla_{e_i}^{T\widetilde{M}}\left(f\widetilde{\Psi}^{\widetilde{M}}\right)\right) - 2\sqrt{-1}f\sum_{i=1}^{\dim G} \widetilde{\Psi}_i \mu^{S\otimes\widetilde{E}}(V_i). \tag{2.20}$$

Let B_f^* be the adjoint of B_f with respect to $h^{S(T\widetilde{M})\otimes\widetilde{E}}$.

From (2.18)–(2.20), we have

$$\left(D_f^{\widetilde{E}}\right)^2 = \left(D^{\widetilde{E}}\right)^2 + B_f - 2\sqrt{-1}f\sum_{i=1}^{\dim G} \widetilde{\Psi}_i L_{V_i} + f^2\left|\widetilde{\Psi}^{\widetilde{M}}\right|^2. \tag{2.21}$$

From (1.5), (2.1) and (2.20), we get on $\partial M \times (-\infty, 0]$,

$$\begin{aligned}B_f =&\sqrt{-1}f\Big(\sum_{i=1}^{n-1} c(e_i)c\left(\nabla_{e_i}^{T\widetilde{M}}\widetilde{\Psi}^{\widetilde{M}}\right) - 2\sum_{i=1}^{\dim G}\widetilde{\Psi}_i\, \mu^{S\otimes\widetilde{E}}(V_i)\Big)\\ &+ \sqrt{-1}\frac{\partial f}{\partial x_n} c(e_n) c\left(\widetilde{\Psi}^{\widetilde{M}}\right).\end{aligned} \tag{2.22}$$

Since $\widetilde{\Psi}^{\widetilde{M}}$, $\nabla_{e_i}^{T\widetilde{M}}\widetilde{\Psi}^{\widetilde{M}} \in T\partial M$ (for $1 \leq i \leq n-1$), $\widetilde{\Psi}_j$ are constant in x_n on $\partial M \times (-\infty, 0]$, (2.22) implies that there exists $C_0 > 0$ such that the following pointwise estimate on $\widetilde{M}$ holds:

$$\frac{1}{2}(B_f + B_f^*) \geqslant -C_0\left(|f| + |df|\right). \tag{2.23}$$

Now, we fix $\gamma \in \Lambda_+^*$.

For $K \in \mathfrak{g}$, we denote by $L_K(\gamma)$ the Lie derivative L_K acting on V_γ^G. Let $\|L_K(\gamma)\|$ be the operator norm of $L_K(\gamma)$ with any (fixed) G-invariant Hermitian norm on V_γ^G. Then the Lie derivative L_K acting on $L^2(\widetilde{M}, S(T\widetilde{M}) \otimes \widetilde{E})^\gamma$ acts only on the factor V_γ^G (cf. (2.13)), and coincides with the linear bounded operator $L_K(\gamma)$. Thus its operator norm is $\|L_K(\gamma)\|$.

From (2.21) and (2.23), there exists $C_0 > 0$ such that for $s \in \mathscr{C}_0^\infty(\widetilde{M}, S(T\widetilde{M}) \otimes \widetilde{E})^\gamma$, we have

$$\left\langle \left(D_f^{\widetilde{E}}(\gamma)\right)^2 s, s\right\rangle_{\widetilde{M}} \geqslant \left\|D^{\widetilde{E}} s\right\|_{\widetilde{M},0}^2 + \left\|f\left|\widetilde{\Psi}^{\widetilde{M}}\right| s\right\|_{\widetilde{M},0}^2 \tag{2.24}$$
$$- C_0 \left\langle (|f| + |df|)\, s, s\right\rangle_{\widetilde{M}} - \|s\|_{\widetilde{M},0} \sum_{i=1}^{\dim G} \|L_{V_i}(\gamma)\| \left\|f\widetilde{\Psi}_i s\right\|_{\widetilde{M},0}.$$

Recall that $\widetilde{\Psi}^{\widetilde{M}}$ is nowhere zero on $\partial M \times (-\infty, 0]$ and is constant in x_n. Moreover $\widetilde{\Psi}_i$ is constant in x_n on $\partial M \times (-\infty, 0]$. Set

$$Q_1 = \inf_{y \in \partial M} \left|\widetilde{\Psi}^{\widetilde{M}}\right|^2 (y, 0) > 0, \quad Q_2 = \sup_{y \in \partial M} \left|\widetilde{\Psi}\right|^2 (y, 0). \tag{2.25}$$

By (1.6), there exists $C_\gamma > 0$ such that on $\partial M \times (-\infty, -C_\gamma]$, the following pointwise estimate holds:

$$\frac{1}{4} f^2 \left|\widetilde{\Psi}^{\widetilde{M}}\right|^2 > C_0(|f| + |df|) + \frac{Q_2}{Q_1} \sum_{i=1}^{\dim G} \|L_{V_i}(\gamma)\|^2. \tag{2.26}$$

By (2.16), the last term of (2.24) can be controlled by $\frac{Q_1}{4Q_2}\|f|\widetilde{\Psi}|s\|_{\widetilde{M},0}^2 + \frac{Q_2}{Q_1}\sum_{i=1}^{\dim G} \|L_{V_i}(\gamma)\|^2 \|s\|_{\widetilde{M},0}^2$. Thus by (2.24) and (2.26), there exists $C_\gamma' > 0$ such that for $s \in \mathscr{C}_0^\infty(\widetilde{M}, S(T\widetilde{M}) \otimes \widetilde{E})^\gamma$,

$$\left\|D_f^{\widetilde{E}}(\gamma) s\right\|_{\widetilde{M},0}^2 \geqslant \left\|D^{\widetilde{E}} s\right\|_{\widetilde{M},0}^2 + \frac{1}{2} Q_1 \|fs\|_{\widetilde{M},0}^2 - C_\gamma' \int_{M \cup \partial M \times [-C_\gamma, 0]} |s|^2 dv_{\widetilde{M}}. \tag{2.27}$$

From (2.27), we can adapt the argument in [6, §3.1]) to know that $D_f^{\widetilde{E}}(\gamma)$ is a Fredholm operator. Here we will show that, by the argument in [12, Prop. 8.2.8], its spectrum is discrete.

We claim that the operator

$$\left(1 + (D_f^{\widetilde{E}}(\gamma))^2\right)^{-1/2} : L^2(\widetilde{M}, S(T\widetilde{M}) \otimes \widetilde{E})^\gamma \to L^2(\widetilde{M}, S(T\widetilde{M}) \otimes \widetilde{E})^\gamma \tag{2.28}$$

is compact.

Equivalently, we need to prove that for any sequence $\omega_k \in L^2(\widetilde{M}, S(T\widetilde{M}) \otimes \widetilde{E})^\gamma$, $\|\omega_k\|_{\widetilde{M},0} = 1$, the sequence $s_k = \left(1 + (D_f^{\widetilde{E}}(\gamma))^2\right)^{-1/2} \omega_k$ has a convergent subsequence in $L^2(\widetilde{M}, S(T\widetilde{M}) \otimes \widetilde{E})^\gamma$.

Take $l > 2C_\gamma'$, for φ_l in (2.9), by (2.11) and (2.12), there exists $C > 0$ such that for any $k \in \mathbb{N}$, we have

$$\left\|D^{\widetilde{E}}(\varphi_l s_k)\right\|_{\widetilde{M},0} + \|\varphi_l s_k\|_{\widetilde{M},0} \leqslant C. \tag{2.29}$$

By Gårding's inequality and Rellich's theorem, (2.29) and the fact that $\{\varphi_l s_k\}$'s have a common compact support, we can select a convergent subsequence of

$\{\varphi_l s_k\}_k$ in $L^2(\widetilde{M}, S(T\widetilde{M}) \otimes \widetilde{E})^\gamma$. Now by a diagonal argument, we have a subsequence $\{s_{m_k}\}_k$ of $\{s_k\}_k$ such that for any $l \in \mathbb{N}$, $s_{m_k}|_{M\cup(\partial M\times[-l,0])}$, the restriction of s_{m_k} to $M \cup (\partial M \times [-l, 0])$, converges in L^2-norm.

By (2.27), $\{f s_{m_k}\}$ is uniformly bounded in $L^2(\widetilde{M}, S(T\widetilde{M}) \otimes \widetilde{E})$, by (1.6), we know $\{s_{m_k}\}$ converges in $L^2(\widetilde{M}, S(T\widetilde{M}) \otimes \widetilde{E})$.

Thus $\left(1 + (D_f^{\widetilde{E}}(\gamma))^2\right)^{-1/2}$ is compact, and the spectrum of $\left(D_f^{\widetilde{E}}(\gamma)\right)^2$ is discrete.

As $D_f^{\widetilde{E}}$ is G-invariant, from (2.13), we get (2.15a). The equation (2.15b) is a consequence of Peter-Weyl theorem (cf. [3, Ch. 2, Theorem 5.7]). The proof of Theorem 2.1 is completed. □

Remark 2.2. *Let v be the function on $\widetilde{M}$ defined by $v = |\widetilde{\Psi}^{\widetilde{M}}| + |\widetilde{\Psi}| + |\nabla^{T\widetilde{M}}\widetilde{\Psi}^{\widetilde{M}}| + |\mu^{S\otimes\widetilde{E}}| + 1$. Recall that a G-invariant real function f on $\widetilde{M}$ is an admissible function for the triple $(S(T\widetilde{M}) \otimes \widetilde{E}, \nabla^{S(T\widetilde{M})\otimes\widetilde{E}}, \widetilde{\Psi})$ in the sense of* [2, Definition 2.6] *if and only if*

$$\lim_{x_n\to-\infty} \frac{f^2|\widetilde{\Psi}^{\widetilde{M}}|^2}{|df||\widetilde{\Psi}^{\widetilde{M}}| + fv + 1}(y, x_n) = +\infty \quad \textit{uniformly for } y \in \partial M. \tag{2.30}$$

Since $\widetilde{\Psi}^{\widetilde{M}}$, $\nabla_{e_i}^{T\widetilde{M}}\widetilde{\Psi}^{\widetilde{M}} \in T\partial M$, $\widetilde{\Psi}_i$ are constant in x_n on $\partial M \times (-\infty, 0]$, if (1.5) *holds, then* (2.30) *is equivalent to* (1.6).

Definition 2.3. *For each $\gamma \in \Lambda_+^*$, the index of $D_{+,f}^{\widetilde{E}}(\gamma)$, thought of as a virtual G-representation, is defined by*

$$\operatorname{Ind}\left(D_{+,f}^{\widetilde{E}}(\gamma)\right) := \operatorname{Ker}\left(D_{+,f}^{\widetilde{E}}(\gamma)\right) - \operatorname{Ker}\left(D_{-,f}^{\widetilde{E}}(\gamma)\right). \tag{2.31}$$

Let $\operatorname{Ind}_\gamma(D_{+,f}^{\widetilde{E}})$ be the multiplicity of V_γ^G in $\operatorname{Ind}\left(D_{+,f}^{\widetilde{E}}(\gamma)\right)$. By Theorem 2.1, we have

$$\operatorname{Ker}\left(D_{+,f}^{\widetilde{E}}\right) - \operatorname{Ker}\left(D_{-,f}^{\widetilde{E}}\right) = \bigoplus_{\gamma\in\Lambda_+^*} \operatorname{Ind}_\gamma\left(D_{+,f}^{\widetilde{E}}\right) \cdot V_\gamma^G \in R[G]. \tag{2.32}$$

Lemma 2.4. *Under the condition in Theorem 2.1, if f_1 is a G-invariant function on $\widetilde{M}$ such that one of the following two conditions holds:*

a) *$f = f_1$ outside a compact subset of $\widetilde{M}$,*
b) *$f_1 = e^{-2x_n} f$ on $\partial M \times (-\infty, 0]$.*

Then for any $\gamma \in \Lambda_+^*$,

$$\operatorname{Ind}_\gamma(D_{+,f}^{\widetilde{E}}) = \operatorname{Ind}_\gamma(D_{+,f_1}^{\widetilde{E}}). \tag{2.33}$$

Proof. Note that C_0 in (2.23) does not depend on f.

If $f = f_1$ outside a compact subset of $\widetilde{M}$, let $f_t = (1-t)f + tf_1$ for $t \in [0, 1]$. Then for $\gamma \in \Lambda_+^*$, there exists $C_\gamma > 0$ such that (2.26) holds, thus (2.27) for

$D^{\widetilde{E}}_{f_t}$ holds uniformly on $t \in [0,1]$. Now $D^{\widetilde{E}}_f(\gamma)\left(1+(D^{\widetilde{E}}_f(\gamma))^2\right)^{-1/2}$ is a continuous family of bounded Fredholm operators, thus $\mathrm{Ind}_\gamma(D^{\widetilde{E}}_{+,f_t})$ does not depend on $t \in [0,1]$. In particular (2.33) holds.

If $f_1 = e^{-2x_n} f$ on $\partial M \times (-\infty, 0]$, set $f_t = e^{-2tx_n} f$ for $t \in [0,1]$. Then again for $\gamma \in \Lambda^*_+$, there exists $C_\gamma > 0$ such that (2.26) holds uniformly on $t \in [0,1]$, thus (2.27) for $D^{\widetilde{E}}_{f_t}$ holds uniformly on $t \in [0,1]$. We conclude as above (2.33) holds. □

The following lemma will be used in the proof of Theorem 3.1.

Lemma 2.5. *If* $(1+|f|)^{-1/2}\omega \in L^2(\widetilde{M}, S(T\widetilde{M}) \otimes \widetilde{E})^\gamma$ *and*

$$\left(D^{\widetilde{E}} + \sqrt{-1} f c(\widetilde{\Psi}^{\widetilde{M}})\right)\omega = 0 \tag{2.34}$$

in the sense of distribution, then for any $m \in \mathbb{N}$,

$$f^m\omega,\ (1-\varphi_1) f^m D^E_{\partial M}\omega,\ (1-\varphi_1) f^m \frac{\partial \omega}{\partial x_n} \in L^2(\widetilde{M}, S(T\widetilde{M}) \otimes \widetilde{E})^\gamma. \tag{2.35}$$

Proof. We assume that ω verifies (2.34) and $(1+|f|)^{-1/2}\omega \in L^2(\widetilde{M}, S(T\widetilde{M}) \otimes \widetilde{E})^\gamma$. Then by the ellipticity of $D^{\widetilde{E}}$, $\omega \in \mathscr{C}^\infty(\widetilde{M}, S(T\widetilde{M}) \otimes \widetilde{E})$.

Take $C_f > 0$ such that f is strictly positive on $\partial M \times (-\infty, -C_f]$. For $k > 1$, $m \in \mathbb{R}$, let $\varphi_{k,m}$ be a G-invariant smooth function on $\widetilde{M}$ such that

$$\varphi_{k,m} = \varphi_k f^m \quad \text{on } \partial M \times (-\infty, -C_f]. \tag{2.36}$$

Then by (2.5), (2.6) and (2.34), on $\mathcal{V}_1 := \partial M \times (-\infty, -C_f]$, we get

$$D^{\widetilde{E}}(\varphi^2_{k,m}\omega) = \varphi_k f^{2m}\Big[2c(e_n)\Big(\varphi'_k + m\varphi_k f^{-1}\frac{\partial f}{\partial x_n}\Big) - \sqrt{-1}\varphi_k f c(\widetilde{\Psi}^{\widetilde{M}})\Big]\omega. \tag{2.37}$$

As in (2.2), we denote by $\|\ \|_{\mathcal{V}_1,0}$, $\langle\,,\,\rangle_{\mathcal{V}_1}$ the L^2-norm and Hermitian product on $L^2(\mathcal{V}_1, S(T\widetilde{M}) \otimes \widetilde{E})$. Note that $dv_{\widetilde{M}} = -dx_n \wedge dv_{\partial M}$ on $\mathcal{V}_1$, thus by (2.21), (2.23), (2.34) and the argument around (2.24), there exists $C > 0$ such that for any $k > 1$, $m \in \mathbb{R}$, we have

$$\begin{aligned} 0 = \Big\langle (D^{\widetilde{E}}_f)^2\omega, \varphi^2_{k,m}\omega\Big\rangle_{\mathcal{V}_1} &\geqslant \mathrm{Re}\,\Big\langle D^{\widetilde{E}}\omega, D^{\widetilde{E}}(\varphi^2_{k,m}\omega)\Big\rangle_{\mathcal{V}_1} \\ &+ \int_{\partial M \times \{-C_f\}} \mathrm{Re}\,\Big\langle c(-e_n)D^{\widetilde{E}}\omega, \varphi^2_{k,m}\omega\Big\rangle\, dv_{\partial M} - C\|f^{m+1/2}\varphi_k\omega\|^2_{\mathcal{V}_1,0} \\ &+ \Big\langle\Big(-C_0(|f| + |df|) + f^2\Big|\widetilde{\Psi}^{\widetilde{M}}\Big|^2\Big)\omega, \varphi^2_{k,m}\omega\Big\rangle_{\mathcal{V}_1}. \end{aligned} \tag{2.38}$$

Here we estimate the Lie derivative term in (2.21) by the term $\|\cdot\|^2_{\mathcal{V}_1,0}$ in (2.38).

Recall that on $\partial M \times (-\infty, 0]$, $\widetilde{\Psi}^{\widetilde{M}}(y, x_n)$ is nowhere zero and constant in x_n. By (1.5), (1.6), (2.34), (2.37) and the assumption that $(1+|f|)^{-1/2}\omega \in$

$L^2(\widetilde{M}, S(T\widetilde{M}) \otimes \widetilde{E})^\gamma$, there exists $C > 0$ such that for any $k \in \mathbb{N}$,

$$\begin{aligned} \mathrm{Re}\left\langle D^{\widetilde{E}}\omega, D^{\widetilde{E}}(\varphi_{k,-1}^2\,\omega)\right\rangle_{\mathcal{V}_1} &= \mathrm{Re}\left\langle -\sqrt{-1}fc(\widetilde{\Psi}^{\widetilde{M}})\omega, D^{\widetilde{E}}(\varphi_{k,-1}^2\,\omega)\right\rangle_{\mathcal{V}_1} \\ &\geqslant \frac{1}{2}\left\|\varphi_k\left|\widetilde{\Psi}^{\widetilde{M}}\right|\omega\right\|_{\mathcal{V}_1,0}^2 - C. \end{aligned} \tag{2.39}$$

From (1.5), (1.6), (2.39) and $(1+|f|)^{-1/2}\omega \in L^2(\widetilde{M}, S(T\widetilde{M}) \otimes \widetilde{E})^\gamma$, when $k \to \infty$, we get $\left|\widetilde{\Psi}^{\widetilde{M}}\right|\omega \in L^2(\widetilde{M}, S(T\widetilde{M}) \otimes \widetilde{E})$ from (2.38) by taking $m = -1$. This implies $\omega \in L^2(\widetilde{M}, S(T\widetilde{M}) \otimes \widetilde{E})$.

By repeating the above process, we know that $f^m\omega \in L^2(\widetilde{M}, S(T\widetilde{M}) \otimes \widetilde{E})$ for any $m \in \mathbb{N}$.

For $k, m \in \mathbb{N}$, by (2.5), as in (2.38), we get

$$\begin{aligned} \left\langle (D^{\widetilde{E}})^2(\varphi_{k,m}\omega), \varphi_{k,m}\omega\right\rangle_{\mathcal{V}_1} &= \left\|\varphi_{k,m} D^E_{\partial M}\omega\right\|_{\mathcal{V}_1,0}^2 \\ &+ \int_{\partial M \times \{-C_f\}} \left\langle \tfrac{\partial}{\partial x_n}(\varphi_{k,m}\omega), \varphi_{k,m}\omega\right\rangle dv_{\partial M} + \left\|\tfrac{\partial}{\partial x_n}(\varphi_{k,m}\omega)\right\|_{\mathcal{V}_1,0}^2. \end{aligned} \tag{2.40}$$

Note that the following pointwise estimate holds:

$$\left|\tfrac{\partial}{\partial x_n}(\varphi_{k,m}\omega)\right|^2 \geqslant \frac{1}{2}\left|\varphi_{k,m}\tfrac{\partial \omega}{\partial x_n}\right|^2 - \left|\tfrac{\partial \varphi_{k,m}}{\partial x_n}\omega\right|^2. \tag{2.41}$$

From (2.34), (2.36) and $f^m\omega \in L^2(\widetilde{M}, S(T\widetilde{M}) \otimes \widetilde{E})$, we know that the left-hand side of (2.40) is uniformly bounded on $k > 1$, and $\frac{\partial \varphi_{k,m}}{\partial x_n}\omega \to mf^{m-1}\frac{\partial f}{\partial x_n}\omega$ in $L^2(\mathcal{V}_1, (S(T\widetilde{M}) \otimes \widetilde{E})|_{\mathcal{V}_1})$ as $k \to +\infty$. From (2.40), (2.41), for any $m \in \mathbb{N}$, when $k \to +\infty$, $f^m D^E_{\partial M}\omega$, $f^m \frac{\partial \omega}{\partial x_n} \in L^2(\mathcal{V}_1, (S(T\widetilde{M}) \otimes \widetilde{E})|_{\mathcal{V}_1})$.

The proof of Lemma 2.5 is completed. □

3. Transversal index and L^2-index

We use the notation in Introduction and Section 2.1.

Let f be a G-invariant function on $\widetilde{M}$ verifying (1.5) and (1.6), and we assume that Ψ^M is nowhere zero on ∂M.

The purpose of this section is to give a self-contained proof of Braverman's following result [2, Theorem 5.5] for manifolds with boundary, which identifies the transversal index in (1.4) and the L^2-index appearing in (2.32).

Theorem 3.1. *For any $\gamma \in \Lambda^*_+$, the following identity holds:*

$$\mathrm{Ind}_\gamma\left(D^{\widetilde{E}}_{+,f}\right) = \mathrm{Ind}_\gamma\left(\sigma^M_{E,\Psi}\right). \tag{3.1}$$

Proof. To prove (3.1), by Lemma 2.4a), we can and we will assume that there exists $C_1 > 0$ such that $f > C_1$ on $\widetilde{M}$.

Set $U = M \cup \partial M \times [-1,0]$ and $\widehat{U} = M \cup \partial M \times (-1,0]$ the interior of U. Let $\rho : (-1,0] \to (-\infty,0]$ be a strictly increasing smooth function such that

$$\rho(t_n) = t_n \text{ for } t_n \in [-1/4,0],\ \rho(t_n) = \log(1+t_n) \text{ for } t_n \in (-1,-1/2]. \tag{3.2}$$

We define the diffeomorphism $\tau : \widehat{U} \to \widetilde{M}$ by

$$\tau(x) = x \text{ for } x \in M, \quad \tau(y,t_n) = (y,\rho(t_n)) \text{ for } (y,t_n) \in \partial M \times (-1,0]. \tag{3.3}$$

Let $g^{T\widehat{U}} := \tau^* g^{T\widetilde{M}}$ be the induced metric on $T\widehat{U}$, then on $\partial M \times (-1,0] \subset \widehat{U}$,

$$g^{T\widehat{U}}(y,t_n) = g_y^{T\partial M} + (\rho'(t_n))^2 (dt_n)^2. \tag{3.4}$$

We define the metric g^{TU} on TU by g^{TM} on M and

$$g^{TU}(y,t_n) = g_y^{T\partial M} + (dt_n)^2 \quad \text{for } (y,t_n) \in \partial M \times (-1,0]. \tag{3.5}$$

From (2.1), the pull-back of $(S(T\widetilde{M}), h^{S(T\widetilde{M})}, \nabla^{S(T\widetilde{M})})$, $(\widetilde{E}, h^{\widetilde{E}}, \nabla^{\widetilde{E}})$ on $\widehat{U}$ are still the pull-back of the corresponding objects on ∂M. Thus they extend naturally to U, and we denote them by $(S(TU), h^{S(TU)}, \nabla^{S(TU)})$, (E, h^E, ∇^E). Moreover, the G-action on $\partial M \times [-1,0] \subset U$ is induced by the G-action on ∂M, and the induced map $\Psi : U \to \mathfrak{g}$ is still constant in t_n on $\partial M \times [-1,0]$.

Since Ψ^M is nowhere zero on $U \setminus M$, by the additivity of the transversal index (cf. [1, Theorem 3.7, §6] and [9, Prop. 4.1]), one has

$$\operatorname{Ind}\left(\sigma^M_{E,\Psi}\right) = \operatorname{Ind}\left(\sigma^U_{E,\Psi}\right) \in R[G]. \tag{3.6}$$

Let $L^2(U, S(TU) \otimes E)$ be the space of L^2-sections of $S(TU) \otimes E$ on U with norm $\| \ \|_0$ associated to $g^{TU}, h^{S(TU)}, h^E$ as in (2.2), and $H^k(U, S(TU) \otimes E)$ the corresponding k^{th} Sobolev space.

We adapt now the idea of [2, §14.2–§14.5] to deform the transversal elliptic symbol $\sigma^U_{E,\Psi}$ in (1.3) and to identify $\operatorname{Ind}_\gamma(D^{\widetilde{E}}_{+,f})$ to $\operatorname{Ind}_\gamma(\sigma^U_{E,\Psi})$.

Let $\vartheta : [0,\infty) \to [1,\infty)$ be a smooth function such that $\vartheta(t) = 1$ for $t \leqslant 1$ and $\vartheta(t) = t$ for $t \geqslant 2$. Let $\vartheta_M : U \to [0,1]$ be a smooth function such that $\vartheta_M \equiv 1$ on M and $\vartheta_M \equiv 0$ on $\partial M \times [-1,-1/2]$. We still denote by $\pi : TU \to U$ the natural projection. Consider the symbol for $x \in U, \xi \in T_x U$,

$$\sigma(x,\xi) := \sqrt{-1}\Big\{\vartheta_M(x)\vartheta(|\xi|_{g^{TU}})^{-1} c(\xi) + c(\Psi^U)\Big\} \otimes \mathrm{Id}_{\pi^* E}\ . \tag{3.7}$$

Then σ is a transversally elliptic symbol of order 0 and is homotopic to $\sigma^U_{E,\Psi}$ on $T_G U$.

As the manifold U with boundary is compact, by [11, Proposition 2.4], there exists a G-vector bundle F over U such that the bundle $S_+(TU) \otimes E \oplus F$ is a trivial G-vector bundle on U. The map $\sqrt{-1}c(\Psi^U) \otimes \mathrm{Id}_E + \mathrm{Id}_F$ defines an isomorphism of the restriction of $S_+(TU) \otimes E \oplus F$ and $S_-(TU) \otimes E \oplus F$ on $\partial M \times [-1,0) = U \setminus M$, and so a trivialization of $S_-(TU) \otimes E \oplus F$ on $U \setminus M$.

Let $\jmath : U \to N$ be a G-equivariant embedding of the manifold U with boundary into a smooth compact G-manifold N with the same dimension (for example, we can take N as the double of U). Then by extending $S_\pm(TU) \otimes E \oplus F$ on U as trivial G-vector bundles on $N \setminus M$, we get G-vector bundles $\widetilde{\mathcal{E}}_\pm$ over N. Moreover we have a natural map $\widetilde{c} : \widetilde{\mathcal{E}}_+ \to \widetilde{\mathcal{E}}_-$ whose restriction to $N \setminus M$ is invertible and

$$\widetilde{c} = \sqrt{-1}c(\Psi^U) \otimes \mathrm{Id}_E + \mathrm{Id}_F \quad \text{on } U. \tag{3.8}$$

We still denote by ϑ_M the extension of ϑ_M on N by taking $\vartheta_M = 0$ on $N \setminus U$, and $\pi : TN \to N$ the natural projection. Then the symbol

$$\widetilde{\sigma}_N(x,\xi) := \sqrt{-1}\vartheta_M(x)\vartheta(|\xi|_{g^{TU}})^{-1}c(\xi) \otimes \mathrm{Id}_{\pi^*E} + \pi^*\widetilde{c}(x) : \pi^*\widetilde{\mathcal{E}}_+ \to \pi^*\widetilde{\mathcal{E}}_- \tag{3.9}$$

for $x \in N, \xi \in T_{G,x}N$, is a transversally elliptic symbol of order 0 on $T_G N$ which extends the transversally elliptic symbol $\widetilde{\sigma} = \sigma + \mathrm{Id}_{\pi^*F}$ on $T_G U$.

The excision theorem [1, Theorem 3.7] tells us that the index $\mathrm{Ind}(\widetilde{\sigma}_N)$ of the transversally elliptic symbol $\widetilde{\sigma}_N$ depends only on $\sigma^U_{E,\Psi}$, but not on the choice of $\jmath$, $\widetilde{\sigma}_N$, and its character is a distribution on G. By the definition of $\mathrm{Ind}(\sigma^U_{E,\Psi})$ (cf. [1, §1], [9, §3]), we have

$$\mathrm{Ind}(\sigma^U_{E,\Psi}) := \mathrm{Ind}(\widetilde{\sigma}_N). \tag{3.10}$$

We construct now a particular zeroth-order transversally elliptic operator P on N whose symbol is homotopic to $\widetilde{\sigma}_N$.

Let g^{TN} be a G-invariant Riemannian metric on N which extends g^{TU} on U, and $h^{\widetilde{\mathcal{E}}} = h^{\widetilde{\mathcal{E}}_+} \oplus h^{\widetilde{\mathcal{E}}_-}$ be a G-invariant Hermitian metric on $\widetilde{\mathcal{E}} = \widetilde{\mathcal{E}}_+ \oplus \widetilde{\mathcal{E}}_-$ on N which extends $h^{S(TU)\otimes E}$ on U. The metrics g^{TN}, $h^{\widetilde{\mathcal{E}}}$ induce a norm $\| \ \|_0$ on $L^2(N, \widetilde{\mathcal{E}})$, the space of L^2-sections of $\widetilde{\mathcal{E}}$ on N, as in (2.2). Let $\mathcal{A} : \mathscr{C}^\infty(N, \widetilde{\mathcal{E}}_+) \to \mathscr{C}^\infty(N, \widetilde{\mathcal{E}}_+)$ be an invertible positive-definite self-adjoint G-invariant second-order differential operator, whose principal symbol is $\sigma(\mathcal{A})(x,\xi) = |\xi|^2_{g^{TN}} \mathrm{Id}_{\widetilde{\mathcal{E}}_+}$.

Recall that we assume that $f > C_1 > 0$ on $\widetilde{M}$. Set

$$\widetilde{f}(x) = \frac{1}{f \circ \tau(x)} \quad \text{if } x \in \widehat{U}; \quad \widetilde{f}(x) = 0 \quad \text{if } x \in N \setminus U. \tag{3.11}$$

By (1.5) and (1.6), $\widetilde{f}$ is a G-invariant $\mathscr{C}^0$ function on N.

We denote by $D^E_{\widehat{U}}$ on $\widehat{U}$ the operator $D^{\widetilde{E}}$ under the map $\tau : \widehat{U} \to \widetilde{M}$. By (2.5) and (3.3), we get

$$D^E_{\widehat{U}} = c(e_n)D^E_{\partial M} + \frac{1}{\rho'(t_n)}c(e_n)\frac{\partial}{\partial t_n} \quad \text{on } \partial M \times (-1, 0], \tag{3.12}$$

and the restriction of $D^E_{\widehat{U}}$ to M is the Spin^c Dirac operator D^E on M associated to $g^{TM}, h^{S(TM)}, h^E$. Set

$$P = \widetilde{c} + \widetilde{f} D^E_{\widehat{U}} \mathcal{A}^{-1/2}. \tag{3.13}$$

By (3.11) and (3.12), P is a well-defined pseudo-differential operator on N.

For $\xi \in TN$, we define $\widehat{c}(\xi) = c(\xi)$ on M, and

$$\widehat{c}(\xi) = c(\xi) \ \text{ for } \ \xi \in T\partial M|_{\partial M \times (-1,0]},$$

$$\widehat{c}\left(\frac{\partial}{\partial t_n}\right)_{(y,t_n)} = \frac{1}{\rho'(t_n)}c(e_n) \ \text{ for } \ (y, t_n) \in \partial M \times (-1, 0].$$

Then by (3.12), under the identification of T^*N and TN by using the metric g^{TN}, the principal symbol of P is

$$\sigma(P)(x,\xi) = \pi^*\widetilde{c} + \sqrt{-1}\widetilde{f}\,\vartheta(|\xi|_{g^{TN}})^{-1}\widehat{c}(\xi), \qquad \text{for } x \in N,\ \xi \in T_xN. \tag{3.14}$$

By (3.9) and (3.14), for $t \in [0,1]$, $x \in N$, $\xi \in T_{G,x}N$, we have

$$\begin{aligned}(t\widetilde{\sigma}_N + (1-t)\sigma(P))(x,\xi)\\ = \sqrt{-1}\vartheta(|\xi|_{g^{TN}})^{-1}\Big(t\,\vartheta_M(x)c(\xi) + (1-t)\widetilde{f}\,\widehat{c}(\xi)\Big) + \pi^*\widetilde{c}(x).\end{aligned} \tag{3.15}$$

Thus for $x \in N \setminus U$, $(t\widetilde{\sigma}_N + (1-t)\sigma(P))(x,\xi)$ is $\pi^*\widetilde{c}(x)$ which is invertible, and for $\xi \in T_{G,x}U$, as $\langle \xi, \Psi^U\rangle = 0$, we know if $\Psi^U(x) \neq 0$, by (3.8), $\widetilde{c}(x)$ is invertible, thus $(t\widetilde{\sigma}_N + (1-t)\sigma(P))(x,\xi)$ is invertible. We conclude finally that for any $t \in [0,1]$,

$$\begin{aligned}\{(x,\xi) \in T_GN : (t\widetilde{\sigma}_N + (1-t)\sigma(P))(x,\xi) \text{ is not invertible}\}\\ = \{(x,0) \in T_GM : \Psi^U(x) = 0\}\end{aligned} \tag{3.16}$$

is compact. In particular, $\sigma(P)$ is a transversally elliptic symbol and homotopic to $\widetilde{\sigma}_N$, thus P is a zeroth-order transversally elliptic operator on N, and by (3.10) and the homotopy invariance of the transversal index (cf. [1, Theorem 3.7, §6] and [9, Prop. 4.1]),

$$\mathrm{Ind}(\sigma^U_{E,\Psi}) = \mathrm{Ind}(\widetilde{\sigma}_N) = \mathrm{Ind}(P). \tag{3.17}$$

For $t \in [0,1]$, consider the family of operators

$$P_t = (1-t)\widetilde{c} + t\,\widetilde{c}\,\mathcal{A}^{-1/2} + \widetilde{f}\,D^E_{\widehat{U}}\,\mathcal{A}^{-1/2} : L^2(N,\widetilde{\mathcal{E}}_+) \to L^2(N,\widetilde{\mathcal{E}}_-). \tag{3.18}$$

For $\gamma \in \Lambda^*_+$, we denote by $P_t(\gamma)$ the restriction of P_t to $L^2(N,\widetilde{\mathcal{E}}_+)^\gamma$, the γ-component of $L^2(N,\widetilde{\mathcal{E}}_+)$. For any $t < 1$, the operator P_t is a transversally elliptic operator depending continuously on t, thus $P_t(\gamma)$ is Fredholm for any $\gamma \in \Lambda^*_+$, and as $P_0 = P$, from (1.4), (3.17),

$$\mathrm{Ind}(P_t(\gamma)) = \mathrm{Ind}(P_0(\gamma)) = \mathrm{Ind}_\gamma(\sigma^U_{E,\Psi}) \cdot V^G_\gamma \qquad \text{for } t < 1. \tag{3.19}$$

Since P_t is a family of bounded operators which depends continuously on t, to show (3.19) holds for $t = 1$, we only need to prove that the operator $P_1(\gamma)$ is Fredholm for any $\gamma \in \Lambda^*_+$.

From (3.18), we have

$$P_1 = \widetilde{c}\,\mathcal{A}^{-1/2} + \widetilde{f}\,D^E_{\widehat{U}}\mathcal{A}^{-1/2} : L^2(N,\widetilde{\mathcal{E}}_+) \to L^2(N,\widetilde{\mathcal{E}}_-). \tag{3.20}$$

Thus $s \in \mathrm{Ker}(P_1)$ if and only if $\omega := \mathcal{A}^{-1/2}s \in H^1(N,\widetilde{\mathcal{E}}_+)$, the 1st Sobolev space on N with values in $\widetilde{\mathcal{E}}_+$ associated to $g^{TN}, h^{\widetilde{\mathcal{E}}_+}$, satisfies that

$$\left(\widetilde{c} + \widetilde{f}\,D^E_{\widehat{U}}\right)\omega = 0. \tag{3.21}$$

As $\widetilde{f} \equiv 0$ and $\widetilde{c}$ is invertible on $N \setminus U$, from (3.21), $\omega = 0$ on $N \setminus U$. Thus (3.21) holds if and only if $\mathrm{supp}(\omega) \subset U$ and (3.21) holds on U. By (3.8) and (3.11), (3.21) is equivalent to $\omega \in H^1(U, S_+(TU) \otimes E)$, $\omega = 0$ on ∂U, and

$$\left(D^E_{\widehat{U}} + \sqrt{-1} f \circ \tau\, c\left(\Psi^U\right)\right) \omega = 0. \tag{3.22}$$

Lemma 3.2. *Assume that* (*cf.* (1.5))

$$\lim_{x_n \to -\infty} \varrho(x_n) e^{x_n} = +\infty. \tag{3.23}$$

If ω verifies (3.22), *then $\omega \in L^2(U, S(TU) \otimes E)^\gamma$ if and only if $\omega \in L^2(\widetilde{M}, S(T\widetilde{M}) \otimes \widetilde{E})^\gamma$. In this case, for any $m \in \mathbb{N}$, $f^m \omega \in L^2(U, S(TU) \otimes E)^\gamma$, and $\omega \in H^1(U, S(TU) \otimes E)$, $\omega = 0$ on ∂U.*

Proof. By (2.1), (3.12) and the discussion after (3.5), (3.22) is equivalent to (2.34) in the sense of distribution.

Let $dv_{\widehat{U}}$, dv_U be the Riemannian volume forms on $(\widehat{U}, g^{T\widehat{U}})$, (U, g^{TU}), respectively. Then by (3.4) and (3.5), we have

$$dv_{\widehat{U}}(y, t_n) = \rho'(t_n)\, dv_U(y, t_n) \quad \text{on } \partial M \times (-1, 0]. \tag{3.24}$$

Note that on $(-1, -1/2]$, $\rho'(t_n) = \frac{1}{1+t_n}$.

By (3.24), if $\omega \in L^2(\widetilde{M}, S(T\widetilde{M}) \otimes \widetilde{E})$, then $\omega \in L^2(U, S(TU) \otimes E)$.

Now assume that ω verifies (3.22) and $\omega \in L^2(U, S(TU) \otimes E)^\gamma$. Then by the ellipticity of $D^{\widetilde{E}}$, $\omega \in \mathscr{C}^\infty(\widetilde{M}, S(T\widetilde{M}) \otimes \widetilde{E})$.

By (3.23), on $\partial M \times (-1, -1/2] \subset \widehat{U}$, $\lim_{t_n \to -1} (f \circ \tau)(t_n)(t_n + 1) = +\infty$. Thus from (3.24),

$$\begin{aligned} &\int_{\partial M \times (-\infty, \log(1/2)]} (1 + |f|)^{-1} |\omega|^2 dv_{\widetilde{M}} \\ &\qquad = \int_{\partial M \times (-1, -1/2]} (1 + |f \circ \tau|)^{-1} |\omega|^2 (t_n + 1)^{-1} dv_U < +\infty. \end{aligned} \tag{3.25}$$

This means $(1 + |f|)^{-1/2} \omega \in L^2(\widetilde{M}, S(T\widetilde{M}) \otimes \widetilde{E})$. Now from Lemma 2.5 and (3.24), we get $\omega, f\omega \in H^1(U, S(TU) \otimes E)$. Thus the restrictions of ω, $f\omega$ on ∂U are well defined. But $f = \infty$ on ∂U, thus $\omega = 0$ on ∂U. □

From Lemma 2.4, we can assume that f is a strictly positive G-invariant smooth function on $\widetilde{M}$ verifying (1.5), (1.6) and (3.23).

From Lemma 3.2, (2.14), (3.21) and (3.22), we know that $\mathrm{Ker}(P_1(\gamma))$ is isomorphic to $\mathrm{Ker}(D^{\widetilde{E}}_{+,f}(\gamma))$, in the same way, $\mathrm{Coker}(P_1(\gamma))$ is isomorphic to $\mathrm{Ker}(D^{\widetilde{E}}_{-,f}(\gamma))$. But we have proved that $D^{\widetilde{E}}_f(\gamma)$ is a Fredholm operator, thus $P_1(\gamma)$ is a Fredholm operator and (3.19) holds for $t = 1$. The proof of Theorem 3.1 is completed. □

References

[1] M.F. Atiyah, *Elliptic operators and compact groups*, Lecture Notes in Mathematics, Vol. 401, Springer-Verlag, Berlin, 1974.

[2] M. Braverman, *Index theorem for equivariant Dirac operators on non-compact manifolds*, K-Theory **27** (2002), no. 1, 61–101.

[3] T. Bröcker and T. Tom Dieck, *Representations of compact Lie groups*, GTM 98, Springer-Verlag, New York, 1985.

[4] P.R. Chernoff, Essential self-adjointness of powers of generators of hyperbolic equations. J. Functional Analysis 12 (1973), 401–414.

[5] H.B. Lawson and M.-L. Michelsohn, *Spin geometry*, Princeton Mathematical Series, vol. 38, Princeton University Press, Princeton, NJ, 1989.

[6] X. Ma and G. Marinescu, *Holomorphic Morse inequalities and Bergman kernels*, Progress in Mathematics **254**, Birkhäuser Boston, MA, 2007, 422 pp.

[7] X. Ma and W. Zhang, *Geometric quantization for proper moment maps*, C. R. Math. Acad. Sci. Paris **347** (2009), 389–394.

[8] ———, *Transversal index and geometric quantization on non-compact manifolds*, arXiv: 0812.3989.

[9] P.-É. Paradan, *Localization of the Riemann-Roch character*. J. Funct. Anal. **187** (2001), no. 2, 442–509.

[10] ———, Spin^c*-quantization and the K-multiplicities of the discrete series*. Ann. Sci. Ecole Norm. Sup. (4) **36** (2003), no. 5, 805–845.

[11] G. Segal, *Equivariant K-theory*. Inst. Hautes Études Sci. Publ. Math. No. **34** (1968), 129–151.

[12] M.E. Taylor, *Partial differential equations. II*, Applied Mathematical Sciences, vol. 116, Springer-Verlag, New York, 1996.

[13] Y. Tian and W. Zhang, *Quantization formula for symplectic manifolds with boundary*, Geom. Funct. Anal. **9** (1999), no. 3, 596–640.

[14] M. Vergne, Quantification géométrique et réduction symplectique. Séminaire Bourbaki, Vol. 2000/2001. Astérisque No. 282 (2002), 249–278.

[15] ———, *Applications of equivariant cohomology*, International Congress of Mathematicians. Vol. I, Eur. Math. Soc., Zürich, 2007, pp. 635–664.

Xiaonan Ma
Université Paris Diderot – Paris 7
UFR de Mathématiques
Case 7012, Site Chevaleret
F-75205 Paris Cedex 13, France
e-mail: `ma@math.jussieu.fr`

Weiping Zhang
Chern Institute of Mathematics & LPMC
Nankai University
Tianjin 300071, P.R. China
e-mail: `weiping@nankai.edu.cn`

The Asymptotics of the Ray-Singer Analytic Torsion of Hyperbolic 3-manifolds

Werner Müller

Dedicated to Jeff Cheeger for his 65th birthday

Abstract. In this paper we consider the analytic torsion of a closed hyperbolic 3-manifold associated with the mth symmetric power of the standard representation of $\mathrm{SL}(2,\mathbb{C})$ and we study its asymptotic behavior as m tends to infinity. The leading coefficient of the asymptotic formula is given by the volume of the hyperbolic 3-manifold. It follows that the Reidemeister torsion associated with the symmetric powers determines the volume of a closed hyperbolic 3-manifold.

Mathematics Subject Classification (2000). Primary: 58J52, Secondary: 11M36.

Keywords. Analytic torsion, hyperbolic manifolds, Ruelle zeta function.

1. Introduction

Let X be a closed, oriented hyperbolic 3-manifold. Then there exists a discrete, torsion free, co-compact subgroup $\Gamma \subset \mathrm{SL}(2,\mathbb{C})$ such that $X = \Gamma\backslash\mathbb{H}^3$, where $\mathbb{H}^3 = \mathrm{SL}(2,\mathbb{C})/\mathrm{SU}(2)$ is the three-dimensional hyperbolic space. Let ρ be a finite-dimensional complex representation of Γ and let $E_\rho \to X$ be the associated flat vector bundle. Choose a Hermitian fiber metric h on E_ρ. Let $T_X(\rho; g, h)$ denote the Ray-Singer analytic torsion of the de Rham complex of E_ρ-valued differential forms [21], where g denotes the hyperbolic metric. If ρ is acyclic, then $T_X(\rho; g, h)$ is metric independent [19, Corollary 2.7]. In this case we denote it by $T_X(\rho)$.

For $m \in \mathbb{N}$ let $\tau_m = \mathrm{Sym}^m$ be the mth symmetric power of the standard representation of $\mathrm{SL}(2,\mathbb{C})$ on $\mathbb{C}^2$ and denote by E_{τ_m} the flat vector bundle associated to $\tau_m|_\Gamma$. It is well known that $H^*(X, E_{\tau_m}) = 0$. This follows, for example, from [4, Chapt. VII, Theorem 6.7]. Hence the restriction of τ_m to Γ is an acyclic representation of Γ. Denote by $T_X(\tau_m)$ the analytic torsion with respect to $\tau_m|_\Gamma$. The purpose of this paper is to study the asymptotic behavior of $T_X(\tau_m)$ as $m \to \infty$.

Our main result is the following theorem.

Theorem 1.1. *Let X be a closed, oriented hyperbolic 3-manifold. Then we have*

$$-\log T_X(\tau_m) = \frac{\operatorname{vol}(X)}{4\pi} m^2 + O(m) \tag{1.1}$$

as $m \to \infty$.

We note that there is an analogous result in the holomorphic setting. In [3] Bismut and Vasserot studied the asymptotic behavior of the holomorphic Ray-Singer torsion for symmetric powers of a positive vector bundle.

Let $\tau_X(\tau_m)$ denote the Reidemeister torsion of X with respect to $\tau_m|_\Gamma$ (see [19]). Then by [19, Theorem 1] we have $T_X(\tau_m) = \tau_X(\tau_m)$. Thus we obtain the following corollary.

Corollary 1.2. *Let X be a closed, oriented hyperbolic 3-manifold. Then we have*

$$-\log \tau_X(\tau_m) = \frac{\operatorname{vol}(X)}{4\pi} m^2 + O(m) \tag{1.2}$$

as $m \to \infty$.

This result has applications to the cohomology of arithmetic hyperbolic 3-manifolds. We will discuss this elsewhere. As an immediate corollary we get

Corollary 1.3. *Let X be a closed, oriented hyperbolic 3-manifold. Then* $\operatorname{vol}(X)$ *is determined by the set of Reidemeister torsions $\{\tau_X(\tau_m) \colon m \in \mathbb{N}\}$.*

Some remarks are in order. The Reidemeister torsion of a compact 3-manifold is known to be a topological invariant [7]. Therefore, it follows from Corollary 1.3 that the volume of a compact, oriented hyperbolic 3-manifold is a topological invariant. This is also a well-known consequence of the Mostow-Prasad rigidity theorem [18, 20].

There are only finitely many closed, oriented hyperbolic 3-manifolds with the same volume [25, Theorem 3.6]. Therefore we get

Corollary 1.4. *A compact, oriented hyperbolic 3-manifold X is determined up to finitely many possibilities by the set $\{\tau_X(\tau_m) \colon m \in \mathbb{N}\}$ of Reidemeister torsion invariants.*

It is known [29] that the number of closed hyperbolic manifolds with a given volume can be arbitrarily large. Therefore the proof of the corollary does not give a uniform bound on the number of closed hyperbolic manifolds with the same set of Reidemeister torsion invariants.

Our approach to prove Theorem 1.1 is based on the expression of $T_X(\tau_m)$ in terms of the twisted Ruelle zeta function attached to τ_m. Recall that for a finite-dimensional complex representation ρ of Γ the twisted Ruelle zeta function $R_\rho(s)$

is defined for $\mathrm{Re}(s) \gg 0$ as the infinite product

$$R_\rho(s) = \prod_{\substack{[\gamma]\neq e\\ \text{prime}}} \det\left(\mathrm{Id} - \rho(\gamma)e^{-s\ell(\gamma)}\right), \tag{1.3}$$

where $[\gamma]$ runs over the nontrivial primitive conjugacy classes of Γ and $\ell(\gamma)$ denotes the length of the unique closed geodesic associated to $[\gamma]$. It follows from [10] that $R_\rho(s)$ admits a meromorphic extension to the whole complex plane. If ρ is unitary and acyclic then $R_\rho(s)$ is regular at $s = 0$ and its value at zero satisfies

$$|R_\rho(0)| = T_X(\rho)^2$$

(see [9]). For an arbitrary unitary representation (which is not necessarily acyclic), the coefficient of the leading term of the Laurent expansion of $R_\rho(s)$ at $s = 0$ is given by the analytic torsion. The corresponding result holds for any compact, oriented hyperbolic manifold of odd dimension $n \geq 3$ [9]. In his thesis [5] U. Bröcker has established a similar result for representations of the fundamental group that are restrictions of finite-dimensional irreducible representations of the isometry group $\mathrm{SO}_0(n, 1)$ of the hyperbolic n-space. Unfortunately, his method is based on elaborate computations which are difficult to verify. This problem has been rectified by Wotzke in his thesis [28]. He gave a different proof which replaces Bröcker's explicite computations by the real version of Kostant's Bott-Borel-Weil theorem [23].

To state the result for $n = 3$ we need to introduce some notation. Let τ be a finite-dimensional, irreducible representation of $\mathrm{SL}(2, \mathbb{C})$, which we regard as real Lie group. Let θ be the Cartan involution of $\mathrm{SL}(2, \mathbb{C})$ with respect to $SU(2)$. Put $\tau_\theta = \tau \circ \theta$. Denote by $R_\tau(s)$ the twisted Ruelle zeta function for the restriction of τ to Γ. Let $E_\tau \to X$ be the flat vector bundle associated to $\tau|_\Gamma$. The flat bundle E_τ can be equipped with a canonical Hermitian fiber metric [14]. Let $\Delta_p(\tau)$ be the corresponding Laplacian on E_τ-valued p-forms and denote by $T_X(\tau)$ the Ray-Singer analytic torsion associated to $\tau|_\Gamma$. Then the main result of [28] for $n = 3$ is the following theorem.

Theorem 1.5. *Let τ be a finite-dimensional, irreducible representation of* $\mathrm{SL}(2, \mathbb{C})$, *regarded as real Lie group. Then we have*

1) *If $\tau_\theta \not\cong \tau$, then $R_\tau(s)$ is regular at $s = 0$ and*

$$|R_\tau(0)| = T_X(\tau)^2.$$

2) *Let $\tau_\theta = \tau$. If $\tau \neq 1$, then the order $h(\tau)$ of $R_\tau(s)$ at $s = 0$ is given by*

$$h(\tau) = -2\sum_{p=1}^{3}(-1)^p p \dim\ker\Delta_p(\tau). \tag{1.4}$$

and if $\tau = 1$, the order equals $4 - 2\dim H^1(X, \mathbb{R})$. The leading term of the Laurent expansion of $R_\tau(s)$ at $s = 0$ is given by

$$T_X(\tau)^2 s^{h(\tau)}.$$

The case of the trivial representation is covered by [9]. In this case the order of $R(s)$ at $s=0$ differs from (1.4).

It follows from Theorem 1.5 that in order to prove Theorem 1.1 it suffices to analyze the asymptotic behavior of $R_{\tau_m}(0)$ as $m\to\infty$. For this purpose we consider another type of twisted Ruelle zeta functions. Let A be the standard split torus of $\mathrm{SL}(2,\mathbb{C})$ and let M be the centralizer of A in $\mathrm{SU}(2)$ (see (2.1) for the explicit description). For $\sigma\in\hat{M}$ let $R(s,\sigma)$ be the Ruelle zeta function defined by (3.6). Using the decomposition of τ_m under the Cartan subgroup MA, it follows that $R_{\tau_m}(s)$ is the product of the twisted Ruelle zeta functions with shifted argument $R(s-(m/2-k),\sigma_{m-2k})$, $k=0,\dots,m$. This reduces the study of the asymptotic behavior of $T_X(\tau_{2m})$ (resp. $T_X(\tau_{2m+1})$) as $m\to\infty$ to the study of the behavior of $|R(k,\sigma_{2k})|$ and $|R(-k,\sigma_{2k})|$, (resp. $|R(k+1/2,\sigma_{2k+1})|$ and $|R(-k-1/2,\sigma_{-(2k+1)})|$), $k>2$, as $k\to\infty$. To analyze the behavior of $|R(k,\sigma_{2k})|$ (resp. $|R(k+1/2,\sigma_{2k+1})|$) as $k\to\infty$ we simply use the infinite product defining it. To deal with the remaining cases we use the functional equation which implies

$$|R(-s,\sigma_k)|=\exp\left(-4\pi^{-1}\operatorname{vol}(X)\operatorname{Re}(s)\right)|R(s,\sigma_{-k})|.$$

This is exactly how the volume of X appears in the asymptotic formula (1.1).

For the sake of completeness we include a proof of Theorem 1.5 which is based on results of [6]. The starting point of the method is the observation that the flat bundle $E_\tau\to X$ is isomorphic to the locally homogeneous vector bundle defined by the restriction of τ to the maximal cocompact subgroup $\mathrm{SU}(2)$ of $\mathrm{SL}(2,\mathbb{C})$ (see [14, Propostion 3.1]). Using this isomorphism the bundle E_τ can be equipped with a canonical Hermitian fiber metric induced from an invariant metric on the corresponding homogeneous vector bundle [14, Lemma 3.1]. We define the Laplacian $\Delta_p(\tau)$ on E_τ-valued p-forms with respect to this metric on E_τ. Then, up to a constant, $\Delta_p(\tau)$ equals $-R(\Omega)$, where $R(\Omega)$ denotes the action of the Casimir operator on sections of the locally homogeneous bundle. This is the key fact which allows us to apply the Selberg trace formula to the heat kernel.

The paper is organized as follows. In Section 2 we summerize some basic facts about hyperbolic 3-manifolds and analytic torsion. In Section 3 we consider twisted Ruelle and Selberg zeta functions and establish some of their basic properties. In Section 4 we introduce certain auxiliary elliptic operators which are needed to derive the determinant formula and to prove the functional equation for the Selberg zeta function. In the next section 5 we establish the functional equation for the Selberg and Ruelle zeta functions. In Section 6 we use the determinant formula of [6] to express the twisted Ruelle zeta function as a ratio of products of regularized determinants of the elliptic operators introduced in Section 5. The determinant formula is one of main tools to study the leading term of the Laurent expansion of $R_\tau(s)$ at $s=0$. In Section 7 we give a proof of Theorem 1.5. In the final Section 8 we prove Theorem 1.1.

2. Preliminaries

2.1. Basic notions

Let $G = \mathrm{SL}(2,\mathbb{C})$ and $K = \mathrm{SU}(2)$. Then K is a maximal compact subgroup of G. We regard G as real Lie group and we recall that G is isomorphic to $\mathrm{Spin}_0(3,1)$ [2]. Under this isomorphism, $\mathrm{SU}(2)$ is mapped to $\mathrm{Spin}(3)$. Thus G acts on the hyperbolic 3-space $\mathbb{H}^3$ and $\mathbb{H}^3 \cong G/K$.

Let $G = NAK$ be the standard Iwasawa decomposition of G and let M be the centralizer of A in K. Then

$$A = \left\{ \begin{pmatrix} \lambda & 0 \\ 0 & \lambda^{-1} \end{pmatrix} : \lambda \in \mathbb{R}^+ \right\}, \quad M = \left\{ \begin{pmatrix} e^{i\theta} & 0 \\ 0 & e^{-i\theta} \end{pmatrix} : \theta \in [0, 2\pi] \right\}. \tag{2.1}$$

We use the natural normalization of the Haar measures for A, N, K and G as in [11, pp. 387–388]. In particular, we choose on K the Haar measure dk of total mass 1.

Let $\mathfrak{g}$, $\mathfrak{k}$, $\mathfrak{a}$, $\mathfrak{m}$ and $\mathfrak{n}$ denote the Lie algebras of G, K, A, M and N, respectively. Let

$$\mathfrak{g} = \mathfrak{k} \oplus \mathfrak{p} \tag{2.2}$$

be the Cartan decomposition of $\mathfrak{g}$. Then $\mathfrak{a}$ is a maximal abelian subspace of $\mathfrak{p}$. Let α be the unique positive root of $(\mathfrak{g}, \mathfrak{a})$. Let $H \in \mathfrak{a}$ be such that $\alpha(H) = 1$. Let $\mathfrak{a}^+ \subset \mathfrak{a}$ be the positive Weyl chamber and let $A^+ = \exp(\mathfrak{a}^+)$. Let $W := W(\mathfrak{g}, \mathfrak{a})$ denote the restricted Weyl group.

Put $\mathfrak{h} = \mathfrak{m} \oplus \mathfrak{a}$. Then $\mathfrak{h}$ is a Cartan subalgebra of $\mathfrak{g}$. We identify $\mathfrak{h}$ with $\mathbb{R}^2$. Then the Weyl group W_G of $(\mathfrak{g}_\mathbb{C}, \mathfrak{h}_\mathbb{C})$ acts on $\mathbb{C}^2$ by sign changes. So W_G has order 4. In a compatible ordering on $\mathfrak{h}_\mathbb{C}^*$ the only positive roots of the pair $(\mathfrak{g}_\mathbb{C}, \mathfrak{h}_\mathbb{C})$ are α_1 and α_2 where $\alpha_1(H) = \alpha_2(H) = \alpha(H) = 1$ and $\alpha_1(iH) = -\alpha_2(iH) = i$. Let $\rho_G = \frac{1}{2}(\alpha_1 + \alpha_2)$.

Let B be the Killing form of $\mathfrak{g}$. Define a symmetric bilinear form on $\mathfrak{g}$ by

$$\langle Y_1, Y_2 \rangle = \frac{1}{4} B(Y_1, Y_2), \quad Y_1, Y_2 \in \mathfrak{g}. \tag{2.3}$$

Then $\langle \cdot, \cdot \rangle$ is positive definite on $\mathfrak{p}$, negative definite on $\mathfrak{k}$ and we have $\langle \mathfrak{k}, \mathfrak{p} \rangle = 0$. The normalization is such that the restriction of $\langle \cdot, \cdot \rangle$ to $\mathfrak{p} \cong T_{eK}(G/K)$ induces the G-invariant Riemannian metric on $\mathbb{H}^3 = G/K$ which has constant curvature -1.

Let $\{Z_i\}$ be a basis of $\mathfrak{g}$ and let $\{Z^j\}$ be the basis of $\mathfrak{g}$ which is determined by $\langle Z_i, Z^j \rangle = \delta_{ij}$. Then the Casimir element $\Omega \in \mathcal{Z}(\mathfrak{g}_\mathbb{C})$ is given by

$$\Omega = \sum_i Z_i Z^i. \tag{2.4}$$

Let $R(\Omega)$ be the differential operator induced by Ω on $C^\infty(\mathbb{H}^3)$. Then by Kuga's lemma we have

$$R(\Omega) = -\Delta,$$

where Δ is the hyperbolic Laplace operator on functions.

2.2. Lattices

Let $\Gamma \subset G$ be a discrete, torsion free, cocompact subgroup. Then $X = \Gamma\backslash\mathbb{H}^3$ is a closed hyperbolic manifold. Given $\gamma \in \Gamma$, we denote by $[\gamma]$ the Γ-conjugacy class of γ. The set of all conjugacy classes of Γ will be denoted by $C(\Gamma)$. Let $\gamma \neq 1$. Then there exist $g \in G$, $m_\gamma \in M$, and $a_\gamma \in A^+$ such that

$$g\gamma g^{-1} = m_\gamma a_\gamma. \tag{2.5}$$

By [27, Lemma 6.6], a_γ depends only on γ and m_γ is determined up to conjugacy in M. By definition there exists $\ell(\gamma) > 0$ such that

$$a_\gamma = \exp(\ell(\gamma)H). \tag{2.6}$$

Then $\ell(\gamma)$ is the length of the unique closed geodesic in X that corresponds to the conjugacy class $[\gamma]$. An element $\gamma \in \Gamma - \{e\}$ is called primitive, if it can not be written as $\gamma = \gamma_0^k$ for some $\gamma_0 \in \Gamma$ and $k > 1$. For every $\gamma \in \Gamma - \{e\}$ there exist a unique primitive element $\gamma_0 \in \Gamma$ and $n_\Gamma(\gamma) \in \mathbb{N}$ such that $\gamma = \gamma_0^{n_\Gamma(\gamma)}$.

2.3. Finite-dimensional representations

Denote by $\hat{M}$ the set of unitary characters of M. Then $\hat{M} \cong \mathbb{Z}$ and the character σ_k that corresponds to $k \in \mathbb{Z}$ is given by

$$\sigma_k\left(\begin{pmatrix} e^{i\theta} & 0 \\ 0 & e^{-i\theta} \end{pmatrix}\right) = e^{ik\theta}. \tag{2.7}$$

Recall that the finite-dimensional irreducible representations of G, regarded as real Lie group, are parametrized by pairs of nonnegative integers [11, p. 32]. For $m \in \mathbb{N}_0$ let

$$\tau_m = \mathrm{Sym}^m \colon G \to \mathrm{GL}(S^m(\mathbb{C}^2))$$

be the mth symmetric power of the standard representation of $G = \mathrm{SL}(2,\mathbb{C})$ on $\mathbb{C}^2$. Denote by $\overline{\tau}_m$ the complex conjugate representation. Then

$$\tau_{m,n} = \tau_m \otimes \overline{\tau}_n \tag{2.8}$$

is the irreducible representation with highest weight (m,n). The restrictions of τ_m to MA decomposes as follows:

$$\tau_m\big|_{MA} = \bigoplus_{k=0}^{m} \sigma_{m-2k} \otimes e^{(\frac{m}{2}-k)\alpha}. \tag{2.9}$$

2.4. Induced representations

Let $P = MAN$ be the standard parabolic subgroup of G. We identify $\mathbb{C}$ with $\mathfrak{a}^*_{\mathbb{C}}$ by $z \to zH$. For $n \in \mathbb{Z}$ and $\lambda \in \mathbb{C}$ let $\pi_{n,\lambda}$ be the induced representation

$$\pi_{n,\lambda} = \mathrm{Ind}_P^G(\sigma_n \otimes e^{i\lambda} \otimes 1). \tag{2.10}$$

Note that this is the parametrization of the principal series used in [11, Chapt. XI, §2]. The representation $\pi_{n,\lambda}$ acts in the Hilbert space $\mathcal{H}_{n,\lambda}$ whose subspace of

C^∞-vectors is given by

$$\begin{aligned}\mathcal{H}^\infty_{n,\lambda} = \{f \in C^\infty(G, V_{\sigma_n}) \colon f(gman) =& e^{-(i\lambda+1)(\log a)}\sigma_n(m)^{-1}f(g),\\ & g \in G,\ man \in P\}.\end{aligned} \tag{2.11}$$

If $\lambda \in \mathbb{R}$, then $\pi_{n,\lambda}$ is unitary. This family of representations is the unitary principal series. All $\pi_{n,\lambda}$, $n \in \mathbb{Z}$, $\lambda \in \mathbb{R}\backslash\{0\}$, are irreducible. They have an explicit realization [11, Chapt. II, §4]. The Casimir eigenvalue $\pi_{n,\lambda}(\Omega)$ is given by

$$\pi_{n,\lambda}(\Omega) = -\lambda^2 + \frac{n^2}{4} - 1. \tag{2.12}$$

This follows from [11, Theorem 8.22]. It can be also verified using the explicit realization of $\pi_{n,\lambda}$. In the latter case one has to take into account that the identification of $\mathfrak{a}_\mathbb{C}$ with $\mathbb{C}$ is differnt from ours.

The nonunitarily induced representations

$$\pi^c_x = \operatorname{Ind}_P^G(1 \otimes e^x \otimes 1), \quad 0 < x < 1, \tag{2.13}$$

are unitarizable. This is the complementary series. The Casimir eigenvalue is given by

$$\pi^c_x(\Omega) = x^2 - 1. \tag{2.14}$$

This also follows from [11, Theorem 8.22] or can be verified using the explicit realization of π^c_x (see [11, Chapt. II, §4]). Denote by $\Theta_{n,\lambda} = \operatorname{tr} \pi_{n,\lambda}$ the character of $\pi_{n,\lambda}$.

2.5. Flat vector bundles

Let $\tau \colon G \to \mathrm{GL}(V_\tau)$ be an irreducible finite-dimensional representation of G. Let E_τ be the flat vector bundle associated to the restriction $\tau|_\Gamma$ of τ to Γ. By [14, Proposition 3.1] E_τ is canonically isomorphic to the locally homogeneous vector bundle associated to $\tau|_K$, i.e.,

$$E_\tau \cong (\Gamma\backslash G \times V_\tau)/K, \tag{2.15}$$

where K acts on $\Gamma\backslash G \times V_\tau$ by $(\Gamma g, v) \cdot k = (\Gamma g k, \tau(k)^{-1}v)$. So we may regard E_τ as locally homogeneous vector bundle equipped with a flat connection which, of course, is different from the canonical invariant connection on the homogeneous bundle.

The vector bundle E_τ can be equipped with a canonical fiber metric. By [14, Lemma 3.1] there exists an inner product $\langle\cdot,\cdot\rangle$ on V_τ which satisfies

$$\begin{aligned}&\text{(a)} \qquad \langle\tau(Y)u, v\rangle = -\langle u, \tau(Y)v\rangle \text{ for all } Y \in \mathfrak{k},\ u, v \in V_\tau;\\ &\text{(b)} \qquad \langle\tau(Y)u, v\rangle = \langle u, \tau(Y)v\rangle \quad \text{for all } Y \in \mathfrak{p},\ u, v \in V_\tau.\end{aligned} \tag{2.16}$$

Such an inner product is called *admissible*. It is unique up to scaling. By (a) the inner product is invariant under $\tau(K)$ and therefore, it defines via (2.15) a Hermitian fiber metric in E_τ. Denote by $\Delta_p(\tau)$ the Laplacian on E_τ-valued p-forms with respect to an admissible metric on E_τ.

2.6. Regularized determinants

Let P be an elliptic differential operator acting on C^∞-sections of a smooth Hermitian vector bundle E over a compact Riemannian manifold X. The metrics g on X and h on E induce an inner product in $C^\infty(X,E)$. Suppose that with respect to this inner product the operator P is symmetric and nonnegative. Then the zeta function $\zeta(s;P)$, $s\in\mathbb{C}$, of P is defined as

$$\zeta(s;P)=\sum_{\lambda\in\operatorname{Spec}(P)\setminus\{0\}} m(\lambda)\lambda^{-s},$$

where $m(\lambda)$ denotes the multiplicity of the eigenvalue λ. The series converges absolutely and uniformly on compact subsets of $\operatorname{Re}(s)>\dim(X)/\operatorname{ord}(P)$. Moreover $\zeta(s;P)$ admits a meromorphic extension to $s\in\mathbb{C}$ which is holomorphic at $s=0$ (see [22, Chapt. II]). Then the regularized determinant $\det P$ of P is defined as

$$\det P=\exp\left(-\frac{d}{ds}\zeta(s;P)\big|_{s=0}\right). \tag{2.17}$$

Assume that P is symmetric and bounded from below. Let $\lambda\in\mathbb{R}$ be such that $P+\lambda>0$. Then $\det(P+\lambda)$ is defined by (2.17). Voros [24] has shown that the function $\lambda\mapsto\det(P+\lambda)$, defined for $\lambda\gg 0$, extends to an entire function $\det(P+s)$ of $s\in\mathbb{C}$ with zeros $-\lambda_j$ where $\lambda_j\in\operatorname{Spec}(P)$.

2.7. Ray-Singer analytic torsion

Finally we recall the definition of the Ray-Singer analytic torsion [21]. Let χ be a finite-dimensional representation of $\pi_1(X)$ and let $E_\chi\to X$ be the associated flat vector bundle over X. Pick a Hermitian fiber metric h in E_χ and let $\Delta_p(\chi)\colon\Lambda^p(X,E_\chi)\to\Lambda^p(X,E_\chi)$ be the Laplacian on the space of E_χ-valued p-forms. Then $\Delta_p(\chi)$ is a nonnegative, second-order elliptic differential operator. So it has a well-defined regularized determinant, defined by (2.17). Then the analytic torsion is defined as the following weighted product of regularized determinants

$$T_X(\chi;g,h)=\prod_{p=1}^{3}\left(\det\Delta_p(\chi)\right)^{(-1)^{p+1}p/2}. \tag{2.18}$$

By definition $T_X(\chi;g,h)$ depends on g and h. However, if $\dim X$ is odd and χ is acyclic, i.e., $H^*(X,E_\chi)=0$, then $T_X(\chi;g,h)$ is independent of g and h [19, Corollary 2.7]. In this case we denote it simply by $T_X(\chi)$.

In this paper we consider the special case where $X=\Gamma\backslash\mathbb{H}^3$ is a closed hyperbolic 3-manifold and χ is the restriction of a representation $\tau\colon G\to\operatorname{GL}(V_\tau)$ to Γ. Then, as explained above, the flat bundle E_τ carries an admissible metric. We denote the analytic torsion attached to $\tau|_\Gamma$ with respect to this metric by $T_X(\tau)$.

3. Twisted Ruelle and Selberg zeta functions

In this section we consider various kinds of twisted geometric zeta functions which are needed for the proof of our main result. We will use the notation introduced in Section 2. First we recall the following estimation of the growth of the length spectrum. For $R > 0$ we have

$$\#\{[\gamma] \in C(\Gamma) \colon \ell(\gamma) \leq R\} \ll e^{2R} \tag{3.1}$$

[6, (1.31)].

If T is an endomorphism of a finite-dimensional vector space, denote by $S^k T$ the kth symmetric power of T. Let $\overline{\mathfrak{n}} = \theta\mathfrak{n}$ be the negative root space. Then for $\sigma \in \hat{M}$ and $s \in \mathbb{C}$ with $\operatorname{Re}(s) > 1$ the twisted Selberg zeta function is defined by

$$Z(s,\sigma) = \prod_{\substack{[\gamma]\neq e\\ \text{prime}}} \prod_{k=0}^{\infty} \det\left(1 - \left(\sigma(m_\gamma) \otimes S^k\left(\operatorname{Ad}(m_\gamma a_\gamma)_{\overline{\mathfrak{n}}}\right)\right) e^{-(s+1)\ell(\gamma)}\right), \tag{3.2}$$

where $[\gamma]$ runs over the non-trivial primitive conjugacy classes in Γ. By [6, (3.6)] we have

$$\log Z(s,\sigma) = -\sum_{[\gamma]\neq e} \frac{\sigma(m_\gamma) e^{-\ell(\gamma)}}{\det(\operatorname{Id} - \operatorname{Ad}(m_\gamma a_\gamma)_{\overline{\mathfrak{n}}}) n_\Gamma(\gamma)}\, e^{-s\ell(\gamma)}. \tag{3.3}$$

It follows from (3.1) that the series converges absolutely and uniformly in the half-plane $\operatorname{Re}(s) > 1$. Therefore the infinite product converges absolutely and uniformly in the half-plane $\operatorname{Re}(s) > 1$. Furthermore by [6, Theorem 3.15] it has a meromorphic extension to the entire complex plane and satisfies a functional equation [6, Theorem 3.18]. To state the functional equation we need some notation. Let $w \in W_A$ be the non-trivial element. Then w acts on $\hat{M}$ by

$$(w\sigma)(m) = \sigma(m_w^{-1} m m_w), \quad m \in M,\, \sigma \in \hat{M},$$

where m_w is a representative of w in the normalizer of $\mathfrak{a}$ in K. Thus $w\sigma_k = \sigma_{-k}$, $k \in \mathbb{Z}$. For each $\sigma \in \hat{M}$ there is an associated Dirac operator $D_X(\sigma)$ acting in a Clifford bundle $E_\sigma \to X$ [6, p. 29]. Let $\eta(D_X(\sigma))$ denote the eta invariant of $D_X(\sigma)$. Let P_σ be the Plancherel polynomial with respect to σ. If the Haar measures are normalized as in [11, pp. 387–388] and $\mathfrak{a}_\mathbb{C}$ is identified with $\mathbb{C}$ by $z \in \mathbb{C} \mapsto zH \in \mathfrak{a}_\mathbb{C}$, then by [11, Theorem 11.2] (up to a minor correction) it is given by

$$P_{\sigma_k}(z) = \frac{1}{4\pi^2}\left(\frac{k^2}{4} - z^2\right), \quad k \in \mathbb{Z}, \tag{3.4}$$

where $\sigma_k \in \hat{M}$ is the character defined by (2.7). We note that our definition of $P_\sigma(z)$ differs from the definition of $P_\sigma(z)$ in [6, p. 56].

Then the functional equation satisfied by $Z(s,\sigma)$ is the following equality

$$Z(s,\sigma) = e^{i\pi\eta(D_X(\sigma))} \exp\left\{-4\pi \operatorname{vol}(X) \int_0^s P_\sigma(r)\,dr\right\} Z(-s, w\sigma) \tag{3.5}$$

[6, Theorem 3.18]. Our formula differs from the formula in [6, Theorem 3.18]. This is due to the different definition of $P_\sigma(z)$. Since the functional equation plays an important role in this paper, we will give a separate proof for the functional equation of the symmetrized Selberg zeta function in Section 5.

A related dynamical zeta function is the twisted Ruelle zeta function $R(s,\sigma)$ which is defined by

$$R(s,\sigma) = \prod_{\substack{[\gamma]\neq e\\ \text{prime}}} \left(1-\sigma(m_\gamma)e^{-s\ell(\gamma)}\right), \tag{3.6}$$

where, as above, $[\gamma]$ runs over the non-trivial primitive conjugacy classes in Γ. Note that $R(s,\sigma_0)$ equals the usual Ruelle zeta function

$$R(s) = \prod_{\substack{[\gamma]\neq e\\ \text{prime}}} \left(1-e^{-s\ell(\gamma)}\right). \tag{3.7}$$

Ruelle zeta functions of this type have been studied by Fried, and Bunke and Olbrich [6]. The two zeta functions are closely related. Namely the Ruelle zeta function can be expressed in terms Selberg zeta functions as follows.

Lemma 3.1. *For every $\sigma \in \hat{M}$ we have*

$$R(s,\sigma) = \frac{Z(s+1,\sigma)Z(s-1,\sigma)}{Z(s,\sigma\otimes\sigma_2)Z(s,\sigma\otimes\sigma_{-2})}. \tag{3.8}$$

Proof. By [6, (3.4)] we have

$$\log R(s,\sigma) = -\sum_{[\gamma]\neq e} \frac{\sigma(m_\gamma)}{n_\Gamma(\gamma)} e^{-s\ell(\gamma)}. \tag{3.9}$$

Using (3.3) we get

$$\begin{aligned}
&\log Z(s+1,\sigma) + \log Z(s-1,\sigma) - \log Z(s,\sigma\otimes\sigma_2) - \log Z(s,\sigma\otimes\sigma_{-2})\\
&\quad = -\sum_{[\gamma]\neq e} \frac{\sigma(m_\gamma)(1-\sigma_2(m_\gamma)e^{-\ell(\gamma)} - \sigma_{-2}(m_\gamma)e^{-\ell(\gamma)} + e^{-2\ell(\gamma)})}{\det(\mathrm{Id}-\mathrm{Ad}(m_\gamma a_\gamma)_{\bar{\mathfrak{n}}})n_\Gamma(\gamma)} e^{-s\ell(\gamma)}\\
&\quad = -\sum_{[\gamma]\neq e} \frac{\sigma(m_\gamma)}{n_\Gamma(\gamma)} e^{-s\ell(\gamma)}.
\end{aligned} \tag{3.10}$$

Together with (3.9) the lemma follows. □

Put

$$\theta_X(\sigma) := 2\eta(D_X(\sigma)) - \eta(D_X(\sigma\otimes\sigma_2)) - \eta(D_X(\sigma\otimes\sigma_{-2})), \quad \sigma\in\hat{M}.$$

We summarize the main properties of $R(s,\sigma)$ by the following proposition.

Proposition 3.2. *For each $\sigma \in \hat{M}$ we have*

1) *The infinite product* (3.6) *is absolutely convergent in the half-plane* $\mathrm{Re}(s) > 2$.

2) $R(s,\sigma)$ *admits a meromorphic extension to whole complex plane.*
3) $R(s,\sigma)$ *satisfies the following functional equation.*

$$R(s,\sigma)=e^{i\pi\theta_X(\sigma)}\exp\left(4\pi^{-1}\operatorname{vol}(\Gamma\backslash\mathbb{H}^3)s\right)R(-s,w\sigma). \tag{3.11}$$

Proof. 1) follows from the estimation (3.1). The meromorphic extension is established in [6, Chap. 4] and the functional equation is proved in [6, Theorem 4.5]. It follows from Lemma 3.1 and the functional equation of the Selberg zeta function. Namely using (3.8) and (3.5) we get

$$\frac{R(s,\sigma)}{R(-s,w\sigma)}=e^{i\pi\theta_X(\sigma)}\exp\Bigg(-4\pi\operatorname{vol}(X)\Bigg\{\int_0^{s+1}P_{\sigma_k}(r)\,dr+\int_0^{s-1}P_{\sigma_k}(r)$$
$$-\int_0^{s}P_{\sigma_{k+2}}(r)\,dr-\int_0^{s}P_{\sigma_{k-2}}(r)\,dr\Bigg\}\Bigg).$$

It follows from (3.4) by a simple computation that

$$\int_0^{s+1}P_{\sigma_k}(r)\,dr+\int_0^{s-1}P_{\sigma_k}(r)q-\int_0^{s}P_{\sigma_{k+2}}(r)\,dr-\int_0^{s}P_{\sigma_{k-2}}(r)\,dr=-\frac{s}{\pi^2}$$

which implies 3). □

Now let $\tau\colon G\to\mathrm{GL}(V)$ be a representation in a finite-dimensional complex vector space V. We fix a norm $\|\cdot\|$ in V. The restriction $\tau|_{MA}$ of τ to MA decomposes into characters:

$$\tau|_{MA}=\bigoplus_{k\in I}\sigma_k\otimes e^{\nu_k\alpha}, \tag{3.12}$$

where $I\subset\mathbb{Z}$ is finite and $\nu_k\in\frac{1}{2}\mathbb{Z}$. Let $c=\max\{|\nu_k|\colon k\in I\}$. Given $g\in G$, we denote by $a(g)\in A^+$ the A^+-component of g with respect to the Cartan decomposition $G=KA^+K$. It follows from (3.12) that there exists $C_1>0$ such that

$$\|\tau(g)\|\le C_1e^{c\alpha(\log a(g))},\quad g\in G.$$

This implies that there exists $c_2>0$ such that

$$\|\tau(\gamma)\|\le Ce^{c_2\ell(\gamma)},\quad \gamma\in\Gamma\setminus\{1\}.$$

Therefore, the infinite product

$$R_\tau(s)=\prod_{\substack{[\gamma]\neq e\\ \text{prime}}}\det\left(\mathrm{I}-\tau(\gamma)e^{-s\ell(\gamma)}\right) \tag{3.13}$$

is absolutely convergent in the half-plane $\operatorname{Re}(s)>c_2+2$. By (3.12) we have

$$\det\left(\mathrm{I}-\tau(\gamma)e^{-s\ell(\gamma)}\right)=\det\left(\mathrm{I}-\tau(m_\gamma a_\gamma)e^{-s\ell(\gamma)}\right)$$
$$=\prod_{k\in I}\det\left(1-\sigma_k(m_\gamma)e^{-(s-\nu_k)\ell(\gamma)}\right).$$

Taking the product of both sides over all non-trivial primitive conjugacy classes, we get

$$R_\tau(s) = \prod_{k\in I} R(s-\nu_k,\sigma_k), \quad \operatorname{Re}(s) > c_2 + 2. \tag{3.14}$$

The right-hand side is a meromorphic function on $\mathbb{C}$. This implies that $R_\tau(s)$ admits a meromorphic continuation to $\mathbb{C}$.

Using (3.8), it follows that $R_\tau(s)$ can also be expressed in terms of twisted Selberg zeta functions. This formula can be simplified using Kostant's Bott-Borel-Weil theorem [13] which we recall next. Let

$$\mu_p \colon MA \to \mathrm{GL}(\Lambda^p \mathfrak{n}_\mathbb{C}), \quad p = 0, 1, 2, \tag{3.15}$$

be the pth exterior power of the adjoint representation of MA on $\mathfrak{n}_\mathbb{C}$. It decomposes into characters as follows

$$\mu_0 = \sigma_0, \quad \mu_1 = (\sigma_2 \otimes e^\alpha) \oplus (\sigma_{-2} \otimes e^\alpha), \quad \mu_2 = \sigma_0 \otimes e^{2\alpha}. \tag{3.16}$$

Denote by $\tilde{\mu}_p$ the contragredient representation of the representation (3.15). Given $(m,n) \in \mathbb{Z}\times\mathbb{Z}$, we define a character $\chi_{(m,n)} \colon MA \to \mathbb{C}^\times$ by

$$\chi_{(m,n)} = \sigma_{m-n} \otimes e^{\frac{m+n}{2}\alpha}. \tag{3.17}$$

Lemma 3.3. *Let τ be an irreducible representation of G with highest weight $\Lambda_\tau \in \mathbb{N}_0 \times \mathbb{N}_0$. We have the following identity of characters of MA.*

$$\sum_{p=0}^{2} (-1)^p \operatorname{tr} \tilde{\mu}_p \cdot \operatorname{tr} \tau = \sum_{w\in W_G} (-1)^{\ell(w)} \chi_{w(\Lambda_\tau + \rho_G) - \rho_G}, \tag{3.18}$$

where $\ell(w)$ denotes the length of w.

Proof. Let $L = MA$. For a finite-dimensional L-module W denote by $\mathrm{ch}_L(W)$ the element in the character ring $R(L)$. Let V_τ be an irreducible G-module with highest weight Λ_τ. By the analog of a result of Kostant [13, Theorem 5.14], for real Lie algebras [4, Theorem III.3.1], [23], we have

$$\sum_{p=0}^{2} (-1)^p \mathrm{ch}_L(H^p(\mathfrak{n}, V_\tau)) = \sum_{w\in W_G} (-1)^{\ell(w)} \chi_{w(\Lambda_\tau + \rho_G) - \rho_G}, \tag{3.19}$$

where $H^p(\mathfrak{n}, V_\tau)$ denotes the Lie algebra cohomology. By the Poincaré principle [13, (7.2.3)] we have

$$\sum_{p=0}^{2} (-1)^p \mathrm{ch}_L(\Lambda^p \mathfrak{n}^* \otimes V_\tau) = \sum_{p=0}^{2} (-1)^p \mathrm{ch}_L(H^p(\mathfrak{n}, V_\tau)). \tag{3.20}$$

Here L acts on $\mathfrak{n}^*$ via the contragredient representation of the adjoint representation. Combining (3.19) and (3.20), the lemma follows. In fact, in the present case the lemma could also be proved by an elementary computation, using the parametrization (2.8). ◻

We are now ready to prove the formula which expresses $R_\tau(s)$ as a fraction of twisted Selberg zeta functions. For $w \in W_G$ write

$$\chi_{w(\Lambda_\tau+\rho_G)-\rho_G} = \sigma_{\tau,w} \otimes e^{(\lambda_{\tau,w}-1)\alpha}, \tag{3.21}$$

where $\sigma_{\tau,w} \in \hat{M}$ and $\lambda_{\tau,w} \in \mathbb{R}$.

Proposition 3.4. *Let τ be an irreducible finite-dimensional representation of G. Then we have*

$$R_\tau(s) = \prod_{w\in W_G} Z(s-\lambda_{\tau,w}, \sigma_{\tau,w})^{(-1)^{\ell(w)}}. \tag{3.22}$$

Proof. Recall that for an endomorphism W of a finite-dimensional vector space we have

$$\det(\mathrm{Id} - W) = \sum_{k=0}^{\infty}(-1)^k \operatorname{tr}(\Lambda^k W). \tag{3.23}$$

Let $m \in M$ and $a \in A$. Note that $\tilde{\mu}_p(ma) = \Lambda^p \operatorname{Ad}(ma)_{\bar{\mathfrak{n}}}$. Hence if we apply (3.23) to $\tilde{\mu}_p$ we get

$$\sum_{p=0}^{2}(-1)^p \operatorname{tr}\tilde{\mu}_p(ma) = \det\left(\mathrm{Id} - \operatorname{Ad}(ma)_{\bar{\mathfrak{n}}}\right).$$

Using (3.18) and (3.21), we get

$$\operatorname{tr}\tau(ma) = \sum_{w\in W_G}(-1)^{\ell(w)}\frac{\sigma_{\tau,w}(m)}{\det\left(\mathrm{Id} - \operatorname{Ad}(ma)_{\bar{\mathfrak{n}}}\right)}e^{(\lambda_{\tau,w}-1)\alpha(\log a)}. \tag{3.24}$$

Next we have

$$\begin{aligned} \log R_\tau(s) &= \sum_{\substack{[\gamma]\neq e\\ \text{prime}}} \operatorname{tr}\log\left(\mathrm{Id} - \tau(\gamma)e^{-s\ell(\gamma)}\right)\\ &= -\sum_{\substack{[\gamma]\neq e\\ \text{prime}}}\sum_{k=1}^{\infty}\frac{\operatorname{tr}\left(\tau(\gamma)e^{-s\ell(\gamma)}\right)^k}{k}\\ &= -\sum_{[\gamma]\neq e}\frac{\operatorname{tr}\tau(\gamma)}{n_\Gamma(\gamma)}e^{-s\ell(\gamma)}. \end{aligned} \tag{3.25}$$

Now let $\gamma \in \Gamma\setminus\{e\}$, $\gamma \sim m_\gamma a_\gamma$. Then $\log a_\gamma = \ell(\gamma)H$. Inserting (3.24) on the right-hand side of (3.25), we get

$$\log R_\tau(s) = \sum_{w\in W_G}(-1)^{\ell(w)+1}\sum_{[\gamma]\neq e}\frac{\sigma_{\tau,w}(m_\gamma)}{\det\left(\mathrm{Id} - \operatorname{Ad}(m_\gamma a_\gamma)_{\bar{\mathfrak{n}}}\right)n_\Gamma(\gamma)}e^{-(s-\lambda_{\tau,w}+1)\ell(\gamma)}$$

By [6, (3.6)], the right-hand side equals

$$\sum_{w\in W_G}(-1)^{\ell(w)}\log Z(s-\lambda_{\tau,w}, \sigma_{\tau,w}),$$

which proves the proposition. $\square$

We also need to consider symmetrized Ruelle and Selberg zeta functions. Recall that the nontrivial element $w_A \in W_A$ acts on $\hat{M}$ by $w_A\sigma_k = \sigma_{-k}$. Let $\sigma \in \hat{M} \setminus \{\sigma_0\}$. Put

$$S(s,\sigma) := Z(s,\sigma)Z(s,w_A\sigma). \tag{3.26}$$

This is the symmetrized Selberg zeta function. Let $\theta\colon G \to G$ be the Cartan involution. Put

$$\tau_\theta = \tau \circ \theta. \tag{3.27}$$

Note that $\tau_p = \mathrm{Sym}^p$ satisfies $\tau_p \circ \theta = \overline{\tau}_p$, and $\overline{\tau}_p \circ \theta = \tau_p$. Thus θ acts on the highest weights by

$$\theta(m,n) = (n,m), \quad (m,n) \in \mathbb{N}_0 \times \mathbb{N}_0. \tag{3.28}$$

By (2.8) it follows that an irreducible finite-dimensional representation τ of G satisfies $\tau_\theta = \tau$, if and only if $\tau = \tau_{m,m}$ for some $m \in \mathbb{N}_0$.

Proposition 3.5. *Let τ be an irreducible finite-dimensional representation of G. Then we have*

$$R_\tau(s)R_{\tau_\theta}(s) = \prod_{w\in W_G} S(s-\lambda_{\tau,w},\sigma_{\tau,w})^{(-1)^{\ell(w)}}, \quad \tau_\theta \not\cong \tau, \tag{3.29}$$

and

$$R_{\tau_{m,m}}(s) = Z(s-(m+1),\sigma_0)Z(s+m+1,\sigma_0)S(s,\sigma_{2m+2})^{-1}, \quad m \in \mathbb{N}_0. \tag{3.30}$$

Proof. Put

$$\Xi(\tau) = \{w(\Lambda_\tau + \rho_G) - \rho_G \colon w \in W_G\}.$$

Let $\tau = \tau_{m,n}$. Then we have

$$\Xi(\tau) = \{(m,n), (-(m+2),n), (m,-(n+2)), (-(m+2),-(n+2))\}.$$

By (3.17) and (3.21), it follows that

$$\begin{aligned}\{(\sigma_{\tau,w},\lambda_{\tau,w})\colon w \in W_G\} \\ = \{(\sigma_{m-n},(m+n)/2+1), (\sigma_{-(m+n+2)},(n-m)/2), \\ (\sigma_{m+n+2},(m-n)/2), (\sigma_{n-m},-(m+n)/2-1)\}.\end{aligned} \tag{3.31}$$

Assume that $m \neq n$. Using that $(\tau_{m,n})_\theta = \tau_{n,m}$ and (3.31), it follows that

$$\begin{aligned}\{(\sigma_{\tau,w},\lambda_{\tau,w})\colon w \in W_G\} \cup \{(\sigma_{\tau_\theta,w},\lambda_{\tau_\theta,w})\colon w \in W_G\} \\ = \{(\sigma_{\tau,w},\lambda_{\tau,w}), (w_A\sigma_{\tau,w},\lambda_{\tau,w})\colon w \in W_G\}.\end{aligned}$$

By (3.22) and (3.26), equality (3.29) follows. Now assume that $\tau_\theta = \tau$. By (3.28) there exists $m \in \mathbb{N}_0$ such that $\tau = \tau_{m,m}$. In this case we get

$$\begin{aligned}\{(\sigma_{\tau,w},\lambda_{\tau,w})\colon w \in W_G\} \\ = \{(\sigma_0,m+1), (\sigma_{-2(m+1)},0), (\sigma_{2(m+1)},0), (\sigma_0,-(m+1)\}.\end{aligned} \tag{3.32}$$

Using again (3.22) and (3.26), we get (3.30). □

4. Bochner-Laplace operators

In this section we study certain auxiliary elliptic operators which are needed to derive the determinant formula and the functional equation for the Selberg zeta function. These operators were first introduced by Bunke and Olbrich [6].

Let $w_A \in W_A$ be the nontrivial element. It acts on $\sigma_k \in \hat{M}$ by $w_A \sigma_k = \sigma_{-k}$. Thus, if $k \neq 0$, then σ_k is not W_A-invariant. For $l \in \mathbb{N}_0$ let $\nu_l \in \hat{K}$ denote the irreducible representation of $K = \mathrm{SU}(2)$ of highest weight l. Then we have

$$\nu_l|_M = \bigoplus_{k=0}^{l} \sigma_{l-2k}. \tag{4.1}$$

Let $R(K)$ and $R(M)$ denote the representation rings of K and M, respectively. The inclusion $i\colon M \to K$ induces the restriction map $i^*\colon R(K) \to R(M)$. From (4.1) we get

$$\begin{aligned} &i^*(\nu_l - \nu_{l-2}) = \sigma_l + \sigma_{-l}, \quad l \in \mathbb{N},\ l \geq 2; \\ &i^*(\nu_1) = \sigma_1 + \sigma_{-1}, \quad i^*(\nu_0) = \sigma_0. \end{aligned} \tag{4.2}$$

It follows from (4.2) that for every $\sigma \in \hat{M}$ there exists a unique $\xi_\sigma \in R(K)$ such that

$$i^*(\xi_\sigma) = \sigma + w_A \sigma. \tag{4.3}$$

Then we have

$$\xi_\sigma = \sum_{\nu \in \hat{K}} m_\nu(\sigma) \nu. \tag{4.4}$$

with $m_\nu(\sigma) \in \{0, \pm 1\}$ for $\sigma \neq \sigma_0$ and $m_{\nu_l}(\sigma_0) = 0$, if $l \neq 0$, and $m_{\nu_0}(\sigma_0) = 2$.

Given $\nu \in \hat{K}$, let $\tilde{E}_\nu$ denote the associated homogeneous vector bundle over G/K and $E_\nu = \Gamma \backslash \tilde{E}_\nu$ the corresponding locally homogeneous bundle over X. For $\sigma \in \hat{M}$ and $\nu \in \hat{K}$ let $m_\nu(\sigma)$ be defined by (4.4). Put

$$E(\sigma) = \bigoplus_{\substack{\nu \\ m_\nu(\sigma) \neq 0}} E_\nu. \tag{4.5}$$

This bundle has a canonical grading

$$E(\sigma) = E^+(\sigma) \oplus E^-(\sigma) \tag{4.6}$$

defined by the sign of $m_\nu(\sigma)$.

Let $\tilde{A}_\nu$ be the elliptic G-invariant differential operator on

$$C^\infty(G/K, \tilde{E}_\nu) \cong (C^\infty(G) \otimes V_\nu)^K$$

which is induced by $-\Omega$, where $\Omega \in \mathcal{Z}(\mathfrak{g}_\mathbb{C})$ is the Casimir element. Let

$$\tilde{\Delta}_\nu = (\nabla^\nu)^* \nabla^\nu$$

be the connection Laplacian associated to the canonical invariant connection ∇^ν of $\tilde{E}_\nu$. By [15, Proposition 1.1] we have

$$\tilde{A}_\nu = \tilde{\Delta}_\nu - \nu(\Omega_K), \tag{4.7}$$

where $\Omega_K \in \mathcal{Z}(\mathfrak{k}_{\mathbb{C}})$ is the Casimir element of K. Being G-invariant, $\tilde{A}_\nu$ descends to an elliptic operator

$$A_\nu \colon C^\infty(X, E_\nu) \to C^\infty(X, E_\nu). \tag{4.8}$$

It follows from (4.7) that A_ν is symmetric and bounded from below. For $l \in \mathbb{Z}$ put

$$c(\sigma_l) = \frac{l^2}{4} - 1. \tag{4.9}$$

Define the operator $A(\sigma)$ acting on $C^\infty(X, E(\sigma))$ by

$$A(\sigma) := \bigoplus_{\substack{\nu \\ m_\nu(\sigma) \neq 0}} A_\nu + c(\sigma). \tag{4.10}$$

Obviously, $A(\sigma)$ preserves the grading of $E(\sigma)$.

Since A_ν is symmetric and bounded from below, the heat operator e^{-tA_ν} is well defined and is a trace class operator. Given $\sigma \in \hat{M}$, put

$$K(t; \sigma) = \sum_{\nu \in \hat{K}} m_\nu(\sigma) \operatorname{Tr}(e^{-tA_\nu}), \tag{4.11}$$

where $m_\nu(\sigma)$ is defined by (4.4). Our next goal is to use the Selberg trace formula to express $K(t, \sigma)$ in terms of the length of the closed geodesics.

Let $\tilde{A}_\nu$ be the lift of A_ν to the universal covering $\tilde{X} = G/K$. It acts in the space of smooth sections of the homogeneous vector bundle $\tilde{E}_\nu$ associated to ν. With respect to the isomorphism $C^\infty(G/K, \tilde{E}_\nu) \cong (C^\infty(G) \otimes V_\nu)^K$ we have

$$\tilde{A}_\nu = -R(\Omega) \otimes \operatorname{Id}_{V_\nu}.$$

Let $e^{-t\tilde{A}_\nu}$, $t > 0$, the heat semigroup generated by $\tilde{A}_\nu$. This is a smoothing operator on

$$L^2(G/K, \tilde{E}_\nu) \cong (L^2(G) \otimes V_\nu)^K$$

which commutes with the action of G. Therefore it is of the form

$$\left(e^{-t\tilde{A}_\nu}\phi\right)(g) = \int_G H_t^\nu(g^{-1}g')\phi(g')\, dg', \quad \phi \in (L^2(G) \otimes V_\nu)^K, \quad g \in G,$$

where the kernel $H_t^\nu \colon G \to \operatorname{End}(V_\nu)$ is C^∞, L^2, and satisfies the covariance property

$$H_t^\nu(k^{-1}gk') = \nu(k)^{-1} \circ H_t^\nu(g) \circ \nu(k'), \quad k, k' \in K,\ g \in G. \tag{4.12}$$

Actually, a much stronger result holds. For $q > 0$ let $\mathcal{C}^q(G)$ be Harish-Cahndra's L^q-Schwartz space. Then we have

$$H_t^\nu \in (\mathcal{C}^q(G) \otimes \operatorname{End}(V_\tau))^{K \times K} \tag{4.13}$$

for all $q > 0$. The proof is similar to the proof of Proposition 2.4 in [1]. By standard arguments it follows that the kernel of the heat operator e^{-tA_ν} is given by

$$H^\nu(t; x, x') = \sum_{\gamma \in \Gamma} H_t^\nu(g^{-1}\gamma g'), \tag{4.14}$$

where $x, x' \in X$ and $x = \Gamma gK$ and $x' = \Gamma g'K$. Therefore the trace of the heat operator e^{-tA_ν} is given by

$$\operatorname{Tr}\left(e^{-tA_\nu}\right) = \int_X \operatorname{tr} H^\nu(t; x, x)\, dx,$$

where tr denotes the trace tr: $\operatorname{End}(E_{\nu,x}) \to \mathbb{C}$ for $x \in X$. Let

$$h_t^\nu(g) = \operatorname{tr} H_t^\nu(g).$$

Using (4.12) and (4.14), it follows that

$$\operatorname{Tr}\left(e^{-tA_\nu}\right) = \int_{\Gamma\backslash G} \sum_{\gamma\in\Gamma} h_t^\nu(g^{-1}\gamma g)\, dg. \tag{4.15}$$

Let R_Γ denote the right regular representation of G on $L^2(\Gamma\backslash G)$. Then (4.15) can be written as

$$\operatorname{Tr}\left(e^{-tA_\nu}\right) = \operatorname{Tr} R_\Gamma(h_t^\nu). \tag{4.16}$$

Let

$$h_t^\sigma = \sum_{\nu\in\hat{K}} m_\nu(\sigma) h_t^\nu. \tag{4.17}$$

Then by (4.11) and (4.16) we get

$$K(t;\sigma) = \operatorname{Tr} R_\Gamma\left(h_t^\sigma\right), \quad t > 0.$$

We can now apply the Selberg trace formula [27]. We use the notation introduced in Section 2. Let $\bar{\mathfrak{n}} = \theta(\mathfrak{n})$, where θ is the Cartan involution. For $\gamma \in \Gamma \setminus \{e\}$ put

$$D(\gamma) = e^{\ell(\gamma)} \det(\operatorname{Id} - \operatorname{Ad}(m_\gamma a_\gamma)_{\bar{\mathfrak{n}}}).$$

Let $\Theta_{n,\lambda}$ denote the character of the principal series representation $\pi_{n,\lambda}$, where $n \in \mathbb{Z}$ and $\lambda \in \mathbb{R}$. Then the Selberg trace formula gives

$$\begin{aligned} K(t;\sigma) &= \operatorname{Vol}(X) h_t^\sigma(e) \\ &\quad + \frac{1}{2\pi} \sum_{[\gamma]\neq e} \frac{\ell(\gamma)}{n_\Gamma(\gamma) D(\gamma)} \sum_{n\in\mathbb{Z}} \overline{\sigma_n(m_\gamma)} \int_{\mathbb{R}} \Theta_{n,\lambda}(h_t^\sigma) e^{-i\ell(\gamma)\lambda}\, d\lambda. \end{aligned} \tag{4.18}$$

Note that $h_t^\sigma(e)$ can also be expressed in terms of characters. By (4.13), each h_t^ν belongs to $\mathcal{C}^q(G)$ for all $q > 0$. Therefore h_t^σ is in $\mathcal{C}^q(G)$. Hence we can apply the Plancherel formula for G (see [11, Theorem 11.2]). With respect to the normalizations of Haar measures used in [11] and the definition of the Plancherel polynom (3.4), we have

$$h_t^\sigma(e) = \sum_{n\in\mathbb{Z}} \int_{\mathbb{R}} \Theta_{n,\lambda}(h_t^\sigma) P_{\sigma_n}(i\lambda)\, d\lambda. \tag{4.19}$$

To continue we need to compute the characters $\Theta_{n,\lambda}(h_t^\sigma)$. First by (4.17) we have

$$\Theta_{n,\lambda}(h_t^\sigma) = \sum_{\nu\in\hat{K}} m_\nu(\sigma) \Theta_{n,\lambda}(h_t^\nu), \tag{4.20}$$

which reduces the problem to the computation of $\Theta_{n,\lambda}(h_t^\nu)$. For any unitary representation π of G on a Hilbert space $\mathcal{H}_\pi$ set

$$\tilde{\pi}(H_t^\nu) = \int_G \pi(g) \otimes H_t^\nu(g)\, dg.$$

This defines a bounded operator on $\mathcal{H}_\pi \otimes V_\nu$. As in [1, pp. 160–161] it follows from (4.12) that relative to the splitting

$$\mathcal{H}_\pi \otimes V_\nu = (\mathcal{H}_\pi \otimes V_\nu)^K \oplus \left[(\mathcal{H}_\pi \otimes V_\nu)^K\right]^\perp,$$

$\tilde{\pi}(H_t^\nu)$ has the form

$$\tilde{\pi}(H_t^\nu) = \begin{pmatrix} \pi(H_t^\nu) & 0 \\ 0 & 0 \end{pmatrix}$$

with $\pi(H_t^\nu)$ acting on $(\mathcal{H}_\pi \otimes V_\tau)^K$. Then it follows as in [1, Corollary 2.2] that

$$\pi(H_t^\nu) = e^{t\pi(\Omega)}\, \mathrm{Id}, \tag{4.21}$$

where Id is the identity on $(\mathcal{H}_\pi \otimes V_\nu)^K$. Let $\{\xi_n\}_{n\in\mathbb{N}}$ and $\{e_j\}_{j=1}^m$ be orthonormal bases of $\mathcal{H}_\pi$ and V_ν, respectively. Then we have

$$\begin{aligned} \operatorname{Tr} \pi(H_t^\nu) &= \sum_{n=1}^\infty \sum_{j=1}^m \langle \pi(H_t^\nu)(\xi_n \otimes e_j), (\xi_n \otimes e_j)\rangle \\ &= \sum_{n=1}^\infty \sum_{j=1}^m \int_G \langle \pi(g)\xi_n, \xi_n\rangle \langle H_t^\nu(g) e_j, e_j\rangle\, dg \\ &= \sum_{n=1}^\infty \int_G h_t^\nu(g) \langle \pi(g)\xi_n, \xi_n\rangle\, dg \\ &= \operatorname{Tr} \pi(h_t^\nu). \end{aligned} \tag{4.22}$$

Together with (4.21) we get

$$\operatorname{Tr} \pi(h_t^\nu) = e^{t\pi(\Omega)} \dim (\mathcal{H}_\pi \otimes V_\nu)^K. \tag{4.23}$$

Now we consider a unitary principal series representation $\pi_{n,\lambda}$ Let $[\nu|_M : \sigma_n]$ denote the multiplicity of $\sigma_n \in \hat{M}$ in $\nu|_M$. It equals 0 or 1. For any representation π of G denote by $\pi^\vee$ the contragredient representation of π. By Frobenius reciprocity [11, p. 208], we have

$$\dim (\mathcal{H}_{n,\lambda} \otimes V_\nu)^K = [\pi_{n,\lambda}^\vee|_K : \nu] = [\pi_{-n,-\lambda}|_K : \nu] = [\nu|_M : \sigma_{-n}] = [\nu|_M : \sigma_n].$$

Combined with (4.23), we obtain

$$\Theta_{n,\lambda}(h_t^\nu) = e^{t\pi_{n,\lambda}(\Omega)} [\nu|_M : \sigma_n].$$

Using (4.20), (4.4) and (4.3), we get

$$\Theta_{n,\lambda}(h_t^\sigma) = e^{t\pi_{n,\lambda}(\Omega)} \sum_{\nu\in\hat{K}} m_\nu(\sigma)[\nu|_M : \sigma_n] = e^{t\pi_{n,\lambda}(\Omega)}[\sigma + w_A\sigma : \sigma_n]. \tag{4.24}$$

The Casimir eigenvalue $\pi_{n,\lambda}(\Omega)$ is given by (2.12). Using the definition of $c(\sigma)$ by (4.9) it can be written as

$$\pi_{n,\lambda}(\Omega) = -\lambda^2 + c(\sigma_n). \tag{4.25}$$

Now we can put our computations together. Let $k \in \mathbb{N}_0$. If we insert (4.24) in (4.18) and (4.19) and use (4.25), we get

$$\begin{aligned} K(t;\sigma_k) = e^{tc(\sigma_k)} \Bigg(& 2\operatorname{vol}(X) \int_{\mathbb{R}} e^{-t\lambda^2} P_{\sigma_k}(i\lambda)\, d\lambda \\ & + \sum_{[\gamma]\neq e} \frac{\ell(\gamma)}{n_\Gamma(\gamma)} L_{\text{sym}}(\gamma;\sigma_k) \frac{e^{-\ell(\gamma)^2/(4t)}}{(4\pi t)^{1/2}} \Bigg), \end{aligned} \tag{4.26}$$

where

$$L_{\text{sym}}(\gamma,\sigma) = \frac{(\sigma(m_\gamma) + (w_A\sigma)(m_\gamma))e^{-\ell(\gamma)}}{\det\left(\operatorname{Id} - \operatorname{Ad}(m_\gamma a_\gamma)_{\overline{\mathfrak{n}}}\right)}. \tag{4.27}$$

Using the definition of $A(\sigma)$ by (4.10) together with (4.11), we finally get

Proposition 4.1. *For every $\sigma \in \hat{M}$ we have*

$$\begin{aligned} \operatorname{Tr}_s\left(e^{-tA(\sigma)}\right) = & 2\operatorname{vol}(X) \int_{\mathbb{R}} e^{-t\lambda^2} P_\sigma(i\lambda)\, d\lambda \\ & + \sum_{[\gamma]\neq e} \frac{\ell(\gamma)}{n_\Gamma(\gamma)} L_{sym}(\gamma;\sigma) \frac{e^{-\ell(\gamma)^2/(4t)}}{(4\pi t)^{1/2}}. \end{aligned} \tag{4.28}$$

5. The functional equation of the Selberg zeta function

One of the main ingredients of the proof of Theorem 1.1 is the functional equation (3.5) satisfied by the Selberg zeta function $Z(s,\sigma)$. In particular, it is important to determine the sign in the exponential factor. We include a proof of the functional equation for the symmetrized Selberg zeta function which suffices for our purpose.

Let $\sigma \in \hat{M}$. Note that $A(\sigma)$ is a second-order elliptic differential operator on a compact manifold. Therefore it is essentially self-adjoint and the unique self-adjoint extension of $A(\sigma)$ has pure point spectrum consisting of a sequence of eigenvalues $\lambda_1 \le \lambda_2 \le \cdots \to \infty$ of finite multiplicities. It follows from Weyl's law that

$$\sum_{\lambda_i > 0} \lambda_i^{-2} < \infty. \tag{5.1}$$

Therefore the resolvent $(A(\sigma)+s^2)^{-1}$, $\operatorname{Re}(s^2) \gg 0$, is a Hilbert-Schmidt operator. Let $\operatorname{Re}(s_i^2) \gg 0$, $i = 1,2$. By the resolvent equation we have

$$(A(\sigma)+s_1^2)^{-1} - (A(\sigma)+s_2^2)^{-1} = (s_2^2 - s_1^2)(A(\sigma)+s_1^2)^{-1} \circ (A(\sigma)+s_2^2)^{-1}.$$

The right-hand side is a product of Hilbert-Schmidt operators and therefore, it is a trace class operator. Hence $(A(\sigma)+s_1^2)^{-1}-(A(\sigma)+s_2^2)^{-1}$ is a trace class operator. Now observe that

$$(A(\sigma)+s^2)^{-1}=\int_0^\infty e^{-ts^2}e^{-tA(\sigma)}\,dt.$$

Furthermore we have the heat expansion

$$\operatorname{Tr}\left(e^{-tA(\sigma)}\right)\sim\sum_{j\geq 0}a_j t^{-3/2+j} \tag{5.2}$$

as $t\to +0$. Let $\operatorname{Re}(s^2),\operatorname{Re}(s_0^2)\gg 0$. Then it follows from (5.2) that

$$\operatorname{Tr}_s\left((A(\sigma)+s^2)^{-1}-(A(\sigma)+s_0^2)^{-1}\right)=\int_0^\infty (e^{-ts^2}-e^{-ts_0^2})\operatorname{Tr}_s\left(e^{-tA(\sigma)}\right)\,dt. \tag{5.3}$$

Now we replace $\operatorname{Tr}_s\left(e^{-tA(\sigma)}\right)$ by the right-hand side of (4.28). First note that for $\operatorname{Re}(s)>0$ we have

$$\int_0^\infty e^{-ts^2}\frac{e^{-\ell(\gamma)^2/(4t)}}{\sqrt{4\pi t}}=\frac{1}{2s}e^{-s\ell(\gamma)} \tag{5.4}$$

Furthermore by Cauchy's theorem we have

$$\begin{aligned}\int_0^\infty &(e^{-ts^2}-e^{-ts_0^2})\left(\int_{\mathbb{R}}e^{-t\lambda^2}P_\sigma(i\lambda)\,d\lambda\right)dt\\ &=\int_{\mathbb{R}}\frac{s_0^2-s^2}{(\lambda^2+s^2)(\lambda^2+s_0^2)}P(i\lambda)\,d\lambda\\ &=\frac{\pi}{s}P_\sigma(s)-\frac{\pi}{s_0}P_\sigma(s_0).\end{aligned} \tag{5.5}$$

For the last equality we used that $P_\sigma(s)$ is an even polynomial. By (5.4) and (5.5), we get

$$\begin{aligned}\operatorname{Tr}_s\left((A(\sigma)+s^2)^{-1}-(A(\sigma)+s_0^2)^{-1}\right)&=2\pi\operatorname{vol}(X)\left(\frac{P_\sigma(s)}{s}-\frac{P_\sigma(s_0)}{s_0}\right)\\ +\frac{1}{2s}\sum_{[\gamma]\neq e}\frac{\ell(\gamma)}{n_\Gamma(\gamma)}L_{\mathrm{sym}}(\gamma;\sigma)e^{-s\ell(\gamma)}&-\frac{1}{2s_0}\sum_{[\gamma]\neq e}\frac{\ell(\gamma)}{n_\Gamma(\gamma)}L_{\mathrm{sym}}(\gamma;\sigma)e^{-s_0\ell(\gamma)}.\end{aligned} \tag{5.6}$$

By (3.3) we have

$$\sum_{[\gamma]\neq e}\frac{\ell(\gamma)}{n_\Gamma(\gamma)}L_{\mathrm{sym}}(\gamma;\sigma)e^{-s\ell(\gamma)}=\frac{Z'(s,\sigma)}{Z(s,\sigma)}+\frac{Z'(s,w_A\sigma)}{Z(s,w_A\sigma)},$$

which is the logarithmic derivative of the symmetrized Selberg zeta function $S(s,\sigma)$ defined by (3.26). Thus we get

$$\begin{aligned}\operatorname{Tr}_s\left((A(\sigma)+s^2)^{-1}-(A(\sigma)+s_0^2)^{-1}\right) =&2\pi\operatorname{vol}(X)\left(\frac{P_\sigma(s)}{s}-\frac{P_\sigma(s_0)}{s_0}\right)\\ &+\frac{1}{2s}\frac{S'(s,\sigma)}{S(s,\sigma)}-\frac{1}{2s_0}\frac{S'(s_0,\sigma)}{S(s_0,\sigma)}.\end{aligned} \tag{5.7}$$

Put

$$\Xi(s,\sigma)=\exp\left(4\pi\operatorname{vol}(X)\int_0^s P_\sigma(r)\,dr\right)\,S(s,\sigma).$$

Then (5.7) can be rewritten as

$$\operatorname{Tr}_s\left((A(\sigma)+s^2)^{-1}-(A(\sigma)+s_0^2)^{-1}\right)=\frac{1}{2s}\frac{\Xi'(s,\sigma)}{\Xi(s,\sigma)}-\frac{1}{2s_0}\frac{\Xi'(s_0,\sigma)}{\Xi(s_0,\sigma)}. \tag{5.8}$$

From this equality one can deduce the existence of the meromorphic extension of $S(s,\sigma)$ and determine the location of the singularities, i.e., zeros and poles of $S(s,\sigma)$. Let $\lambda_1<\lambda_2<\cdots$ be the eigenvalues of $A(\sigma)$. For each λ_j let $\mathcal{E}(\lambda_j)$ be the eigenspace of $A(\sigma)$ with eigenvalue λ_j. Put

$$m_s(\lambda_j,\sigma)=\dim_{\mathrm{gr}}\mathcal{E}(\lambda_j).$$

If $\lambda_j<0$, we choose the square root $\sqrt{\lambda_j}$ which has positive imaginary part. Put

$$s_j^\pm=\pm i\sqrt{\lambda_j},\quad j\in\mathbb{N}.$$

Proposition 5.1. *The Selberg zeta function $S(s,\sigma)$, defined by* (3.26), *has a meromorphic extension to $s\in\mathbb{C}$. The set of singularities of $S(s,\sigma)$ equals $\{s_j^\pm : j\in\mathbb{N}\}$. If $\lambda_j\neq 0$, then the order of $S(s,\sigma)$ at both s_j^+ and s_j^- is equal to $m_s(\lambda_j,\sigma)$. The order of the singularity at $s=0$ is $2m_s(0,\sigma)$.*

Proof. The left-hand side of (5.8) equals

$$\sum_{j=1}^\infty m_s(\lambda_j,\sigma)\left\{\frac{1}{s^2+\lambda_j}-\frac{1}{s_0^2+\lambda_j}\right\}.$$

By (5.1) the series converges absolutely and uniformly on compact subsets which shows that it is a meromorphic function of $s\in\mathbb{C}$ and the only poles are simple and occur exactly at the points $\{s_j^\pm : j\in\mathbb{N}\}$. Hence the logarithmic derivative of $\Xi(s,\sigma)$ is a meromorphic function with the same poles. Let $\lambda_j\neq 0$. Then

$$\frac{2s}{s^2+\lambda_j}=\frac{1}{s-s_j^+}+\frac{1}{s-s_j^-}.$$

It follows that $s_j^\pm$ are simple poles of $\Xi'(s,\sigma)\cdot\Xi(s,\sigma)^{-1}$ with residue $m_s(\lambda_j,\sigma)$. Hence $\Xi(s,\sigma)$ has a meromorphic extension to $\mathbb{C}$ and the order of $\Xi(s,\sigma)$ at $s_j^\pm$ equals $m_s(\lambda_j,\sigma)$. In the same way it follows that the order of $\Xi(s,\sigma)$ at $s=0$ equals $2m_s(0,\sigma)$. □

Now subtract from (5.8) the same equation for $-s$ and multiply by $2s$. Then we get

$$\frac{\Xi'(s,\sigma)}{\Xi(s,\sigma)} + \frac{\Xi'(-s,\sigma)}{\Xi(-s,\sigma)} = 0,$$

which shows that the logarithmic derivative of $\Xi(s)\cdot\Xi(-s)^{-1}$ equals zero. Therefore $\Xi(s)\cdot\Xi(-s)^{-1}$ is constant. By Proposition 5.1 the order of $S(s,\sigma)$ at zero is even. Hence

$$\lim_{s\to 0}\frac{\Xi(s)}{\Xi(-s)} = 1.$$

This implies $\Xi(s) = \Xi(-s)$. Since $P_\sigma(z)$ is even, we obtain the following functional equation for $S(s,\sigma)$:

$$S(s,\sigma) = \exp\left(-8\pi \operatorname{vol}(X)\int_0^s P_\sigma(r)\,dr\right) S(-s,\sigma). \tag{5.9}$$

Note that $\overline{Z(s,\sigma_m)} = Z(\overline{s},\sigma_{-m})$. Hence for $s\in\mathbb{R}$ we have

$$S(s,\sigma) = |Z(s,\sigma)|^2.$$

Then (5.9) is reduced to

$$|Z(s,\sigma)| = \exp\left(-4\pi \operatorname{vol}(X)\int_0^s P_\sigma(r)\,dr\right)|Z(-s,\sigma)|, \quad s\in\mathbb{R}. \tag{5.10}$$

6. The determinant formula

By [6, Theorem 3.19] the twisted Selberg zeta function can be expressed as a graded regularized determinant of $A(\sigma)$. We include a simple proof of this formula for our case.

First we recall the notion of the graded regularized determinant of an elliptic self-adjoint operator. Let $E = E^+\oplus E^-$ be a $\mathbb{Z}/2\mathbb{Z}$-graded Hermitian vector bundle over a compact Riemannian manifold. Let $P\colon C^\infty(Y,E)\to C^\infty(Y,E)$ be an elliptic differential operator which is symmetric and bounded from below. Assume that P preserves the grading, i.e., assume that with respect to the decomposition

$$C^\infty(Y,E) = C^+(Y,E^+)\oplus C^\infty(Y,E^-)$$

P takes the form

$$P = \begin{pmatrix} P^+ & 0 \\ 0 & P^- \end{pmatrix}.$$

Then we define the graded determinant $\det_{\mathrm{gr}}(P)$ of P by

$$\det{}_{\mathrm{gr}}(P) = \frac{\det(P^+)}{\det(P^-)}. \tag{6.1}$$

Given $\sigma\in\hat{M}$, let $A(\sigma)$ be the elliptic operator defined by (4.10). It acts in a graded vector bundle. Hence the graded determinant $\det_{\mathrm{gr}}(s^2 + A(\sigma))$ is defined. Let $P_\sigma(r)$ be the Plancherel polynomial (3.4). By [6, Theorem 3.19] the twisted symmetrized Selberg zeta function $S(s,\sigma)$ can be expressed by the graded determinant as follows.

Proposition 6.1. *We have*

$$S(s;\sigma)=\det\nolimits_{\mathrm{gr}}\left(s^2+A(\sigma)\right)\exp\left(-4\pi\operatorname{vol}(X)\int_0^s P_\sigma(r)\,dr\right),\quad \sigma\neq\sigma_0,\tag{6.2}$$

and

$$Z(s;\sigma_0)=\det\left(s^2-1+\Delta\right)\exp\left((6\pi)^{-1}\operatorname{vol}(X)s^3\right),\tag{6.3}$$

where Δ is the Laplace operator on $C^\infty(X)$ and det *is the usual regularized determinant.*

Proof. We give a simple proof of this formula. Let $\sigma\in\hat{M}$. For $\operatorname{Re}(s^2)\gg 0$ let

$$\zeta(z,s)=\int_0^\infty e^{-ts^2}\operatorname{Tr}_s\left(e^{-tA(\sigma)}\right)t^{z-1}\,dt.\tag{6.4}$$

The integral converges absolutely and uniformly on compact subsets of the half-plane $\operatorname{Re}(z)>3/2$. It admits an extension to a meromorphic function of $z\in\mathbb{C}$ which is differentiable in s. It is regular at $z=0$ and we have

$$\zeta(z,s)=-\log\det\nolimits_{\mathrm{gr}}(A(\sigma)+s^2)+O(z).\tag{6.5}$$

Furthermore for $\operatorname{Re}(z)>3/2$ we have

$$-\frac{1}{2s}\frac{d}{ds}\zeta(z,s)=\int_0^\infty e^{-ts^2}\operatorname{Tr}_s\left(e^{-tA(\sigma)}\right)t^z\,dt.\tag{6.6}$$

Thus by (6.5) we get

$$\begin{aligned}\frac{1}{2s}\frac{d}{ds}\log\det\nolimits_{\mathrm{gr}}\left(A(\sigma)+s^2\right)&-\frac{1}{2s_0}\frac{d}{ds}\log\det\nolimits_{\mathrm{gr}}\left(A(\sigma)+s^2\right)\big|_{s=s_0}\\ &=\lim_{z\to0}\left(-\frac{1}{2s}\frac{d}{ds}\zeta(z,s)+\frac{1}{2s_0}\frac{d}{ds}\zeta(z,s)\big|_{s=s_0}\right)\\ &=\int_0^\infty(e^{-ts^2}-e^{-ts_0^2})\operatorname{Tr}_s\left(e^{-tA(\sigma)}\right)\,dt.\end{aligned}\tag{6.7}$$

Assume that $\sigma\neq\sigma_0$. Together with (5.3) and (5.7) we get

$$\frac{d}{ds}\log\det\nolimits_{\mathrm{gr}}\left(A(\sigma)+s^2\right)=\frac{d}{ds}\log S(s,\sigma)+4\pi\operatorname{vol}(X)P_\sigma(s)+bs\tag{6.8}$$

for some $b\in\mathbb{C}$. Integrating this equality gives

$$\log S(s,\sigma)=-4\pi\operatorname{vol}(X)\int_0^s P_\sigma(r)\,dr+\log\det\nolimits_{\mathrm{gr}}\left(A(\sigma)+s^2\right)+\frac{b}{2}s^2+c\tag{6.9}$$

for some $c\in\mathbb{C}$. In order to determine the constants b and c we take $s\in\mathbb{R}$ and consider the asymptotic behavior of both sides of (6.9) as $s\to\infty$. By (3.3) it follows that $\log S(s,\sigma)\to0$ as $s\to\infty$. Next consider the behavior of $\log\det_{\mathrm{gr}}\left(A(\sigma)+s^2\right)$ for $s\in\mathbb{R}$ and $s\to\infty$. By (6.5) we have

$$\log\det\nolimits_{\mathrm{gr}}\left(A(\sigma)+s^2\right)=-\frac{d}{dz}\left(z\zeta(z,s)\right)\big|_{z=0}.\tag{6.10}$$

Denote the first term on the right-hand side of (4.28) by $I(t,\sigma)$ and the second by $H(t,\sigma)$. Let $\mathrm{Re}(z)>3/2$. Then by (4.28) and (6.4) we get

$$\zeta(z,s)=\int_0^\infty e^{-ts^2}I(t,\sigma)t^{z-1}\,dt+\int_0^\infty e^{-ts^2}H(t,\sigma)t^{z-1}\,dt. \tag{6.11}$$

It follows from the definition of $H(t,\sigma)$ that the integral

$$\int_0^\infty e^{-ts^2}H(t,\sigma)t^{z-1}\,dt$$

is an entire function of $z\in\mathbb{C}$ and for every compact subset $\omega\subset\mathbb{C}$ there exist $C,c>0$ such that

$$\left|\frac{d}{dz}\int_0^\infty e^{-ts^2}H(t,\sigma)t^{z-1}\,dt\right|\leq C\,e^{-cs^2}\quad z\in\omega,\ s\geq 0. \tag{6.12}$$

To deal with the first integral on the right-hand side of (6.11), we note that

$$\int_0^\infty e^{-ts^2}\left(\int_{\mathbb{R}}e^{-t\lambda^2}\lambda^{2j}\,d\lambda\right)dt=\Gamma(j+1/2)\Gamma(-j-1/2+z)s^{2j-2z+1}. \tag{6.13}$$

By (3.4) the Plancherel polynomial $P_\sigma(z)$ is of the form $P_\sigma(z)=a_1+a_2z^2$. Using the definition of $I(t,\sigma)$ and (6.13), we get

$$\begin{aligned}\frac{d}{dz}\left(z\int_0^\infty e^{-ts^2}I(t,\sigma)t^{z-1}\,dt\right)\bigg|_{z=0}&=-4\pi\operatorname{vol}(X)\left(a_1s-\frac{a_2}{3}s^3\right)\\&=-4\pi\operatorname{vol}(X)\int_0^s P_\sigma(r)\,dr.\end{aligned} \tag{6.14}$$

Together with (6.10), (6.11), and (6.12) we obtain

$$\log\det\nolimits_{\mathrm{gr}}\left(A(\sigma)+s^2\right)=4\pi\operatorname{vol}(X)\int_0^\infty P_\sigma(r)\,dr+O(e^{-cs^2})$$

for $s\in\mathbb{R}$, $s\to\infty$. This implies that the constants b and c in (6.9) are zero. Exponentiating (6.9), we get (6.2). The proof of (6.3) is similar. □

Remark. From the statement of Theorem 3.19 in [6] it is not apparent that the determinant is the graded determinant. However, it is the general understanding in [6] that the trace of a trace class operator on a graded bundle is the super trace corresponding to the grading (see [6, p. 29]). Consequently regularized determinants of elliptic operators on graded bundles are always understood in [6] as graded determinants.

Now let τ be an irreducible, finite-dimensional representation of G with highest weight $\Lambda_\tau=(m,n)$. For $w\in W_G$ let $\sigma_{\tau,w}\in\hat{M}$ and $\lambda_{\tau,w}$ be defined by (3.21). Let

$$\Delta(w)=\bigoplus_{\substack{\nu\\ m_\nu(\sigma_{\tau,w})\neq 0}}A_\nu+\tau(\Omega). \tag{6.15}$$

This is an elliptic operator acting on $C^\infty(X,E(\sigma_{\tau,w}))$.

Using (3.31), an explicit computation shows that for all $w \in W_G$ we have

$$\lambda_{\tau,w}^2 + c(\sigma_{\tau,w}) = \frac{1}{2}\left(m(m+2) + n(n+2)\right) = \tau(\Omega). \tag{6.16}$$

Using (6.16), and (6.15), it follows that

$$A(\sigma_{\tau,w}) + \lambda_{\tau,w}^2 = \Delta(w) \tag{6.17}$$

as operators on $C^\infty(X, E(\sigma_{\tau,w}))$. Then it follows from (6.2) that

$$\begin{aligned} S(s - \lambda_{\tau,w}; \sigma_{\tau,w}) = &\det_{\mathrm{gr}}(s^2 - 2\lambda_{\tau,w}s + \Delta(w)) \\ &\cdot \exp\left(-4\pi \operatorname{vol}(X) \int_0^{s-\lambda_{\tau,w}} P_{\sigma_{\tau,w}}(r)\, dr\right), \end{aligned}$$

if $\sigma_{\tau,w} \neq \sigma_0$. If $\sigma_{\tau,w} = \sigma_0$, we use (6.3), which leads to a similar formula

Proposition 6.2. *Let $\tau_\theta \not\cong \tau$. There is a constant $c = c(\tau)$ such that*

$$R_\tau(s) R_{\tau_\theta}(s) = e^{c \operatorname{vol}(X) s} \prod_{w \in W_G} \det_{\mathrm{gr}}(s^2 - 2\lambda_{\tau,w}s + \Delta(w))^{(-1)^{\ell(w)}}. \tag{6.18}$$

Proof. By assumption we have $\tau = \tau_{m,n}$ with $m \neq n$. It follows from (3.31) that $\sigma_{\tau,w} \not\cong \sigma_0$ for all $w \in W_G$. Put

$$F(s) = \sum_{w \in W_G} (-1)^{\ell(w)+1} \int_0^{s-\lambda_{\tau,w}} P_{\sigma_{\tau,w}}(r)\, dr. \tag{6.19}$$

Then it follows from (3.29) and (6.2) that

$$R_\tau(s) R_{\tau_\theta}(s) = e^{-4\pi \operatorname{vol}(X) F(s)} \prod_{w \in W_G} \det_{\mathrm{gr}}\left(s^2 - 2s\lambda_{\tau,w} + \Delta(w)\right)^{(-1)^{\ell(w)}}. \tag{6.20}$$

Using (3.4) and (3.31), an explicite computation gives

$$F(s) = -\pi^{-2}(m+1)(n+1)s.$$

□

Now we consider the case $\tau_\theta = \tau$. Then $\tau = \tau_{m,m}$ for some $m \in \mathbb{N}_0$.

Proposition 6.3. *Let $m \in \mathbb{N}_0$. There exists a constant $c = c(m)$ such that*

$$\begin{aligned} R_{\tau_{m,m}}(s) = &e^{c \operatorname{vol}(X) s} \\ &\cdot \frac{\det\left((s+m+1)^2 - 1 + \Delta\right) \det\left((s-m-1)^2 - 1 + \Delta\right)}{\det_{\mathrm{gr}}\left(s^2 + A(\sigma_{2m+2})\right)}. \end{aligned} \tag{6.21}$$

Proof. Put

$$F_m(s) = \frac{1}{6\pi}\left((s+m+1)^3 + (s-m-1)^3\right) + 4\pi \int_0^s P_{\sigma_{2m+2}}(r)\, dr.$$

Using (3.30), (6.2) and (6.3), it follows that

$$R_{\tau_{m,m}}(s) = e^{\operatorname{vol}(X) F_m(s)} \frac{\det\left((s+m+1)^2 - 1 + \Delta\right) \det\left((s-m-1)^2 - 1 + \Delta\right)}{\det_{\mathrm{gr}}\left(s^2 + A(\sigma_{2m+2})\right)}.$$

Using (3.4), it follows that $F_m(s) = 2\pi^{-1}(m+1)^2 s$. □

7. Proof of Theorem 1.5

Since [28] has not been published yet, we include a proof of Theorem 1.5. Let $\tau\colon G \to \mathrm{GL}(V_\tau)$ be an irreducible finite-dimensional representation with associated flat bundle E_τ equipped with an admissible metric. Let $\Delta_p(\tau)$ be the Laplacian on E_τ-valued p-forms. Let

$$K(t,\tau) := \sum_{p=1}^{3}(-1)^p p \operatorname{Tr}\left(e^{-t\Delta_p(\tau)}\right) \tag{7.1}$$

and

$$q(\tau) = \sum_{p=1}^{3}(-1)^p p \dim\ker H^p(X, E_\tau).$$

Then by definition of the analytic torsion we have

$$\log T_X(\tau) = \frac{1}{2}\frac{d}{ds}\left(\frac{1}{\Gamma(s)}\int_0^\infty (K(t,\tau) - q(\tau))\, t^{s-1}\, dt\right)\bigg|_{s=0}, \tag{7.2}$$

where the right-hand side is defined near $s = 0$ by analytic continuation of the Mellin transform. The first step of the proof is to apply the trace formula to express $K(t,\tau)$ in terms of the length of closed geodesics. This is the basis for the relation between analytic torsion and the twisted Ruelle zeta function.

Let $\mathfrak{p}$ be the orthogonal complement of $\mathfrak{k}$ in $\mathfrak{g}$ with respect to the Killing form. Let $x_0 = eK$. Recall that there is a canonical isomorphism $T_{x_0}(G/K) \cong \mathfrak{p}$. Let R_Γ denote the right regular representation of G on $L^2(\Gamma\backslash G)$ (resp. $C^\infty(\Gamma\backslash G)$). Using (2.15), we get a canonical isomorphism

$$\Lambda^p(X, E_\tau) \cong \left(C^\infty(\Gamma\backslash G) \otimes \Lambda^p\mathfrak{p}^* \otimes V_\tau\right)^K, \tag{7.3}$$

where K acts by $k \in K \mapsto R_\Gamma(k)\otimes\Lambda^p \operatorname{Ad}_{\mathfrak{p}}^*(k)\otimes\tau(k)$. There is a similar isomorphism for the space $L^2\Lambda^p(X, E_\tau)$ of L^2-sections of $\Lambda^p T^*X \otimes E_\tau$. With respect to the isomorphism (7.3), we have the following generalization of Kuga's lemma

$$\Delta_p(\tau) = -R_\Gamma(\Omega)\otimes \mathrm{Id} + \tau(\Omega)\,\mathrm{Id}, \tag{7.4}$$

(see [14, (6.9)]), where Ω is the Casimir element and $\tau(\Omega)$ is the Casimir eigenvalue of τ.

Let $\tilde{\Delta}_p(\tau)$ be the lift of $\Delta_p(\tau)$ to the universal covering $\tilde{X} = G/K$. Let $e^{-t\tilde{\Delta}_p(\tau)}$, $t > 0$, be the corresponding heat semigroup. This is a smoothing operator on

$$L^2\Lambda^p(\tilde{X}; \tilde{E}_\tau) \cong \left(L^2(G)\otimes\Lambda^p\mathfrak{p}^*\otimes V_\tau\right)^K,$$

which commutes with the action of G. Therefore, it is of the form

$$\left(e^{-t\tilde{\Delta}_p(\tau)}\phi\right)(g) = \int_G H_t^{\tau,p}(g^{-1}g')\phi(g')\, dg', \quad g \in G,$$

where $\phi \in (L^2(G)\otimes\Lambda^p\mathfrak{p}^*\otimes V_\tau)^K$, and the kernel $H_t^{\tau,p}\colon G \to \operatorname{End}(\Lambda^p\mathfrak{p}^*\otimes V_\tau)$ belongs to $C^\infty \cap L^2$ and satisfies the covariance property

$$H_t^{\tau,p}(k^{-1}gk') = \nu_p(\tau)(k)^{-1} H_t^{\tau,p}(g)\nu_p(\tau)(k'), \tag{7.5}$$

with respect to the representation

$$\nu_p(\tau) := \Lambda^p \operatorname{Ad}_K^* \otimes \tau \colon K \to \operatorname{GL}(\Lambda^p \mathfrak{p}^* \otimes V_\tau). \tag{7.6}$$

Moreover, for all $q > 0$ we have

$$H_t^{\tau,p} \in (\mathcal{C}^q(G) \otimes \operatorname{End}(\Lambda^p \mathfrak{p}^* \otimes V_\tau))^{K \times K}, \tag{7.7}$$

where $\mathcal{C}^q(G)$ denotes Harish-Chandra's L^q-Schwartz space. The proof is similar to the proof of Proposition 2.4 in [1]. Let

$$h_t^{\tau,p}(g) = \operatorname{tr} H_t^{\tau,p}(g).$$

Repeating the arguments which we used to prove (4.16), we get

$$\operatorname{Tr}\left(e^{-t\Delta_p(\tau)}\right) = \operatorname{Tr} R_\Gamma(h_t^{\tau,p}). \tag{7.8}$$

Put

$$k_t^\tau = \sum_{p=1}^{3} (-1)^p p\, h_t^{\tau,p}. \tag{7.9}$$

By (7.1) we have

$$K(t,\tau) = \operatorname{Tr} R_\Gamma(k_t^\tau).$$

We can now apply the Selberg trace formula [27]. Let the notation be as in (4.18). Then we get

$$\begin{aligned} K(t,\tau) &= \operatorname{Vol}(X) k_t^\tau(e) \\ &\quad + \frac{1}{2\pi} \sum_{[\gamma]\neq e} \frac{\ell(\gamma)}{n_\Gamma(\gamma) D(\gamma)} \sum_{n\in\mathbb{Z}} \overline{\sigma_n(m_\gamma)} \int_{\mathbb{R}} \Theta_{n,\lambda}(k_t^\tau) e^{-i\ell(\gamma)\lambda}\, d\lambda, \end{aligned} \tag{7.10}$$

where the notation is the same as in (4.18). The characters $\Theta_{n,\lambda}(k_t^\tau)$ can be computed in the same way as in Section 4. Let π be a unitary representation of G on a Hilbert space $\mathcal{H}_\pi$. Set

$$\tilde{\pi}(H_t^{\tau,p}) = \int_G \pi(g) \otimes H_t^{\tau,p}(g)\, dg.$$

This defines a bounded operator on $\mathcal{H}_\pi \otimes \Lambda^p \mathfrak{p}^* \otimes V_\tau$. As in [1, pp. 160–161] it follows from (7.5) that relative to the splitting

$$\mathcal{H}_\pi \otimes \Lambda^p \mathfrak{p}^* \otimes V_\tau = (\mathcal{H}_\pi \otimes \Lambda^p \mathfrak{p}^* \otimes V_\tau)^K \oplus \left[(\mathcal{H}_\pi \otimes \Lambda^p \mathfrak{p}^* \otimes V_\tau)^K\right]^\perp,$$

$\tilde{\pi}(H_t^{\tau,p})$ has the form

$$\tilde{\pi}(H_t^{\tau,p}) = \begin{pmatrix} \pi(H_t^{\tau,p}) & 0 \\ 0 & 0 \end{pmatrix}$$

with $\pi(H_t^{\tau,p})$ acting on $(\mathcal{H}_\pi \otimes \Lambda^p \mathfrak{p}^* \otimes V_\tau)^K$. Using (7.4) it follows as in [1, Corollary 2.2] that

$$\pi(H_t^{\tau,p}) = e^{t(\pi(\Omega)-\tau(\Omega))} \operatorname{Id}$$

on $(\mathcal{H}_\pi \otimes \Lambda^p \mathfrak{p}^* \otimes V_\tau)^K$. As in (4.22) we get

$$\operatorname{tr} \pi(h_t^{\tau,p}) = \operatorname{tr} \pi(H_t^{\tau,p}) = e^{t(\pi(\Omega)-\tau(\Omega))} \dim(\mathcal{H}_\pi \otimes \Lambda^p \mathfrak{n}^* \otimes V_\tau)^K. \tag{7.11}$$

Now let π be a unitary principal series representation $\pi_{n,\lambda}$ acting in the Hilbert space $\mathcal{H}_{n,\lambda}$. Using (7.11) and (2.12) we get

$$\Theta_{n,\lambda}(h_t^{\tau,p}) = e^{-t(\lambda^2+1-n^2/4+\tau(\Omega))} \dim \left(\mathcal{H}_{n,\lambda} \otimes \Lambda^p \mathfrak{p}^* \otimes V_\tau\right)^K. \tag{7.12}$$

Denote by $\mathbb{C}_n$ the M-module defined by σ_n. By Frobenius reciprocity [11, p. 208] we have

$$\dim \left(\mathcal{H}_{n,\lambda} \otimes \Lambda^p \mathfrak{p}^* \otimes V_\tau\right)^K = \dim \left(\mathbb{C}_n \otimes \Lambda^p \mathfrak{p}^* \otimes V_\tau\right)^M.$$

and by (7.9) we get

$$\Theta_{n,\lambda}(k_t^\tau) = e^{-t(\lambda^2+1-n^2/4+\tau(\Omega))} \sum_{p=1}^{3} (-1)^p p \dim \left(\mathbb{C}_n \otimes \Lambda^p \mathfrak{p}^* \otimes V_\tau\right)^M.$$

Choose an orthonormal basis of $\mathfrak{p}$ as in [16, p. 9]. Using this basis it follows that as M-modules, $\mathfrak{p}$ and $\mathfrak{a} \oplus \mathfrak{n}$ are equivalent. Thus we get

$$\sum_{p=1}^{3} (-1)^p p \, \Lambda^p \mathfrak{p}^* = \sum_{p=1}^{3} (-1)^p p \left(\Lambda^p \mathfrak{n}^* + \Lambda^{p-1} \mathfrak{n}^*\right) = \sum_{p=0}^{2} (-1)^{p+1} \Lambda^p \mathfrak{n}^*. \tag{7.13}$$

Therefore, the Fourier transformation of k_t^τ is given by

$$\Theta_{n,\lambda}(k_t^\tau) = e^{-t(\lambda^2+1-n^2/4+\tau(\Omega))} \sum_{p=0}^{2} (-1)^{p+1} \dim(\mathbb{C}_n \otimes \Lambda^p \mathfrak{n}^* \otimes V_\tau)^M. \tag{7.14}$$

This formula can be simplified using the real version of Konstant's Bott-Borel-Weil theorem [23]. We apply Lemma 3.3 to determine the $n \in \mathbb{Z}$ for which $\dim(\mathbb{C}_n \otimes \Lambda^p \mathfrak{n}^* \otimes V_\tau)^M \neq 0$. We decompose the characters on the right-hand side of (3.18) according to (3.21). Let $\sigma_{\tau,w} \in \hat{M}$ and $\lambda_{\tau,w} \in \frac{1}{2}\mathbb{Z}$ be defined by (3.21). Using (3.18) and (6.16), we get

$$\begin{aligned} \sum_{n \in \mathbb{Z}} \overline{\sigma_n(m_\gamma)} \int_{\mathbb{R}} \Theta_{n,\lambda}(k_t^\tau) e^{-i\ell(\gamma)\lambda} \, d\lambda \\ = \sum_{w \in W_G} (-1)^{\ell(w)+1} \sigma_{\tau,w}(m_\gamma) e^{-t\lambda_{\tau,w}^2} \frac{e^{-\ell(\gamma)^2/(4t)}}{(4\pi t)^{1/2}}. \end{aligned} \tag{7.15}$$

Next we consider the contribution of the identity to (7.10). By (7.7), k_t^τ is in $\mathcal{C}^q(G)$ for all $q > 0$. Therefore we can apply the Plancherel formula for G (see [11, Theorem 11.2]). With respect to the normalizations of Haar measures used in [11], we have

$$k_t^\tau(e) = \sum_{n \in \mathbb{Z}} \int_{\mathbb{R}} \Theta_{n,\lambda}(k_t^\tau) P_{\sigma_n}(i\lambda) \, d\lambda,$$

where $P_{\sigma_n}(z)$ is the Plancherel polynom (3.4). Repeating the arguments that led to (7.15), we get

$$k_t^\tau(e) = \sum_{w \in W_G} (-1)^{\ell(w)+1} e^{-t\lambda_{\tau,w}^2} \int_{\mathbb{R}} e^{-t\lambda^2} P_{\sigma_{\tau,w}}(i\lambda) \, d\lambda. \tag{7.16}$$

Combined with (7.10) and (7.15), we obtain

$$K(t,\tau)=\sum_{w\in W_G}(-1)^{\ell(w)+1}e^{-t\lambda^2_{\tau,w}}\Bigg(\operatorname{vol}(X)\int_{\mathbb{R}}e^{-t\lambda^2}P_{\sigma_{\tau,w}}(i\lambda)\,d\lambda \qquad (7.17)$$
$$+\sum_{\{\gamma\}\neq\{e\}}\frac{\ell(\gamma)}{n_\Gamma(\gamma)}L(\gamma;\sigma_{\tau,w})\frac{e^{-\ell(\gamma)^2/(4t)}}{(4\pi t)^{1/2}}\Bigg),$$

where $L(\gamma,\sigma)$ is defined by

$$L(\gamma,\sigma)=\frac{\sigma(m_\gamma)e^{-\ell(\gamma)}}{\det\left(\operatorname{Id}-\operatorname{Ad}(m_\gamma a_\gamma)_{\overline{\mathfrak{n}}}\right)}. \qquad (7.18)$$

Unfortunately, the constants $\lambda_{\tau,w}$ appearing in the exponential factors prevent us from applying the Mellin transform to this formula directly. This problem occurred already in [9]. To overcome this problem we use the auxiliary operators introduced in Section 4.

Using (4.26) and (6.16), it follows that for $w\in W_G$ we have

$$e^{\tau(\Omega)t}K(t,\sigma_{\tau,w})=e^{-t\lambda^2_{\tau,w}}\Bigg(2\operatorname{vol}(X)\int_{\mathbb{R}}e^{-t\lambda^2}P_{\sigma_{\tau,w}}(i\lambda)\,d\lambda$$
$$+\sum_{[\gamma]\neq 1}\frac{\ell(\gamma)}{n_\Gamma(\gamma)}\left(L(\gamma;\sigma_{\tau,w})+L(\gamma;w_A(\sigma_{\tau,w}))\right)\frac{e^{-\ell(\gamma)^2/(4t)}}{(4\pi t)^{1/2}}\Bigg). \qquad (7.19)$$

Next observe that by (3.31) there exists a decomposition

$$W_G=W_0\sqcup W_1,$$

with $|W_i|=2$, $i=1,2$, and a bijection $j\colon W_0\to W_1$ such that for $w\in W_0$ we have

$$\sigma_{\tau,j(w)}=w_A(\sigma_{\tau,w}),\quad \lambda_{\tau,j(w)}=-\lambda_{\tau,w}.$$

Hence by (7.17) and (7.19) we get

$$K(t,\tau)=\frac{1}{2}\sum_{w\in W_G}(-1)^{\ell(w)+1}e^{\tau(\Omega)t}K(t;\sigma_{\tau,w}). \qquad (7.20)$$

This equality can be expressed in a slightly different way as follows. Denote by Tr_s the supertrace with respect to the grading of $E(\sigma_{\tau,w})$. Using the definition of $K(t,\sigma_{\tau,w})$ by (4.11) and the definition of $\Delta(w)$ by (6.15), we get

$$K(t,\tau)=\frac{1}{2}\sum_{w\in W_G}(-1)^{\ell(w)+1}\operatorname{Tr}_s\left(e^{-t\Delta(w)}\right). \qquad (7.21)$$

To continue we need to determine the location of the spectrum of the operators $A(\sigma)$.

Lemma 7.1. *For $\sigma\in\hat{M}$ we have $A(\sigma)\geq -1$. Moreover, if $k\notin\{0,\pm 2\}$, then $A(\sigma_k)>-1$.*

Proof. Let $\hat{G}$ denote the unitary dual of G. Let

$$L^2(\Gamma\backslash G) = \widehat{\bigoplus}_{\pi\in\hat{G}} m_\Gamma(\pi)\mathcal{H}_\pi \tag{7.22}$$

be the spectral decomposition of the right regular representation of G on $L^2(\Gamma\backslash G)$. Let (ν, V_ν) be an irreducible unitary representation of K. Then $L^2(X, E_\nu) \cong \left(L^2(\Gamma\backslash G) \otimes V_\nu\right)^K$. Using (7.22), we get

$$\left(L^2(\Gamma\backslash G) \otimes V_\nu\right)^K = \widehat{\bigoplus}_{\pi\in\hat{G}} m_\Gamma(\pi) \left(\mathcal{H}_\pi \otimes V_\nu\right)^K. \tag{7.23}$$

This decomposition corresponds to the spectral resolution of A_ν as follows. Assume that we have $m_\Gamma(\pi)\dim(\mathcal{H}_\pi \otimes V_\nu)^K \neq 0$. Then $m_\Gamma(\pi)(\mathcal{H}_\pi \otimes V_\nu)^K$ is an eigenspace of A_ν with eigenvalue $-\pi(\Omega)$. Note that $\hat{G}$ is the union of the trivial representation, the unitary principal series $\pi_{k,\lambda}$ with $k \in \mathbb{Z}$ and $\lambda \in \mathbb{R}$, and the complementary series π_x^c with $0 < x < 1$ [12, Proposition 49], [11, Theorem 16.2] (where for the latter reference the different parametrization of the induced representations has to be taken into account).

First consider the principal series $\pi_{n,\lambda}$. By Frobenius reciprocity [11, p. 208] we have for $l \in \mathbb{N}_0$

$$\dim(\mathcal{H}_{\pi_{k,\lambda}} \otimes V_{\nu_l})^K = [\pi_{k,\lambda}^\vee|_K : \nu_l] = [\nu_l|_M : \sigma_{-k}] = [\nu_l|_M : \sigma_k]. \tag{7.24}$$

By (4.1) it follows that $(\mathcal{H}_{\pi_{k,\lambda}} \otimes V_{\nu_l})^K \neq 0$ implies $l \geq k$. Moreover, by (4.2) it follows that $m_{\nu_l}(\sigma_m) \neq 0$ implies $m \geq l$. Thus if $m_\nu(\sigma_m) \neq 0$ and $(\mathcal{H}_{\pi_{k,\lambda}} \otimes V_\nu)^K \neq 0$, then we have $m \geq k$. Hence if $\nu \in \hat{K}$ and $\sigma \in \hat{M}$ are such that $m_\nu(\sigma) \neq 0$ and $(\mathcal{H}_{\pi_{k,\lambda}} \otimes V_\nu)^K \neq 0$, then it follows from (2.12) and (4.9) that

$$-\pi_{k,\lambda}(\Omega) + c(\sigma) \geq 0. \tag{7.25}$$

Next consider the complementary series. By (4.9) we have $c(\sigma) \geq -1$ for all $\sigma \in \hat{M}$. Since $0 < x < 1$, it follows from (2.14) that

$$-\pi_x^c(\Omega) + c(\sigma) > -1. \tag{7.26}$$

for all $\sigma \in \hat{M}$. Finally, the trivial representation of G occurs in (7.23) only if ν is the trivial representation ν_0. Moreover, by (4.2) we have $m_{\nu_0}(\sigma_l) \neq 0$, only if $l = 0$ or $l = 2$. Thus by (7.25), (7.26), and the definition of $A(\sigma)$ by (4.10), the statement of the Lemma follows. □

We apply this lemma to study the kernel of the operator $\Delta(w)$, $w \in W_G$, which is defined by (6.15).

Lemma 7.2. *Let τ be an irreducible, finite-dimensional representation of G. Assume that $\tau_\theta \not\cong \tau$. Then* $\ker\Delta(w) = \{0\}$ *for all $w \in W_G$.*

Proof. Let $\tau = \tau_{m,n}$ with $m \neq n$. We use (6.17) to express $\Delta(w)$ in terms of $A(\sigma_{\tau,w})$. By (3.31) we have $\lambda_{\tau,w} \in \frac{1}{2}\mathbb{Z} \setminus \{0\}$ for all $w \in W_G$. If $|\lambda_{\tau,w}| > 1$, it follows from (6.17) and Lemma 7.1 that $\Delta(w) > 0$. It remains to consider the cases $\lambda_{\tau,w} = \pm 1$ and $\lambda_{\tau,w} = \pm 1/2$. In the first case we have $|m - n| = 2$. Then it follows from (3.31) that $\sigma_{\tau,w} = \sigma_{2l}$ with $|l| \geq 2$. By Lemma 7.1 we get $\Delta(w) > 0$. In the second case we have $|m-n| = 1$. By (3.31) it follows that $\sigma_{\tau,w} = \sigma_{2l+1}$ for some $l \in \mathbb{Z}$. Let $\nu \in \hat{K}$ such that $m_\nu(\sigma_{2l+1}) \neq 0$. By (4.2) there exists $p \in \mathbb{N}_0$ such that $\nu = \nu_{2p+1}$. Since π_x^c is induced from the trivial representation and $[\nu_{2p+1}|_M : \sigma_0] = 0$, Frobenius reciprocity [11, p. 208] implies

$$\dim(\mathcal{H}_{\pi_x^c} \otimes V_{\nu_{2p+1}})^K = [\pi_x^c|_K : \nu_{2p+1}] = [\nu_{2p+1}|_M : \sigma_0] = 0. \tag{7.27}$$

Thus in this case the complementary series does not occur in (7.23). Also the trivial representation does not occur. By (7.25) it follows that $A(\sigma_{\tau,w}) \geq 0$. Using (6.17) we get $\Delta(w) > 0$. $\square$

Now we can turn to the proof of Theorem 1.5. First assume that $\tau \not\cong \tau_\theta$. Then it follows from [4, Chapt. VII, Theorem 6.7] that $H^*(X, E_\tau) = 0$. Hence $\Delta_p(\tau) > 0$ for all p, $0 \leq p \leq 3$. By Lemma 7.2 we also have $\Delta(w) > 0$, $w \in W_G$. Hence $K(t,\tau)$ and $\operatorname{Tr}(e^{-t\Delta(w)})$, $w \in W_G$, are exponentially decreasing as $t \to \infty$. Therefore we can take the Mellin transform of both sides of (7.21) and we get

$$\frac{1}{\Gamma(s)} \int_0^\infty K(t,\tau) t^{s-1}\, dt = \frac{1}{2} \sum_{w \in W_G} (-1)^{\ell(w)+1} \frac{1}{\Gamma(s)} \int_0^\infty \operatorname{Tr}_s\left(e^{-t\Delta(w)}\right) t^{s-1}\, dt,$$

which holds for $\operatorname{Re}(s) > 3/2$. After analytic continuation we compare the derivatives at $s = 0$ of both sides. Using (7.2) we get

$$T_X(\tau)^4 = \prod_{p=1}^{3} \det\left(\Delta_p(\tau)\right)^{2(-1)^{p+1}p} = \prod_{w \in W_G} \det\nolimits_{\mathrm{gr}}\left(\Delta(w)\right)^{(-1)^{\ell(w)}}. \tag{7.28}$$

Now we use the determinant formula (6.2) to relate the right-hand side to the value at zero of the Ruelle zeta function. Since $\Delta(w) > 0$, it follows that $\det_{\mathrm{gr}}(s^2 - 2s\lambda_{\tau,w} + \Delta(w))$ is regular at $s = 0$ and its value at $s = 0$ is equal to $\det_{\mathrm{gr}}(\Delta(w)) \neq 0$. Hence

$$\lim_{s \to 0} \prod_{w \in W_G} \det\nolimits_{\mathrm{gr}}\left(s^2 - 2s\lambda_{\tau,w} + \Delta(w)\right)^{(-1)^{\ell(w)}} = \prod_{w \in W_G} \det\nolimits_{\mathrm{gr}}\left(\Delta(w)\right)^{(-1)^{\ell(w)}}. \tag{7.29}$$

By (6.20) it follows that $R_\tau(s) R_{\tau_\theta}(s)$ is regular at zero. Now observe that $\overline{\tau} = \tau_\theta$. Furthermore by (3.13) we have

$$\overline{R_\tau(s)} = R_{\tau_\theta}(\overline{s}).$$

This implies that $R_\tau(s)$ is regular at $s = 0$ and by (6.20) we get

$$|R_\tau(0)|^2 = \prod_{w \in W_G} \det\nolimits_{\mathrm{gr}}\left(\Delta(w)\right)^{(-1)^{\ell(w)}}. \tag{7.30}$$

Combining (7.28) and (7.30), the first statement of Theorem 1.5 follows.

Next assume that $\tau_\theta = \tau$. Then there exists $m \in \mathbb{N}_0$ such that $\tau = \tau_{m,m}$. We use (6.21) to determine the order of $R_\tau(s)$ at $s = 0$. For $m \in \mathbb{N}_0$ let

$$h_m = \dim\nolimits_{\mathrm{gr}} \ker(A(\sigma_{2m+2})), \tag{7.31}$$

where $\dim_{\mathrm{gr}}$ denotes the graded dimension of a graded vector space, i.e., if $V = V^+ \oplus V^-$ is a graded finite-dimensional vector space, then $\dim_{\mathrm{gr}} V = \dim V^+ - \dim V^-$. Assume that $m \geq 1$. Then $\det\left((s \pm (m+1))^2 - 1 + \Delta\right)$ is regular and nonzero at $s = 0$. Furthermore, $\det_{\mathrm{gr}}\left(s^2 + A(\sigma_{2m+2})\right)$ has order $2h_m$ at $s = 0$ and

$$\lim_{s\to 0} s^{-2h_m} \det_{\mathrm{gr}}\left(s^2 + A(\sigma_{2m+2})\right) = \det_{\mathrm{gr}}\left(A(\sigma_{2m+2})\right). \tag{7.32}$$

By (6.21), it follows that $R_\tau(s)$ has order $-2h_m$ at $s = 0$ and we have

$$\lim_{s\to 0} s^{2h_m} R_{\tau_{m,m}}(s) = \frac{\det\left((m+1)^2 - 1 + \Delta\right)^2}{\det_{\mathrm{gr}}\left(A(\sigma_{2m+2})\right)}. \tag{7.33}$$

On the other hand, using (3.32), it follows from (7.21) that

$$K(t,\tau) = \operatorname{Tr}_s\left(e^{-tA(\sigma_{2m+2})}\right) - 2e^{-t((m+1)^2-1)} \operatorname{Tr}\left(e^{-t\Delta}\right). \tag{7.34}$$

Taking the limit $t \to \infty$ of both sides of (7.34), we get

$$h_m = \sum_{p=1}^{3} (-1)^p p \dim\left(\ker \Delta_p(\tau)\right). \tag{7.35}$$

Moreover (7.34) also implies

$$T_X(\tau)^2 = \prod_{p=1}^{3} \det\left(\Delta_p(\tau)\right)^{(-1)^{p+1}p} = \frac{\det\left((m+1)^2 - 1 + \Delta\right)^2}{\det_{\mathrm{gr}}\left(A(\sigma_{2m+2})\right)}.$$

Combining this equality with (7.33) and (7.35), we obtain the second statement of Theorem 1.5 for $m \geq 1$. Finally consider the case $\tau = 1$. In this case we need to take into account the simple zero of $\det(s^2 \pm 2s + \Delta)$ at $s = 0$. Thus we get

$$\lim_{s\to 0} s^{2h_0 - 2} R_1(s) = \frac{\det(\Delta)^2}{\det_{\mathrm{gr}}\left(A(\sigma_2)\right)}. \tag{7.36}$$

Furthermore, (7.34) gives.

$$h_0 = \sum_{p=1}^{3} (-1)^p p \dim H^p(X,\mathbb{R}) + 2 = \dim H^1(X,\mathbb{R}) - 1.$$

This implies that the order of $R_1(s)$ at $s = 0$ equals $4 - 2\dim H^1(X,\mathbb{R})$.

8. Proof of Theorem 1.1

We are now ready to prove our main result. We consider the representation τ_m. Using (2.9), it follows from (3.14) that

$$R_{\tau_m}(s) = \prod_{k=0}^{m} R\left(s - (m/2 - k), \sigma_{m-2k}\right). \tag{8.1}$$

We distinguish the cases where m is odd and even. Let $m \geq 3$. Then we get

$$\begin{aligned} R_{\tau_{2m}}(s) &= \prod_{k=0}^{4} R(s-(2-k), \sigma_{4-2k}) \prod_{k=3}^{m} R(s-k, \sigma_{2k}) R(s+k, \sigma_{-2k}) \\ &= R_{\tau_4}(s) \prod_{k=3}^{m} R(s-k, \sigma_{2k}) R(s+k, \sigma_{-2k}). \end{aligned} \tag{8.2}$$

Similarly, for $m \geq 2$ we get

$$R_{\tau_{2m+1}}(s) = R_{\tau_3}(s) \prod_{k=2}^{m} R(s-k-1/2, \sigma_{2k+1}) R(s+k+1/2, \sigma_{-(2k+1)}). \tag{8.3}$$

Now recall that by Proposition 3.2, 1), each $R(s, \sigma_l)$, $l \in \mathbb{Z}$, is regular in the half-plane $\operatorname{Re}(s) > 2$ and does not vanish in this half-plane. By the functional equation (3.11) the same holds in the half-plane $\operatorname{Re}(s) < -2$. Therefore the products on the right-hand side of (8.2) and (8.3) are regular at $s = 0$. Furthermore it follows from (3.6) that

$$|R(s, \sigma_l)| = |R(\overline{s}, \sigma_{-l})|. \tag{8.4}$$

Using (8.2), (8.3) and Theorem (1.5) we get

$$T_X(\tau_{2m})^2 = T_X(\tau_4)^2 \prod_{k=3}^{m} |R(k, \sigma_{2k})| \cdot |R(-k, \sigma_{2k})|, \quad m \geq 3. \tag{8.5}$$

and

$$T_X(\tau_{2m+1})^2 = T_X(\tau_3)^2 \prod_{k=2}^{m} |R(k+1/2, \sigma_{2k+1})| \cdot |R(-k-1/2, \sigma_{2k+1})|, \quad m \geq 2. \tag{8.6}$$

By the functional equation (3.11) and (8.4) we get

$$|R(-k, \sigma_{2k})| = \exp\left(-\frac{4}{\pi} \operatorname{vol}(\Gamma\backslash\mathbb{H}^3) k\right) |R(k, \sigma_{2k})|$$

Together with (8.5) this leads to

$$T_X(\tau_{2m}) = T_X(\tau_4) \prod_{k=3}^{m} \exp\left(-\frac{2}{\pi} \operatorname{vol}(\Gamma\backslash\mathbb{H}^3) k\right) |R(k, \sigma_{2k})|. \tag{8.7}$$

Similarly

$$T_X(\tau_{2m+1}) = T_X(\tau_3) \prod_{k=2}^{m} \exp\left(-\frac{2}{\pi} \operatorname{vol}(\Gamma\backslash\mathbb{H}^3)(k+1/2)\right) |R(k+1/2, \sigma_{2k+1})|. \tag{8.8}$$

To continue we need the following estimation.

Lemma 8.1. *There exists $C > 0$ such that for all $m \in \mathbb{N}$, $m \geq 3$, we have*

$$\sum_{k=3}^{m} \big| \log |R(k, \sigma_{2k})| \big| \leq C, \quad \sum_{k=2}^{m} \big| \log |R(k+1/2, \sigma_{2k+1})| \big| \leq C.$$

Proof. We consider the first case. Since $|\sigma_{2k}(m_\gamma)| = 1$, we have

$$1 - e^{-k\ell(\gamma)} \leq \left|1 - \sigma_{2k}(m_\gamma)e^{-k\ell(\gamma)}\right| \leq 1 + e^{-k\ell(\gamma)}.$$

Let $k \geq 3$. Using that the infinite product (3.6) is absolutely convergent for $\operatorname{Re}(s) > 2$, we get

$$\sum_{\substack{[\gamma]\neq e\\ \text{prime}}} \log\left(1 - e^{-k\ell(\gamma)}\right) \leq \log|R(k,\sigma_{2k})| \leq \sum_{\substack{[\gamma]\neq e\\ \text{prime}}} \log\left(1 + e^{-k\ell(\gamma)}\right),$$

which implies

$$\big|\log|R(k,\sigma_{2k})|\big| \leq \sum_{\substack{[\gamma]\neq e\\ \text{prime}}} \sum_{n=1}^{\infty} \frac{1}{n} e^{-nk\ell(\gamma)}. \tag{8.9}$$

Let $\delta = \inf\{\ell(\gamma)\colon \gamma \in \Gamma \setminus \{e\}\}$. Put $C_1 = (1 - e^{-\delta})^{-1}$. Using (8.9) we get

$$\begin{aligned}
\sum_{k=3}^{m} \big|\log|R(k,\sigma_{2k})|\big| &\leq \sum_{n=1}^{\infty} \frac{1}{n} \sum_{\substack{[\gamma]\neq e\\ \text{prime}}} \sum_{k=3}^{m} e^{-\ell(\gamma)nk}\\
&= \sum_{n=1}^{\infty} \frac{1}{n} \sum_{\substack{[\gamma]\neq e\\ \text{prime}}} \left(\frac{1 - e^{-n(m+1)\ell(\gamma)}}{1 - e^{-n\ell(\gamma)}} - \left(1 + e^{-n\ell(\gamma)} + e^{-2n\ell(\gamma)}\right)\right)\\
&= \sum_{n=1}^{\infty} \frac{1}{n} \sum_{\substack{[\gamma]\neq e\\ \text{prime}}} \frac{e^{-3n\ell(\gamma)} - e^{-(m+1)n\ell(\gamma)}}{1 - e^{-n\ell(\gamma)}} \leq C_1 \sum_{\substack{[\gamma]\neq e\\ \text{prime}}} \sum_{n=1}^{\infty} \frac{e^{-3n\ell(\gamma)}}{n}\\
&= C_1 \log R(3,\sigma_0)^{-1} = C.
\end{aligned}$$

The other case is similar. □

Taking the logarithm of both sides of (8.7) and (8.8), respectively, we obtain

$$\log T_X(\tau_{2m}) = \log T_X(\tau_4) + \sum_{k=3}^{m} \log|R_{2k}(k)| - \frac{1}{\pi} \operatorname{vol}(\Gamma\backslash\mathbb{H}^3)\,(m(m+1) - 6)\,.$$

and

$$\begin{aligned}
\log T_X(\tau_{2m+1}) = \log T_X(\tau_3) &+ \sum_{k=2}^{m} \log\left|R_{2k+1}\left(k + \frac{1}{2}\right)\right|\\
&- \frac{1}{\pi} \operatorname{vol}(\Gamma\backslash\mathbb{H}^3)(m(m+2) - 3).
\end{aligned}$$

Applying Lemma 8.1 we get

$$-\log T_X(\tau_m) = \frac{1}{4\pi} \operatorname{vol}(\Gamma\backslash\mathbb{H}^3) m^2 + O(m)$$

as $m \to \infty$. This completes the proof of Theorem 1.1.

Acknowledgement

I would like to thank Jonathan Pfaff for a careful reading of the manuscript and for pointing out some minor mistakes.

References

[1] D. Barbasch, H. Moscovici, *L^2-index and the trace formula*, J. Funct. Analysis **53** (1983), 151–201.

[2] R. Berndt, *Representations of linear groups. An introduction based on examples from physics and number theory.* Vieweg, Wiesbaden, 2007.

[3] J.-M. Bismut, E. Vasserot, *The asymptotics of the Ray-Singer analytic torsion of the symmetric powers of a positive vector bundle.* Ann. Inst. Fourier (Grenoble) **40** (1990), no. 4, 835–848.

[4] A. Borel, N. Wallach, *Continuous cohomology, discrete subgroups, and representations of reductive groups*, Second edition. Mathematical Surveys and Monographs, 67. Amer. Math. Soc., Providence, RI, 2000.

[5] U. Bröcker, *Die Ruellesche Zetafunktion für G-induzierte Anosov-Flüsse*, Ph.D. thesis, Humboldt-Universität Berlin, Berlin, 1998.

[6] U. Bunke and M. Olbirch, *Selberg zeta and theta functions*, A differential operator approach, Akademie Verlag, Berlin, 1995.

[7] T.A. Chapman, *Topological invariance of Whitehaed torsion*, Amer. J. Math. **96** (1974), 488–497.

[8] J. Elstrodt, F. Grunewald, J. Mennicke, *Groups acting on hyperbolic space. Harmonic analysis and number theory.* Springer Monographs in Mathematics. Springer-Verlag, Berlin, 1998.

[9] D. Fried, *Analytic torsion and closed geodesics on hyperbolic manifolds*, Invent. math. **84** (1986), 523–540.

[10] D. Fried, *Meromorphic zeta functions of analytic flows*, Commun. Math. Phys. **174** (1995), 161–190.

[11] A.W. Knapp, *Representation theory of semisimple groups*, Princeton University Press, Princeton and Oxford, 2001.

[12] A.W. Knapp and E.M. Stein, *Intertwining operators for semisimple Lie groups*, Annals of Math. **93** (1971), 489–578.

[13] B. Kostant, *Lie algebra cohomology and the generalized Borel-Weil theorem.* Ann. of Math. (2) **74** (1961) 329–387.

[14] Matsushima, Murakami, *On vector bundle valued harmonic forms and automorphic forms on symmetric riemannian manifolds*, Ann. of Math. **78** (1963), 365–416.

[15] R.J. Miatello, *The Minakshisundaram-Pleijel coefficients for the vector-valued heat kernel on compact locally symmetric spaces of negative curvature.* Trans. Amer. Math. Soc. **260** (1980), 1–33.

[16] J.J. Millson, *Closed geodesics and the η-invariant*, Annals of Math. **108** (1978), 1–39.

[17] J. Milnor, *Whitehead torsion*, Bull. Amer. Math. Soc. **72** (1966), 358–426.

[18] G.D. Mostow, *Strong rigidity of locally symmetric spaces*, Princeton Univ. Press and Univ. of Tokyo Press, 1973.

[19] W. Müller, *Analytic torsion and R-torsion for unimodular representations*, J. Amer. Math. Soc. **6** (1993), 721–753.

[20] G. Prasad, *Strong rigidity of* $\mathbb{Q}$*-rank* 1 *lattices,* Invent. math. **21** (1973), 255–286.

[21] D.B. Ray, I.M. Singer; *R-torsion and the Laplacian on Riemannian manifolds*, Advances in Math. 7, 145–210. (1971).

[22] M.A. Shubin, *Pseudodifferential operators and spectral theory,* Second edition. Springer-Verlag, Berlin, 2001.

[23] J. Silhan, *A real analog of Kostant's version of the Bott-Borel-Weil theorem*, J. of Lie theory **14** (2004), 481–499.

[24] A. Voros, *Spectral functions, special functions and the Selberg zeta function*, Commun. Math. Phys. **110** (1987), 439–465.

[25] W. Thurston, *Three-dimensional manifolds, Kleinian groups and hyperbolic geometry,* Bull. Amer. Math. Soc. **6** (1982), no. 3, 357–381.

[26] W. Thurston, *Three-dimensional geometry and topology,* Vol. 1. Princeton Mathematical Series, 35. Princeton University Press, Princeton, NJ, 1997.

[27] N.R. Wallach, *On the Selberg trace formula in the case of compact quotient*, Bull. Amer. Math. Soc. **82** (1976), 171–195.

[28] A. Wotzke, *Die Ruellsche Zetafunktion und die analytische Torsion hyperbolischer Mannigfaltigkeiten*, Ph.D. thesis, Bonn, 2008, Bonner Mathematische Schriften, Nr. 389.

[29] B. Zimmermann, *A note on hyperbolic* 3*-manifolds of the same volume*, Monatsh. Math. **117**, (1994), no. 1-2, 139–143.

Werner Müller
Universität Bonn
Mathematisches Institut
Endenicher Allee 60
D-53115 Bonn, Germany
e-mail: `mueller@math.uni-bonn.de`

Differential Characters for K-theory

James Simons and Dennis Sullivan

Dedicated to Jeff Cheeger for his 65th birthday

Abstract. We describe a sequence of results that begins with the introduction of differential characters on singular cycles in the seventies motivated by the search for invariants of geometry or more generally bundles with connections. The sequence passes through an Eilenberg-Steenrod type uniqueness result for ordinary differential cohomology using these characters and a construction of a differential K-theory using Grothendieck's construction on classes of complex bundles with connection. The last element of the sequence returns full circle with a differential character definition of differential K-theory. The cycles in this definition of characters for differential K-theory are closed smooth manifolds provided with complex structures and hermitian connections on their stable tangent bundles.

Mathematics Subject Classification (2000). 53Cxx.

Keywords. Connections on bundles, differential K-theory, differential K-characters.

Background

Maps from smooth singular cycles in a manifold X to $\mathbb{R}/\mathbb{Z}$ satisfying a differential form homology variation property, termed Differential Characters and denoted by $\hat{H}^k(M,\mathbb{R}/\mathbb{Z})$, were introduced and studied by Jeff Cheeger and one of us in the early 70s [1]. Precisely,

$$1) \qquad \hat{H}^i(M,\mathbb{R}/\mathbb{Z}) = \left\{ f \in \mathrm{Hom}(Y_{i-1}(X),\mathbb{R}/\mathbb{Z}) \mid f(\partial a) = \int_a \omega_f \mod \mathbb{Z} \right\}$$

where $Y_{i-1}(X)$ denotes smooth cycles of dim $i-1$. It follows from the definition that ω_f is uniquely determined by f and is contained in $\wedge^i_{\mathbb{Z}}$, closed k-forms with integral periods. $\hat{H}(\mathbb{R}/\mathbb{Z})$ comprises a functor from the smooth category into $\mathbb{Z}$-graded rings and may be shown to satisfy the following commutative diagram of natural transformations:

2) 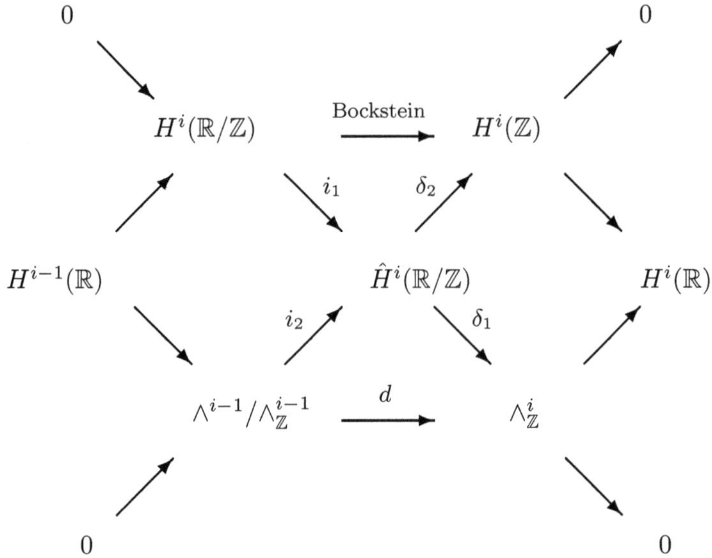

where the upper outside sequence is the Bockstein exact sequence, and the lower outside sequence is easily derived from the deRham theorem.

In one application it was shown that in a principal G-bundle with connection, the pair consisting of an invariant polynomial on G whose Chern-Weil form defines a real class with integral periods, together with the choice of an integral characteristic class with the same real image determines a differential character on the base manifold. δ_1 and δ_2 map the character into the form and the integral class respectively. The diagram in 2) shows these data to determine the character modulo an element in a torus of dimension equal to the one lower odd Betti number. Should both the form and the class vanish the character lies in this torus and may be non-zero even if the bundle is locally flat.

Other constructions arising in different contexts have been made and then shown equivalent to differential characters in the smooth context. See [20] for a discussion of Deligne cohomology somewhat in the spirit of the summary here. Other examples related to Harvey-Lawson spark complexes and Lawson homology can be found in the references and discussion of [21]. A third set of examples appears in the circle of ideas combining algebraic geometry, arithmetic and analysis discussed in [22] and its references. In the smooth category all these functors satisfied 2) and are naturally equivalent to $\hat{H}(\mathbb{R}/\mathbb{Z})$. This observation inspired the uniqueness result in [2], discussed below.

Differential cohomology

In the modern viewpoint differential characters are one instantiation of a contravariant functor called ordinary differential cohomology, a fibre product functor that combines differential forms with ordinary integer cohomology defined, say, using singular cochains in a manner essentially as depicted in 2). Exotic or generalized differential cohomology theory also makes sense as a fibre product functor combining differential forms with any exotic or generalized cohomology theory [1] defined, say, using spectra [19]. This possibility is based on two neat facts: the first is that any finite type generalized cohomology theory tensor a field k of characteristic zero is canonically isomorphic to ordinary cohomology theory with graded coefficients in the exotic theory of a point tensor k. The second is the well-known fact that real cohomology for manifolds is canonically described by the complex of differential forms. See Hopkins and Singer [5] for the general construction using spectra of the fibre product of differential forms with any finite type generalized cohomology theory.

In a second paper [2] the present authors verified the current viewpoint on differential characters by showing ordinary differential cohomology as a contravariant functor on the smooth category is uniquely characterized by the diagram in 2). Specifically it is shown that any abstract theory satisfying 2) is canonically isomorphic to differential characters.

Differential K-theory via structured bundles

In a third paper [3] the present authors constructed a differential geometric instantiation of differential K-theory in even degrees. This was defined abstractly in [5] by combining total even-dimensional differential forms with the spectrum of the generalized cohomology theory associated to K-theory. The geometric construction of [3] was made using stable equivalence classes of complex hermitian bundles with unitary connections. The equivalence relation in this construction combines strict isomorphism together with stabilization and a further equivalence (termed CS) whereby one connection can be changed to another connection if and only if the Chern-Simons difference form is exact. Such objects are called Structured Bundles, and their isomorphism classes over a given manifold naturally form a commutative semi-ring. Applying the Grothedieck construction yields a commutative ring-valued functor $\hat{K}$. This construction should be contrasted with the first reference in [22] and the fibre product definition of [11]. $\hat{K}$ satisfies the following commutative diagram of natural transformations:

[1] A contravariant functor satisfying three out of the four Eilenberg-Steenrod axioms: a homotopy functor on pairs with the exact sequence and excision but omitting the point axiom.

3) 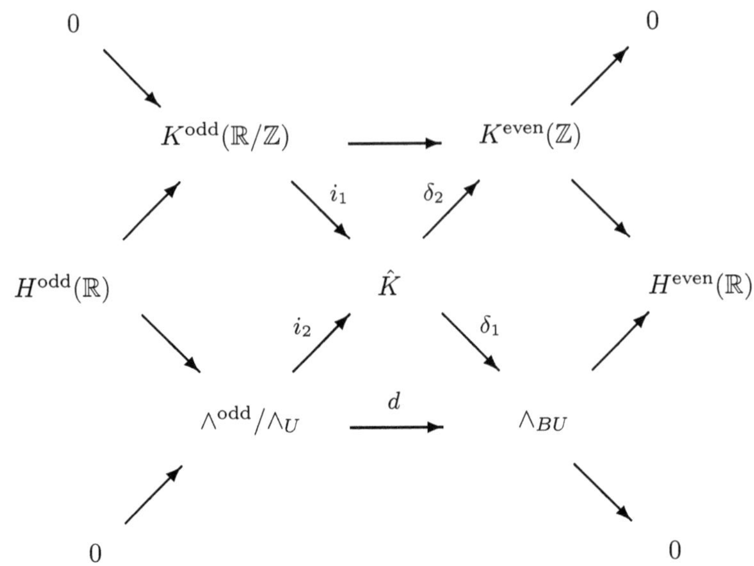

In the above, $\wedge_{BU}$ is the ring of total even closed forms cohomologous to the Chern characters of complex vector bundles over X. $\wedge_U = \{g^*(\Theta)\} + \wedge^{\text{odd}}_{exact}$, where Θ is the total odd bi-invariant form on U representing the universal transgression of the Chern character and $g : X \to U$ runs through all smooth maps. δ_2 is the map which forgets the connection, and δ_1 maps an element of $\hat{K}(X)$ into its Chern character form (well defined for a *CS* equivalence class). That $\text{Im}(i_1)$ is the kernel of δ_1 follows fairly easily from work in [23], and the construction and properties of i_2 are exposed in [3]. The upper exact sequence is the Bockstein in K-theory, and the lower exact sequence is easily derived from the deRham theorem. Again, should both the Chern character form and the element of $K(X)$ vanish, the element of $\hat{K}(X)$ lies in a torus of dimension the sum of the odd Betti numbers. Here, the lattice defining the torus is the cohomological image of $\wedge_U$ in $H^{\text{odd}}(\mathbb{R})$, commensurate but distinct from the lattice in the ordinary case. Changing lattices and torsion is the essential difference between differential K-theory and the appropriate sum of ordinary differential cohomologies.

K-characters

In a fourth paper [4] the present authors construct a character instantiation of differential K-theory, analogous to that of differential characters in ordinary differential cohomology. In place of singular cycles one uses representatives of complex bordism endowed with unitary connections on their stable tangent bundles.

To be precise, in complex bordism a Cycle in a smooth manifold X is a closed compact stably almost complex manifold M (termed an SAC) together with a smooth map of M into X. A compact SAC, S, with boundary diffeomorphic to M

induces a SAC structure on M. Should a map of M into X be extendable to a map of such an S into X, M is termed a Boundary. Of course all boundaries are cycles. A cycle together with a complex hermitian structure with unitary connection on a stabilized version of its tangent bundle will be termed an Enriched Cycle. A boundary M is called an Enriched Boundary (of S) if the hermitian structure and unitary connection on its stabilized tangent bundle is induced from similar data on the stabilized tangent bundle of S which data is of product form [2] in a collar neighborhood of M. Under disjoint union, isomorphism classes over X of enriched cycles and boundaries mapping into X form semi-groups, $EC(X)$ and $EB(X)$.

We also note that if M is an enriched cycle or boundary in X, and Q is an enriched cycle mapping to a point, then $Q \times M$ is an enriched cycle or boundary in X. Finally, $ES(X)$ and $EB(X)$ are graded by dimension of M, and it is important to distinguish between the even and odd cases. Thus ES^{even} and ES^{odd} have the obvious meaning. We define K-characters to be the abelian groups

$$\widehat{Kch}^{\text{even}}(X) = \{f \in \text{Hom}(EC^{\text{odd}}(X), \mathbb{R}/\mathbb{Z})\}$$

satisfying:

1. $f(\partial S) = \int_S \omega_f \wedge \text{Todd}(S) \bmod Z$, where $\text{Todd}(S)$ is its total Todd form with respect to the connection on its stabilized tangent bundle, and ω_f is a total even form on X pulled back to S.
2. $f(Q \times M) = \text{Todd}(Q) f(M)$, where $Q \in EC^{\text{even}}(\text{point})$.

$\widehat{Kch}^{\text{odd}}$ has the same definition but the cycles are even and ω_f is odd. In either case it is straightforward to show that

a) ω_f is closed, and uniquely determined by f, and
b) the pulled back cohomology class of ω_f cupped with the Todd class of an even cycle (even case) or odd cycle (odd case) takes integral value on the fundamental homology class of the cycle.

One may associate a K-character to a bundle E with unitary connection ∇. The value of the character on an odd cycle M in the base is the reduction mod 1 of the integral as computed in b) where ω_f is $ch(E)$, the Chern form of the bundle E with unitary connection ∇ and the manifold S with boundary M is constructed abstractly outside X via algebraic topology considerations. The integrality of the integral over the closed manifold V constructed by glueing two such choices of S together,

$$\int_V ch(E)\,\text{Todd}(V) \in Z$$

means the character associated to (E, ∇) by this construction is well defined.

[2]By product form we mean a combing of the collar neighborhood such that parallel transport along the strands is an isomorphism (including hermitian structure and unitary connection) between the stabilized version of the tangent bundle of S along its boundary with that over each interior slice of the collar neighborhood.

From the character of a bundle with connection one may compute invariants of the bundle E itself. These are homomorphisms of the complex bordism with $\mathbb{Q}/\mathbb{Z}$ coefficients into $\mathbb{Q}/\mathbb{Z}$ which assign values in $\mathbb{Z}/k$ to $\mathbb{Z}/k$-manifolds.

In more detail if ϕ denotes the character and if k times a cycle M in X bounds S in X, one forms the quantity $\phi(M) - [1/k \int_S ch(E)\,\mathrm{Todd}(S)]$ in $\mathbb{R}/\mathbb{Z}$. This actually lies in $\mathbb{Q}/\mathbb{Z}$ and is well defined on $\mathbb{Z}/k$-bordism, cf. [7].

Theorem T (topology). *The $\mathbb{Q}/\mathbb{Z}$ periods and the $\mathbb{Q}$ periods: $\int_V ch(E)\,\mathrm{Todd}(V)$ for even cycles V in X satisfy the* Todd *product rule (second property in the description of cycle above), they are compatible in that the complex bordism diagram commutes*

$$\begin{array}{ccc} \Omega(X,\mathbb{Q}) & \to & \mathbb{Q} \\ \downarrow & & \downarrow \\ \Omega(X,\mathbb{Q}/\mathbb{Z}) & \to & \mathbb{Q}/\mathbb{Z} \end{array}$$

and the invariants determine the complex bundle E up to stable isomorphism.

Moreover, any such compatible system of bordism periods satisfying the Todd *product rule comes from a complex bundle, cf.* [7].

Corollary. *The ω_f in the definition of K-character lies in $\wedge_{BU}$ in the even case and $\wedge_U$ in the odd case.*

Remark. Thus just as the original differential characters of a bundle with connection determine integral characteristic classes, the K-characters determine the bundle in integral K-theory itself.

A second application of K-characters is differential geometric. One may observe that the K-character of a bundle with connection is unchanged by strict isomorphism, stabilization, or by changing the connection by CS equivalence. Thus there is a well-defined map from differential theory $\hat{K}$ as defined in [3] to the set of K-characters denoted $\widehat{Kch}$.

Theorem G (geometry). *The K-character of a complex bundle with unitary connection determines its position in $\hat{K}$. Moreover any K-character comes from a complex bundle with unitary connection. In other words we have a canonical equivalence $\hat{K} \to \widehat{Kch}$.*

The proofs of Theorem T and Theorem G use two remarkable properties related to bordism and duality of complex K-theory itself. The first is the sixties result of Conner-Floyd [6] expressing K-homology of a space X (the homology theory Alexander dual to K-theory) by stably almost complex bordism of X tensored with the integers regarded as a module over the bordism of a point using the Todd genus ring homomorphism. The second property, also noted in the sixties [7], is that the homology theory constructed algebraically from K cohomology theory via Pontryagin duality by applying the functor $\mathrm{Hom}(\,,\mathbb{R}/\mathbb{Z})$ agrees with the Alexander dual homology theory. As a consequence one may show that the subgroup of even K-characters with $\omega_f = 0$ may be identified with $K(\mathbb{R}/\mathbb{Z})$. This leads to the exact sequence

$$4) \quad 0 \to K^{\text{odd}}(\mathbb{R}/\mathbb{Z}) \to \widehat{Kch}^{\text{even}} \to \wedge_{BU} \to 0$$

and the theorem follows from the Five Lemma using part of diagram 3).

The argument in the proof of Theorem G, together with a proof of the Mayer-Vietoris property for any abstract functor fitting into 3), (see arXiv 2010) immediately yields the K^{even} theory analog of the uniqueness theorem in [2]. Namely that in the even case any functor satisfying 3) is canonically isomorphic to K-characters[3].

Theorem G shows that $\widehat{Kch}^{\text{even}}$ must fit the diagram in 3), and indeed i_1, i_2 and δ_1 are obviously and intrinsically defined.

As an immediate consequence of the map from $\hat{K}$ to $\widehat{Kch}^{\text{even}}$ and the Atiyah-Patodi-Singer theorem we get

Theorem A (analysis). *Let V be a complex hermitian vector bundle over X with unitary connection, and let M be an enriched odd cycle in X, with the property that the unitary connection on its stabilized tangent bundle is consistent with the Levi-Civita connection on the tangent bundle itself associated with the Riemannian metric induced by the stabilized hermitian structure in the sense that they define the same $\hat{A}$ forms. Then the value on M of the K-character associated to V is equal to the η invariant of the spin $\mathbb{C}$ Dirac operator on M with coefficients in the pull back of V defined using the spin^c structure and line bundle connection coming from the SAC structure on M.*

A consequence of the character instantiation of even differential K-theory is a simple definition of the **wrong way map** for families. Namely, let $Y \to X$ be a projection, the fibres of which are compact even-dimensional SAC manifolds provided with a smooth family of hermitian structures and unitary connections on their stabilized tangent bundles. If M is an enriched odd cycle in X, one can show that the pre-image of M may itself be regarded as an enriched odd cycle in Y. This map from enriched cycles in X to enriched cycles in Y immediately provides a map from $\widehat{Kch}^{\text{even}}(Y)$ to $\widehat{Kch}^{\text{even}}(X)$. With this map in hand we have

Question (families index theorem for $\hat{K}$). *Can the wrong way map in $\hat{K}^{\text{even}}$ for a metrized family of stably almost complex manifolds of even dimension defined above using topology and differential geometry be computed analytically as follows: for each complex bundle with unitary connection (E, ∇) in the total space and for each odd enriched cycle M in the base form the pulled back cycle $\bar{M}$ in the total space and restrict (E, ∇) to $\bar{M}$ to obtain $(\bar{E}, \bar{\nabla})$. The value of the character of the push forward of (E, ∇) on the cycle M is the eta invariant mod 1 of the spin $\mathbb{C}$ Dirac operator of $\bar{M}$ with coefficients in $(\bar{E}, \bar{\nabla})$.*

[3]At the abstract level of spectra this last uniqueness result was achieved independently, more generally and somewhat earlier by Bunke and Schick [9].

Remark. Thanks to Bismut, this is true whenever the connection on the tangent bundle used to define the pull back cycle is equivalent as structured bundles to a Bismut metric connection: a metric connection for the direct sum metric on the pullback cycle so that the three form obtained by cyclic symmetrization of the torsion is closed. We are in the process of analyzing this possibility.

References

[1] Cheeger, J. and Simons, J. "Differential Characters and Geometric Invariants". Notes of Stanford Conference 1973, Lecture Notes in Math. No. 1167. Springer-Verlag, New York. 1985. pp. 50–90.

[2] Simons, James and Sullivan, Dennis. "Axiomatic Characterization of Ordinary Differential Cohomology". arxiv:math.0701077v1. Journal of Topology 1 no. 1. 2007.

[3] Simons, James and Sullivan, Dennis. "Bundles With Connections and Differential K-theory". arxiv;0810.4935v1[mathAT] to appear in Proceedings of Clay Conference celebrating Alain Connes' sixtieth birthday.

[4] Simons, James and Sullivan, Dennis. "Differential K Characters". arxiv (to appear).

[5] Hopkins, M.J. and Singer, I.M. "Quadratic functions in Geometry, Topology, and M-theory". J. Diff. Geom 70. 2005. pp. 329–452.

[6] Conner, Pierre and Floyd, Ed. "The Relation of Cobordism to K theory". 1966 Springer Lecture Notes.

[7] Sullivan, Dennis. "Geometric Topology: Localization,Periodicity and Galois Symmetry, The 1970 MIT Notes". Springer 2005.

[8] Morgan, J.W. and Sullivan, D.P. "The Transversality Characteristic Class and Linking Cycles in Surgery Theory". Ann. of Math. 99. 1974. pp. 461–544.

[9] Bunke, Ulrich and Schick, Thomas. "Uniqueness of smooth extensions of generalized cohomology theories". arxiv submitted Jan. 29, 2009.

[10] Freed, D.S. "Z/k manifolds and families of Dirac operators". Invent. Math 92. 1988. pp. 243–254.

[11] Freed D.S. and Lott, John. "An index theorem in differential K theory". arxiv0907.3508v.

[12] Freed, D.S. "On Determinant Line Bundles". Mathematical Aspects of String Theory. Ed. S.T. Yau. World Scientific. 1987. pp. 189–238.

[13] Bismut, J.M. and D.S. Freed. "The Analysis of Elliptic Familes II". Comm. Math. Phys. 107. 1986. pp. 103–163.

[14] Cheeger, Jeff. "Eta-invariants, the adiabatic approximation and conical singularities". Journal of Differential Geometry 26. 1987. pp. 175–221.

[15] Cheeger, Jeff. "On the formulas of Atiyah-Patodi-Singer and Witten". ([16], [17].)

[16] Atiyah, M.F., Patodi, V.K. and Singer, I.M. "Spectral Asymmetry and Riemannian Geometry III". Math. Proc. Camb. Phil. Soc. 79. 1976. pp. 71–99.
[17] Witten, Edward. "Global Gravitational Anomalies". Commun. Math. Phys. 100. 1985. pp. 197–229.
[18] Freed, Dan S. and Melrose, Richard B. "A mod k Index Theorem". Invent. Math. 107. 1992. pp. 283–299.
[19] Brown, E.H. "Cohomology Theories". Ann. of Math. 75. 1962. pp. 467–484.
[20] Gajer, Pawel. "Geometry of Deligne Cohomology". arxiv alg geom 9601025v1.
[21] Hu, Wenchuan. "A map from Lawson homology to Deligne cohomology". arxiv.0810.0442 [math.AG]
[22] Gillet, Henri and Soule', Christophe. "Characteristic classes for algebraic vector bundles with hermitian metrics". Annals of Math. 131. 1990. pp. 163–203, 205–238 and "An arithmetic Riemann Roch theorem", Inventiones Math. 110. 1992. pp. 473–543.
[23] Lott, John. "$\mathbb{R}/\mathbb{Z}$ Index Theory". Comm Anal Geom 2. 1994. pp. 279–311.

James Simons
Renaissance Technologies LLC
600 Route 25A
East Setauket, NY 11733, USA

Dennis Sullivan
The CUNY Graduate Center
365 Fifth Avenue, Room 4208
New York, NY 10016-4309, USA
e-mail: dsullivan@gc.cuny.edu

MIX
Papier aus verantwortungsvollen Quellen
Paper from responsible sources
FSC® C105338
FSC
www.fsc.org

If you have any concerns about our products,
you can contact us on
ProductSafety@springernature.com

In case Publisher is established outside the EU,
the EU authorized representative is:
Springer Nature Customer Service Center GmbH
Europaplatz 3, 69115 Heidelberg, Germany

Printed by Libri Plureos GmbH
in Hamburg, Germany